Microbiomes of Extreme Environments
Volume 1
Biodiversity and Biotechnological Applications

Editors

Ajar Nath Yadav

Department of Biotechnology
Dr. Khem Singh Gill Akal College of Agriculture
Eternal University, Baru Sahib
Sirmour, Himachal Pradesh, India

Ali Asghar Rastegari

Falavarjan Branch, Islamic Azad University
Isfahan, Iran

Neelam Yadav

Gopi Nath P.G. College
Veer Bahadur Singh Purvanchal University
Ghazipur, Uttar Pradesh, India

CRC Press
Taylor & Francis Group
Boca Raton London New York

CRC Press is an imprint of the
Taylor & Francis Group, an **Informa** business

A SCIENCE PUBLISHERS BOOK

First edition published 2021
by CRC Press
6000 Broken Sound Parkway NW, Suite 300, Boca Raton, FL 33487-2742

and by CRC Press
2 Park Square, Milton Park, Abingdon, Oxon, OX14 4RN

© 2021 Taylor & Francis Group, LLC

CRC Press is an imprint of Taylor & Francis Group, LLC

Library of Congress Cataloging-in-Publication Data
Names: Yadav, Ajar Nath, editor.
Title: Microbiomes of extreme environments. Volume 1. Biodiversity and
  biotechnological applications / editors, Ajar Nath Yadav, Department of
  Biotechnology ,Dr. Khem Singh Gill Akal College of Agriculture, Eternal
  University, Baru Sahib, Sirmour, Himachal Pradesh, India, Ali Asghar
  Rastegari, Falavarjan Branch, Islamic Azad University, Isfahan, Iran,
  Neelam Yadav, Gopi Nath P.G. College, Veer Bahadur Singh Purvanchal
  University, Ghazipur, Uttar Pradesh, India.
Description: First edition. | Boca Raton : CRC Press, 2021. | Includes
  bibliographical references and index.
Identifiers: LCCN 2020038815 | ISBN 9780367342746 (hardcover)
Subjects: LCSH: Extreme environments--Microbiology. |
  Extremozymes--Ecology. | Microbial diversity.
Classification: LCC QR100.9 .M55 2021 | DDC 577.5/8--dc23
LC record available at https://lccn.loc.gov/2020038815

ISBN: 978-0-367-34274-6 (hbk)
ISBN: 978-0-367-68268-2 (pbk)

Typeset in Times New Roman
by Radiant Productions

Preface

Extreme microbiomes are those microorganisms thriving under severe conditions where no other living being will have any chance to survive. The extraordinary habitats are manifest with high temperatures (thermophiles), low temperature (psychrophiles), hypersaline environments (halophiles), low and high pH (Acidophiles/alkaliphiles), high pressure (Piezophiles) distributes worldwide. The extreme habitats offer a unique reservoir of genetic diversity and biological source of extremophiles. The extremophilic microbial diversity and their biotechnological potential use in agricultural and industrial applications will be a milestone for future needs. The extremophilic microbes and bioactive compounds produced by extremophilic microbes will serve as potentials products for sustainable agriculture and environments. The enzymes for extremophilic microbes could be applicable for different industrial processes. The study of extremophiles also helps to understands diversity and microbial ecology, microbe-environment interactions, adaptation and evolution, element cycling under the extreme ecosystems. Besides, the extreme nature of those interactions and the physiological capabilities developed in extremophiles is attracting the attention of agriculture and industrial applications. Recent developments clearly show that cell components of extremophilic archaea and bacteria are unique and deliver a valuable source of new biocatalysts and compounds. Since many industrial enzymes are required to function under extreme conditions, there is also a considerable commercial pressure to discover stable biocatalysts in modern microbial biotechnology. Extremophiles and their cell components, therefore, are expected to play an important role in the chemical, food, pharmaceutical, paper and textile industries as well as environmental biotechnology.

The present book on "*Microbiomes of Extreme Environments, Vol. 1: Biodiversity and Biotechnological Applications*" covers biodiversity of extremophilic microbes and their role in agriculture, environment, medicine and allied sectors. This book will be immensely useful to biological sciences, especially to microbiologists, microbial biotechnologists, biochemists, researchers, and scientists of microbial and microbial biotechnology. We are honoured that leading scientists who have extensive, in-depth experience and expertise in extreme microbiology and biotechnology took the time and effort to develop these outstanding chapters. Each chapter is written by internationally recognized researchers/scientists so the reader is given an up-to-date and detailed account of our knowledge of the extreme microbiology and their potential biotechnological applications in agriculture, environmental and allied sectors.

We are grateful to many people who helped to bring this book to light. Dr. Ajar Nath Yadav gives special thanks to his wife for her constant support and motivation in putting everything together. Dr. Yadav also gives special thanks to his esteemed friends, well-wishers, colleagues and senior faculty members of Eternal University, Baru Sahib, India.

Ajar Nath Yadav
Baru Sahib, Himachal Pradesh, India

Ali Asghar Rastegari
Isfahan, Iran

Neelam Yadav
Mau, Uttar Pradesh, India

Acknowledgements

All authors sincerely acknowledge the contribution of up-to-date information on extremophilic microbes, their biodiversity and biotechnological applications in agriculture and environment and allied sectors. The editors are thankful to all authors for their valuable contributions.

The editors would like to thank their families who were very patient and supportive during this journey. We are grateful to the many people who helped to bring this book to light.

The editor, Dr. Ajar Nath Yadav is grateful to his Ph.D. research scholars Tanvir Kaur, Rubee Devi, Divjot Kour, and colleagues for their support and motivation during this project.

We are sure that this book will interest scientists, graduates, undergraduates and post doctorates who are investigating "Extreme Microbiology and Biotechnology".

Contents

Chapter 1

Extremophilic Microbes and their Extremozymes for Industry and Allied Sectors

Hiran Kanti Santra and *Debdulal Banerjee**

Introduction

Extremophiles are ubiquitous in nature and are found in unique extreme environments representing a diverse variety according to their area of occurrence and functioning. Naturally extremophiles are simple organisms, either single celled or filamentous with prokaryotic nucleus and are abundant in places of odd conditions where natural microbiota is not available (MacElroy 1974). The term is anthropocentric in nature and was coined about 40 years ago and their occurrence have been proved from 800 meter deep beneath the Antarctic Sheet (subglacial lakes) to deep sea hydrothermal vents; 2 km from oceanic surface (occurrence of metazoan pompei) (Christner et al. 2014; Fox 2014; Desbruyères and Laubier 1980). Extremophiles are of distinct biochemical machinery (DNA, lipids, enzymes) and physiology that allow them to survive in extreme habitats and these are the basis of their biotechnological exploitation.

The first hurdle for researchers was the proper isolation and appropriate maintenance in simulated situations of laboratory. Molecular biology tools and approaches of metagenomics have solved these problems to some extent. Extremophiles are separated in two groups according to their response to changing environment; obligate extremophiles—those can grow only in extreme conditions and the other one is facultative extremophiles—which can withstand extreme conditions and grow uninterruptedly but develops at its normal optimum conditions also. The idea of extremophily is flexible in three different domains of life (prokaryotes, archaea and eukaryotes), i.e., for thermophilic bacteria the maximum withstanding temperature capacity is 95°C where-as for archaea this temperature is 122°C and unicellular eukaryotes represent the highest tolerance upto 62°C.

Recent biotechnological advances have allowed scientists to explore sea, sky, deep-sea vents, hot springs, the upper troposphere and stratosphere, mines, industries and outer space for extremophiles isolation (Wilson and Brimble 2009; Bhojiya and Joshi 2012). Other than extremophiles some organisms called polyextremophiles (predominantly prokaryote and a few are also eukaryote) that can grow in two or more of these situations are known as polyextremophiles (Horikoshi and Bull 2011; Yadav et al. 2019a). Examples include the red alga *Cyanidioschyzon* sp. which is one way

Microbiology and Microbial Biotechnology Laboratory, Department of Botany and Forestry, Vidyasagar University, Midnapore–721102, West Bengal, India.
* Corresponding author: debu33@gmail.com; db@mail.vidyasagar.ac.in

acidophilic and in other way moderately thermophilic, i.e., can grow in pH ranges of 0.2–3.5 and temperatures of 38°C–57°C. Such a type of adaptation of these microorganisms to the multiple stress factors are attributed to various factors (Weber et al. 2007). Extremozymes are able to catalyze reactions in non-aqueous environments, in water–solvent mixtures, at extremely high pressures, high and low temperatures or even below the water freezing point, at acidic and alkaline pH conditions (Adams et al. 1995).

Biotechnology being a significant tool for both industrial and daily life (food and power generation, biofuels, pharmaceutically valuable products) use enzymes as biocatalyst and the essential qualities of the enzymes like substrate affinity, solvent tolerance, temperature stability, or selectivity, are modified using genetic engineering as a weapon (Elleuche et al. 2014). The main hindrance for the achievement of such targets includes the poor stability of enzymes of mesophilic sources against extreme values of temperature, pH, and ionic strength (Hough and Danson 1999; Eichler 2001). To solve the problem of poor stability of enzymes at extreme conditions, biocatalysts from extreme environments are needed to be explored. Extremophiles grow in environments of extreme conditions such as high and low temperature (–2°C to 15°C in cold environments, 60°C–110°C in case of hot environments), ionic strength of 2–5 M NaCl, and pH ranges of < 4.0 to > 9.0. The majority of them are bacteria and archaea with their separate metabolic pathways that are effective in extreme conditions. Due to these credentials they are known as extremozymes having a broad range of biotechnological applications (Yadav et al. 2017a; Yadav et al. 2020). Archaea represents 20% of the earth's biomass (DeLong and Pace 2001).

There are innumerable differences in structural, genetical and biochemical properties between these three groups of organisms. The nexus between molecular biology and bioprocess custom design has improved the performance of archaeal bio-products. This could result in significant savings, and therefore allow industrial applications of these unique biomaterials (Alquéres et al. 2007). In the present scenario, from the biotechnological point of view archaea derived materials are of immense demand. Examples include the extremely stable membrane lipids of these organisms representing a new and novel drug delivery system (Patel and Sprott 1999; Schiraldi et al. 2002; Oren 2010; Zhao et al. 2015). Self-assembling components like S-layer glycoprotein and bacteriorhodopsin obtained from archae are of nanotechnological importance (Oesterhelt et al. 1991; Sleytr et al. 1997). Haloarchaeal polysaccharides and polymeric substances are components of the oil industry (Rodriguez-Valera 1992) and biodegradable plastics respectively (Fernández-Castillo et al. 1986).

Methanogenic archaea represents the source for clean and low-cost energy generation (Reeve et al. 1997). Another example in molecular biology includes the use of Taq DNA polymerase from *Thermus aquaticus*, which is randomly used in Polymerase Chain Reactions (PCRs) (Canganella and Wiegel 2011). Agarases that are capable of hydrolyzing agar have a large application at the laboratory and industrial level for liberating DNA and other bio-molecules stuck in agarose. They are also effective tools for the bioremediation of agar used daily for laboratory purposes and extraction of bioactive or medicinal compounds from algae and seaweed. Bacteriostatic neoagarosaccharides are also the product of agarase activity. They slow down the process of starch degradation, promote anticancer activity and have antioxidative potentials (Giordano et al. 2006; Elleuche et al. 2014). A very authentic source of agarase is the salt-tolerant extremophile *Pseudoalteromonads*, *Pseudomonas*, and *Vibrio* (DasSarma et al. 2010). Methanogens are not as widely characterized compared to other extremophiles. But methanogens are used in some processes such as, biogas production or organic waste decomposition by anaerobic fermentation (Zhang et al. 2011; Zhu et al. 2011).

A large number of enzymes are of daily importance in human life. Table 1.1 represents the extremophiles with their diverse extremozyme productivity. One such example includes proteases or peptidases from halophiles with a higher degree of stability on organic solvent leading to a sharp decrease in salt concentration and finally the problem of metal corrosion caused by the salt settled on it (Kim and Dordick 1997). Another enzyme of importance is starch hydrolyzing α-amylase with thermal stability obtained from *Bacillus licheniformis*, *Bacillus amyloliquefaciens*, *Bacillus subtilis*,

Table 1.1 Representation of different types of extremozymes from extremophiles.

Extremophile	Extremozyme	Reference
Micrococcus sp.	Nuclease	Kamekura et al. (1982)
Thermus aquaticus	DNA polymerase	Jones and Foulkes (1989)
P. furiosus	DNA polymerase	Lundberg et al. (1991)
Sclerotinia borealis	Pectinase	Takasawa et al. (1997)
Thermococcus aggregans	Hyperthermophilic pullunase	Niehaus et al. (2000)
Pseudoalteromonas haloplanktis	DNA ligase	Georlette et al. (2000)
Bacillus subtilis	Cellulase	Mawadza et al. (2000)
Pseudomonas fluorescens	Alanine racemase	Yokoigawa et al. (2001)
Haloferax mediterranei	Glucose dehydrogenase	Pire et al. (2001)
Thermoplasma acidophilum	Glucoamylase	Serour and Antranikian (2002)
Haloferax volcanii	Isocitrate dehydrogenase	Camacho et al. (2002)
Thermococcus litoralis	L-Aminoacylase	Toogood et al. (2002)
Pseudoalterimonas sp.	Amylase	Matsumoto et al. (2003)
Sulfolobus solfataricus	Alpha-glucosidase	Giuliano et al. (2004)
Halococcus sp.	Amylase	Fukishima et al. (2005)
Haloferax mediterranei	Glutamate dehydrogenase	Díaz et al. (2006)
Pyrococcus furiosus	Alcohol dehydrogenase	Kube et al. (2006)
Sulfolobus solfataricus MT4	Maltooligosyl-trehalose synthase	Cimini et al. (2008)
Haloferax volcanii	Cysteine desulfurase	Zafrilla et al. (2010)
Halobacterium sp.	Protease	Akolkar et al. (2010)
Micrococcus sp.	Glutaminase	Yoshimune et al. (2010)
Virgibacillus sp.	Chitinase	Essghaier et al. (2011)
Bacillus sp.	Xylanase	Prakash et al. (2011)
Marinimicrobium sp.	Inulinase	Li et al. (2011)
Geomicrobium sp.	Protease	Karan et al. (2011)
Marinobacter sp.	Lipase	Pérez et al. (2011)
Haloferax mediterranei	Cu-nitrite reductase	Esclapez et al. (2013)
Halobacterium sp. NRC-1	Alcohol dehydrogenase	Liliensiek et al. (2013)
Sulfolobus solfataricus	Lactonases	Rémy et al. (2016)
Thermus thermophiles	α-galactosidase	Aulitto et al. (2017)
Halorubrum lacusprofundi	β-galactosidase	Laye et al. (2017)
Marinomonas sp. BSi20414	β-1,3-Galactosidase	Ding et al. (2017)
Sulfolobus acidocaldarius	Phospho-triesterase like lactonases	Restaino et al. (2018)

and *Bacillus stearothermophilus* (Kour et al. 2019). The process of starch breaking includes the exposure of enzymes to high temperatures and pH for that reason optimization using extremophiles is required (Elleuche et al. 2014). Geneticists applied traditional protein engineering approaches for the production of extremozymes using mesophiles but without success. So other than direct extraction of extremozymes from extremophiles; expression of extremophilic genes on *E. coli* is of utmost importance. Misfolding and codon usage differences can affect the functional extremozyme expression systems (generally used for *Escherichia coli* and *Bacillus* sp.). So there is an urgent need of research in extremophiles to search for proper hosts for appropriate gene expression, efficient transformations and also suitable expression vectors (Elleuche et al. 2014).

Diversity of extremophiles

Thermophilic microorganisms that are capable of growing at temperatures between 41°C and 122°C are potent producers of extremozymes such as, amylases, cellulases, chitinases, esterases, lipases, mannanase, pullulanases, pectinases, proteases, phytases, xylanases (Bertoldo and Antranikian 2002; Van der Maarel et al. 2002; Madigan and Marrs 1997; Sunna and Bergquist 2003; Verma et al. 2019). They can withstand proteolysis, adverse effects of denaturing agents, organic solvents, high salinity, etc. and the risk of contamination is also reduced, low adhesiveness is maintained, greater solubility of substrates are achieved (Raddadi et al. 2015). Their physical property, configurations and electrostatic interactions are the key features for maintaining their thermostability and function.

Thermostable lipases are used in several bioprocesses like paper industry, milk industry, in processing of dyed products, leather industry, pharmaceuticals, esterification, grease hydrolysis, interesterification, organic biosynthesis, transesterification, optical nanosensors and analytes (Eichler 2001; Haki and Rakshit 2003; Staiano et al. 2005). Proteases from thermozymes are useful for the synthesis of dipeptides and starch-processing (Bruins et al. 2001; Jayakumar et al. 2012). Bioremediation strategies for controlling environmental pollution and bleaching of papers are mediated by cellulases, hemicellulases and xylanases (De Pascale et al. 2008; Unsworth et al. 2007; Rosenbaum et al. 2012). Thermophilic enzymes (amylase, cellulose and protease) are useful in the laundry and detergent industries. *Pyrococcus furiosus* produces a thermozyme (amylase) with great application in mutational studies (Rosenbaum et al. 2012).

Piezophiles are able to survive at high hydrostatic pressures particularly in deep-sea environments, deep-sea and volcano areas, etc. *Pyrococcus horikoshii* produces peptidase which is stable at high pressure. Other examples include *Pyrococcus abyssi* and *Sulfolobus solfataricus* with piezophilic adaptation (Fusi et al. 1995; Mombelli et al. 2002). *Pyrococcus furiosus*, a hyperthermophilic archaeon is known to be the source of Pfu polymerase. A well known molecular biology tool for PCR processes (Lundberg et al. 1991). Piezophiles and thermopiezophiles are separate according to their tolerance level; the latter one can withstand high temperatures as well as great pressure, while the former one can survive in only high pressure conditions, but not in high temperatures (Gomes and Steiner 1998; Gomes et al. 2000). This is the reason that thermopieophiles can be considered as polyextremophiles.

Abe and Horikoshi reported that α-amylase from a piezophilic source acts on maltooligosaccharides and produces trisaccharides, tetrasaccharides at great pressure and small amount of energy (Abe and Horikoshi 2001). This particular reaction is of high importance in the food industry (Cannio et al. 2004; Giuliano et al. 2004). Other than this piezophilic proteins are of great use in the detergent and food industries (Cavicchioli et al. 2011). Acidophiles are micro-organisms that grow at optimum pH of below 3–4 (Jaenicke 1981). Endoß-glucanase from *Sulfolobus solfataricus* demonstrated higher stability at optimum pH of 1.8 (Huang et al. 2005). Pikuta et al. demonstrated that carboxylesterase in *Ferroplasma acidiphilum* has a pH optimum of approximately 2 (Pikuta et al. 2007). Enzymes produced by acidophiles have great potential for biotechnological and industrial applications for the production of biofuel and ethanol. Halophilic microbes are able to survive in excessive salt concentrations (minimum 1 M NaCl) and the establishment of different chemical, structural, physiological modifications have allowed the selectivity as well as stability of proteins with their unique physicochemical properties (Delgado-García et al. 2012; Jackson et al. 2001). Halophiles are subdivided in three groups; extreme halophiles—capable of growing at 2,500–5,200 mM NaCl, moderate halophiles—survives at 500–2,500 mM NaCl concentrations and the slightly halophilic one-grows at 200–500 mM NaCl. Halophilic extremozymes are able to survive at low water activity and in the presence of organic solvents (Datta et al. 2010; Madern et al. 1995).

Halophilic extremozymes (amylases, proteases, lipases, xylanases) are reported from *Acinetobacter* sp., *Bacillus* sp., *Haloferax* sp., *Halobacterium* sp., *Halorhabdus* sp., *Marinococcus* sp., *Micrococcus* sp., *Natronococcus* sp., *Halobacillus* sp., and *Halothermothrix* sp. (Sutrisno et al. 2004; Taylor et al. 2004; Woosowska and Synowiecki 2004). Lipases and esterases are of potent importance in polyunsaturated fatty acids and food production along with biodiesel production (Litchfield 2011; Schreck and Grunden 2014). The halophiles are extremely salt specific; e.g., *Haloferax volcanii* can grow in the presence of KCl but cannot grow in NaCl concentrations (Ortega et al. 2011). Halophilic extremozymes are characterized with low solubility in aqueous/organic and non-aqueous media (Serour and Antranikian 2002; Suzuki et al. 2001). Some examples include the activity in non-aqueous media; protease from *Halobacterium halobium* (Kim and Dordick 1997), *Saliniovibrio* sp. strain AF-2004 (Karbalaei-Heidari et al. 2007), *Natrialba magadii* (Ruiz and De Castro 2007) and organic solvent-tolerant; amylase from *Haloarcula* sp. strain S-1 (Fukushima et al. 2005), *Nesterenkonia* sp. strain F (Shafiei et al. 2011), *Salimicrobium halophilum* strain LY20 (Li and Yu 2012) and glutamate dehydrogenase from *Halobacterium salinarum* strain NRC-36014 (Munawar and Engel 2012). Enzyme production varies according to the salt concentrations (10 to 15% NaCl), e.g., *Halogeometricum* sp. TSS101 (Vidyasagar et al. 2006). Extremozymes from halophilic microorganisms present great opportunities for the industries of food, bioremediation and biosynthetic processes.

Indian Himalaya as a store house of extremophiles

Extremophilic communities are not only necessary for their biotechnological exploitation but also relevant to understand the utility of primitive analogues of bioactive molecules of early environments on earth (Raddadi et al. 2015; Saxena et al. 2016; Yadav et al. 2015c). A large portion of terrestrial and aquatic ecosystems of this planet are temporarily or permanently dominated by subzero temperatures and extremophiles are the dominant microflora of those regions. Oceans represent 71% of the earth's surface and 90% of that are of below 4°C (Stibal et al. 2012). Psychrophilic microorganisms thrive well at low temperatures close to the freezing point of water and exhibit higher growth rate at subzero temperatures (Shivaji et al. 2011). The psychrophiles could be used as biofertilizers, biocontrol agents, and bioremediators (Yadav et al. 2019b; Yadav et al. 2018). There have been some remarkable works in past few years from the cold deserts of different areas of the world but in the Indian context, the Indian Himalayas represents a niche of psychrophilic microbes and their proper isolation along with biotechnological exploitation could open up new horizons for their effective use in agriculture, medicine and bioengineering processes (Singh et al. 2016; Yadav et al. 2017b).

The Indian subcontinent has been benefitted with the presence of the great Himalayan mountain range and in particular Leh, Ladakh represents one of the coldest places of this mountain valley and scientists have proliferated their search to a great extent for the detailed study of psychrophiles from the cold deserts of the Indian Himalaya. Not only cold active enzymes but the Indian Himalayas represent a store house of hot extremozymes. Extremophiles diversity of the cold deserts of Indian Himalaya is moderate (Sahay et al. 2013; Saxena et al. 2014; 2015; Yadav et al. 2015a, b). Diversity of extremophiles from high Arctic snow has already been explored (Harding et al. 2011). Some well-known hot springs of the Indian Himalaya are the Soldhar and Ringigad hot springs, Uttaranchal Himalaya (Kumar et al. 2004), Chumathang hot spring, NW Indian Himalayas (Yadav 2015a), Vashisht, Khirganga and Kasol hot springs (Shirkot and Verma 2015) and Tattapani hot spring (Priya et al. 2016). These niches are hotspots of diverse microbes with novel genes and bioactive molecules with immense applications in industry and allied sectors (Saxena et al. 2016).

Himalayan hot springs

Hot springs are known to be the most potent sites for the residence of thermostable extremozyme producing thermophilic extremophiles with valuable industrial and medicinal importance. Especially hot springs situated at high elevations are of immense importance in this respect. Himalayan hotsprings situated at higher altitudes in the lap of the great Himalayan mountain ranges have been explored for thermo-stable enzymes (Sahay et al. 2017). They investigated the culturable thermophilic flora of bacteria and identified those using 16S rRNA technology. They performed isolation, phylogenetic profiling and also screening for hydrolytic enzyme production by the extremophiles from the hot-springs of Manikaran (elevation of 1760 meters from the mean sea level) and Yumthang (3564 meters from mean sea level) situated at Beas and Parvati valley geothermal system and North Sikkim district of Sikkim of India respectively. The temperature and pH of hot spring water varied from 89°C to 95°C and 43°C to 63°C, 7.8 to 8.2 and 6.5 to 6.8 for Manikaran and Yumthang hot-springs respectively. Six separate samples were collected for screening of extremozyme containing microbial flora from each hot springs and sealed in a sterilized thermostatic container (flask) for transportation to the laboratory for further analysis.

In total 140 thermophilic bacteria from 12 samples were phylogenetically constructed and 51 isolates (positive result showing taxa in screening for enzyme production; 28 from Manikaran and 23 from Yumthang) were identified in 37 distinct species belonging to 14 genera of *Anoxybacillus, Bacillus, Brevibacillus, Brevundimonas, Burkholderia, Geobacillus, Paenibacillus, Planococcus, Pseudomonas, Rhodanobacter, Thermoactinomyces, Thermobacillus, Thermonema* and *Thiobacillus*. The strains are *Bacillus megaterium* NBY38, *Bacillus pumilus* NBY4, *Bacillus simplex* NBY1, *Bacillus* sp. NBY16, *Bacillus subtilis* NBY44, *Bacillus thermoamylovorans* NBY26, *Brevibacillus* sp. NBY28, *Brevundimonas bullata* NBY27, *Burkholderia glathei* NBY12, *Paenibacillus* sp. NBY33, *Paenibacillus thiaminolyticus* NBY3, *Pseudomonas putida* NBY55, *Pseudomonas* sp. NBY30, *Rhodanobacter* sp. NBY11, *Rhodobacter capsulatus* NBY31, *Thermoactinomyces* sp. NBY60, *Thermobacillus* sp. NBY36, *Thiobacillus denitrificans* NBY57, *Bacillus amyloliquefaciens* NBM82, *Bacillus cibi* NBM33, *Bacillus licheniformis* NBM60, *Bacillus megaterium* NBM1, *Bacillus mycoides* NBM19, *Bacillus nealsonii* NBM23, *Bacillus pumilus* NBM31, *Bacillus* sp., NBM43, *Bacillus subtilis* NBM45, *Bacillus subtilis* NBM48, *Bacillus thermoamylovorans* NBM38, *Bacillus thuringiensis* NBM12, *Brevibacillus choshinensis* NBM66, *Brevibacillus reuszeri* NBM83, *Brevundimonas bullata* NBM35, *Geobacillus* sp. NBM49, *Paenibacillus dendritiformis* NBM41, *Paenibacillus ehimensis* NBM24, *Paenibacillus glycanilyticus* NBM30, *Paenibacillus glycanilyticus* NBM75, *Paenibacillus jamilae* NBM25, *Paenibacillus popilliae* NBM68, *Paenibacillus thiaminolyticus* NBM40, *Paenibacillus thiaminolyticus* NBM71, *Planococcus* sp. NBM37, *Rhodobacter capsulatus* NBM2, *Thermobacillus* sp. NBM6, *Thermonema lapsum* NBM28, *Anoxybacillus* sp. NBY46, *Bacillus amyloliquefaciens* NBY37, *Bacillus cereus* NBY23, *Bacillus circulans* NBY8, *Bacillus licheniformis* NBY5.

They were tested for their ability to degrade substrates such as protein, amylum starch, xylan and cellulose and other similar compounds. Out of 51 positive hydrolase producing isolates 24 were found to be stable at a wide variation of pH and temperature. The outcomes are impressive as they have isolated three bacteria (*Thermobacillus* sp. NBM6, *Paenibacillus ehimensis* NBM24 and *Paenibacillus popilliae* NBM68) that produce cellulose free xylanase. These extracellular hydrolytic enzymes (extremozymes) from thermophilic bacterial flora come with great economic benefit in sectors like industries, medical purposes and also in the agricultural sectors. Xylanases are used in pulp pre-bleaching, better digestion of animal feed stock, improvement of cereal based foods, development of fermented products from lignocellulosic and agro waste matters, plant fibers degumming and finally clarity enhancement of fruit juices in the beverage industries. As treatment using xylanase at higher temperature disrupts the structures of cell wall facilitating the removal of lignin in the various stages of bleaching of paper. So cellulose free xylanase are of utmost importance. Table 1.2 represents the recent works on extremophilic microflora of hot springs of the Indian Himalaya with their biotechnological prospects of extremozymes.

Cold deserts

The bio-control ability of the microbial flora ranging from bacteria to fungus is a widely accepted popular concept of sustainable agriculture. Problems arise in case of bio-control of pests or pathogenic fungus especially in regions of high altitude where these crops are not supported with the biological control conditions. So isolation of bio-control agents with a wide range of host susceptibility from

Table 1.2. Extremophilic microbes and their extremozymes from hot springs of the Indian Himalayas.

Profile of extremophiles	References
Extremophiles: Bacteria, yeast and filamentous organisms were present. Soils were dominated by spore forming rods. Out of 58 aerobic isolates, 53 were gram positive bacilli. Gram positive anaerobic oval rods population was also found. Bacilli were of endospore producing in nature **Application:** Phosphate solubilizing ability and antagonistic activity **Area of isolation:** Soldhar and Ringigad hotsprings of Garhwal region of Uttaranchal Himalaya	Kumar et al. (2004)
Extremophiles: *Bacillus aryabhattai* (JQ904723), *Brevibacillus* sp. (AJ313027), *Bacillus licheniformis* (GQ280087), *Bacillus pumilus* (JQ435673), *Bacillus subtilis* (EF442670), *Bacillus pumilus* (HF536558), *Bacillus* sp. (KC121051), *Bacillus cereus* (GU011948), *Rhodococcus* sp. (DQ285075), *Kocuria* sp. (DQ448711), *Planococcus* sp. (JX312584), *Micrococcus* sp. (JN866765), *Staphylococcus haemolyticus* (JQ624771), *Bacillus arbutinivorans Bacillus* sp. (FN397517) *Bacillus niacini* (AB680904), *Bacillus vireti* (HQ397585), *Aneurinibacillus danicus* (NR_028657), *Brevibacillus* sp. (FJ529026), *Staphylococcus succinus* (HQ423378), *Bacillus* sp. (JF901703), Uncultured *Klebsiella* sp. (GQ416648), *Pseudomonas* sp. (EU680995), *Microbacterium oxydans* (EU714340), *Pseudomonas psychrophila* (JQ782895), *Exiguobacterium acetylicum* (JX307688), *Rhodococcus baikunurensis* (JX683682), *Pseudomonas fluorescens* (JX127246), *Lysinibacillus xylaniticus* (JQ739716), *Staphylococcus hominis* (JN644561), *Bacillus megaterium* (EU931553), *Chelatococcus daeguensis* (NR_044297), *Bacillus flexus* (JQ936679), *Bacillus* sp. (HE821233), *Bacillus subtilis* (JX845578), *Lysinibacillus* sp. (FN397524), *Brevibacillus agri* (HQ222834), *Bacillus beijingensis* (JQ799102), *Bhargavaea cecembensis* (JQ071510), *Microbacterium* sp. (DQ339613) **Application:** Some of them are extremely thermotolerant and are potent producers of amylase and protease **Area of isolation:** Manikaran hot springs	Kumar et al. (2014)
Extremophiles: *Brevibacillus thermoruber, Paenibacillus barengoltzii, Bacillus licheniformis, Geobacillus thermoleovorans, Geobacillus caldoxylosilyticus, Thermus aquaticus, Geobacillus vulcani, G. thermocatenulatus, G. lituanicus, G. toebii, Paenibacillus barengoltzii, Aneurinibacillus thermoaerophilus* **Area of isolation:** Five hot water springs of Himachal Pradesh (Manikaran, Vashisht, Khirganga, Kasol, Kullu)	Shirkot and Verma (2015)
Extremophiles: *Geobacillus thermodenitrificans* IP_WH1(KP842609), *Bacillus licheniformis* IP_WH2(KP842 610), *B. aerius* IP_WH3(KP842611), *B. licheniformis* IP_WH4(KP842612), *B. licheniformis* IP_60Y(KP842613), *G. thermodenitrificans* IP_60A1(KP842614), *Geobacillus* sp. IP_60A2(KP842615) and *Geobacillus* sp. IP_80TP(KP842 616) **Application:** Thermotolerant cellulase production by *Geobacillus thermodenitrificans* **Area of isolation:** Tattapani Hot Spring Sediment in North West Himalayas	Priya et al. (2016)

regions of higher altitude or subzero temperature could be an alternative solution to this problem. Thus the importance of extremophilic microorganisms growing in subzero temperatures known as psychrotrophic bacteria are greatly increasing (Verma et al. 2017; Yadav 2015). In this respect Yadav et al. (2015b) studied the diversity of the culturable bacteria found in the subzero temperature conditions of the cold desert of Leh Ladakh of Jammu and Kashmir including Khardungla Pass, Indus, Zanskar River conflux, Pangong Lake and Chumathang.

A total number of 325 bacterial isolates were obtained from soil and water samples of those regions using 10 different nutrient compositions and characterized using 16S rDNA amplified ribosomal DNA restriction analysis with three restriction endonucleases AluI (from *Arthrobacter luteus*), MspI (from *Moraxella* sp.), and HaeIII (from *Haemophilus influenzae*) leading to the formation of 23–40 groups for different sites having 75% similarity index and adding up-to 175 groups. 16S rRNA gene sequencing identified 175 bacteria belonging to four phyla Actinobacteria (16%), Bacteroidetes (3%), Firmicutes (54%), and Proteobacteria (28%) including 9 different genera with 57 different species.

Almost 39% of the total morphotypes belonged to *Bacillus* and allied genus (BBDG – *Bacillus* derived genera) followed by *Pseudomonas* (14%), *Arthrobacter* (9%), *Exiguobacterium* (8%), *Alishewanella* (4%), *Brachybacterium, Providencia, Planococcus* (3%), *Janthinobacterium, Sphingobacterium, Kocuria* (2%) and *Aurantimonas, Citricoccus, Cellulosimicrobium, Brevundimonas, Desemzia, Flavobacterium, Klebsiella, Paracoccus, Psychrobacter, Sporosarcina, Staphylococcus, Sinobaca, Stenotrophomonas, Sanguibacter,* Vibrio (1%). The identified isolates are *Arthrobacter* sp., *Arthrobacter sulfonivorans, Arthrobacter sulfureus, Brachybacterium* sp., *Cellulosimicrobium cellulans, Citricoccus* sp., *Kocuria kristinae, Kocuria palustris, Sanguibacter antarcticus, Flavobacterium antarcticum, Sphingobacterium* sp., *Bacillus anthracis, Bacillus baekryungensis, Bacillus cereus, Bacillus firmus, Bacillus flexus, Bacillus licheniformis, Bacillus marisflavi, Bacillus mojavensis, Bacillus muralis, Bacillus pumilus, Bacillus simplex, Bacillus* sp., *Bacillus subtilis, Desemzia incerta, Exiguobacterium antarcticum, Exiguobacterium* sp., *Exiguobacterium undae, Lysinibacillus fusiformis, Lysinibacillus* sp., *Lysinibacillus sphaericus, Paenibacillus* sp., *Paenibacillus terrae, Paenibacillus xylanexedens, Planococcus antarcticus, Planococcus donghaensis, Planococcus kocurii, Pontibacillus* sp., *Sinobaca beijingensis, Sporosarcina aquimarina, Staphylococcus arlettae, Aurantimonas ltamirensis, Brevundimonas terrae, Paracoccus* sp., *Janthinobacterium* sp., *Alishewanella* sp., *Klebsiella* sp., *Providencia* sp., *Pseudomonas peli, Pseudomonas putida, Pseudomonas reactans, Pseudomonas* sp., *Pseudomonas stutzeri, Psychrobacter glacincola, Stenotrophomonas maltophilia, Vibrio metschnikovii.* Representatives of each cluster were screened at low temperatures (5°C to 15°C) for their PGP (Plant Growth Promoting) characters.

Their biotechnological potential includes their use as a plant growth promoter in terms of ammonia, hydrogen cyanide, IAA (Indole-3-acetic acid), gibberelic acid, siderophore production, phosphate solubilization, ACC (1-aminocyclopropane-1-carboxylate) deaminase activity. Isolates were also potent in their bio-control ability of the pathogenic microorganisms like *Rhizoctonia solani* and *Macrophomina phaseolina* (root rot pathogens) that causes world-wide global destruction of economically important food products in plains and high altitudes (cold climatic) of agricultural lands. The species richness was highest in case of Pangon Lake followed by Khardungla Pass, Chumathang, Indus River and Zanskar River. Evenness was highest in the Indus River where as Shanon and Simpson's index was highest for Pangon Lake followed by Khardungla Pass (H-3.40 and D-0.96) and Chumathang regions (H-3.27 and D-0.96). In terms of representation of niche specific bacteria, the Pangon Lake represents the highest value of 7 followed by the Chumathang where-as Khardungla Pass had the lowest value of 2. Table 1.3 represents the other remarkable reports of extremophiles and their extremozymes from the Himalayan cold-deserts.

Extremophilic pigments with biotechnological applications

Generally microorganisms synthesize a wide range of colored (pigmented) molecules useful for their growth and metabolism (Madigan et al. 2012), energy production by photosynthesis (Siefirmann-Harms 1987), defense against microorganisms and finally provide protection during stress conditions (Martin-Cerezo et al. 2015) (Fig. 1.1). In this respect, extremophiles are selected as they offer growth in high acidic or saline conditions, in high atmospheric pressure situations even in extreme temperature of both cold and hot conditions for the biotechnological exploitation of the microorganisms for the production and application of pigments for human welfare instead of eubacterial species arcahebacterial species (García-López et al. 2017). During biotechnological exploitation these extremophiles can withstand the radiation and exposure to aggressive chemicals as they are used to such stress.

Radiophiles

Radiation is a common cause of mutation leading to cell (eukaryotic or prokaryotic) death. Pigments produced intracellular by the microbes residing in extreme environments are able to

Table 1.3 Extremophilic microbes and their extremozymes from cold deserts of the Indian Himalayas.

Extremophiles and their profile	References
Extremophiles: *Arthrobacter* sp., *Arthrobacter sulfonivorans, Arthrobacter sulfureus, Brachybacterium* sp., *Cellulosimicrobium cellulans, Citricoccus* sp., *Kocuria kristinae, Kocuria palustris, Sanguibacter antarcticus, Flavobacterium antarcticum, Sphingobacterium* sp., *Bacillus anthracis, Bacillus baekryungensis, Bacillus cereus, Bacillus firmus, Bacillus flexus, Bacillus licheniformis, Bacillus marisflavi, Bacillus mojavensis, Bacillus muralis, Bacillus pumilus, Bacillus simplex, Bacillus* sp., *Bacillus subtilis, Desemzia incerta, Exiguobacterium antarcticum, Exiguobacterium* sp., *Exiguobacterium undae, Lysinibacillus fusiformis, Lysinibacillus* sp., *Lysinibacillus sphaericus, Paenibacillus* sp., *Paenibacillus terrae, Paenibacillus xylanexedens, Planococcus antarcticus, Planococcus donghaensis, Planococcus kocurii, Pontibacillus* sp., *Sinobaca beijingensis, Sporosarcina aquimarina, Staphylococcus arlettae, Aurantimonas ltamirensis, Brevundimonas terrae, Paracoccus* sp., *Janthinobacterium* sp., *Alishewanella* sp., *Klebsiella* sp., *Providencia* sp., *Pseudomonas peli, Pseudomonas putida, Pseudomonas reactans, Pseudomonas* sp., *Pseudomonas stutzeri, Psychrobacter glacincola, Stenotrophomonas maltophilia, Vibrio metschnikovii* **Applications:** Plant growth promoting abilities in terms of ammonia, hydrogen cyanide, IAA (Indole-3-acetic acid) siderophore production, phosphate solubilization ACC (1-aminocyclopropane-1-carboxylate) deaminase activity. Bio-control ability of the pathogenic microorganisms like *Rhizoctonia solani* and *Macrophomina phaseolina* **Area of isolation:** Cold desert of Leh Ladakh of Jammu and Kashmir including Khardungla Pass, Indus, Zanskar River conflux, Pangong Lake and Chumathang	Yadav et al. (2015b)
Extremophiles: 140 bacteria (psychrophiles, psychrotrophs and Psychrotolerant) belonging to Firmicutes, Proteobacteria (α, β, and γ) or Actinobacteria. **Applications:** Producers of cold-active hydrolases such as protease, amylase, xylanase and cellulase **Area of isolation:** Sub-glacial freshwater (Gurudongmar) and brackish water (Pangong) lakes of the Himalayas	Sahay et al. (2013)
Extremophiles: 1540 bacteria, 157 archaea, 260 fungi and 200 actinomycetes **Applications:** Eubacteria from saline soils help to overcome salt stress and improve crop plants, Bacteria from Sambar salt lake improves growth on wheat in saline soils, increase proline content and total soluble sugar content. *B. pumilus* inoculation cause 43% increase in wheat yield, acid tolerant bacteria improves growth of horticultural species **Area of isolation Soil:** Plant and water samples from the cold desert of Leh and Rohtang, from the mangrove regions of Sunderbans, West Bengal and Bhitarkanika, Orissa; thermal springs of Rajgir, Manikaran, Bakreswar, Balrampur and Vashisht Sambar salt lake, hypersaline soils of Rann of Kutch Pullicat Lake (Tamilnadu) and Chilka Lake, acidic soils of Manipur, Kerala, Meghalaya and Mizoram	Saxena et al. (2014)
Extremophiles: Total 557 bacteria from 85 samples belonging to Actinobacteria, Bacteroidetes, Firmicutes and Proteobacteria **Applications:** Ability to produce cold active hydrolytic enzymes at low temperatures (amylase, a-glucosidase, pectinase, protease, cellulose, a-galactosidase, laccase, chitinase, lipase). Seventeen isolates (Arthrobacter, Exiguobacterium, Janthinobacterium, providencia, Pseudomonas, Sporosarcina) show antifreezing activity (cryoprotectant, organic acids, amino acids production). Lignocellulolytic activity by Bacillus atropheus, B. sppp., *Eupenecillium crustaceum* and *Penicillium citrinum*. Enhancement of K and Zn *in-vitro* in wheat and soybean after successful inoculation **Area of isolation:** High mountain pass (Khardungla and Rohtang), Rivers (Indus, Zanskar, I-Z confluence and Beas), subglacial lakes (Pangog, Dashair and Chandratal) of the Indian Himalayas	Saxena et al. (2015)

Table 1.3 Contd. ...

...Table 1.3 Contd.

Extremophiles and their profile	References
Extremophiles: *Aeromicrobium* sp., IARI-R-30 *Kocuria* sp., IARI-R-31 *Janibacter* sp., IARI-R-45 *Arthrobacter* sp., IARI-R-46 *A. psychrolactophilus*, IARI-R-60 *Plantibacter* sp., IARI-R-69 *Arthrobacter pascens*, IARI-R-98 *A. psychrochitiniphilus*, IARI-R-100 *A. sulfonivorans*, IARI-ABR-40 *Arthrobacter nicotianae*, IARI-R-138 *Rhodococcus qingshengii*, IARI-R-139 *Microbacterium oxydans*, IARI-R-141 *Microbacterium* sp., IARI-R-142 *Rhodococcus* sp., IARI-R-7 IARI-ABR-10 *Citricoccus* sp., IARI-R-2 *B. psychrosaccharolyticus*, IARI-R-3 *Bacillus simplex*, IARI-R-4 *Exiguobacterium homiense*, IARI-R-8 *Sanguibacter suarezii*, *Lysinibacillus fusiformis*, IARI-R-11 *Lysinibacillus sphaericus*, IARI-R-25 *Bacillus amyloliquefaciens*, IARI-R-26 *Bacillus thuringiensis*, IARI-ABR-41 *Bacillus altitudinis*, IARI-ABR-44 *Bacillus megaterium*, IARI-ABR-49 *Bacillus subtilis*, IARI-ABR-36 *Paenibacillus tylopili*, IARI-ABR-43 *Paenibacillus pabuli*, IARI-ABR-39 *Paenibacillus terrae*, IARI-R-27 *Paenibacillus lautus*, IARI-R-28 *Bacillus muralis*, IARI-R-29 *Staphylococcus cohnii*, IARI-R-40 *Exiguobacterium marinum*, IARI-R-110 *Sporosarcina psychrophila*, IARI-R-111 *Sporosarcina globispora*, IARI-ABR-37 *Sporosarcina pasteurii*, IARI-R-137 *Exiguobacterium indicum*, IARI-R-140 *Exiguobacterium* sp., IARI-ABR-1 Staphylococcus xylosus, ARI-ABR-2 *S. saprophyticus*, IARI-ABR-3 *Virgibacillus* sp., IARI-ABR-5 *J. halotolerans*, IARI-ABR-18 *V. halodenitrificans*, IARI-ABR-33 *Brevundimonas terrae*, IARI-ABR-47 *Bosea* sp., IARI-ABR-48 *Methylobacterium* sp., IARI-R-9 *Variovorax ginsengisoli*, IARI-R-50 Janthinobacterium lividum, IARI-R-70 *Janthinobacterium* sp., IARI-ABR-46 *Burkholderia* sp., IARI-ABR-51 *Burkholderia cepacia*, IARI-R-6 *Aeromonas hydrophila*, IARI-R-10 *Pseudomonas stutzeri*, IARI-R-12 *P. xanthomarina*, IARI-R-48 *P. extremaustralis*, IARI-R-53 *Pseudomonas cedrina*, IARI-R-57 *Pseudomonas fragi*, IARI-R-59 *Pseudomonas tolaasii*, IARI-R-64 *Pseudomonas* sp., IARI-R-83 *Providencia* sp., IARI-R-87 *Pantoea agglomerans*, IARI-R-89 *Yersinia intermedia*, IARI-R-91 *Providencia rustigianii*, IARI-R-92 *Yersinia aleksiciae*, IARI-R-97 *P. deceptionensis*, IARI-R-112 *Yersinia massiliensis*, IARI-R-113 *Yersinia kristensenii*, IARI-R-125 *Psychrobacter marincola*, IARI-R-127 *Psychrobacter frigidicola*, IARI-R-128 *Pseudomonas orientalis*, IARI-R-129 *Yersinia ruckeri*, IARI-R-131 *Pseudomonas putida*, IARI-R-132 *Pseudomonas moraviensis*, IARI-R-134 *P. psychrophila*, IARI-R-143 *Pseudomonas jessenii*, IARI-ABR-38 *Pseudomonas reactans*, IARI-ABR-42 *Pseudomonas fluorescens*, IARI-ABR-45 *Pseudomonas gessardii*, IARI-ABR-50 *Pseudomonas peli*, IARI-R-25 *Flavobacterium* sp., IARI-ABR-31 *F. psychrophilum*, IARI-ABR-27 *Sphingobacterium* sp. **Application:** *In vitro* antifungal activity against *Rhizoctonia solani* and *Macrophomina phaseolina*. This is the first report of *Arthrobacter nicotianae*, *Brevundimonas terrae*, *Paenibacillus tylopili* and *Pseudomonas cedrina* in cold deserts exhibiting multifunctional PGP applications (ammonia, HCN, gibberellic acid, IAA and siderophore; solubilization of phosphorus and activity of ACC deaminase) at low temperatures **Area of isolation:** Water, soil and sediment samples from Chandratal Lake, Rohtang Pass, Dashair Lake, Beas River	Yadav et al. (2015a)
Extremophiles: 35 Bacilli grouped in seven families [Bacillaceae (48%), Staphylococcaceae (14%), Bacillales incertae sedis (13%), Planococcaceae (12%), Paenibacillaceae (9%), Sporolactobacillaceae (3%), Carnobacteriaceae (1%))]. *Bacillus amyloliquefaciens*, IARI-ABL-36 *Bacillus baekryungensis*, IARI-L-73 *Bacillus cereus*, IARI-L-21 *Bacillus firmus*, IARI-ABL-37 *Bacillus flexus*, IARI-ABL-38 *Bacillus licheniformis*, IARI-ABL-39 *Bacillus marisflavi*, IARI-ABL-40 *Bacillus mojavensis*, IARI-L-74 *Bacillus muralis*, IARI-R-2 *B. psychrosaccharolyticus*, IARI-L-54 *Bacillus pumilus*, IARI-L-118 *Bacillus simplex*, IARI-L-69 *Bacillus subtilis*, IARI-L-46 *Desemzia incerta*, IARI-L-70 *Exiguobacterium antarcticum*, IARI-R-137 *Exiguobacterium indicum*, IARI-R-40 *Exiguobacterium marinum*, IARI-L-116 *Exiguobacterium undae*, IARI-ABR-5 *Jeotgalicoccus halotolerans*, IARI-R-8 *Lysinibacillus fusiformis*, IARI-R-27 *Paenibacillus lautus*, IARI-L-127 *Paenibacillus terrae*, IARI-L-76 *Paenibacillus xylanexedens*, IARI-ABL-9 *Planococcus antarcticus*, IARI-L-39 *Planococcus donghaensis*, IARI-ABL-3 *Planococcus kocurii*, IARI-ABL-41 *Pontibacillus* sp., IARI-ABL-18 *Sinobaca beijingensis*, IARI-L-77 *Sporosarcina aquimarina*, IARI-R-111 *Sporosarcina globispora*, Sporosarcina Psychrophila, IARI-ABL-33 *Staphylococcus arlettae*, IARI-R-29 *Staphylococcus cohnii*, IARI-ABR-1 *Staphylococcus xylosus*, IARI-ABR-18 *Virgibacillus halodenitrificans* **Application:** cold-active enzyme (Amylase, b-glucosidase, pectinase, protease, cellulose, xylanase, b-galactosidase, laccase, chitinase, and lipase) production having immense importance on medical, industrial and agricultural sectors **Area of isolation:** Subglacial lakes (Chandratal Lake, Dashair Lake, and Pangong Lake) of North West Indian Himalayas	Yadav et al. (2016a)

Table 1.3 Contd. ...

...Table 1.3 Contd.

Extremophiles and their profile	References
Extremophiles: 247 morphotypes grouped into 43 clusters. 43 Bacilli were of different species of 11 genera known as Desemzia, Exiguobacterium, Jeotgalicoccus, Lysinibacillus, Paenibacillus, Planococcus, Pontibacillus, Sinobaca, Sporosarcina, Staphylococcus, Virgibacillus **Applications:** At low temperatures some of the isolates (*Bacillus licheniformis, B. muralis, Desemzia incerta, Paenibacillus tylopili, Sporosarcina globispora*) are able to produce indole-3-acetic acid, ammonia, siderophore, gibberellic acid, hydrogen cyanide, solubilization of zinc, phosphate and potassium, 1-aminocyclopropane-1-carboxylate deaminase activity also biocontrol ability against *Rhizoctonia solani* and *Macrophomina phaseolina*. **Area of isolation:** Water, soil and sediments of North Western Indian Himalayas (Khardungla Pass, Chumathang Pass and Rotang Pass).	Yadav et al. (2016b)

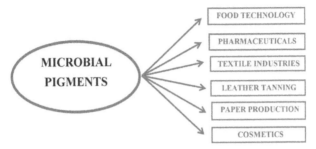

Figure 1.1 Multidimensional applications of microbial pigments.

protect themselves from this type of harmful radiation (gamma rays, UV radiation radio waves, X-rays, etc.) are known as radiophiles (resistant to radiation). The most harmful is considered as UV radiation in extreme environments (deserts or ice poles). Bacterial genus Deinococcus, Pyrococcus, Thermococcus are found to be radiophiles and *Deinococcus radiodurans* can resist UV radiation and ionizing radiation upto 20 times and 200 times respectively in comparison to *Escherichia coli* (Chandi et al. 2009; Singh and Gabani 2011). Radiation directly affects the biochemical machinery of the cell might be DNA or photosystems (Harm 1980; Karentz et al. 1991; Cockell and Rothschild 1999). Out of three types of UV radiation UV-B, UV-C and UV-A with a wavelength of 290 nm to 320 nm, 190 nm to 290 nm and 320 nm to 400 nm respectively, UV-B (solar) and UV-C (germicidal) have lower wavelengths and higher energy are the most harmful ones (Cordoba-Jabonero et al. 2003).

A large amount of UV-C radiation is eliminated by the stratospheric shield of ozone layer however UV-A and UV-B radiation act on DNA leading to pyrimidine (co-valent bond between thymine-thymine or cytosine-cytosine) dimer formation on the same strand causing a misreading of the DNA polymerase and direct damage of DNA (at a wavelength of 254 nm UV breaks molecular bonds of DNA), proteins and lipids by producing reactive oxygen species (Madigan et al. 2012). The entire bacterial community is affected leading to alteration of their composition and biomass (Cordoba-Jabonero et al. 2003). To cope with these situations microorganisms are equipped with protective measures in the forms of pigments especially in case of microbial species of ice poles (the growth withstands the harsh conditions on snow and ice surface) where nutrient contents are low and radiation is high in amount (Muller et al. 1998). Usually species of algae like unicellular members of family Chlamydomonaceae; *Chlamydomonas nivalis* grows as a green mat over the snow surface. The color changes to red (carotenoid rich) when the young cells turns to their resting phases (Lutz et al. 2014) and known as red snow. These carotene rich pigments help the algal species to survive against photolytic and also photodynamic stress. Other than these, another pigment xanthomonadin found in *Xanthomonas* sp. protects the organism (Rajagopal et al. 1997).

Halophiles and halotolerants

In some halophiles residing in the salty waters of Santa Pola salterns instead of chlorophyll, other light sensitive proteins like red and orange colored pigment carotenoids (also known as bacterioruberins) are present (Oren and Rodriguez-Valera 2001). In very negligible oxygen concentrations some species of haloarchaea are able to synthesize integral membrane protein known as bacteriorhodopsin (Inoue et al. 2014). The retinal aldehyde associated with bacteriorhodopsin gives a purple coloration to this protein. The bacteriorhodopsin synthesis is stimulated by low O_2 concentration and as a result when the cells are transferred to such an environment the haloarcheal cells change their color from orange red to purple red and uptake the bacteriorhodopsin into their cytoplasmic membrane. Bacteriorhodopsin usually absorbs green light and converts to cis state from its transconfiguration leading to the transfer of proton across the plasma membrane and finally the retinal returns to it stable transstate. This is how the cis-trans interchange mediates the transfer of proton across the membrane leading to completion of the cycle. On successful accumulation of protons outside the cell membrane a proton motive force is generated coupled with synthesis of ATP (Adenosine tri-phosphate) with the help of ATPase. This type of system is seen in *Halobacterium salinarum* and the light-stimulated proton pump releases Na+ out of the cell through the Na+-H+ antiport system and mediates the entry of the nutrients including K+ required for osmotic balance maintenance. Amino acid entries to the bacterial cells are done indirectly by light mediated amino acid-Na+ symporter (Madigan et al. 2012).

Piezophiles and piezotolerants

Piezotolerants can withstand higher hydrostatic pressure under water but piezophiles grow optimum at high pressure environments (Madigan et al. 2012). Such an example of a piezophilic organism is Moritella found in Mariana Trench of the Pacific Ocean at a depth of 10,000 meter. Violet colored pigment production has been reported by a strain of *Shewanella violacea* found in the Ryuku Trench at a depth of 5110 meter at a hydrostatic pressure of 0.1 to 70.0 MPa at a temperature of 4°C to 15°C (Kato et al. 1995).

Acidophiles and alkaliphiles

Stierle et al. in the year of 2015 isolated an acidophilic fungus from the Berkeley Pit acid mine waste lakes and identified it as a species of *Pleurostomophora* sp. that can produce berckchaetorubramine, a red pigment. Biotechnological use includes its application in the medical industry as an inhibitor of interleukin 1 β (IL-1β), Tumor Necrosis Factor alpha (TNFα), IL-6 production. The results are confirmed after several human cell line trials (Stierle et al. 2015).

Alkaliphiles generally grows in a pH range of 10–13, i.e., above the value of 7 (Horikoshi 1999). They are the products of useful metabolites like carotenoids and antibiotics (Preiss et al. 2015). Carotenoids are responsible for the yellow coloration of the Bacillus shaped bacterial cell (Aono and Horikoshi 1991). They have the systems of proton pumps and synthases that are able to scavenge ROS (Reactive Oxygen Species) protecting the bacterial cells from oxygen toxicity (Aono and Horikoshi 1991; Steiger et al. 2015). Other than this gray colored cyanobacterial species are also found in stromatolites of Australian lagoons.

Psychrotolerants

Psychrotolerant (cold adapted species of microorganisms) *Sphingobacterium antarcticus* can produce zeaxanthin (a type of polar carotenoid), β-carotene (Jagannadham et al. 2000). Another example includes the production of xanthorhodopsin and red cytochrome c3 by psychrophilic bacterial species of Antarctica; *Octadecabacter arcticus*, *Octadecabacter antarcticus* and *Shewanella frigidimarina* respectively (Vollmers et al. 2013).

Thermophiles and hyper thermophiles

Thermophiles are also potent producers of pigments. Examples include production of carotenoid pigments by purple sulfur bacterium *Thermochromatium tepidum* isolated from the hot springs of Yellowstone National Park (Madigan 1984; Suzuki et al. 2007). Bacterioruberin and bacteriorhodopsin were isolated from halophilic archaea found in Uzon Caldera and Geyser Valley of Kamchatka.

Variety of microbial pigments

Carotenoids and phycobilins

Carotenoids protect the bacterial cell from photooxidative damage and acts as a photoprotectant (Becker-Hapak et al. 1997; Goodwin 1980). They also protect cells from cold stress by modulating the membrane fluidity in case of psychrophilic bacteria. Other than β-carotene (non-polar carotenoid), zeaxanthin (polar carotenoid) increases rigidity of the membrane (Subczynski et al. 1992). Russell and Hamamoto (1998) reported that microorganisms when grown in suboptimal temperatures synthesize larger amounts of unsaturated fatty acids than microorganisms grown at higher temperatures and maintain their fluidity of the membrane (Russell and Hamamoto 1998). A study reveals that in case of psychrotrophic bacteria from the Antarctica have carotenoid pigments and the rate of synthesizing this pigment increases when the organisms are transferred to lower or subzero temperatures (Shivaji et al. 1988; 1989a; 1989b; Chattopadhyay et al. 1997). Carotenoid pigments can bind to synthetic membranes and increase their fluidity (Jagannadham et al. 1996a; 1996b; 2000). Psychrophilic algae like *Chlamydomonas nivalis* contains astaxanthin, chemically known as 3, 3-dihidroxi-β, β'-caroteno-4, 4-diona; a xanthophyll. The presence astaxanthin is also reported from *Haematococcus pluvialis*, *Chlorella zofingiensis*, *Chlorococcum*, and *Phaffia rhodozyma*.

Violacein and prodigiosin

Violacein is a water insoluble purple color pigment having antibacterial and antioxidative properties, produced by *Chromobacterium violaceum* that grows specifically on a medium containing tryptophan (Soliev et al. 2011). This bacterium is a common habitat of soil and water. Prodigiosin (red pyrrole-containing pigment) is produced by actinobacterial and gamma-proteobacterial members like *Streptomyces coelicolor*, *Hahella chejuensis*, *Pseudomonas magnesiorubra*, *Vibrio psychroerythreus*, and *Serratia* sp. (Williamson et al. 2006). Some strains of psychrophilic *Janthinobacterium lividum* are found to produce red colored prodigiosin and purple colored violacein. Purple and red colored strains exhibited antibiosis in the terms of high level of resistance against β-lactam antibiotics (Schloss et al. 2010). Prodigiosin are of clinical importance as they act as immunosuppressive and anticancer agents (Williamson et al. 2006).

Pyocyanin and pzaphilones

Pseudomonas aeruginosa (a gammaproteobacteria) produces pyocyanin, an electrochemically active metabolite involved in biological activities like gene expressions, fitness maintenance of bacterial cell and biofilm formation, which is responsible for the blue-green characteristic coloration of the *Pseudomonas* sp. and also have antimicrobial (antibacterial and antifungal) activity (Jayaseelan et al. 2014).

Azaphilones are known to be a class of diverse metabolites found in fungal families known as fungal polyketide with an oxygenated bicyclic core and chiral quaternary center and is of immense importance for their diverse biological (antimicrobial, cytotoxic and anti-inflammatory) activity. Examples include production of red pigment berckchaetorubramine by the acidophilic fungi *Pleurostomophora* sp. (Stierle et al. 2015).

Multidimensional biotechnological application

Extremophilic microorganisms and their products have a major impact on biotechnology fields (Angelaccio 2013). Pigments of microbial origin are advantageous over synthetic pigments (Malik et al. 2012; Venil et al. 2013). The positive aspects of using microbial pigments for biotechnological purposes include their easy manipulation of genes, cheaper cost of substrates and simple techniques for culture maintenance in bioreactors (Venil et al. 2013). At present synthetic pigments are of great importance due to their large use in tanning of leather, production of paper, food and hair coloration, preparation of photoelectrochemical cells, light-harvesting arrays and also in textile industries (Venil et al. 2013) (Fig. 1.2).

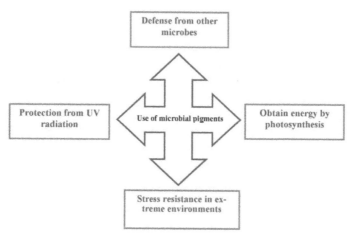

Figure 1.2 Role of microbial pigments in microbial life cycle.

Applications in the food industry

Colored food is the new trend of the food industry in the 21st century in order to meet the interest of consumers on the one hand and on the other hand they acts as additives, antioxidants and also color intensifiers (Malik et al. 2012). The use of food grade colorants is of safer over artificial additives with harmful effect on the environment and health hazards. In order to meet this need pigments from microorganisms' source has become the new area of interest as natural compounds have positive health benefits (Malik et al. 2012). Natural food colors used in breakfast cereals, baby food products, pastas, sauces, processed cheeses, fruit drinks, vitamin-enriched milk products, and in some energy drinks are not only environment friendly but also improve the visual appearance of foods with enhanced probiotic health (Nagpal et al. 2011). Some examples include isolation of astaxanthin from *Agrobacterium aurantiacum*, *Paracoccus carotinifaciens*, *Xanthophyllomyces dendrorhos* and from extremophiles of red snow of the Antarctica, etc. and its use in feed supplement (Fujii et al. 2010). Canthaxanthin used in food, beverages and also in pharmaceutical sectors are obtained from *Bradyrhizobium* sp. and *Haloferax alexandrines*. *Blakeslea trispora* and *Asbhya gossypii* are exploited for the production of food-grade riboflavin and carotene, respectively (Venil et al. 2013). Zeaxanthin from microbial sources are also known to possess applications in food, feed and pharmaceutical industries (Abdel-Aal et al. 2013; Baiao et al. 1999; Nishino et al. 2009).

Applications in the pharmaceutical industry

Microbial pigments have a multidimensional role in the pharmaceutical industry as a nutritional supplement and for the maintenance of human health. Like, β-carotene and astaxanthin (a xanthophyll) are obtained from microbial sources and are photoprotective in nature maintaining the yellow color of the retinal macula and blocking the sun rays to protect some parts of the retina (Venil and Lakshmanaperumalsamy 2009; Soliev et al. 2011). General bacterial pigments such as prodigiosin, carotene and xanthophylls have carcinogens preventing properties coupled with antioxidative and free radical properties (Soliev et al. 2011). Antibiotic properties are also reported from prodigiosin by Hassankhani et al. (2014) and Lapenda et al. (2015). Melanin is found in eukaryotic as well as prokaryotic cells and is responsible for the black, brown or gray colors of many bacteria with potent antioxidative property (Delgado-Vargas et al. 2000). Melanin is also used as an agent of sunblock to protect the skin from UV radiation (Soliev et al. 2011). Another live saving biomolecule is an antibiotic obtained from bacterial or fungal origins that can potentially kill or inhibit other disease causing microorganisms (Madigan et al. 2012). Extremophiles are also able to produce antibiotics, e.g., *Nonomuraea* sp., a yellow colored actinomycete isolated from cave soils of Thailand producing a glycopeptide antibiotic (Gaballo et al. 2006). More research of this type is needed as the development of antibiotic resistance is a grave problem of this era and antibiotics from new and novel unexplored extremophilic micro-organisms' source could be a better alternative.

Applications in the textile industry

Dyes from extremophiles are also used in textile industries for the purposes of coloring natural fibers (wool, cotton, silk), fur, and leather (Venil et al. 2013). Red pigment prodigiosin has been used to dye many fibers like wool, nylon, acrylics, silk and is obtained from *Vibrio* sp. and *Serratia marcescens* (Ahmad et al. 2012). Bacterial species like *Janthinobacterium lividum* and *Chromobacterium violaceum* produces violacein. Due to the increasing interest on natural dyes in the recent past it has been proved to be a major dye used in textile industries and applied in various fabrics like pure silk, pure rayon, jacquard rayon, acrylic, cotton, silk satin, and polyester (Malik et al. 2012; Ahmad et al. 2012; Shirata et al. 2000). It provides separate color tones according to the nature of the fiber. A bluish purple color is found in case of natural fibers and a dark blue color in case of synthetic ones. Other than dyeing properties deoxyviolacein shows antibiotic potentials (Jiang et al. 2012).

Applications as laboratory tools

Bacteria with a fluorescent ability are used for disease detection assays in medical laboratories. They are used to label antibodies and can determine the degree of progress of any specific reaction. Such an example is phycoerythrin which is used as a fluorescence based indicator for the detection of the rate of damage due to peroxy radicals. As the radicals react with phycoerythrin it is easy to make an assay that is indicative of the concentration of the pigment over time (DeLange and Glazer 1989). At first during the beginning of the reaction the pigment will appear on the assay as a zone of a complete fluorescent glow but as the reaction progresses by the addition of the peroxy radicals dark spots will start to appear in the regions where the pigment has already reacted with it and finally leading to the overall decrease of fluorescence over time (DeLange and Glazer 1989). Another pigment of this technique that could be exploited is R-phycoerthyrin for the labeling of the antibodies like IgG by the help of a linker protein SPDP (Mahmoudian et al. 2010). Pathogen of amoebiasis; *Entamoeba histolytica* could be detected using this technique (Thammapalerd et al. 1996). Fluorescent dyes are also used to detect the microorganisms in fluorescent microscopy. *Methanosarcina barkeri* and *Methanobacterium formicicum* (methanogenic archaea), *Chloracidobacterium thermophilum* (acidobacterium) are identified based on this principle due to their blue green autofluorescence and presence of bacteriochlorophyll C (responsible for the red color) inside the chlorosomal vesicles respectively (Madigan et al. 2012). A microeukaryotic diatom *Stephanodiscus* sp. shows red colored autofluorescence (Madigan et al. 2012).

Applications in bioremediation

Bioremediation is the process of environmental clean-up by means of microorganisms. Microorganisms which are able to reduce metal compounds are now a hot topic of research and a weapon for cleaning land contaminated with heavy metals and radionuclides. Some microorganisms

(*Geobacter metallireducens* and *Desulfovibrio desulfuricans*) use U (VI) as their terminal electron acceptor and help in removing uranium from contaminated soils. Removal of selenium from Se contaminated water through precipitation is mediated by *Clostridium* sp. (Kauffman et al. 1986), *Citrobacter* sp., *Flavobacterium* sp., and *Pseudomonas* species (Burton et al. 1987) and Se (VI) (selenite; SeO_4^{-2}) is reduced to Se (O) (elemental selenium). Reduction of Cr (VI) to Cr (III) is done enzymatically and Cr loses its mobility and toxicity by the use of *Bacillus cereus*, *Bacillus subtilis*, *Pseudomonas aeruginosa*, *Achromobacter eurydice*, *Micrococcus roseus*, and *Escherichia coli* (Gvozdyak et al. 1987), as well as *Pseudomonas ambigua* (Horitsu et al. 1987), *Pseudomonas fluorescens* (Bopp and Ehrlich 1988), *Enterobacter cloacae* (Wang et al. 1989), *Streptomyces* spp. (Das and Chandra 1990), *Pseudomonas putida* (Ishibashi et al. 1990), *D. desulfuricans*, and *D. vulgaris*. *Geobacter metallireducens* reduces the soluble Hg (II) to Hg (O) in its volatile form and might be used as a remediation strategy (Lovley 1993).

For microbial fuel cells

Microbial fuel cells use bacteria as a catalyst and oxidize organic and inorganic matters for the generation of an electric current. The principle is that the electrons released by the substrates that are degraded by bacterial species are transferred directly to a negative terminal and flow to a positive terminal of cathode and finally linked using a conductive material with either a resistor or directly operated under a load (generate electricity and run a device) (Logan et al. 2006). These fuel cells are usually operated using mixed bacterial cultures and the results are better than the use of pure cultures (Kour et al. 2019). The species of mixed culture used for this purpose includes *Alcaligenes faecalis* and *Enterococcus gallinarum* (Facultative anaerobic bacteria able to produce hydrogen), *Pseudomonas aeruginosa* (Rabaey et al. 2004). Other classes of bacterial species are from Alphaproteobacteria, Betaproteobacteria, Deltaproteobacteria, Actinobacteria, Acidobacteria, Chloroflexi, and Verrucomicrobia were found in microbial communities (Phung et al. 2004).

Lipases from extremophiles

Lipases (EC 3.1.1.3) are autocatalytic triacylglycerol acyl hydrolases that catalyzes the release of free fatty acids and glycerol, monoacylglycerol and diacylglycerol from triglycerides (Elleuche et al. 2016). Lipases use water-insoluble long-chain acyl esters (> 10 carbon atoms) as substrates and esterases mediate the hydrolysis and synthesis of water-soluble short-chain fatty acids (< 10 carbon atoms) (Bornscheuer 2002). Psychrophilic microorganisms that grow in cold environments such as sea ice, deep sea and snow fields are able to synthesize lipases to fulfill their energy requirement. A well-known lipase from psychrophilic source is yeast; *Candida antarctica* Ca1B (a member of hemiascomycetes) (Joseph et al. 2008).

Diversity of lipases

Inspite of the fact that a large percentage of the earth's biosphere is cold and inhabited by psychrophilic microorganisms, a few cold-active esterases and lipases have been reported from bacterial strains residing permanently in cold sea ice, soil or glaciers and in the deep sea or mountain regions (Al Khudary et al. 2008; Wu et al. 2013; Wi et al. 2014; Parra et al. 2015; Tchigvintsev et al. 2015). The lypolytic enzymes are optimum at their function around 30°C and gradually lose its use at 45°C and above, and are known as cold active. The lipases that are optimum at temperatures above 25°C, can retain their activity upto 45°C are called thermostable lipase. One exception of this definition is the lipase obtained from the Alaskan psychrothrophic bacterium *Pseudomonas* sp. B11-1 (optimum activity at 45°C) (Choo et al. 1998). Cold active lipases retain their activity even at subzero temperatures (Elend et al. 2007). Lipases from extremophiles are listed in Table 1.4. Genomic library screening of deep sea sediments in the Gulf of Mexico resulted in the isolation of cold-adapted and salt-tolerant esterase (Est10) from *Psychrobacter pacificensis* which works best at 25°C and retains 55% of its activity even at 0°C and this is notably increased when salt (2 M to

Table 1.4 Lipase production by extremophiles.

Extremophiles	pH	Temp.	Reference
Moraxella sp. TA 144	8.0	20°C	Feller et al. (1991)
Pseudomonas sp. B11-1 (LipP)	8.0	5–35°C	Choo et al. (1998)
Pyrococcus furiosus	7.6	100°C	Ikeda and Clark (1998)
Pseudomonas sp. KB700A (KB-Lip)	8.0–8.5	35°C	Rashid et al. (2001)
Bacillus thermoleovorans ID-1	8.0–9.0	60°C	Lee et al. (2001)
Aeropyrum pernix K1	8.0	90°C	Gao et al. (2003)
Archaeoglobus fulgidus DSM 4304	11.0	70°C	Rusnak et al. (2005)
Pseudomonas fluorescens (LipB68)	8.0	20°C	Luo et al. (2006)
Photobacterium lipolyticum M37	8.0	25°C	Ryu et al. (2006)
Corynebacterium paurometabolum MTCC 6841	8.5	25°C	Joshi et al. (2006)
Psychrobacter sp. 7195 (LipA1)	9.0	30°C	Zhang et al. (2007)
Uncultured bacterium (LipCE)	7.0	30°C	Elend et al. (2007)
Geobacillus thermoleovorans YN	9.5	70°C	Soliman et al. (2007)
Pseudomonas sp. 7323 (LipA)	9.0	30°C	Zhang and Zeng (2008)
Thermoanaerobacter thermohydrosulfuricus	8.0	75°C	Royter et al. (2009)
Vibrio fischeri	8.0	30°C	Mohankumar et al. (2010)
Uncultured bacterium (rEML1)	8.0	25°C	Jeon et al. (2009)
Marinobacter lipolyticus strain SM19	8.0	80°C	Pérez et al. (2011)
Pseudomonas sp. TK-3 (LipTK-3)	8.0	25–30°C	Tanaka et al. (2012)
Chromohalobacter sp. strain LY7-8	9.0	60°C	Li and Yu (2012)
Pelagibacterium halotolerans B2 T	7.5	45°C	Jiang et al. (2012)
Salimicrobium sp. strain LY19	7.0	50°C	Xin and Hui-Ying (2013)
Colwellia psychrerythraea 34H (CpsLip)	7.0	27°C	Do et al. (2013)
Pseudomonas sp. YY31 (LipYY31)	8.0	25°C	Yamashiro et al. (2013)
Thermotoga maritima	5.0–5.5	70°C	Tao et al. (2013)
Thermus thermophiles HB27	6.3	58.2°C	Fuciños et al. (2014)
Halobacillus sp. AP-MSU8	9.0	40°C	Esakkiraj et al. (2014)
Pseudomonas sp.	8.0	20°C	Katiyar et al. (2017)
Geobacillus sp. EPT9	8.5	55°C	Zhu et al. (2015)
Geobacillus stearothermophilus	9.0	50°C	Alabras et al. (2017)
Bacillus subtilis strain KT985358	8.0	37°C	Saraswat et al. (2017)
Proteus mirabilis	6.0	37°C	Jaiswal et al. (2017)
Pseudomonas sp. LSK25	7.0	10°C	Salwoom et al. (2019)

5 M NaCl) is added to the medium (Wu et al. 2013). Another example of cold active enzyme is EstC isolated from *Streptomyces coelicolor* A3. This extremophile has 50 genes engaged for coding of lipolytic enzyme. The optimum values are of 35°C and 25% of the activity still remains when the temperature is reduced to 10°C (Brault et al. 2012).

Biotechnological applications of cold active lipase

Optimum activity in lower temperatures has allowed researchers to exploit the biotechnological applications of lipase obtained from psychrophiles. Some of the popular uses include:

- Synthesis (organic) of fragile chiral compounds and their application as food additives in the manufacture of cheese and bakery.

- Used as detergents and also in environmental bio-transformations (bioremediations), production of biomedicine (pharmaceuticals) or as a tool of molecular biology (Joseph et al. 2008; Joshi and Satyanarayana 2015).

- Provides protection to the host microorganisms by inhibiting formation of inclusion bodies in psychrophilic bacteria (Feller et al. 1996).

- As a modifier of oils and fats and helps in production of sugar-based polymers, surfactants, emulsifiers, structured lipids, wax esters, synthesize fragrant compounds (Joseph et al. 2008).

- Lipases are potent ingredients in detergent solutions for laundry and in household dishwaters. Lipolase from the fungus *Thermomyces lanuginosa* produced in the filamentous ascomycete *Aspergillus oryzae* is applied in the laundry industry for removal of stains (fats, butter, lipstick, sauces and salad oil) from clothes.

- Washing enzyme Lipomax®, an engineered variant of lipase from *Pseudomonas alcaligenes* that was introduced by Gist-Brocades in 1995 (Joshi and Satyanarayana 2015).

- Applied as additives in bleaching compositions, for cleaning of contact lens and liquid leather (Joseph et al. 2008).

- Acts as food aroma enhancer. Lipase from psychrotrophic *Pseudomonas* strain P38 catalyzes esterification of a flavoring compound at low temperatures (Tan et al. 1996).

- Biodiesel production using the cold active lypolytic enzymes from *Pseudomonas fluorescens* B68, LipB68 is mediated using soybean oil as the substrate and incubating at 20°C for 120 hours. Ninety two percent biodiesel is obtained by this trans-esterification reaction and occurrence of this process in low temperature saves a lot of energy suitable for industrial purposes (Luo et al. 2006).

- Naproxen and ibuprofen are two non-steroidal medicines with an anti-inflammatory property obtained by the action of lypolytic enzymes (Hess et al. 2008).

- Lipase from *S. marcescensis* is used for the production of calcium channel blocker drug Diltiazem (Wicka et al. 2013).

Application of lipases from thermophiles and hyper-thermophiles

Lipases are necessary enzymes with multidimensional application on food, pharmaceutical, textile industries, etc. and can also play a role in bio-conversion of products, generation of volatile compounds or green synthesis by an enzymatic action (Fig. 1.3).

- As most of the industrial processes are carried out at a temperature higher than 45°C use of thermostable lipase contributes to higher rates of reaction, increases substrate solubility, reduces viscosity and contamination risks (Sharma et al. 2002).

- Lipases obtained from thermophiles or hyper-thermophiles show greater stability under harsh denaturing conditions (Mozhaev 1993). Use of lipolytic enzymes from a hyper thermophilic source reduces the requirement of pretreatment steps using chemicals, steam or pressure, use of solvents (in case of T1 lipase mediated methyl butyrate production; Wahab et al. 2014) and intervenes with the elimination of unwanted productions by avoiding necessary steps for separation saving money and man power.

- Lipase isolated from *T. thermohydrosulfuricus* SOL1 and *Caldanaerobacter subterraneus* subsp. *tengcongensis* were resistant to solvents or detergents up to a concentration of 99% (Royter et al. 2009). Esterase from *T. thermohydrosulfuricus* SOL1 show high preference for esters of secondary alcohols thus important for conversion of pharmaceutically valuable

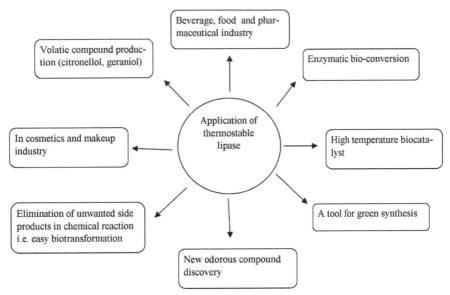

Figure 1.3 Diverse applications of thermostable lipases.

substrates and since it has high temperature tolerance and resistance against organic solvent it mediates biotransformation (Royter et al. 2009).

- Esterase from *Acidicaldus* sp. is applied for the synthesis of medically important S-enantiomers of naproxen and ibuprofen esters (Lopez et al. 2014).

- In the food and beverage sectors inter-esterification processes are important for the conversion of palm oil into cocoa butter fat substitute with a higher melting point and serves as an attractive food, confectionery and cosmetic supplement.

- Quality enhancement of cheap oils by enzymatic bio-conversion. High demanding products like human milk fat substitute or cocoa butter equivalents are commercially made available by the action of lipase (Hasan et al. 2005).

- Formation of flavor precursors by selective hydrolysis of triacylglycerides (Jaeger and Reetz 1998).

- Application of terpene esters with short-chain fatty acids like terpinyl esters, geraniol esters or citronellol esters obtained through enzymatic conversion are high value products of the beverage, food and pharmaceutical industry.

Proteases from extremophiles

Proteases also known as peptidases or proteolytic enzymes are the hydrolases (EC 3) acting on peptide bonds (subclass 4, EC 3.4) and have been the pioneers in industrial application of enzymes. Proteolytic enzymes obtained from microorganisms are of immense importance in a wide range of industrial processes varying from food and feed, textile and leather industry, pharmaceutical and medicinal purposes to laundry and detergent industry which is known to be the largest sector of protease application to date (Rao et al. 1998; Kumar et al. 2008; Khan 2013). Proteases are ubiquitous in living members of nature covering viruses, bacteria and human beings. Almost 2% of the total human genome is considered to be functionally coding for the protease enzyme (Li et al. 2013). They are potent agents for maintaining many processes such as, dietary protein digestion, turnover of cellular protein, metabolic and regulatory processes which are major events for proper continuation of life (Bialkowska et al. 2016). Protease helps us to understand the structure and functional relationship of proteins and represent the large percentage of industrially valuable enzymes on the global market (Sarethy et al. 2011).

Protease from high temperature tolerant bacteria (Thermophiles)

Thermophiles are generally found in high temperature regions and are adapted to thrive at higher temperatures. They are generally classified into several groups based on their optimal activity at different temperature conditions (Table 1.5). Protease producing thermophiles are ubiquitous in nature and found from archaea, bacteria and filamentous fungi. They are found in soils of tropical forests (De Azeredo et al. 2004; Jaouadi et al. 2010a), geothermal springs of marine ecosystems (Klingeberg et al. 1995), hydrothermal vents of deep sea (Matsuzawa et al. 1988; van den Burg et al. 1991), geothermal sediments and hot-streams (Jang et al. 2002), volcanic lava, composts (Hasbay and Ogel 2002), hot water pipelines and boiling outflows of thermal power plants (Dib et al. 1998), thermophilic digesters (Majeed et al. 2013) (Fig. 1.4). They not only expand at temperatures above 60–70°C but are also more resistant to organic solvents, extreme pH and detergents (Jaouadi et al. 2010a; Lagzian and Asoodeh 2012).

Table 1.5 Classification of thermophilic extremophiles Kikani et al. (2010).

S.N.	Thermophiles	Temperature of survival
1	Facultative thermophile	60–65°C and also 37°C
2	Obligate thermophile	65–70°C (growth stops at temp. < 40°C)
3	Extreme thermophile	40–70°C
4	Hyperthermophile	80–115°C

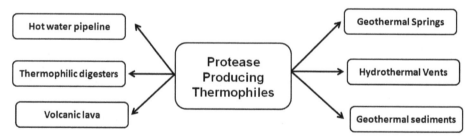

Figure 1.4 Naturally occurring sources of protease producing extremophiles.

Cold-adapted proteases

Psychrophilic and psychrotolerant microorganisms produce peptide bond-breaking enzymes known as proteases. These organisms grow at 0°C. The optimum temperature for these microorganisms does not exceed 15°C and ranges within 20°C for true psychrophiles. In the case of psychrotolerant microbes the optimum temperature varies between 20°C to 35°C (Morita 1975). They are predominantly found in sea waters, arctic soils, glaciers and alpine soils.

Alkaliphilic protease

Alkaliphilic proteases produced by thermo alkaliphiles, psychro alkaliphiles and haloalkaliphiles, grow at pH values of 9.0 but are stunted at pH of neutral 7.0 or lower than that, and are stable in alkaline environments (Horikoshi 1999; Sarethy et al. 2011). Alkaliphiles are abundant in various environments like bathroom tile joints, compost, feather samples found at lake shores, mine-water containment dam, muds, soda lakes, salt lakes, etc. (Gessesse et al. 2003; Saeki et al. 2002;

Kobayashi et al. 2007; Karan et al. 2011; Dastager et al. 2008; Deng et al. 2010; Mitsuiki et al. 2002; Bakhtiar et al. 2002; Raval et al. 2014).

Halophilic proteases

Halophilic proteases are produced by halophilic organisms in the presence of NaCl for their catalytic activity. Halotolerant proteins, are not obtained from halophiles alone, but they are active in a wide range of NaCl concentrations, without any specific requirement on NaCl (Graziano and Merlino 2014). Microorganisms that produces salt-dependent extracellular proteases belong to archaea, bacteria and eukaryotes isolated from brine springs of underground salt deposits, deep-sea sediments, saline and hypersaline lakes, salt pans, salt marshes, solar salterns, salt mines, soda lakes and sea water (Yin et al. 2014; Setati 2010; Sinha and Khare 2013). Some halophiles with very low requirement of NaCl concentration (0.5 M) are known as slight halophiles while other classification of halophiles are completely based on their salt requirements (Table 1.6) (Setati 2010).

Table 1.6 Classification of halophiles.

S.N.	Halophiles	NaCl concentration
1	Archaea	3.4–5.1 M
2	Slight halophiles	0.5 M
3	Moderate halophiles	0.5–2.6 M
4	Extreme halophiles	4.3 M

Application of proteases from extremophilic sources

Proteases are universally used for their use in the detergent industry, textile and leather industries, food and pharmaceuticals and bioremediation of keratinaceous waste generated from poultry farms. However the limitations of enzymes obtained from mesophilic sources have led researchers to search for them from extremophiles and applied in sectors where earlier the acceptability was minimum due to their low tolerance parameters. Table 1.7 represents the recent discoveries of protease from extremophilic microorganisms.

Detergent industry

Proteases are useful in industries such as detergents, degumming of silks, food and feed industry, de-hairing of leathers, cosmetics and pharmaceuticals. They are widely used as washing powders and automatic dishwashing detergents for the degradation of proteinaceous stains caused by blood, milk, egg, grass and also sauces. Gupta et al. (2002) reported the use of subtilopeptidase as an optical lens cleaner. Other commercial proteases of use in several industrial processes are: Alkazym (Novodan, Copenhagen, Denmark), Tergazyme (Alconox, New York, USA), Ultrasil (Henkel, Dusseldorf, Germany) and P3-paradigm (Henkel-Ecolab, Dusseldorf, Germany)—these are of importance in the membrane-cleaning process and other enzymes are prepared for the same purpose. Pronod 153L is made to remove stain of bloods on surgical instruments. Another example of thermophile protease in the detergent industry is alcalase, a serine endopeptidase with optimal activity at 60°C and pH 8.3 (Biotex) obtained from alkalitolerant bacteria *Bacillus licheniformis* (Maurer 2004). Halophilic protease also remains stable in organic solvents and this uniqueness has made them a valuable tool for their use in the detergent industry. Some halophiles with protease as their extremozyme are *Bacillus* sp. EMB9 (Sinha and Khare 2013), *Geomicrobium* sp. EMB2 (Karan and Khare 2010) or *Virgibacillus* sp. EMB13 (Sinha and Khare 2012).

Table 1.7 Proteases from extremophiles and their applications.

Extremophile	Applications	Reference
Coprothermobacter proteolyticus (Thermophilic bacterium)	Proteolysin is a novel highly thermostable and cosolvent-compatible protease with a wide range of pH tolerance and is active at temperatures of up to 80°C	Toplak et al. (2013)
Microbacterium sp. (from alkaline saline environments)	Probably the first report of alkaline proteases from this genus with commercial and educational value. Optimum pH 7.0 and 9.0 with mild NaCl concentration	Lü et al. (2014)
Meiothermus taiwanensis WR-220 (thermophilic bacterium)	Produces keratinolytic protease useful in poultry waste bioremediation. Optimum temperature 65°C and pH 10	Wu et al. (2017)
Bacillus licheniformis, Thermomonas hydrothermalis (From hot springs of Jordan)	Produces thermostable amylase, protease, cellulose, gelatins, and lecithin	Balsam et al. (2017)
Pyrobaculum neutrophilum (Hyperthermophilic archaeon)	A thermophilic serine protease inhibitor (product name is Pnserpin) with anti-inflammatory activity	Fei et al. (2017)
Halobacillus andaensis	Produce bioactive peptide production from fish muscle protein	Delgado-García et al. (2019)
Pseudomonas aeruginosa (from *Mauritia flexuosa* palm swamps soil samples in Peruvian Amazon)	The serine metallo protease and serine protease hydrolyze distillers dried grains with soluble proteins. Optimal temp. is 60°C and pH range of 8 and 11	Flores-Fernández et al. (2019)

Food industry

The higher specificity of enzyme coupled with lower risk of contaminations by mesophiles has allowed the use of extremophilic proteases in the food industry. Alkaliphilic proteases cause the formation of protein hydrolystaes by the degradation of a large number of substrates like; casein (Miprodan; MD Foods, Germany), whey (Lacprodan; MD Foods, Germany), soy (Proup; Novo Nordisk, Denmark) and meat (Flavourzyme, Novo Nordisk, Denmark) (Gupta et al. 2002). They are useful in baby food formulations, preparation of fruit juices or soft drinks, meat tenderization and recovery of protenaceous materials from fish, meat and crustacean shell waste at the time of chitin production (Singhal et al. 2012; Synowiecki 2010).

Leather industry

Proteases have been used in the leather industry for a long time substituting harmful substances used during the time of soaking, dehairing. Alkaline proteases are involved in the elimination of non-collagenous components of the skin and other globular (albumin, globulin) proteins. These processes in one way save time, reduce pollution and on the other hand increase the quality of the leather. Alkaline proteases from *B. subtilis* HQDB32, *B. amyloliquefaciens* are used for the removal of hair from sheep skin and hide respectively (Varela et al. 1997; George et al. 1995).

Poultry industry

The waste generated due to the disposal of the poultry materials worldwide causes a higher degree of environmental pollution that could be minimized by the use of keratinase obtained from mesophilic or extremophilic microorganisms. Thermostable keratinases decompose poultry waste like meat or poultry industry byproducts (Suzuki et al. 2006). Conventional procedures do not produce any nutritive byproduct but keratinase mediated breakdown leads to the formation of hydrolysates which has potential application as fertilizers. Other than this bioremediation property keratinases are used for dehairing of leather, a component of cosmetic industry, detergents and edible films.

Thermophilic protease from *Fervidobacterium islandicum* AW-1 is active at a temperature of 100°C and pH of 9.0 (Nam et al. 2002). *Bacillus tequilensis* hsTKB2, a thermoalkali tolerant halophilic keratinase producer is in one way alkaliphilic (pH 9.0–10.5), thermostable (50–80°C) and halophilic (0–30% NaCl) (Paul et al. 2014).

Cold-active β-galactosidases

β-D-Galactosidase (EC 3.2.1.23) with the systematic name of β-D-galactoside galactohydrolase is an exoglycosidase that hydrolysizes the terminal non reducing β-D-galactoside residue in β-D-galactosides (Wanarska and Kur 2005). The common β-D-galactoside is a lactose made up of D-glucose and D-galactose. In high lactose concentration glucosyl transferase activity is also exhibited (Cruz et al. 1999). Onishi and Tanaka (1996) reported β-D-glucosidase, β-D-fucosidase and α-L-arabinosidase activity of β-D-galactosidase of *Rhodotorula minuta* IFO897.

It was explored in a broad range of bacteria, fungi, plants and animals. Generally Lac Z β-D-galactosidase is isolated from *E. coli* for industrial application but toxicity from *E. coli* to be used as a source has minimized the use of this enzyme. *Bifidobacterium adolescentis* DSM 20083 (Van Laere et al. 2000) and *Bacillus circulans* (Vetere and Paoletti 1998) produces two and three different β-D-galactosidases, respectively. But for LLM (Low Lactose Milk) production yeast *Kluyveromyces lactis* (Maxilact® product line from DSM, Lactozym® Pure from Novozymes and GODO-YNL2 Lactase from DuPont) and the fungus *Aspergillus oryzae* (Tolerase TM L from DSM) are used as safe sources (GRAS status).

Two cold-active β-D-galactosidases are reported to date from eukaryotic microorganisms; the yeast-like fungus *Guehomyces pullulans* (Song et al. 2010; Nakagawa et al. 2006b). But due to their mesophilic origin they are not suitable products to be used for low temperature reactions. As a result, β-D-galactosidases from cold adaptive microorganisms are of popular demand. In a comparison with their mesophilic partner, cold-active enzymes can carry out hydrolysis of lactose at low temperatures (10°C) that has allowed the hydrolysis of milk during cold storage leading to save time, energy and less risk of contamination by mesophilic enzyme producers (Cieśliński et al. 2016).

The additional benefits of using cold active enzymes over mesophiles include avoidance of nonenzymatic products generation that can cause browning at high temperatures. The main sources of cold-active β-D-galactosidases are psychrophilic and psychrotolerant microorganisms isolated from low temperature areas. *Arthrobacter* sp. C2-2 was obtained from fellfield soil (Karasová-Lipovová et al. 2003), *Planococcus* sp. SOS Orange from a hypersaline pond situated at the Antarctica (Sheridan and Brenchley 2000), *Carnobacterium piscicola* BA was isolated from a soil sample taken in late winter from a field treated with whey in Pennsylvania, USA (Coombs and Brenchley 1999), *Pseudoalteromonas* sp. 22b, which was isolated from the digestive tract of Antarctic krill, *Thysanoessa macrura* (Cieśliński et al. 2005) and *Alkalilactibacillus ikkense,* which was isolated from the Ikka columns in South-West Greenland (Schmidt and Stougaard 2010). Due to the problem of creating proper laboratory condition for maintenance, scientists have created the metagenomic library and identified the gene (zd410) that encodes cold-active β-D-galactosidase (Wang et al. 2010).

Biotechnological applications of β-D-galactosidase

Cold-active enzymes with their high catalytic efficiency at low temperatures and thermolability at high temperatures are the primary basis for their wide range of biotechnological implications (Fig. 1.5).

- Enzymatic degradation leads to galactose-glucose syrup (dairy syrup) formation, used in the food industry as a sweetener.
- Lactose hydrolysis could avoid the problems and reduce the costs of transport and storage of whey; that arise due to its concentration by crystallization as a result of evaporation.

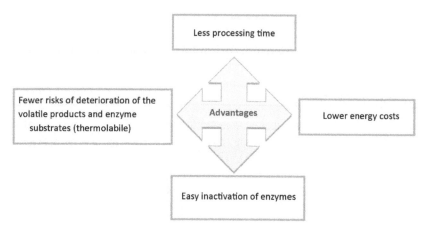

Figure 1.5 Benefits of cold active enzymes.

- Low lactose milk production: This is generally used in the dairy industry for the production of LLM (Low Lactose containing Milk) or the hydrolysis of lactose in whey and whey permeate. It also helps in the screening of the cloned cells in genetic engineering processes by indicating the blue and white colors (Juers et al. 2012). LLM are useful for consumption by people with low lactose tolerance, persons who suffer from abdominal pain, bloating borborygmi and loose stools due to consumption of milk and allied products (Juers et al. 2012). A huge percentage of Asian (more than 90%), African (80–100%), Native American (more than 90%) and Southern European (more than 80%) populations suffer from these issues (Mlichova and Rosenberg 2006). The two products of enzymatic hydrolysis of lactose; galactose and glucose are used as a carbon source by microorganisms and leads to the production of helpful products such as, lactates, acetates, ethanol, butanediol and biopolymers (Guimaraes et al. 1992; Coté et al. 2004; Mehaia et al. 1993; Nath et al. 2014). Two patents have been found regarding cold active β-D-galactosidases from *Alkalilaclibacillus ikkense* (Stougaard and Shmidt, Patent No.: US 8,288,143 B2) and *Pseudoalteromonas haloplanktis* (Hoyoux et al. Patent No.: US 6,727,084 B1) for the hydrolysis of lactose in dairy products and cow's milk in processes conducted at low temperatures.

- Production of Lactulose: β-D-galactosidases produce heterooligosaccharides (HOSs) by transferring the galactosyl moiety to another sugar (other-than lactose, glucose or galactose). 4-*O*-β-D-galactopyranosyl-β-D-fructofuranose (commonly known as lactulose) is used in the food industry as prebiotics, in medicine as the treatment of constipation, hepatic encephalopathy. A large amount of oligosaccharides could be obtained by the activity of β-D-galactosidase (Pawlak-Szukalska et al. 2014). One such example includes the synthesis of heterooligosaccharides like lactulose (i.e., galactosylfructose), galactosyl-xylose and galactosyl-arabinose from lactose and the appropriate monosaccharide.

- Galacto-oligosaccharides production: Galactosyltransferase activity of this enzyme leads to the synthesis of Galacto-Oligosaccharides (GOSs) with two or more galactose units by breaking lactose. Rahim and Lee 1991 reported the production of GOSs by psychrotolerant *Bacillus subtilis* KL88. Karasová-Lipovová et al. (2003) isolated psychrotolerant Antarctic bacterium *Arthrobacter* sp. C2-2 and evaluated its GOSs synthesizing abilities by using milk at low temperatures. Nakagawa et al. (2006a) found the cold-active rBglAp β-D-galactosidase from *Arthrobacter psychrolactophilus* F2 for the synthesis of GOS trisaccharides at 10°C. Pawlak-Szukalska et al. (2014) noted the formation of tri- and tetrasaccharides in lactose solutions (292–584 mM) as an action of cold-active β-D-galactosidase from *Arthrobacter* sp. 32cB. GOSs act as a prebiotic and stimulate the growth of bifidobacteria that leads to a decrease in the population of putrefactive bacteria in the gut of humans and other animals. GOSs have some influence on consumers in terms of reduction in bad cholesterol in blood, less chances of development of colon cancer incidences, absorption of Ca^{2+} from consumed diet, etc. Not only with pharmaceutical applications, but they are used in the cosmetics and food industries as low calorie sweetening agents (Mlichová and Rosenberg 2006).

- The galactosyltransferase synthesizes β-glycosides where D-galactose residue is bond by β-1→4 linkages to such aglycons as different alcohols (Stevenson et al. 1993), antibiotics (Scheckermann et al. 1997), ergot alkaloids (Kren et al. 1992) and flavonol glycoside myricitrin (Shimizu et al. 2006).

- Acts as the biocatalysts for food ingredients, cosmetics, pharmaceuticals production and glycosylates thermolabile chemicals.

- Lactose acts as the galactosyl donor in the glycosylation of aromatic glycoside salicin (2-(hydroxymethyl) phenyl-β-D-glucopyranoside) catalyzed by the β-D-galactosidase from *Arthrobacter* sp. 32cB.

- Cold-active β-D-galactosidase from Antarctic *Pseudoalteromonas* sp. 22b demonstrates the ability for the synthesis of alkyl galactosidase (Makowski et al. 2009).

- D-Tagatose production: D-Tagatose, a low calorie monosaccharide (a GRAS product; 2001, by US FDA) is known as an isomer of the aldohexose D-galactose (or a natural ketohexose) is used as a substitute of sucrose in beverages, dietary candies, and dietary products with innumerable health benefits l such as probiotic, low-glycaemic sweetener, anti-cariogenic and antioxidant potentials (Levin 2002; Oh 2007). It can check fasting blood glucose levels and glycosylated HbA1c in case of Type-2 DM patients; reduce LDL-cholesterols and obesity issues (Lu et al. 2008; Ensor et al. 2015). Hydrolysis of pure lactose by β-D-galactosidase leads to the production of D-galactose and then D-glucose is separated and pure D-galactose is isomerized to D-tagatose. Cold-active β-D-galactosidase reported from the marine *Pseudoalteromonas haloplanktis* is able to hydrolyze (96%) lactose in whey permeate at 23°C and pH of 7.0. Recombinant strain of *Pichia pastoris* (incorporation of β-D-galactosidase gene from psychrotolerant bacterium *Arthrobacter chlorophenolicus*) produces β-D-galactosidase on growing on the whey permeate and utilizes the product D-glucose, while D-galactose remains in the medium. About 90% efficiency of the reaction is obtained after 168 hours of incubation at 30°C temperature in whey permeate producing upto 120 g L^{-1} lactose. Here unlike *Pseudoalteromonas haloplanktis* the inhibitory effect of D-glucose was overcome (Wanarska and Kur 2012).

Phospho-triestarase like lactonases from extremophiles

OPs (Organophosphate compounds) are the core of pesticides (38% of the leading pesticides such as, coumaphos, chlorpyrifos, diazinon, dimethoate, paraoxon, parathion) and nerve based toxic agents (chemical warfare agents used in World War II like sarin, soman, tabun) (Singh 2009; Raushel 2002). Pesticides are of great importance in agricultural sectors for the purpose of high product (in terms of quality and quantity) generation but their negative impact includes water pollution leading to human health disorders. Chemically OPs are esters of phosphorus with various combinations of carbon, nitrogen, oxygen, and sulfur. They are divided in six different subclasses like phosphates, phosphonates, phosphorothioates, phosphorodithioates, phosphorothiolates and phosphoramidates (Can 2014). OP toxicity works by blocking the key enzyme acetylcholineesterase by phosphorylation of the serine residue in its active site and leads to the accumulation of acetylcholine (neurotransmitter) in neuromuscular or neural junction.

The problems include ataxia, deterioration of semen quality and sperm chromatin, bronchial hypersecretion, bronchoconstriction, convulsions, chromosomal aberrations, cancerogenesis, DNA damage, endocrine disorders, gene mutations, hypothermia, hyper salivation, insulin resistance, lacrimation, respiratory failure, skeletal muscle fasciculation and twitching, and finally death (Gupta

2005; Carey et al. 2013; Ojha and Gupta 2015; Salazar-Arredondo et al. 2008; Lasram et al. 2014). As OPs cause hazards to community health there is an urgent need to find remediation strategies to overcome this problem. Therefore the main way to avoid this problem is bioremediation of OPs by microorganisms. There is a great demand for finding highly resistant and stable enzymes for this purpose. Organophosphorus hydrolases (OPHs), Methyl Parathion Hydrolases (MPHs), Serum paraoxonases (PONs) and Microbial prolidases/OP acid anhydrolases are enzymes that are able to hydrolyze these OP components.

Phosphotriesterase-Like Lactonases (PLLs)

The Phosphotriesterase-Like Lactonase (PLL) family includes a group of enzymes that have the main lactonase activity on lactones and Acyl-Homoserin Lactones (AHLs), phosphotriesterase activity towards organophosphates. They are also found in mesophiles; from *Deinococcus radiodurans* (*Dr* OPH/*Dr* 0930), *Mycobacterium tuberculosis* (PPH), *Mycobacterium avium* subsp. *Paratuberculosis* K-10 (MCP) (Chow et al. 2009), *Rhodococcus erythropolis* (AhlA) (Afriat et al. 2006; Hawwa et al. 2009b). Thermostable PLLs are obtained from *Geobacillus stearothermophilus* (GsP) (Hawwa et al. 2009a), *Geobacillus kaustophilus* (GKL/*Gka* P) (Chow et al. 2010). PLLs are reported in the hyperthermophilic crenearchaeon *Sulfolobus solfataricus* (*Sso* Pox) and *Sulfolobus acidocaldarius* (*Sac* Pox) (Afriat et al. 2006; Merone et al. 2005; Porzio et al. 2007). Other sources of PLLs include *Sulfolobus islandicus* (*Sis* Lac) (Hiblot et al. 2012b), *Vulcanisaeta moutnovskia* (*Vmo* Lac/VmutPLL) (Kallnik et al. 2014). Phosphotriesterase activity has been found in several hydrolases such as prolidase, PLL and MBL enzymes extremophilic origins. So the new trend is to use extremophilic PLL enzymes as a tool for bioremediation.

Other than OP degradation these extremozymes could be exploited for the elimination of undesired microorganisms from water purifying plants by using lactonases immobilized on membrane filters (Ng et al. 2011). This trait of these enzymes was achieved due to their ability to interfere with bacterial quorum sensing property by the Sso Pox lactonase activity on acylhomoserine lactones. Another report includes the use of mutant *Sso* Pox (W263F) as a detoxification tool. The enzyme was dissolved in several buffered aqueous solvents (like 30% ethanol, 30% or 50% methanol and 0.1% sodium-dodecyl-sulfate) to evaluate its catalytic activity under stressing denaturing conditions. One such example includes the toxin extraction from contaminated soils (Merone et al. 2010; Hiblot et al. 2012a). The results were compared with those obtained with *bd* PTE from *B. diminuta*. It was revealed that W263F outperforms *bd* PTE in most of the tested situations. After a 15 minute treatment at room temperature 99.5% of paraoxon was hydrolized in 30% methanol and ethanol and 0.1% sodium-dodecyl-sulfate (Merone et al. 2010). PLL enzymes have proved to be successful for OP detoxification. It is a fact that *p*-nitrophenol produced by paraoxon and parathion hydrolysis is toxic in nature but lesser toxicity is reported than the original OP and there are soil bacteria that are able to degrade *p*-nitrophenol by utilizing it as carbon source (Munnecke 1979) offers complete removal of these xenobiotics from nature (Spain and Gibson 1991).

Role of carbonic anhydrase for reducing global warming issues

Global warming is one of the major issues of the 21st century leading to environmental deterioration. CO_2 is the major contributor of the global warming process causing the rise of the earth's average temperature over the past few decades and the average CO_2 concentration in air has risen from 280 ppm to 398 ppm due to industrialization (Shakun et al. 2012). These situations could be a result of uncontrolled fossil fuel combustion and over exploitation of fossil fuels, extreme emissions of thermal power plants and also rapid uncontrolled deforestation (Solomona et al. 2009). While during the modern age people have benefited with industrialization events, however a negative impact is seen in terms of melting of Antarctic ices, thinning of green land masses and finally leading to an increase in the ocean water levels over the past 30 years than in the mid-19th century (Bose and Satyanarayan 2018).

As oceans are the main agent of absorption of CO_2 (30%) rapid acidification (26% higher than the normal) of oceanic water is another grave problem and leads to the gradual fall in the concentration of O_2 in coastal water bodies. Other affects include the change in the normal water cycle, imbalance of ecosystem, shifts in abundance and distribution patterns of marine flora and fauna, poisoning of terrestrial or oceanic ecosystem, over occurrence of uncontrollable microbial diseases, extinction of beneficial or rare species, qualitative and quantitative reduction of crop yield, rise in price of essential commodities and lifesaving medicines, safe and drinkable water crisis (Houghton 2005; Adams et al. 1998). So it is the right time to seek some solutions in this respect and carbon capturing followed by storing it in an immobilized form to any geological formations from where it cannot be leaked by means of physical (using chemicals, geological carbon sequestration) and biological methods (using plants and microbes) are the best solutions. Other methods include the scrubbing of CO_2 from air and fixing it as a mineral carbonate.

Carbon sequestration by extremozyme carbonic anhydrase

Inspite of the physical and biological means of planting trees or using chemicals to reduce carbon emissions from popular sources, a great trust is put forward on microbial sources. Microbes from mesophilic or extremophilic origin are able to produce enzymes known as carbonic anhydrase that can cause CO_2 breakdown. Carbonic anhydrase is a ubiquitous zinc metalloenzyme (EC No. 4.2.1.1) that mediates the inter-conversion of carbon dioxide to bicarbonates (Smith and Ferry 2000). The process of carbonic anhydrase mediated breakdown of CO_2 includes the supply of flue gas on a bioreactor earlier provided with immobilized enzymes set on appropriate beads. Enzymatic action splits CO_2 and releases HCO_3^- and H^+ which are used further for several purposes. The selection of microorganisms as a source of carbonic anhydrase is shifted to organisms of extreme environments as they can thrive well in adverse realms of nature. Using thermoalkali stable (can withstand high temperatures and high pH; known as polyextremophile) carbonic anhydrases is an alternative cost effective and highly valued process of mitigating global warming issues. Reports include the presence of all the three necessary genes in prokaryotic organisms; extremophilic archaebacteria and eubacteria (Smith and Ferry 2000). The list of extremophiles as contributors of carbonic anhydrase is listed in Table 1.8. It is a new concept of using extremozymes to solve environmental issues like global warming and hopefully can contribute to open new paths in this respect.

Conclusion and future prospects

Though the research approaches on extremophiles and extremozymes from them have drastically changed over the past few years and nowadays the reliability on extremozymes as a substitute of mesophilic counterparts has increased due to their diverse array of application in food, pharmaceuticals, laboratory techniques, enzyme production industry and also in environmental clean-up projects, etc. The developments of patents on D-Tagatose production using β-D-galactosidase is a proof of this success (Wanarska and Kur 2012). Still hindrances regarding their isolation and culture in simulated laboratory conditions are a grave problem concerning their biotechnological exploitation but the aggressive development of molecular biology tools and metagenomic approaches are opening up new horizons of extremophiles research. Genetic engineering approaches are making the task much easier. As they are present in one and all extreme conditions ranging from high temperature to extreme salinity, the gene pool is also of a large variety. Therefore easy exploitation could be made from these sources to facilitate functions where mesophilic or conventional enzymes have failed to establish a proper procedure or demands high money and processing costs. In the 21st century the research and application purposes of enzymes are no doubt at their peak but still a big percentage of this untapped knowledge is yet to be revealed for the welfare of mankind. Hopefully in the near future extremozymes will solve the issues of environment cleanup by bioremediation or organophosphate like xenobiotic compounds or carbon sequestration and by other ways to develop

Table 1.8 Carbonic anhydrases from extremophiles.

Extremophiles	Authors
Methanosarcina thermophila (γ-CAs)	Alber et al. (1994)
Methanosarcina thermophila	Alber et al. (1996)
Methanobacterium thermoautotrophicum	Smith and Ferry (1999)
Pyrococcus horikoshii	Jeyakanthan et al. (2008)
Sulfurihydrogenibium yellowstonense	Luca et al. (2012)
Sulfurihydrogenibium yellowstonense (α-CAs)	Capasso et al. (2012)
Sulfurihydrogenibium azorense (α-CAs)	Luca et al. (2013)
Thiobacillus thioparus strain THI115	Ogawa et al. (2013)
Sulfurihydrogenibium azorense	De Luca et al. (2013)
Persephonella marina EX-H1T	Kanth et al. (2014)
Persephonella marina EX-H1	Jo et al. (2014)
Thalassiosira weissflogii	Vullo et al. (2014)
Thermovibrio ammonificans	James et al. (2014)
Nostoc sp.	De Luca et al. (2015)
Pseudoalteromonas haloplanktis	Vullo et al. (2015)
Nostoc commune	Vullo et al. (2015)
Photobacterium profundum (α-CAs)	Somalinga et al. (2016)
Colwellia psychrerythraea	Luca et al. (2016)
Bacillus halodurans	Faridi and Satyanarayana (2016)

techniques to enhance the sectors of food, pharmaceuticals, bio-transformation, bio-synthesis, textiles, fuel sources, diagnosis of disease, beverage industry, etc.

Acknowledgement

The authors are thankful to Department of Botany and Forestry, Vidyasagar University for their support. The are also grateful to the Department of Science and Technology Biotechnology for financial support in the form of SMCMF (Swami Vivekananda Merit Cum Means Fellowship).

References

Abdel-Aal, E.S.M., H. Akhtar, K. Zaheer and R. Ali. 2013. Dietary sources of lutein and zeaxanthin carotenoids and their role in eye health. Nutrients 5: 1169–1185.

Abe, F. and K. Horikoshi. 2001. The biotechnological potential of piezophiles. Trend. Biotechnol. 19: 102–108.

Adams, M.R., H.B. Hurd, S. Lenhart and N. Leary. 1998. Effects of global climate change on agriculture: an interpretative review. Climat. Res. 11: 19–30.

Adams, M.W., F.B. Perler and R.M. Kelly. 1995. Extremozymes: Expanding the limits of biocatalysis. Nat. Biotechnol. 13: 662–668.

Afriat, L., C. Roodveldt, G. Manco and D.S. Tawfik. 2006. The latent promiscuity of newly identified microbial lactonase is linked to a recently diverged phosphotriesterase. Biochemistry 45: 13677–13686.

Ahmad, W.A., W.Y.W. Ahmad, Z.A. Zakaria and N.Z. Yusof. 2012. Application of Bacterial Pigments as Colorant: The Malaysian Perspective. N.Y. Springer 29–38: 59–70.

Akolkar, A.V., D. Durai and A.J. Desai. 2010. *Halobacterium* sp. SP1 (1) as a starter culture for accelerating fish sauce fermentation. J. Appl. Microbiol. 109: 44–53.

Alabras, R., K. Alashkar and A. Almariri. 2017. Production and optimization of extracellular lipase enzyme produced by locally strain of *Geobacillus stearothermophilus*. J. Che. Phar. Sci. 10: 1.

Alber, B.E. and J.G. Ferry. 1994. A carbonic anhydrase from the archaeon *Methanosarcina thermophila*. Proc. Nati. Acad. Sci. U.S.A. 91: 6909–6913.

Alber, B.E. and J.G. Ferry. 1996. Characterization of heterologously produced carbonic anhydrase from *Methanosarcina thermophila*. J. Bacteriol. 178: 3270–3274.

Al-Khudary, R., N.I. Stösser, F. Qoura and G. Antranikian. 2008. *Pseudoalteromonas arctica* sp. nov., an aerobic, psychrotolerant, marine bacterium isolated from Spitzbergen. Int. J. Syst. Evol. Microbiol. 58: 2018–2024.

Alquéres, S.M.C., R.V. Almeida, M.M. Clementino, R.P. Vieira, W.I. Almeida, A.M. Cardoso et al. 2007. Exploring the biotechnological applications in the archaeal domain. Braz. J. Microbiol. 38: 398–405.

Angelaccio, S. 2013. Extremophilic SHMTs: from structure to biotechnology. B.M.C. Res. Int. 1–10.

Aono, R. and K. Horikoshi. 1991. Carotenes produced by alkaliphilic yellow-pigmented strains of *Bacillus*. Agric. Biol. Chem. 55: 2643–2645.

Aulitto, M., S. Fusco, G. Fiorentino, D. Limauro, E. Pedone, S. Bartolucci et al. 2017. *Thermus thermophilus* as source of thermozymes for biotechnological applications: homologous expression and biochemical characterization of an α-galactosidase. Microb. Cell. Fact. 16: 28.

Baiao, N.C., J. Mendez, J. Mateos, M. García and G.G. Mateos. 1999. Pigmenting efficacy of several oxycarotenoids on egg yolk. J. Appl. Poult. Res. 8: 472–479.

Bakhtiar, S., M.M. Andersson, A. Gessesse, B. Mattiasson and R. Hatti-Kaul. 2002. Stability characteristics of a calcium-independent alkaline protease from *Nesterenkonia* sp. Enz. Microb. Technol. 32: 525–531.

Balsam, T., Al. Mohammad, H.I. Daghistani, A. Jaouani, S. Abdel-Latif and C. Kennes. 2017. Isolation and characterization of thermophilic bacteria from jordanian hot springs: *Bacillus licheniformis* and *Thermomonas hydrothermalis* isolates as potential producers of thermostable enzymes. Int. J. Microbiol. 6943952: 12.

Becker-Hapak, M., E. Troxtel, J. Hoerter and A. Eisenstark. 1997. RpoS dependent overexpression of carotenoids from *Erwinia herbicola* in OXYR-deficient *Escherichia coli*. Biochem. Biophys. Res. Commun. 239: 305–309.

Bertoldo, C. and G. Antranikian. 2002. Starch-hydrolyzing enzymes from thermophilic archaea and bacteria. Curr. Opin. Chem. Biol. 6: 151–60.

Bhojiya, A.A. and H. Joshi. 2012. Isolation and characterization of zinc tolerant bacteria from Zawar Mines Udaipur, India. Int. J. Environ. Eng. Manage. 3: 239–242.

Białkowska, A., E. Gromek, T. Florczak, J. Krysiak, K. Szulczewska and M. Turkiewicz. 2016. Extremophilic proteases: developments of their special functions, potential resources and biotechnological applications. pp. 399–444. *In*: Rampelotto, P.H. (ed.). Biotechnology of Extremophiles Advances and Challenges. Springer International Publishing, Switzerland.

Bopp, L.H. and H.L. Ehrlich. 1988. Chromate resistance and reduction in *Pseudomonas fluorescens* strain LB300. Arch. Microbiol. 150: 426–31.

Bornscheuer, U.T. 2002. Microbial carboxyl esterases: classification, properties and application in biocatalysis. FEMS Microbiol. Rev. 26: 73–81.

Bose, H. and T. Satyanarayana. 2018. Carbonic anhydrases of extremophilic microbes and their applicability in mitigating global warming through carbon sequestration. pp. 227–248. *In*: Durvasula, R.V. and D.V. Subbarao (ed.). Extremophiles From Biology to Biotechnology. CRC Press Taylor and Francis Group.

Brault, G., F. Shareck, Y. Hurtubise, F. Lepine and N. Doucet. 2012. Isolation and characterization of EstC, a new cold-active esterase from *Streptomyces coelicolor* A3(2). Plos One 7: 32041.

Bruins, M.E., A.E. Janssen and R.M. Boom. 2001. Thermozymes and their applications: a review of recent literature and patents. Appl. Biochem. Biotechnol. 90: 155–186.

Burton, G.A. Jr., T.H. Giddings, P. DeBrine and R. Fall. 1987. High incidence of selenite-resistant bacteria from a site polluted with selenium. Appl. Environ. Microbiol. 53: 185–88.

Camacho, M., A. Rodríguez-Arnedo and M.J. Bonete. 2002. NADP-dependent isocitrate dehydrogenase from the halophilic archaeon *Haloferx volcanii*: Cloning, sequence determination and overexpression in *Escherichia coli*. FEMS Microbiol. Lett. 209: 155–160.

Can, A. 2014. Quantitative structure-toxicity relationship (QSTR) studies on the organophosphate insecticides. Toxicol. Lett. 230: 434–443.

Canganella, F. and J. Wiegel. 2011. Extremophiles: From abyssal to terrestrial ecosystems and possibly beyond. Naturwissenschaften 98 253–279.

Cannio, R., N.Di. Prizito, M. Rossi and A. Morana. 2004. A xylan degrading strain of *Sulfolobus solfataricus*: isolation and characterization of the xylanase activity. Extremophiles 8: 117–124.

Capasso, C., V. De Luca, V. Carginale, R. Cannio and M. Rossi. 2012. Biochemical properties of a novel and highly thermostable bacterial α-carbonic anhydrase from *Sulfurihydrogenibium yellowstonense* YO3AOP1. J. Enz. Inhb. Medcn. Chem. 27: 892–897.

Carey, J.L., C. Dunn and R.J. Gaspari. 2013. Central respiratory failure during acute organophosphate poisoning. Respir. Physiol. Neurobiol. 189: 403–410.

Cavicchioli, R., D. Amils and T.Mc. Genity. 2011. Life and applications of extremophiles. Environ. Microbiol. 13: 1903–1907.

Chandi, C.R., H.S. Mohanta and S.K. Dash. 2009. Extremophiles as novel cell factories. pp. 282–299. *In*: Mishra, C.S.K. (ed.). Biotechnology Applications. I.K. International Publications, New Delhi, India.

Chattopadhyay, M.K., M.V. Jagannadham, M. Vairamani and S. Shivaji. 1997. Carotenoid pigments of an Antarctic psychrotrophic bacterium *Micrococcus roseus*: temperature dependent biosynthesis, structure and interaction with synthetic membranes. Biochem. Biophys. Res. Commun. 239: 85–90.

Choo, D.W., T. Kurihara, T. Suzuki, K. Soda and N. Esaki. 1998. A cold-adapted lipase of an Alaskan psychrotroph, *Pseudomonas* sp. strain B11-1: gene cloning and enzyme purification and characterization. Appl. Environ. Microbiol. 64: 486–491.

Chow, J.Y., L. Wu and W.S. Yew. 2009. Directed evolution of a quorum-quenching lactonase from *Mycobacterium avium* subsp. *paratuberculosis* K-10 in the amidohydrolase superfamily. Biochemistry 48: 4344–4353.

Chow, J.Y., B. Xue, K.H. Lee, A. Tung, L. Wu, R.C. Robinson et al. 2010. Directed evolution of a thermostable quorum-quenching lactonase from the amidohydrolase superfamily. J. Biol. Chem. 285: 40911–40920.

Christner, B.C., J.C. Priscu, A.M. Achberger, C. Barbante, S.P. Carter, K. Christianson et al. 2014. A microbial ecosystem beneath the West Antarctic ice sheet. Natr. 512: 310–313.

Cieśliński, H., J. Kur, A. Białkowska, I. Baran, K. Makowski and M. Turkiewicz. 2005. Cloning, expression, and purification of a recombinant cold-adapted β-galactosidase from antarctic bacterium *Pseudoalteromonas* sp. 22B. Protein. Expr. Purif. 39: 27–34.

Cieśliński, H., M. Wanarska, A. Pawlak-Szukalska, E. Krajewska, M. Wicka and J. Kur. 2016. Cold-active β-galactosidases: sources, biochemical properties and their biotechnological potential. pp. 445–470. *In:* Rampelotto, P.H. (ed.). Biotechnology of Extremophiles Advances and Challenges. Springer International Publishing, Switzerland.

Cimini, D., M.De. Rosa, A. Panariello, V. Morelli and C. Schiraldi. 2008. Production of a thermophilic maltooligosyltrehalose synthase in *Lactococcus lactis*. J. Ind. Microbiol. Biotechnol. 35: 1079–1083.

Cockell, C.S. and L.J. Rothschild. 1999. The effects of ultraviolet radiation on diurnal photosynthetic patterns in the three taxonomically diverse microbial mats. Photochem. Photobiol. 69: 203–210.

Coombs, J.M. and J.E. Brenchley. 1999. Biochemical and phylogenetic analyses of cold-active β-galactosidase from the lactic acid bacterium *Carnobacterium piscicola* BA. Appl. Environ. Microbiol. 65: 5443–5450.

Cordoba-Jabonero, C., L.M. Lara, M. Mancho, A. Marquez and R. Rodrigo. 2003. Solar ultraviolet transfer in the Martian atmosphere: biological and geological implications. Planet Space Sci. 51: 399–410.

Coté, A., W.A. Brown, D. Cameron and G.P. van Walsum. 2004. Hydrolysis of lactose in whey permeate for subsequent fermentation to ethanol. J. Dairy. Sci. 87: 1608–1620.

Cruz, R., V.D. Cruz, J.G. Belote, M.D. Khenayfes, C. Dorta and L.H.D. Oliveira. 1999. Properties of a new fungal beta-galactosidase with potential application in the dairy industry. Rev. Microbiol. 30: 265–271.

Das, S. and A.L. Chandra. 1990. Chromate reduction in Streptomyces. Experientia. 46: 731–33.

DasSarma, P., J.A. Coker, V. Huse and S. Dassarma. 2010. Halophiles, industrial applications. pp. 1–9. *In*: Flickinger, M.C. (ed.). Encyclopedia of Industrial Biotechnology: Bioprocess, Bioseparation, and Cell Technology. Hoboken, NJ: John Wiley & Sons.

Dastager, S.G., A. Dayanand, W.J. Li, C.J. Kim, J.C. Lee, D.J. Park et al. 2008. Proteolytic activity from an alkali-thermotolerant *Streptomyces gulbargensis* sp. nov. Curr. Microbiol. 57: 638–642.

Datta, S., B. Holmes, J. Park, Z. Chen, D.C. Dibble, M. Hadi et al. 2010. Ionic liquid tolerant hyperthermophilic cellulases for biomass pretreatment and hydrolysis. Green. Chem. 12: 338345.

De Azeredo, L., D. Freire, R. Soares, S. Leite and R. Coelho. 2004. Production and partial characterization of thermophilic proteases from *Streptomyces* sp. isolated from Brazilian cerrado soil. Enzyme. Microb. Technol. 34: 354–358.

De Pascale, D., A.M. Cusano, F. Author, E. Parrilli, G. di Prisco, G. Marino et al. 2008. The cold-active Lip1 lipase from the Antarctic bacterium *Pseudoalteromonas haloplanktis* TAC125 is a member of a new bacterial lipolytic enzyme family. Extremophiles 12: 311–323.

DeLange, R.J. and A.N. Glazer. 1989. Phycoerythrin fluorescence-based assay for peroxy radicals: a screen for biologically relevant protective agents. Anal. Biochem. 177: 300–306.

Delgado-García, M., B. Valdivia-Urdiales, C.N. Aguilar-González, J.C. Contreras-Esquivel and R. Rodríguez-Herrera. 2012. Halophilic hydrolases as a new tool for the biotechnological industries. J. Sci. Food. Agric. 92: 2575–2580.

Delgado-García, M., A.C. Flores-Gallegos, M. Kirchmayr, J.A. Rodríguez, J.C. Mateos-Díaz, N. Aguilar et al. 2019. Bioprospection of proteases from *Halobacillus andaensis* for bioactive peptide production from fish muscle protein. Electr. J. Biotechnol. 39: 52–60.

Delgado-Vargas, F., A.R. Jiménez and O. Paredes-López. 2000. Natural pigments: carotenoids, anthocyanins, and betalains: characteristics, biosynthesis, processing, and stability. Crit. Rev. Food. Sci. Nutr. 40: 173–289.

DeLong, E.F. and N.R. Pace. 2001. Environmental diversity of bacteria and archaea. System. Biol. 50: 470–478.

Deng, A., J. Wu, Y. Zhang, G. Zhang and T. Wen. 2010. Purification and characterization of a surfactant stable high-alkaline protease from *Bacillus* sp. B001. Bioresour. Technol. 101: 7100–7106.

Desbruyères, D. and L. Laubier. 1980. *Alvinella pompejana* gen. sp. nov., Ampharetidae aberrant des sources hydrothermales de la ride Est-Pacifique. Oceanolo, Act. 3: 267–274.

Díaz, S., F. Pérez-Pomares, C. Pire, J. Ferrer and M.J. Bonete. 2006. Gene cloning, heterologous overexpression and optimized refolding of the NAD-glutamate dehydrogenase from *Haloferax mediterranei*. Extremophiles 10: 105–115.

Dib, R., J.M. Chobert, M. Dalgalarrondo, G. Barbier and T. Haertlé. 1998. Purification, molecular properties and specificity of a thermoactive and thermostable proteinase from *Pyrococcus abyssi*, strain st 549, hyperthermophilic archaea from deep-sea hydrothermal ecosystem. FEBS. Lett. 431: 279–284.

Ding, H., Q. Zeng, L. Zhou, Y. Yu and B. Chen. 2017. Biochemical and structural insights into a novel thermostable β-1, 3-galactosidase from *Marinomonas* sp. BSi20414. Mar. Drug. 15: 13.

Do, H., J.H. Lee, M.H. Kwon, H.E. Song, J.Y. An, S.H. Eom et al. 2013. Purification, characterization and preliminary X-ray diffraction analysis of a cold-active lipase (CpsLip) from the psychrophilic bacterium *Colwellia psychrerythraea* 34H. Acta. Crystallogr. Sect. F. Struct. Biol. Cryst. Commun. 69: 920–924.

Eichler, J. 2001. Biotechnological uses of archaeal extremozymes. Biotechnol. Adv. 19: 261–278.

Elend, C., C. Schmeisser, H. Hoebenreich, H.L. Steele and W.R. Streit. 2007. Isolation and characterization of a metagenome-derived and cold-active lipase with high stereospecificity for (R)-ibuprofen esters. J. Biotechnol. 130: 370–377.

Elleuche, S., C. Schröder, K. Sahm and G. Antranikian. 2014. Extremozymes—Biocatalysts with unique properties from extremophilic microorganisms. Curr. Opin. in Microbiol. 29: 116–123.

Elleuche, S., C. Schröder and G. Antranikian. 2016. Lipolytic extremozymes from psychro and (hyper-) thermophilic prokaryotes and their potential for industrial applications. pp. 351–374. *In*: Rampelotto, P.H. (ed.). Biotechnology of Extremophiles Advances and Challenges. Springer International Publishing, Switzerland.

Ensor, M., A.B. Banfield, R.R. Smith, J. Wiliams and R.A. Lodder. 2015. Safety and efficacy of D–tagatose in glycemic control in subjects with type 2 diabetes. J. Endocrinol. Diab. Obes. 3: 1065.

Esakkiraj, P., G. Prabakaran, T. Maruthiah, G. Immanuel and A. Palavesam. 2014. Purification and characterization of halophilic alkaline lipase from *Halobacillus* sp. Proc. Natl. Acad. Sci. India Sect. B Biol. Sci. 86: 309–314.

Esclapez. J., B. Zafrilla, R.M. Martínez-Espinosa and M.J. Bonete. 2013. Cu-NirK from *Haloferax mediterranei* as an example of metalloprotein maturation and exportation via Tat system. Biochimica. Biophysica. Acta. 1834: 1003–1009.

Espliego, J.M.E., V.B. Saiz, J.T. Crespo, A.V. Luque, L. Mónica. C. Carrasco et al. 2018. Extremophile enzymes and biotechnology. pp. 248–276. *In*: Durvasula, R.V. and D.V. Subbarao (eds.). Extremophiles From Biology to Biotechnology. CRC Press, Taylor and Francis Group.

Essghaier, B., A. Hedi, M. Bejji, H. Jijakli, A. Boudabous and N. Sadfi-Zouaoui. 2011. Characterization of a novel chitinase from a moderately halophilic bacterium, *Virgibacillus marismortui* strain M3-23. Annl. Microbiol. 62: 835–841.

Faridi, S. and T. Satyanarayana. 2016. Novel alkalistable α-carbonic anhydrase from the polyextremophilic bacterium *Bacillus halodurans*: Characteristics and applicability in flue gas CO_2 sequestration. Environ. Sci. Pollut. Res. 23: 15236–15249.

Fei, R., H. Zhang, S. Zhong, B. Xue, Y. Gao and X. Zhou. 2017. Anti-inflammatory activity of a thermophilic serine protease inhibitor from extremophile *Pyrobaculum neutrophilum*. Europ. J. Inflamm. 15: 143–151.

Feller, G., M. Thiry, J.L. Arpigny and C. Gerday. 1991. Cloning and expression in *Escherichia coli* of three lipase-encoding genes from the psychrotrophic antarctic strain *Moraxella* TA144. Gen. 102: 111–115.

Feller, G., E. Narinx, J.L. Arpigny, M. Aittaleb, E. Baise, S. Genicot et al. 1996. Enzymes from psychrophilic organisms. FEMS. Microbiol. Rev. 18: 189–202.

Fernández-Castillo, R.F., F. Rodriguez-Valera, J. González-Ramos and F. Ruiz-Berraquero. 1986. Accumulation of poly (beta-hydroxybutyrate) by halobacteria. Appl. Environ. Microbiol. 51: 214–216.

Flores-Fernández, C.N., M. Cárdenas-Fernández, D. Dobrijevic, K. Jurlewicz, A.I. Zavaleta, J.M. Ward et al. 2019. Novel extremophilic proteases from *Pseudomonas aeruginosa* M211 and their application in the hydrolysis of dried distiller's grain with solubles. Biotechnol. Prog. 35: e2728.

Fox, D. 2014. Lakes under the ice: Antarctica's secret garden. Nat. 512: 244–246.

Fuciños, P., E. Atanes, O. Lopez-lopez, M. Solaroli, M.E. Cerdan, M.I. Gonzalez-Siso et al. 2014. Cloning, expression, purification and characterization of an oligomeric His-tagged thermophilic esterase from *Thermus thermophilus* HB27. Process. Biochem. 49: 927–935.

Fujii, M., Y. Takano, H. Kojima, T. Hoshino, R. Tanaka and M. Fukui. 2010. Microbial community structure, pigment composition, and nitrogen source of red snow in Antarctica. Microb. Ecol. 59: 466–475.

Fukishima, T., T. Mizuki, A. Echigo, A. Inoue and R. Usami. 2005. Organic solvent tolerance of halophilic α-amylase from a haloarchaeon, *Haloarcula* sp. strain S-1. Extremophiles 9: 321–331.

Fukushima, T., T. Mizuki, A. Echigo, A. Inoue and R. Usami. 2005. Organic solvent tolerance of halophilic a-amylase from a haloarchaeon, *Haloarcula* sp. strain S-1. Extremophiles 9: 85–89.

Fusi, P., M. Grisa, E. Mombelli, R. Consonni, P. Tortora and M. Vanoni. 1995. Expression of a synthetic gene encoding P2 ribonuclease from the extreme thermoacidophilic archaebacterium *Sulfolobus solfataricus* in mesophylic hosts. Gene 154: 99–103.

Gaballo, A., A. Abbrescia, L.L. Palese, L. Micelli, R.Di. Summa, P. Alifano et al. 2006. Structure and expression of the atp operon coding for F1Fo-ATP synthase from the antibiotic-producing actinomycete *Nonomuraea* sp. ATCC 39727. Res. Microbiol. 157: 675–683.

Gao, R., Y. Feng, K. Ishikawa, H. Ishida, S. Ando, Y. Kosugi et al. 2003. Cloning, purification and properties of a hyperthermophilic esterase from archaeon *Aeropyrum pernix* K1. J. Mol. Catal. B. Enzym. 24–25: 1–8.

García-López, E., A. Alcázar, A.M. Moreno and C. Cristina. 2017. Color-producing extremophiles. pp. 61–80. *In*: Singh, O.V. (ed.). Bio-pigmentation and Biotechnological Implementations. Wiley Blackwell John Willey & Sons. Inc.

George, S., V. Raju, M.R.V. Krishnan, T.V. Subramanian and K. Jayaraman. 1995. Production of protease by *Bacillus amyloliquefaciens* in solid-state fermentation and its application in the unhairing of hides and skins. Process. Biochem. 30: 457–462.

Georlette, D., Z.O. Jonsson, F.V. Petegem, J.P. Chessa, J.V. Beeumen, U. Hubscher et al. 2000. A DNA ligase from the psychrophile *Pseudoalteromonas haloplanktis* gives insight into the adaptation of proteins to low temperature. Europ. J. Biochem. 267: 3502–3512.

Gessesse, A., R. Hatti-Kaul, B.A. Gashe and B. Mattiasson. 2003. Novel alkaline proteases from alkaliphilic bacteria grown on chicken feather. Enzym. Microb. Technol. 32: 519–524.

Giordano, A., G. Andreotti, A. Tramice and A. Trincone. 2006. Marine glycosyl hydrolases in the hydrolysis and synthesis of oligosaccharides. Biotechnol. J. 1: 511–530.

Giuliano, M., C. Schiraldi, M.R. Marotta, J. Hugenholtz and M.De. Rosa. 2004. Expression of *Sulfolobus solfataricus* alpha-glucosidase in *Lactococcus lactis*. Appl. Microbiol. Biotechnol. 64: 829–832.

Gomes, J. and W. Steiner. 1998. Production of a high activity of an extremely thermostable ß-mannanase by the thermophilic eubacterium *Rhodothermus marinus*. Biotechnol. Lett. 20: 729733.

Gomes, J., I. Gomes, K. Terler, N. Gubala, G. Ditzelmuller and W. Steiner. 2000. Optimisation of culture medium and conditions for a-L-arabinofuranosidase production by the extreme thermophilic eubacterium *Rhodothermus marinus*. Enzyme. Microb. Technol. 27: 414–422.

Goodwin, T.W. 1980. The Biochemistry of Carotenoids, vol. 1. London: Chapman and Hall.

Graziano, G. and A. Merlino. 2014. Molecular bases of protein halotolerance. Biochim. Biophys. Acta. 4: 850–858.

Guimaraes, W.V., G.L. Dudey and L.O. Ingram. 1992. Fermentation of sweet whey by ethanologenic *Escherichia coli*. Biotechnol. Bioeng. 40: 41–45.

Gupta, R., Q.K. Beg and P. Lorenz. 2002. Bacterial alkaline proteases: molecular approaches and industrial application. Appl. Microbiol. Biotechnol. 59: 15–32.

Gupta, R.C. 2005. Toxicology of organophosphate and carbamate compounds. Academic, London.

Gvozdyak, P.I., N.F. Mogilevich, A.F. Ryl'skii and N.I. Grishchenko. 1987. Reduction of hexavalent chromium by collection strains of bacteria. Microbiol. 55: 770–73.

Haki, G.D. and S.K. Rakshit. 2003. Developments in industrially important thermostable enzymes: a review. Bioresour. Technol. 89: 7–34.

Harding, T., A.D. Jungblut, C. Lovejoy and W.F. Vincent. 2011. Microbes in high arctic snow and implications for the cold biosphere. Appl. Environ. Microbiol. 77: 3234–3243.

Harm, W. 1980. Biological Effects of Ultraviolet Radiation. IUPAB Biophysics Series I. Cambridge: Cambridge University Press.

Hasan, F., A.A. Shah and A. Hameed. 2005. Industrial applications of microbial lipases. Enzyme. Microb. Technol. 139: 235–251.

Hasbay, I. and Z.B. Ogel. 2002. Production of neutral and alkaline extracellular proteases by the thermophilic fungus, *Scytalidium thermophilum*, grown on microcrystalline cellulose. Biotechnol. Lett. 24: 1107–1110.

Hassankhani, R., M.R. Sam, M. Esmaeilou and P. Ahangar. 2014. Prodigiosin isolated from cell wall of *Serratia marcescens* alters expression of apoptosis-related genes and increases apoptosis in colorectal cancer cells. Med. Oncol. 32: 366.

Hawwa, R., J. Aikens, R.J. Turner, B.D. Santarsiero and A.D. Mesecar. 2009a. Structural basis for thermostability revealed through the identification and characterization of a highly thermostable phosphotriesterase-like lactonase from *Geobacillus stearothermophilus*. Arch. Biochem. Biophys. 488: 109–120.

Hawwa, R., S.D. Larsen, K. Ratia and A.D. Mesecar. 2009b. Structure-based and random mutagenesis approaches increase the organophosphate-degrading activity of a phosphotriesterase homologue from *Deinococcus radiodurans*. J. Mol. Biol. 393: 36–57.

Hess, M., M. Katzer and G. Antranikian. 2008. Extremely thermostable esterases from the thermoacidophilic euryarchaeon *Picrophilus torridus*. Extremoph.12: 351–364.

Hiblot, J., G. Gotthard, E. Chabriere and M. Elias. 2012a. Characterisation of the organophosphate hydrolase catalytic activity of *Sso* Pox. Sci. Rep. 2: 779.

Hiblot, J., G. Gotthard, E. Chabriere and M. Elias. 2012b. Structural and enzymatic characterization of the lactonase SisLac from *Sulfolobus islandicus*. P. One. 7: 47028.

Horikoshi, K. 1999. Alkaliphiles: some applications of their products for biotechnology. Microbiol. Mol. Bio. Rev. 63: 735–750.

Horikoshi, K. and A.T. Bull. 2001. Prologue: Definition, categories, distribution, origin and evolution, pioneering studies, and emerging fields of extremophiles. *In*: Extremophiles Handbook. Berlin: Springer.

Horitsu, H., S. Futo, Y. Miyazawa, S. Ogai and K. Kawai. 1987. Enzymatic reduction of hexavalent chromium by hexavalent chromium tolerant *Pseudomonas ambigua* G-1. Agric. Bioi. Chem. 51: 2417–20.

Hough, D.W. and M.J. Danson. 1999. Extremozymes. Crr. Opn. Chem. Biol. 3: 39–46.

Houghton, T.J. 2005. Global Warming. Rep. Prog. Phys. 68: 1343–1403.

Hoyoux, A., J.M. Francois, P. Dubois, E. Baise, I. Jennes, S. Genicot et al. 2004. Cold-active beta-galactosidase, the process for its preparation and the use thereof. US Patent # 6,727,084 B1.

Huang, Y., G. Krauss, H. Cottaz, H. Driguez and G. Lipps. 2005. A highly acid-stable and thermostable endoß-glucanase from the thermoacidophilic archaeon *Sulfolobus solfataricus*. Biochem. J. 385: 581–588.

Ikeda, M. and D.S. Clark. 1998. Molecular cloning of extremely thermostable esterase gene from hyperthermophilic archaeon *Pyrococcus furiosus* in *Escherichia coli*. Biotechnol. Bioeng. 57: 624–629.

Inoue, K., T. Tsukamoto and Y. Sudo. 2014. Molecular and evolutionary aspects of microbial sensory rhodopsins. Biochim. Biophys. Acta. 1837: 562–577.

Ishibashi, Y., C. Cervantes and S. Silver. 1990. Chromium reduction in *Pseudomonas putida*. Appl. Environ. Microbioi. 56: 2268–70.

Jackson, C.R., H.W. Langner, J. Donahoe-Christiansen, W.P. Inskeep and T.R. McDermott. 2001. Molecular analysis of microbial community structure in an arsenite-oxidizing acidic thermal spring. Environ. Microbiol. 3: 532–542.

Jaeger, K.E. and M.T. Reetz. 1998. Microbial lipases form versatile tools for biotechnology. Trend. Biotechnol. 16: 396–403.

Jaenicke, R. 1981. Enzymes under extreme of physical conditions. Annu. Rev. Biophys. Bioeng. 10: 1–67.

Jagannadham, M.V., M.K. Chattopadhyay and S. Shivaji. 1996a. The major carotenoid pigment of a psychrotrophic *Micrococcus roseus* strain: fluorescence properties of the pigment and its binding to membranes. Biochem. Biophys. Res. Commun. 220: 724–728.

Jagannadham, M.V., K. Narayanan, R.C. Mohan and S. Shivaji. 1996b. *In vivo* characteristics and localization of carotenoid pigments in psychrotrophic and mesophilic *Micrococcus roseus* using photoacoustic spectroscopy. Biochem. Biophys. Res. Commun. 227: 221–226.

Jagannadham, M.V., M.K. Chattopadhyay, C. Subbalakshmi, M. Vairamani, K. Narayanan, C.M. Rao et al. 2000. Carotenoids of an Antarctic psychrotolerant bacterium, *Sphingobacterium antarcticus*, and a mesophilic bacterium, *Sphingobacterium multivorum*. Arch. Microbiol. 173: 418–424.

Jaiswal, A., M. Preet and B. Tripti. 2017. Production and optimization of lipase enzyme from mesophiles and thermophiles. J. Microb. Biochem. Technol. 9: 3.

James, P., M.N. Isupov, C. Sayer, V. Saneei, S. Berg, M. Lioliou et al. 2014. The structure of a tetrameric alpha-carbonic anhydrase from *Thermovibrio ammonificans* reveals a core formed around intermolecular disulfides that contribute to its thermostability. Acta Crystallogr. D. Biol. Crystallogr. 70: 2607–2618.

Jang, H.J., B.C. Kim, Y.R. Pyun and Y.S. Kim. 2002. A novel subtilisin-like serine protease from *Thermoanaerobacter yonseiensis* KB-1: its cloning, expression, and biochemical properties. Extremophiles 6: 233–243.

Jaouadi, B., B. Abdelmalek, D. Fodil, F.Z. Ferradji, H. Rekik, N. Zaraim et al. 2010a. Purification and characterization of a thermostable keratinolytic serine alkaline proteinase from *Streptomyces* sp. strain AB1 with high stability in organic solvents. Bioresour. Technol. 101: 8361–8369.

Jayakumar, R., S. Jayashree, B. Annapurna and S. Seshadri. 2012. Characterization of thermostable serine alkaline protease from an alkaliphilic strain *Bacillus pumilus* MCAS8 and its applications. Appl. Biochem. Biotechnol. 168: 1849–1866.

Jayaseelan, S., D. Ramaswamy and S. Dharmaraj. 2014. Pyocyanin: production, applications, challenges and new insights. World J. Microbiol. Biotechnol. 30: 1159–1168.

Jeon, J.H., J.T. Kim, Y.J. Kim, H.K. Kim, H.S. Lee, S.G. Kang et al. 2009. Cloning and characterization of a new cold-active lipase from a deep-sea sediment metagenome. Appl. Microbiol. Biotechnol. 81: 865–874.

Jeyakanthan, J., S. Rangarajan, P. Mridula, S.P. Kanaujia, Y. Shiro, S. Kuramitsu et al. 2008. Observation of a calcium-binding site in the gamma-class carbonic anhydrase from *Pyrococcus horikoshii*. Acta Crystallogr. D. Biol. Crystallogr. 64: 1012–1019.

Jiang, P., H. Wang, S. Xiao, M. Fang, R. Zhang, S. He et al. 2012. Pathway redesign for deoxyviolacein biosynthesis in *Citrobacter freundii* and characterization of this pigment. Appl. Microbiol. Biotechnol. 94: 1521–1532.

Jiang, X., Y. Huo, H. Cheng, X. Zhang, X. Zhu and M. Wu. 2012. Cloning, expression and characterization of a halotolerant esterase from a marine bacterium *Pelagibacterium halotolerans* B2 T. Extremophiles 16: 427–435.

Jo, B.H., J.H. Seo and J.C. Cha. 2014. Bacterial extremo-α-carbonic znhydrases from deep-sea hydrothermal vents as potential biocatalysts for CO_2 sequestration. J. Mol. Cataly. Enzymatic 109: 31–39.

Jones, M. and N. Foulkes. 1989. Reverse transcription of mRNA by *Termus aquaticus* DNA polymerase. Nucl. Acid. Res. 17: 8387–8388.

Joseph, B., P.W. Ramteke and G. Thomas. 2008. Cold active microbial lipases: some hot issues and recent developments. Biotechnol. Adv. 26: 457–470.

Joshi, G.K., S. Kumar, B.N. Tripathi and V. Sharma. 2006. Production of alkaline lipase by *Corynebacterium paurometabolum*, MTCC 6841 isolated from Lake Naukuchiatal, Uttaranchal State, India. Curr. Microbiol. 52: 354–8.

Joshi, S. and T. Satyanarayana. 2015. *In vitro* engineering of microbial enzymes with multifarious applications: prospects and perspectives. Bioresour. Technol. 176: 273–283.

Juers, D.H., B.W. Matthews and R.E. Huber. 2012. LacZ beta-galactosidase: structure and function of an enzyme of historical and molecular biological importance. Prot. Sci. 21: 1792–1807.

Kallnik, V., A. Bunescu, C. Sayer, C. Bräsen, R. Wohlgemuth, J. Littlechild et al. 2014. Characterization of a phosphotriesterase-like lactonase from the hyperthermoacidophilic crenarchaeon *Vulcanisaeta moutnovskia*. J. Biotechnol. 190: 11–17.

Kamekura, M., T. Hamakawa and H. Onishi. 1982. Application of halophilic nuclease H of *Micrococcus varians* subsp. *halophilus* to commercial production of flavoring agent 5′-GMP. Appl. Environ. Microbiol. 44: 994–995.

Kanth, B.K., S.Y. Jun and S. Kumari. 2014. Highly thermostable carbonic anhydrase from *Persephonella marina* EX-H1: Its expression and characterization for CO_2 sequestration applications. Proc. Biochem. 49: 2114–2121.

Karan, R and S.K. Khare. 2010. Purification and characterization of a solvent-stable protease from *Geomicrobium* sp. EMB2. Environ. Technol. 31: 1061–1072.

Karan, R., S.P. Singh, S. Kapoor and S.K. Khare. 2011. A novel organic solvent tolerant protease from a newly isolated *Geomicrobium* sp. EMB2 (MTCC 10310): production optimization by response surface methodology. New. Biotechnol. 28: 138–145.

Karasová-Lipovová, P., H. Strnad, V. Spiwok, S. Malá, B. Králová and N.J. Russell. 2003. The cloning, purification and characterisation of a cold-active b-galactosidase from the psychrotolerant Antarctic bacterium *Arthrobacter* sp. C2-2. Enzy. Microbiol. Technol. 33: 836–844.

Karbalaei-Heidari, H.R., A.A. Ziaee and M.A. Amoozegar. 2007. Purification and biochemical characterization of a protease secreted by the *Salinivibrio* sp. strain AF-2004 and its behavior in organic solvents. Extremophiles 11: 237–243.

Karentz, D., J.E. Cleaver and D.L. Mitchell. 1991. Cell survival characteristics and molecular responses of Antarctic phytoplankton to ultraviolet-B radiation. J. Phycol. 27: 326–341.

Katiyar, P., V. Hare and V.S. Baghel. 2017. Isolation, partial purification and characterization of a cold active lipase from *Pseudomonas* sp., isolated from Satopanth Glacier of Western Himalaya, India. Int. J. Sci. Res. and Man. 5: 6106–6112.

Kato, C., T. Sato and K. Horikoshi. 1995. Isolation and properties of barophilic and barotolerant bacteria from deep-sea mud samples. Biodivers. Conserv. 4: 1–9.

Kauffman, J.W., W.C. Laughlin and R.A. Baldwin. 1986. Microbiological treatment of uranium mine waters. Environ. Sci. Technol. 20: 243–48.

Khan, F. 2013. New microbial proteases in leather and detergent industries. Innov. Res. Chem. 1: 1–6.

Kikani, B.A., R.J. Shukla and S.P. Singh. 2010. Biocatalytic potential of thermophilic bacteria and actinomycetes. vol. 2. pp. 1000–1007. *In*: Méndez-Vilas, A. (ed.). Current Research, Technology and Education. Topics in Applied Microbiology and Microbial Biotechnology. Formatex Research Center, Badajoz, Spain.

Kim, J. and S. Dordick. 1997. Unusual salt and solvent dependence of a protease from an extreme halophile. Biotechnol. Bioeng. 55: 471–479.

Klingeberg, M., B. Galunsky, C. Sjoholm, V. Kasche and G. Antranikian. 1995. Purification and properties of a highly thermostable, sodium dodecyl sulfate-resistant and stereospecific proteinase from the extremely thermophilic archaeon *Thermococcus stetteri*. Appl. Environ. Microbiol. 61: 3098–3104.

Kobayashi, T., J. Lu, Z. Li, V.S. Hung, A. Kurata, Y. Hatada et al. 2007. Extremely high alkaline protease from a deep-subsurface bacterium, *Alkaliphilus transvaalensis*. Appl. Microbiol. Biotechnol. 75: 71–80.

Kour, D., K.L. Rana, T. Kaur, B. Singh, V.S. Chauhan, A. Kumar et al. 2019. Extremophiles for hydrolytic enzymes productions: biodiversity and potential biotechnological applications. pp. 321–372. *In*: Molina, G., V.K. Gupta, B. Singh and N. Gathergood (eds.). Bioprocessing for Biomolecules Production. doi:10.1002/9781119434436.ch16.

Kren, V., P. Sedmera, V. Havlicek and A. Fiserova. 1992. Enzymatic galactosylation of ergotalkaloids. Tetrahedron. Lett. 47: 7233–7236.

Kube, J., C. Brokamp, R. Machielsen, J. van der Oost and H. Markl. 2006. Influence of temperature on the production of an archaeal thermoactive alcohol dehydrogenase from *Pyrococcus furiosus* with recombinant *Escherichia coli*. Extremophiles 10: 221–227.

Kumar, B., P. Trivedi, A.K. Mishra, A. Pandey and Lok Man S. Palni. 2004. Microbial diversity of soil from two hot springs in Uttaranchal Himalaya. Microbiol. Res. 141–146.

Kumar, D., N Savitri, N. Thakur, R. Verma and T.C. Bhalla. 2008. Microbial proteases and application as laundry detergent additive. Res. J. Microbiol. 3: 661–672.

Kumar, M., A.N. Yadav, R. Tiwari, R. Prasanna and A.K. Saxena. 2014. Deciphering the diversity of culturable thermotolerant bacteria from Manikaran hot springs. Ann. Microbiol. 64: 741–751.

Lagzian, M. and A. Asoodeh. 2012. An extremely thermotolerant, alkaliphilic subtilisin-like protease from hyperthermophilic *Bacillus* sp. MLA64. Int. J. Biol. Macromol. 51(5): 960–967.

Lapenda, J.C., P.A. Silva, M.C. Vicalvi, K.X. Sena and S.C. Nascimento. 2015. Antimicrobial activity of prodigiosin isolated from *Serratia marcescens* UFPEDA 398. World J. Microbiol. Biotechnol. 31: 399–406.

Lasram, M.M., I.B. Dhouib, A. Annabi, S.E. Fazaa and N. Gharbi. 2014. A review on the molecular mechanisms involved in insulin resistance induced by organophosphorus pesticides. Toxicology 322: 1–13.

Laye, V.J., R. Karana, J.M. Kima, W.T. Pechera, P. DasSarmaa and S. DasSarmaa. 2017. Key amino acid residues conferring enhanced enzyme activity at cold temperatures in an Antarctic polyextremophilic β-galactosidase. Proc. Natl. Acad. Sci. USA 114: 12530–12535.

Lee, D.W., H.W. Kim, K.W. Lee, B.C. Kim, E.A. Choe, H.S. Lee et al. 2001. Purification and characterization of two distinct thermostable lipases from the gram-positive thermophilic bacterium *Bacillus thermoleovorans* ID-1. Enzym. Microb. Technol. 29: 363–371.

Levin, G.V. 2002. Tagatose, a new GRAS sweetener and health product. J. Med. Food. 5: 23–36.

Li, A.X., L.Z. Guo and W.D. Lu. 2011. Alkaline inulinase production by a newly isolated bacterium *Marinimicrobium* sp. LS-A18 and inulin hydrolysis by the enzyme. World J. Microbiol. Biotechnol. 28: 81–89.

Li, Q., L. Yi, P. Marek and B.L. Iverson. 2013. Commercial proteases: present and future. FEBS. Lett. 587: 1155–1163.

Li, X. and H.Y. Yu. 2012. Characterization of a novel extracellular lipase from a halophilic isolate, *Chromohalobacter* sp. LY7-8. Afr. J. Microbiol. Res. 14: 3516–3522.

Li, X. and H.Y. Yu. 2012. Purification and characterization of novel organic solvent-tolerant β-amylase and serine protease from a newly isolated *Salimicrobium halophilum* strain LY20. FEMS Microbiology Letters. 329(2): 204–11.

Liliensiek, A.K., J. Cassidy, G. Gucciardo, C. Whitely and F. Paradisi. 2013. Heterologous overexpression, purification and characterisation of an alcohol dehydrogenase (ADH2) from *Halobacterium* sp. NRC-1. Mol. Biotechnol. 55: 143–149.

Litchfield, C.D. 2011. Potential for industrial products from the halophilic archaea. J. Ind. Microbiol. Biotechnol. 38: 1635–1647.

Logan, B., B. Hamelers, R. Rozendal, U. Schroeder, J. Keller, S. Freguia et al. 2006. Microbial fuel cells: methodology and technology. Environ. Sci. Technol. 40: 5181–5192.

Lopez, G., J. Chow, P. Bongen, B. Lauinger, J. Pietruszka, W.R. Streit et al. 2014. A novel thermoalkalostable esterase from *Acidicaldus* sp. strain USBA-GBX-499 with enantioselectivity isolated from an acidic hot springs of Colombian Andes. Appl. Microbiol. Biotechnol. 98: 8603–8616.

Lovley, D.R. 1993. Dissimilatory metal reduction. Annu. Rev. Microbiol. 47: 263–290.

Lü, J., X. Wu, Y. Jiang, X. Cai, L. Huang, Y. Yang et al. 2014. An extremophile *Microbacterium* strain and its protease production under alkaline conditions. J. Bas. Microbiol. 54: 378–385.

Lu, Y., G.V. Levin and T.W. Donner. 2008. Tagatose, a new antidiabetic and obesity control drug. Diabet. Obes. Metab. 10: 109–134.

Luca, V.D., D. Vullo, A. Scozzafava, V. Carginale, M. Rossi, C.T. Supuran et al. 2012. Anion inhibition studies of an alpha-carbonic anhydrase from the thermophilic bacterium *Sulfurihydrogenibium yellowstonense* YO3AOP1. Bioorg. Med. Chem. Lett. 22: 5630–5634.

Luca, V.D., V. De Vullo, D. Scozzafava, A. Carginale, C. Vincenzo, M. Rossi et al. 2013. α-carbonic anhydrase from the thermophilic bacterium *Sulphurihydrogenibium azorense* is the fastest enzyme known for the CO$_2$ hydration reaction. Bioorg. Med. Chem. 21: 1465–1469.

Luca, V.D., D. Vullo, S. Del Prete, V. Carginale, S.M. Osman, Z. AlOthman et al. 2016. Cloning, characterization and anion inhibition studies of γ-carbonic anhydrase from the Antarctic bacterium *Colwellia psychrerythraea*. Bioorg. Med. Chem. 24: 835–40.

Luca, V.D., S. Del Prete, D. Vullo, V. Carginale, P. Di Fonzo, S.M. Osman et al. 2016. Expression and characterization of a recombinant psychrophilic γ-carbonic anhydrase (NcoCA) identified in the genome of the Antarctic cyanobacteria belonging to the genus *Nostoc*. J. Enzy. Inhib. and Medicin. Chem. 31: 810–817.

Lundberg, K., D. Shoemaker, M. Adams, J. Short, J. Sorge and E. Marthur. 1991. High-fidelity amplification using a thermostable polymerase isolated from *Pyrococcus furiosus*. Gen. 108: 1–6.

Luo, Y., Y. Zheng, Z. Jiang, Y. Ma and D. Wei. 2006. A novel psychrophilic lipase from *Pseudomonas fluorescens* with unique property in chiral resolution and biodiesel production via transesterification. Appl. Microbiol. Biotechnol. 73: 349–355.

Lutz, S., A.M. Anesio, J. Villar and L.G. Benning. 2014. Variations of algal communities cause darkening of a Greenland glacier. FEMS. Microbiol. Ecol. 89: 402–414.

MacElroy, R.D. 1974. Some comments on evolution of extremophiles. Biosystems. 6: 74–75.

Madern, D., C. Pfister and G. Zaccai. 1995. Mutation at a single acidic amino acid enhances the halophilic behaviour of malate dehydrogenase from *Haloarcula marismortui* in physiological salts. Eur. J. Biochem. 3: 1088–1095.

Madigan, M.T. 1984. A novel photosynthetic purple bacterium isolated from a Yellowstone hot spring. Sci. 225: 313–315.

Madigan, M.T. and B.L. Marrs. 1997. Gliestremofili. Le Sci. 346: 78–85.

Madigan, M.T., J.M. Martinko, D.A. Stahl and D.P. Clark. 2012. Brock Biology of Microorganisms, 13th edn. San Francisco, CA: Pearson Education.

Mahmoudian, J., M. Jeddi-Tehrani, H. Rabbani, A.R. Mahmoudi, M.M. Akhondi, A.H. Zarnani et al. 2010. Conjugation of R-phycoerythrin to a polyclonal antibody and F (ab') 2 fragment of a polyclonal antibody by two different methods. Avicenna J. Med. Biotech. 2: 87–91.

Majeed, T., R. Tabassum, W.J. Orts and C.C. Lee. 2013. Expression and characterization of *Coprothermobacter proteolyticus* alkaline serine protease. Sci. World. J. Article ID 396156.

Makowski, K., A. Białkowska, J. Olczak, J. Kur and M. Turkiewicz. 2009. Antarctic, cold-adapted β-galactosidase of *Pseudoalteromonas* sp. 22b as an effective tool for alkyl galactopyranosides synthesis. Enzyme. Microbial. Technol. 44: 59–64.

Malik, K., J. Tokkas and S. Ghoyal. 2012. Microbial pigments: a review. Int. J. Microbial. Res. Technol. 1: 361–365.

Martin-Cerezo, M.L., E. Garcia-Lopez and C. Cid. 2015. Isolation and identification of a red pigment from the Antarctic bacterium *Shewanella frigidimarina*. Protein. Pept. Lett. 22: 1076–1082.

Matsumoto, M., H. Yokouchi, N. Suzuki, H. Ohata and T. Matsunaga. 2003. Saccharification of marine microalgae using marine bacteria for ethanol production. Appl. Biochem. Biotechnol. 108: 247–254.

Matsuzawa, H., K. Tokugawa, M. Hamaoki, M. Mizoguchi, H. Taguchi, I. Terada et al. 1988. Purification and characterization of aqualysin I (a thermophilic alkaline serine protease) produced by *Thermus aquaticus* YT-1. Eur. J. Biochem. 171: 441–447.

Maurer, K.H. 2004. Detergent proteases. Curr. Opin. Biotechnol. 15: 330–334.

Mawadza, C., R. Hatti-Kaul, R. Zvauya and Bo Mattiasson. 2000. Purification and characterization of cellulases produced by two *Bacillus* strains. J. Biotechnol. 83: 177–187.

Mehaia, M.A., J. Alverez and M. Cheryan. 1993. Hydrolysis of whey permeate lactose in a continuous stirred tank membrane reactor. Int. Dairy. J. 3: 179–192.

Merone, L., L. Mandrich, M. Rossi and G. Manco. 2005. A thermostable phosphotriesterase from the archaeon *Sulfolobus solfataricus*: cloning, overexpression and properties. Extremophiles 9: 297–305.

Merone, L., L. Mandrich, E. Porzio, M. Rossi, S. Müller, G. Reiter et al. 2010. Improving the promiscuous nerve agent hydrolase activity of a thermostable archaeal lactonase. Bioresour. Technol. 101: 9204–9212.

Mitsuiki, S., M. Sakai, Y. Moriyama, M. Goto and K. Furukawa. 2002. Purification and some properties of a keratinolytic enzyme from an alkaliphilic *Nocardiopsis* sp. TOA-1. Biosci. Biotechnol. Biochem. 66: 164–167.

Mlichová, Z. and M. Rosenberg. 2006. Current trends of β-galactosidase application in food technology. J. Food. Nutr. Res. 45: 47–54.

Mohankumar, A. and P. Ranjitha. 2010. Purification and characterization of esterase from marine *Vibrio fischeri* isolated from squid. Ind. J. Mar. Sci. 39: 262–269.

Mombelli, E., E. Shehi, P. Fusi and P. Tortora. 2002. Exploring hyperthermophilic proteins under pressure: theoretical aspects and experimental findings. Biochem. Biophys. Acta. 1595: 392–396.

Morita, R.Y. 1975. Psychrophilic bacteria. Bacteriol. Rev. 39: 144–167.

Mozhaev, V.V. 1993. Mechanism-based strategies for protein thermostabilization. Trends Biotechnol. 11: 88–95.

Muller, T., W. Bleiss, C.D. Martin, S. Rogaschewski and G. Fuhr. 1998. Snow algae from northwest Svalbard: their identification, distribution, pigment and nutrient content. Polar. Biol. 20: 14–32.

Munawar, N. and P.C. Engel. 2012. Overexpression in a non-native halophilic host and biotechnological potential of NAD dependent glutamate dehydrogenase from *Halobacterium salinarum* strain NRC-36014. Extremophiles 16: 463–476.

Munnecke, D.M. 1979. Hydrolysis of organophosphate insecticides by an immobilized-enzyme system. Biotechnol. Bioeng. 21: 2247–2261.

Nagpal, N., N. Munjal and S. Chatterjee. 2011. Microbial pigments with health benefits—a mini review. Trends. Biosci. 4: 157–160.

Nakagawa, T., Y. Fujimoto, R. Ikehata, T. Miyaji and N. Tomizuka. 2006a. Purification and molecular characterization of cold-active beta-galactosidase from Arthrobacter psychrolactophilus strain F2. Appl. Microbiol. Biotechnol. 72: 720–725.

Nakagawa, T., R. Ikehata, M. Uchino, T. Miyaji, K. Takano and N. Tomizuka. 2006b. Cold-active acid β-galactosidase activity of isolated psychrophilic-basidiomycetous yeast *Guehomyces pullulans*. Microbiol. Res. 161: 75–79.

Nam, G.W., D.W. Lee, H.S. Lee, N.J. Lee, B.C. Kim, E.A. Choe et al. 2002. Native-feather degradation by *Fervidobacterium islandicum* AW-1, a newly isolated keratinase-producing thermophilic anaerobe. Arch. Microbiol. 178: 538–547.

Nath, A., S. Mondal, S. Chakraborty, C. Bhattacharjee and R. Chowdhury. 2014. Production, purification, characterization, immobilization, and application of β-galactosidase: a review. Asia.-Pac. J. Chem. Eng. 9: 330–348.

Ng, F.S.W., D.M. Wright and S.Y.K. Seah. 2011. Characterization of a phosphotriesterase-like lactonase from *Sulfolobus solfataricus* and its immobilization for disruption of quorum sensing. Appl. Environ. Microbiol. 77: 1181–1186.

Niehaus, F., A. Peters, T. Groudieva and G Antranikian. 2000. Cloning, expression and biochemical characterization of a unique thermostable pullulan-hydrolysing enzyme from the hyperthermophilic archaeon *Thermococcus aggregans*. FEMS. Microbiol. Lett. 190: 223–229.

Nishino, H., M. Murakoshi, H. Tokuda and Y. Satomi. 2009. Cancer prevention by carotenoids. Arch. Biochem. Biophys. 483: 165–168.

Oesterhelt, D., C. Brauchle and N. Hampp. 1991. Bacteriorhodopsin: A biological material for information processing. Q. Rev. Biophys. 24: 425–78.

Ogawa, T., K. Noguchi, M. Saito, Y. Nagahata, H. Kato, A. Ohtaki et al. 2013. Carbonyl sulfide hydrolase from *Thiobacillus thioparus* strain THI115 is one of the β-carbonic anhydrase family enzymes. J. Am. Chem. Soc. 135: 3818–25.

Oh, D.K. 2007. Tagatose: properties, applications, and biotechnological processes. Appl. Microbiol. Biotechnol. 76: 1–8.

Ojha, A. and Y.K. Gupta. 2015. Evaluation of genotoxic potential of commonly used organophosphate pesticides in peripheral blood lymphocytes of rats. Hum. Exp. Toxicol. 34: 390–400.

Onishi, N. and T. Tanaka. 1996. Purification and properties of a galacto- and gluco-oligosaccharide producing beta-glycosidase from *Rhodotorula minuta* IFO879. J. Ferment. Bioeng. 82: 439–443.

Oren, A. and F. Rodríguez-Valera. 2001. The contribution of halophilic bacteria to the red coloration of saltern crystallizer ponds. FEMS. Microbiol. Ecol. 36: 123–130.

Oren, A. 2010. Industrial and environmental applications of halophilic microorganisms. Environ. Technol. 31: 825–834.

Ortega, G., A. Laín, X. Tadeo, B. López-Méndez, D. Castano, O. Millet. 2011. Halophilic enzyme activation induced by salts. Scientific Reports. 14(1): 6.

Parra, L.P., G. Espina, J. Devia, O. Salazar, B. Andrews and J.A. Asenjo. 2015. Identification of lipase encoding genes from Antarctic seawater bacteria using degenerate primers: expression of a cold-active lipase with high specific activity. Enzyme. Microb. Technol. 68: 56–61.

Patel, G.B. and G.D. Sprott. 1999. Archaeobacterial ether lipid liposomes (archaeosomes) as novel vaccine and drug delivery systems. Critic. Rev. Biotechnol. 19: 317–357.

Paul, T., A. Das, A. Mandal, A. Jana, S.K. Halder, P.K. Das Mohapatra et al. 2014. Smart cleaning properties of a multi tolerance keratinolytic protease from an extremophilic *Bacillus tequilensis* hsTKB2: prediction of enzyme modification site. Waste Biomass Valori. 5: 931–945.

Pawlak-Szukalska, A., M. Wanarska, A.T. Popinigis and J. Kur. 2014. A novel cold-active β-D-galactosidase with transglycosylation activity from the Antarctic *Arthrobacter* sp. 32cB–gene cloning, purification and characterization. Proc. Biochem. 49: 2122–2133.

Pérez, D., S. Martín, G. Fernández-Lorente, M Filice, J.M. Guisán and A. Ventosa. 2011. A novel halophilic lipase, LipBL, showing high efficiency in the production of eicosapentaenoic acid (EPA). P. One 6: e23325.

Phung, N.T., J. Lee, K.H. Kang, I.S. Chang, G.M. Gadd and G.H. Kim. 2004. Analysis of microbial diversity in oligotrophic microbial fuel cells using 16SrDNA sequences. FEMS. Microbiol. Lett. 233: 77–82.

Pikuta, E.V., R.B. Hoover and J. Tang. 2007. Microbial extremophiles at the limits of life. Crit. Rev. Microbiol. 33: 183–209.

Pire, C., J. Esclapez, J. Ferrer and M.J. Bonete. 2001. Heterologous expression of glucose dehydrogenase from the halophilic archaeon *Haloferax mediterranei*, an enzyme of the medium chain dehydrogenase family. FEMS. Microbiol. Lett. 200: 221–227.

Porzio, E., L. Merone, L. Mandrich, M. Rossi and G. Manco. 2007. A new phosphotriesterase from Sulfolobus acidocaldarius and its comparison with the homologue from *Sulfolobus solfataricus*. Biochimie 89: 625–636.

Prakash, P., S.K. Jayalakshmi, B. Prakash, M. Rubul and K. Sreeramulu. 2011. Production of alkaliphilic, halotolerent, thermostable cellulase free xylanase by *Bacillus halodurans* PPKS-2 using agro waste: Single step purification and characterization. World J. Microbiol. Biotechnol. 28: 183–192.

Preiss, L., D.B. Hicks, S. Suzuki, T. Meier and T.A. Krulwich. 2015. Alkaliphilic bacteria with impact on industrial applications, concepts of early life forms, and bioenergetics of ATP synthesis. Front. Bioeng. Biotechnol. 3: 75.

Priya, I., M. Dhar, B. Bajaj, S. Koul and J. Vakhlu. 2016. Cellulolytic activity of thermophilic bacilli isolated from Tattapani hot spring sediment in North West Himalayas. Ind. J. Microbiol. 56: 228–231.

Rabaey, K., N. Boon, S.D. Siciliano, M. Verhaege and W. Verstraete. 2004. Biofuel cells select for microbial consortia that self-mediate electron transfer. Appl. Environ. Microbiol. 70: 5373–5382.

Raddadi, N., A. Cherif, D. Daffonchio, M. Neifar and F. Fava. 2015. Biotechnological applications of extremophiles, extremozymes and extremolytes. Appl. Microbiol. Biotechnol. 99: 7907–7913.

Rahim, K.A. and B.H. Lee. 1991. Specificity, inhibitory studies, and oligosaccharide formation by beta-galactosidase from psychrotrophic *Bacillus subtilis* KL88. J. Dairy Sci. 74: 1773–1778.

Rajagopal, L., C.S. Sundari, D. Balasubramanian and R.V. Sonti. 1997. The bacterial pigment xanthomonadin offers protection against photodamage. FEBS. Lett. 415: 125–128.

Rao, M.B., A.M. Tanksale, M.S. Ghatge and V.V. Deshpande. 1998. Molecular and biotechnological aspects of microbial proteases. Microbiol. Mol. Biol. Rev. 62: 597–635.

Rashid, N., Y. Shimada, S. Ezaki, H. Atomi and T. Imanaka. 2001. Low-temperature lipase from psychrotrophic *Pseudomonas* sp. strain KB700A. Appl. Environ. Microbiol. 67: 4064–4069.

Raushel, F.M. 2002. Bacterial detoxification of organophosphate nerve agents. Curr. Opin. Microbiol. 5: 288–295.

Raval, V.H., S. Pillai, C.M. Rawal and S.P. Singh. 2014. Biochemical and structural characterization of a detergent-stable serine alkaline protease from seawater haloalkaliphilic bacteria. Proc. Biochem. 49: 955–962.

Reeve, J.N., J. Nölling, R.M. Morgan and D.R. Smith. 1997. Methanogenesis: Genes, genomes, and who's on first? J. Bacteriol. 179: 5975–5986.

Rémy, B., L. Laure Plener, L. Poirier, M. Elias, D. Daudé and Eric Chabrière. 2016. Harnessing hyperthermostable lactonase from *Sulfolobus solfataricus* for biotechnological applications. Sci. Rep. 6: 37780.

Restaino, O.F., M.G. Borzacchiello, I. Scognamiglio, L. Fedele, A. Alfano, E. Porzio et al. 2018. High yield production and purification of two recombinant thermostable phosphotriesterase-like lactonases from *Sulfolobus acidocaldarius* and *Sulfolobus solfataricus* useful as bioremediation tools and bioscavengers. B.M.C. Biotechnol. 18: 18.

Rodriguez-Valera, F. 1992. Biotechnological potential of halobacteria. Biochem. Soci. Sympo. 58: 135–147.

Rosenbaum, E., F. Gabel, M.A. Durá, S. Finet, C. Cléry-Barraud, P. Masson et al. 2012. Effects of hydrostatic pressure on the quaternary structure and enzymatic activity of a large peptidase complex from *Pyrococcus horikoshii*. Arch. Biochem. Biophys. 517: 104–110.

Royter, M., M. Schmidt, C. Elend, H. Hobenreich, T. Schafer, U.T. Bornscheuer et al. 2009. Thermostable lipases from the extreme thermophilic anaerobic bacteria *Thermoanaerobacter thermohydrosulfuricus* SOL1 and *Caldanaerobacter subterraneus* subsp. *tengcongensis*. Extremophiles 13: 769–783.

Ruiz, D.M. and R.E. De Castro. 2007. Effect of organic solvents on the activity and stability of an extracellular protease secreted by the haloalkaliphilic archaeon *Natrialba magadii*. J. Ind. Microbiol. Biotechnol. 34: 111–115.

Rusnak, M., J. Nieveler, R.D. Schmid and R. Petri. 2005. The putative lipase, AF1763, from *Archaeoglobus fulgidusis* is a carboxylesterase with a very high pH optimum. Biotechnol. Lett. 27: 743–748.

Russell, N.J. and T. Hamamoto. 1998. Psycrophiles. pp. 25–45. *In*: Horikoshi, K. and W.D. Grant (eds.). Extermophiles: Microbial Life in Extreme Environments. Wilmington, DE: Wiley-Liss.

Ryu, H.S., H.K. Kim, W.C. Choi, M.H. Kim, S.Y. Park, N.S. Han et al. 2006. New cold adapted lipase from *Photobacterium lipolyticum* sp. nov. that is closely related to filamentous fungal lipases. Appl. Microbiol. Biotechnol. 70: 321–326.

Saeki, K., J. Hitomi, M. Okuda, Y. Hatada, Y. Kageyama, M. Takaiwa et al. 2002. A novel species of alkaliphilic Bacillus that produces an oxidatively stable alkaline serine protease. Extremophiles 6: 65–72.

Sahay, H., B.K. Babu, S. Singh, R. Kaushik, A.K. Saxena, D.K. Arora et al. 2013. Cold-active hydrolases producing bacteria from two different sub-glacial Himalayan lakes. J. Basic Microbiol. 53: 703–714.

Sahay, H., A.N. Yadav, A.K. Singh, S. Singh, R. Kaushik and A.K. Saxena. 2017. Hot springs of Indian Himalayas: potential sources of microbial diversity and thermostable hydrolytic enzymes. 3 Biotech 7: 118.

Salazar-Arredondo, E., M.J. Solis-Heredia, E. Rojas-Garcia, I.H. Ochoa and B.V. Quintanilla. 2008. Sperm chromatin alteration and DNA damage by methyl-parathion, chlorpyrifos and diazinon and their oxon metabolites in human spermatozoa. Reprod. Toxicol. 25: 455–460.

Salwoom, L., R.N.Z. Raja Abd Rahman, A.B. Salleh, F. Mohd Shariff, P. Convey, D. Pearce et al. 2019. Isolation, characterisation, and lipase production of a cold-adapted bacterial strain *Pseudomonas* sp. LSK25 isolated from Signy Island, Antarctica. Mol. 16: 24.

Saraswat, P.R., V. Verma, S. Sistla and I. Bhushan. 2017. Evaluation of alkali and thermotolerant lipase from an indigenous isolated *Bacillus* strain for detergent formulation Author links open overlay. Electr. J. Biotechnol. 30: 33–38.

Sarethy, I.P., Y. Saxena, A. Kapoor, M. Sharma, S.K. Sharma, V. Gupta et al. 2011. Alkaliphilic bacteria: applications in industrial biotechnology. J. Ind. Microbiol. Biotechnol. 38: 769–790.

Saxena, A.K., A.N. Yadav, R. Kaushik, S. Tyagi, M. Kumar, R. Prasanna et al. 2014. Use of microbes from extreme environments for the benefits of agriculture. *In*: Afro-Asian Congress on Microbes for Human and Environmental Health. Amity University, Noida New Delhi.

Saxena, A.K., A.N. Yadav, R. Kaushik, S.P. Tyagi and L. Shukla. 2015. Biotechnological applications of microbes isolated from cold environments in agriculture and allied sectors. *In*: International Conference on Low Temperature Science and Biotechnological Advances. Soc. Low Temp. Biol. (vol. 104).

Saxena, A.K., A.N. Yadav, M. Rajawat, R. Kaushik, R. Kumar, M. Kumar et al. 2016. Microbial diversity of extreme regions: an unseen heritage and wealth. Ind. J. Plant Genet. Resour. 29: 246–248.

Scheckermann, C., F. Wagner and L. Fischer. 1997. Galactosylation of antibiotics using the betagalactosidase from *Aspergillus oryzae*. Enzym. Microb. Technol. 20: 629–634.

Schiraldi, C., M. Giuliano and M.De. Rosa. 2002. Perspectives on biotechnological applications of archaea. Archaea. 1: 75–86.

Schloss, P.D., H.K. Allen, A.K. Klimowicz, C. Mlot, J.A. Gross, S. Savengsuksa et al. 2010. Psychrotrophic strain of *Janthinobacterium lividum* from a cold Alaskan soil produces prodigiosin. DNA Cell Biol. 29: 533–541.

Schmidt, M. and P. Stougaard. 2010 Identification, cloning an expression of a cold-active β-galactosidase from a novel arctic bacterium, *Alkalilactibacillus ikkense*. Environ. Technol. 3: 1107–1114.

Schreck, S.D. and A.M. Grunden. 2014. Biotechnological applications of halophilic lipases and thioesterases. Appl. Microbiol. Biotechnol. 98: 1011–1021.

Serour, E. and G. Antranikian. 2002. Novel thermoactive glucoamylase from the thermoacidophilic archaea *Thermoplasma acidophilum, Picrophilus torridus* and *Picrophilus oshimae*. Anton. Van. Leewen. 81: 73–83.

Setati, M.E. 2010. Diversity and industrial potential of hydrolase-producing halophilic/halotolerant eubacteria. Afr. J. Biotechnol. 9: 1555–1560.

Shafiei, M., A.A. Ziaee and M.A. Amoozegar. 2011. Purification and characterization of an organic-solvent-tolerant halophilic a-amylase from the moderately halophilic *Nesterenkonia* sp. strain F. J. Ind. Microbiol. Biotechnol. 38: 275–281.

Shakun, J.D., P.U. Clark, F. He, S.A. Marcott, A.C. Mix, Z. Liu et al. 2012. Global warming preceded by increasing carbon dioxide concentrations during the last deglaciation. Nat. 48: 49–54.

Sharma, R., S.K. Sona, R.M. Vohra, L.K. Gupta and J.K. Gupta. 2002. Purification and characterisation of a thermostable alkaline lipase from a new thermophilic *Bacillus* sp. RSJ-1. Proc. Biochem. 37: 1075–1084.

Sheridan, P.P. and J.E. Brenchley. 2000. Characterization of a salt-tolerant family 42 β-galactosidase from a psychrophilic antarctic Planococcus isolate. Appl. Environ. Microbiol. 66: 2438–2444.

Shimizu, R., H. Shimabayashi and M. Moriwaki. 2006. Enzymatic production of highly soluble myricitrin glycosides using beta-galactosidase. Biosci. Biotechnol. Biochem. 70: 940–948.

Shirata, A., T. Tsukamoto, H. Yasui, T. Hata, S. Hayasaka, A. Kojima et al. 2000. Isolation of bacteria producing bluish-purple pigment and use for dyeing. Jpn. Agric. Res. Q. 34: 131–140.

Shirkot, P. and A. Verma. 2015. Assessment of thermophilic bacterial diversity of thermal springs of Himachal Pradesh. ENVIS. Bull. Himal. Ecol. 23: 27–34.

Shivaji, S., R.N. Shyamala, L. Saisree, S. Vipula, G.S.N. Reddy and P.M. Bhargava. 1988. Isolation and identification of *Micrococcus roseus* and *Planococcus* sp. from Schirmacher Oasis, Antarctica. J. Biosci. 13: 409–414.

Shivaji, S., R.N. Shyamala, L. Saisree, G.S.N. Reddy, G.S. Kumar and P.M. Bhargava. 1989a. Isolates of Arthrobacter from the soils of Schirmacher Oasis, Antarctica. Polar. Biol. 10: 225–229.

Shivaji, S., N. Shyamala Rao, L. Saisree, S. Vipula, G.S.N. Reddy and P.M. Bhargava. 1989b. Isolation and identification of *Pseudomonas* spp. from Schirmacher Oasis, Antarctica. Appl. Environ. Microbiol. 55: 767–770.

Shivaji, S., M. Pratibha, B. Sailaja, K.H. Kishore, A.K. Singh, Z. Begum et al. 2011. Bacterial diversity of soil in the vicinity of Pindari glacier, Himalayan mountain ranges, India, using culturable bacteria and soil 16S rRNA gene clones. Extremophiles 15: 1–22.

Siefirmann-Harms, D. 1987. The light-harvesting and protective functions of carotenoids in photosynthetic membranes. Physiol. Plant. 69: 501–568.

Singh, B.K. 2009. Organophosphorus-degrading bacteria: ecology and industrial applications. Nat. Rev. Microbiol. 7: 156–164.

Singh, O.V. and P. Gabani. 2011. Extremophiles: radiation resistance microbial reserves and therapeutic implications. J. Appl. Microbiol. 110: 851–861.

Singh, R.N., S. Gaba, A.N. Yadav, P. Gaur, S. Gulati, R. Kaushik et al. 2016. First, high quality draft genome sequence of a plant growth promoting and cold active enzymes producing psychrotrophic *Arthrobacter agilis* strain L77. Stand Genomic Sci. 11: 54. doi:10.1186/s40793-016-0176-4.

Singhal, P., V.K. Nigam and A.S. Vidyarthi. 2012. Studies on production, characterization and applications of microbial alkaline proteases. Int. J. Adv. Biotechnol. Res. 3: 653–669.

Sinha, R. and S.K. Khare. 2012. Isolation of a halophilic *Virgibacillus* sp. EMB13: characterization of its protease for detergent application. Ind. J. Biotechnol. 11: 416–426.

Sinha, R. and S.K. Khare. 2013. Characterization of detergent compatible protease of a halophilic *Bacillus* sp. EMB9: differential role of metal ions in stability and activity. Bioresour. Technol. 145: 357–361.

Sleytr, U.B., D. Pum and M. Sára. 1997. Advances in S-layer nanotechnology and biomimetics. Adv. Biophys. 34: 71–79.

Smith, K.S. and J.G. Ferry. 1999. A plant-type (β-Class) carbonic anhydrase in the thermophilic methanoarchaeon *Methanobacterium thermoautotrophicum*. J. Bacteriol. 181: 6247–6253.

Smith, K.S. and J.G. Ferry. 2000. Prokaryotic carbonic anhydrases. FEMS. Microbiol. Rev. 24: 335–366.

Soliev, A.B., K. Hosokawa and K. Enomoto. 2011. Bioactive pigments from marine bacteria: applications and physiological roles. Evid. Based. Complem. Altern. Med. (3): 670349.

Soliman, N.A., M. Knoll, Y.R. Abdel-Fattah, R.D. Schmid and S. Lange. 2007. Molecular cloning and characterization of thermostable esterase and lipase from *Geobacillus thermoleovorans* YN isolated from desert soil in Egypt. Proc. Biochem. 42: 1090–1100.

Solomona, S., G.K. Plattner, R. Knuttic and P. Friedlingstein. 2019. Irreversible climate change due to carbon dioxide emissions. Proc. Nat. Aca. Sci. the U. S. A. 106: 1704–1709.

Somalinga, V., G. Buhrman, A. Arun, R.B. Rose and A.M. Grunden. 2016. A high-resolution crystal structure of a psychrohalophilic α-carbonic anhydrase from *Photobacterium profundum* reveals a unique dimer interface. Plos One 11: e0168022.

Song, C., G.L. Liu, J.L. Xu and Z.M. Chi. 2010. Purification and characterization of extracellular β-galactosidase from the psychrotolerant yeast *Guehomyces pullulans* 17–1 isolated from sea sediment in Antarctica. Proc. Biochem. 45: 954–960.

Spain, J.C. and D.T. Gibson. 1991. Pathway for biodegradation of p-nitrophenol in a Moraxella sp. Appl. Environ. Microbiol. 57: 812–819.

Staiano, M., P. Bazzicalupo, M. Rossi, and S.D. Auria. 2005. Glucose biosensors as models for the development of advanced protein-based biosensors. Mol. Biosyst. 1: 354–362.

Steiger, S., L. Perez-Fons, S.M. Cutting, P.D. Fraser and G. Sandmann. 2015. Annotation and functional assignment of the genes for the C30 carotenoid pathways from the genomes of two bacteria: *Bacillus indicus* and *Bacillus firmus*. Microbiology. 161: 194–202.

Stevenson, D.E., R.A. Stanley and R.H. Furneaux. 1993. Optimization of alkyl β-D-galactopyranoside synthesis from lactose using commercially available β-galactosidases. Biotechnol. Bioeng. 42: 657–666.

Stibal, M., F. Hasan, J.L. Wadham, M.J. Sharp and A.M. Anesio. 2012. Prokaryotic diversity in sediments beneath two polar glaciers with contrasting organic carbon substrates. Extremophiles 16: 255–265.

Stierle, A.A., D.B. Stierle, T. Girtsman, T.C. Mou, C. Antczak and H. Djaballah. 2015. Azaphilones from an acid mine extremophile strain of a *Pleurostomophora* sp. J. Nat. Prod. 78: 2917–2923.

Stougaard, P. and M. Schmidt. 2012. Cold-active beta-galactosidase, a method of producing same and use of such enzyme. U.S. Patent # 8,288,143 B2.

Subczynski, W.K., E. Markowska, W.I. Gruszecki and J. Sielewiesiuk. 1992. Effect of polar carotenoids on dimyristoylphosphatidylcholine membranes: a spin-label study. Biochim. Biophys. Acta. 1105: 97–108.

Sunna, A. and P.L. Bergquist. 2003. A gene encoding a novel extremely thermostable 1, 4-beta-xylanase isolated directly from an environmental DNA sample. Extremophiles 7: 63–70.

Sutrisno, A., M. Ueda, Y. Abe, M. Nakazawa and K. Miyatake. 2004. A chitinase with high activity toward partially Nacetylated chitosan from a new, moderately thermophilic, chitin-degrading bacterium, *Ralstonia* sp. A-471. Appl. Microbiol. Biotechnol. 63: 398–406.

Suzuki, H., Y. Hirano, Y. Kimura, S. Takaichi, M. Kobayashi, K. Miki et al. 2007. Purification, characterization and crystallization of the core complex from thermophilic purple sulfur bacterium *Thermochromatium tepidum*. Biochim. Biophys. Acta. 1767: 1057–1063.

Suzuki, T., T. Nakayama, T. Kurihara, T. Nishino and N. Esaki. 2001. Cold-active lipolytic activity of psychrotrophic *Acinetobacter* sp. strain no. 6. J. Biosci. Bioeng. 92: 144–148.

Suzuki, Y., Y. Tsujimoto, H. Matsui and K. Watanabe. 2006. Decomposition of extremely hard-to-degrade animal proteins by thermophilic bacteria. J. Biosci. Bioeng. 102: 73–81.

Synowiecki, J. 2010. Some application of thermophiles and their enzymes for protein processing. Afr. J. Biotechnol. 9: 7020–7025.

Takasawa, T., K. Sagisaka, K. Yagi, K. Uchiyama, A. Aoki, K. Takaoka et al. 1997. Polygalacturonase isolated from the culture of the psychrophilic fungus *Sclerotinia borealis*. Can. J. Microbiol. 43: 417–424.

Tan, S., R.K. Owusu Apenten and J. Knapp. 1996. Low temperature organic phase biocatalysis using cold-adapted lipase from psychrotrophic *Pseudomonas* P38. Food. Chem. 57: 415–418.

Tanaka, D., S. Yoneda, Y. Yamashiro, A. Sakatoku, T. Kayashima, K. Yamakawa et al. 2012. Characterization of a new cold-adapted lipase from *Pseudomonas* sp. TK-3. Appl. Biochem. Biotechnol. 168: 327–338.

Tao, W., F. Shengxue, M. Duobin, Y. Xuan, D. Congcong and W. Xihua. 2013. Characterization of a new thermophilic and acid tolerant esterase from *Thermotoga maritima* capable of hydrolytic resolution of racemic ketoprofen ethyl ester. J. Mol. Catal. B Enzym. 85–86: 23–30.

Taylor, I.N.R., C. Brown, M. Rycroft, G. King, J.A. Littlechild, M.C. Lloyd et al. 2004. Application of thermophilic enzymes in commercial biotransformation processes. Biochem. Soc. Trans. 32: 290–292.

Tchigvintsev, A., H. Tran, A. Popovic, F. Kovacic, G. Brown, R. Flick et al. 2015. The environment shapes microbial enzymes: five cold-active and salt-resistant carboxylesterases from marine metagenomes. Appl. Microbiol. Biotechnol. 99: 2165–2178.

Thammapalerd, N., T. Supasiri, S. Awakairt and S. Chandrkrachang. 1996. Application of local products R-phycoerythrin and monoclonal antibody as a fluorescent antibody probe to detect *Entamoeba histolytica* trophozoites. Southeast. Asia. J. Trop. Med. Pub. Heal. 27: 297–303.

Toogood, H.S., E.J. Hollingsworth, R.C. Brown, I.N. Taylor, S.J. Taylor, R. McCague et al. 2002. A thermostable L-aminoacylase from *Thermococcus litoralis*: Cloning, overexpression, characterization and applications in biotransformations. Extremophi. 6: 111–122.

Toplak, A., B. Wu, F. Fusetti, P.J.L.M. Quaedflieg and D.B. Janssen. 2013. Proteolysin, a novel highly thermostable and cosolvent-compatible protease from the thermophilic bacterium *Coprothermobacter proteolyticus*. Appl. Environ. Microbiol. 79: 5625–5632.

Unsworth, L.D., O.J. Van Der and S. Koutsopoulos. 2007. Hyperthermophilic enzymes-stability, activity and implementation strategies for high temperature applications. FEBS. J. 274: 4044–4056.

Van den Burg, B., H.G. Enequist, M.E. van den Haar, V.G. Eijsink, B.K. Stulp and G. Venema. 1991. A highly thermostable neutral protease from *Bacillus caldolyticus*: cloning and expression of the gene in *Bacillus subtilis* and characterization of the gene product. J. Bacteriol. 173: 4107–4115.

Van der Maarel, M.J., B. van der Veen, J.C. Uitdehaag, H. Leemhuis and L. Dijkhuizen. 2002. Properties and application of starch-converting enzymes of the a-amylase family. J. Biotechnol. 94: 137–155.

Van Laere, K.M.J., T. Abee, H.A. Schols, G. Beldman and A.G.J. Voragen. 2000. Characterization of a novel beta-galactosidase from *Bifidobacterium adolescentis* DSM 20083 active towards trans-galacto oligosaccharides. Appl. Environ. Microbiol. 66: 1379–1384.

Varela, H., M.D. Ferrari, L. Belobrajdic, A. Vazquez and M.L. Loperena. 1997. Skin unhairing proteases of *Bacillus subtilis*: production and partial characterization. Biotechnol. Lett. 19: 755–758.

Venil, C.K. and P. Lakshmanaperumalsamy. 2009. An insightful overview on microbial pigment, prodigiosin. Elect. J. Biol. 5: 49–61.

Venil, C.K., Z.A. Zakaira and W.A. Ahmad. 2013. Bacterial pigments and their applications. Proc. Biochem. 48: 1065–1079.

Verma, P., A.N. Yadav, V. Kumar, D.P. Singh and A.K. Saxena. 2017. Beneficial plant-microbes interactions: biodiversity of microbes from diverse extreme environments and its impact for crop improvement. pp. 543–580. *In*: Singh, D.P., H.B. Singh and R. Prabha (eds.). Plant-Microbe Interactions in Agro-Ecological Perspectives: Volume 2: Microbial Interactions and Agro-Ecological Impacts. Springer Singapore, Singapore. doi:10.1007/978-981-10-6593-4_22.

Verma, P., A.N. Yadav, K.S. Khannam, S. Mishra, S. Kumar, A.K. Saxena et al. 2019. Appraisal of diversity and functional attributes of thermotolerant wheat associated bacteria from the peninsular zone of India. Saudi J. Biol. Sci. 26: 1882–1895. doi:https://doi.org/10.1016/j.sjbs.2016.01.042.

Vetere, A. and S. Paoletti. 1998. Separation and characterization of three beta-galactosidases from *Bacillus circulans* BBA. Gen. Subj. 1380: 223–231.

Vidyasagar, M., S. Prakash and K. Sreeramulu. 2006. Optimization of culture conditions for the production of haloalkaliphilic thermostable protease from an extremely halophilic archaeon *Halogeometricum borinquense* sp. TSS 101. Lett. Appl. Microbiol. 43: 385–391.

Vollmers, J., S. Voget, S. Dietrich, K. Gollnow, M. Smits, K. Meyer et al. 2013. Poles apart: Arctic and Antarctic Octadecabacter strains share high genome plasticity and a new type of xanthorhodopsin. P. One. 8: e63422.

Vullo, D., P.S. Del, S.M. Osman. V.D. Luca, A. Scozzafava, Z. Alothman et al. 2014. Sulfonamide inhibition studies of the δ-carbonic anhydrase from the diatom *Thalassiosira weissflogii*. Bioorg. Med. Chem. Lett. 24: 275–279.

Vullo, D., D.V. Luca, S.D. Prete, V. Carginale, A. Scozzafava, C. Capasso et al. 2015. Sulfonamide inhibition studies of the c-carbonic anhydrase from the Antarctic cyanobacterium *Nostoc commune*. Bioorg. Med. Chem. 23: 1728–34.

Vullo, D., V.D. Luca, S. Del Prete, V. Carginale, A. Scozzafava, C. Capasso et al. 2015. Sulfonamide inhibition studies of the c-carbonic anhydrase from the Antarctic bacterium *Pseudoalteromonas haloplanktis*. Bioorg. Med. Chem. Lett. 25: 3550–5.

Wahab, R.A., M. Basri, R.N.Z.R. Abdul Rahman, A.B. Salleh, M.B. Abdul Rahman, N. Chaibakhsh et al. 2014. Enzymatic production of a solvent-free menthyl butyrate via response surface methodology catalyzed by a novel thermostable lipase from *Geobacillus zalihae*. Biotechnol. Biotechnol. Equip. 28: 1065–1072.

Wanarska, M. and J. Kur. 2005. β-D-Galactosidases—sources, properties and applications. Biol. Technol. 4: 46–62.

Wanarska, M. and J. Kur. 2012. A method for the production of D-tagatose using a recombinant Pichia pastoris strain secreting β-D-galactosidase from Arthrobacter chlorophenolicus and a recombinant L-arabinose isomerase from *Arthrobacter* sp. 22c. Microbial Cell Factories. 11(1): 113.

Wang, K., G. Li, Y. Qin, C.T. Znang and Y.H. Liu. 2010. A novel metagenome-derived β-galactosidase: gene cloning, overexpression, purification and characterization. Appl. Microbiol. Biotechnol. 88: 155–165.

Wang, P., T. Mori, T.K. Komori, M. Sasatsu, K. Toda and H. Ohtake 1989. Isolation and characterization of an Enterobacter cloacae strain that reduces hexavalent chromium under anaerobic conditions. Appl. Environ. Microbiol. 55: 1665–69.

Weber, P.M.A., J.R. Horst, G.G. Barbier and C. Oesterhe. 2007. Metabolism and metabolomics of eukaryotes living under extreme conditions. Internat. Rev. Cytol. 256: 1–34.

Wi, A.R., S.J. Jeon, S. Kim, H.J. Park, D. Kim, S.J. Han et al. 2014. Characterization and a point mutational approach of a psychrophilic lipase from an arctic bacterium, *Bacillus pumilus*. Biotechnol. Lett. 36: 1295–1302.

Wicka, M., E. Krajewska and A. Pawlak. 2013. Cold-adapted bacterial lipolytic enzymes and their applications. PhD Interdisp. J. 2: 107–112.

Williamson, N.R., P.C. Fineran, F.J. Leeper and G.P. Salmond. 2006. The biosynthesis and regulation of bacterial prodiginines. Nat. Rev. Microbiol. 4: 887–899.

Wilson, Z.E. and A.M. Brimble. 2009. Molecules derived from the extremes of life. Nat. Prod. Rep. 26: 44–71.

Woosowska, S. and J. Synowiecki. 2004. Thermostable glucosidase with broad substrate specificity suitable for processing of lactose-containing products. Food Chem. 85: 181–187.

Wu, G., G. Wu, T. Zhan, Z. Shao and Z. Liu. 2013. Characterization of a cold-adapted and salt-tolerant esterase from a psychrotrophic bacterium *Psychrobacter pacificensis*. Extremophi. 17: 809–819.

Wu, W., M. Chen, I. Tu, Y. Lin, N.E. Kumar, M. Chen et al. 2017. The discovery of novel heat-stable keratinases from *Meiothermus taiwanensis* WR-220 and other extremophiles. Sci. Rep. 7: 4658.

Xin, L. and Y. Hui-Ying. 2013. Purification and characterization of an extracellular esterase with organic solvent tolerance from a halotolerant isolate, *Salimicrobium* sp. LY19. BMC Biotechnol. 13: 108.

Yadav, A.N., S.G. Sachan, P. Verma and A.K. Saxena. 2004. Bioprospecting of plant growth promoting psychrotrophic Bacilli from the cold desert of north western Indian Himalaya. Microbiol. Res. 159: 141–146.

Yadav, A.N. 2015. Bacterial diversity of cold deserts and mining of genes for low temperature tolerance. Ph.D. Thesis, IARI, New Delhi/BIT, Ranchi pp. 234, DOI: 10.13140/RG.2.1.2948.1283/2.

Yadav, A.N., S.G. Sachan, P. Verma and A.K. Saxena. 2015a. Prospecting cold deserts of north western Himalayas for microbial diversity and plant growth promoting attributes. J. Biosci. Bioeng. 119: 683–693.

Yadav, A.N., S.G. Sachan, P. Verma, S.P. Tyagi, R. Kaushik and A.K. Saxena. 2015b. Culturable diversity and functional annotation of psychrotrophic bacteria from cold desert of Leh Ladakh (India). World J. Microbiol. Biotechnol. 31: 95–108.

Yadav, A.N., P. Verma, M. Kumar, K.K. Pal, R. Dey, A. Gupta et al. 2015c. Diversity and phylogenetic profiling of niche-specific Bacilli from extreme environments of India. Ann. Microbiol. 65: 611–629.

Yadav, A.N., S.G. Sachan, P. Verma, R. Kaushik and A.K. Saxena. 2016a. Cold active hydrolytic enzymes production by psychrotrophic Bacilli isolated from three sub-glacial lakes of NW Indian Himalayas. Journal of Basic Microbiology. 56(3): 294–307.

Yadav, A.N., S.G. Sachan, P. Verma and A.K. Saxena. 2016b. Bioprospecting of plant growth promoting psychrotrophic Bacilli from the cold desert of north western Indian Himalayas.

Yadav, A.N., R. Kumar, S. Kumar, V. Kumar, T. Sugitha, B. Singh et al. 2017a. Beneficial microbiomes: biodiversity and potential biotechnological applications for sustainable agriculture and human health. J. Appl. Biol. Biotechnol. 5: 45–57.

Yadav, A.N., P. Verma, V. Kumar, S.G. Sachan and A.K. Saxena. 2017b. Extreme cold environments: A suitable niche for selection of novel psychrotrophic microbes for biotechnological applications. Adv. Biotechnol. Microbiol. 2: 1–4.

Yadav, A.N., P. Verma, S.G. Sachan, R. Kaushik, A.K. Saxena. 2018. Psychrotrophic microbiomes: molecular diversity and beneficial role in plant growth promotion and soil health. pp. 197–240. *In*: Panpatte, D.G., Y.K. Jhala, H.N. Shelat and R.V. Vyas (eds.). Microorganisms for Green Revolution-Volume 2: Microbes for Sustainable Agro-ecosystem. Springer, Singapore. doi:10.1007/978-981-10-7146-1_11.

Yadav, A.N., S. Gulati, D. Sharma, R.N. Singh, M.V.S. Rajawat, R. Kumar et al. 2019a. Seasonal variations in culturable archaea and their plant growth promoting attributes to predict their role in establishment of vegetation in Rann of Kutch. Biologia 74: 1031–1043. doi:10.2478/s11756-019-00259-2.

Yadav, A.N., D. Kour, S. Sharma, S.G. Sachan, B. Singh, V.S. Chauhan et al. 2019b. Psychrotrophic microbes: biodiversity, mechanisms of adaptation, and biotechnological implications in alleviation of cold stress in plants. pp. 219–253. *In*: Sayyed, R.Z., N.K. Arora and M.S. Reddy (eds.). Plant Growth Promoting Rhizobacteria for Sustainable Stress Management: Volume 1: Rhizobacteria in Abiotic Stress Management. Springer Singapore, Singapore. doi:10.1007/978-981-13-6536-2_12.

Yadav, A.N., J. Singh, A.A. Rastegari and N. Yadav. 2020. Plant Microbiomes for Sustainable Agriculture. Springer International Publishing, Cham.

Yamashiro, Y., A. Sakatoku, D. Tanaka and S. Nakamura. 2013. A cold-adapted and organic solvent tolerant lipase from a psychrotrophic bacterium *Pseudomonas* sp. strain YY31: identifi cation, cloning, and characterization. Appl. Biochem. Biotechnol. 171: 989–1000.

Yin, J., J.C. Chen, Q. Wu and G.Q. Chen. 2014. Halophiles, coming stars for industrial biotechnology. Biotechnol. Adv. 33: 1433–1442.

Yokoigawa, K., Y. Okubo, H. Kawai, N. Esaki and K. Soda. 2001. Structure and function of psychrophilic alanine racemase. J. Mol. Cataly. B: Enzym. 12: 27–35.

Yoshimune, K., Y. Shirakihara, M. Wakayama and I. Yumoto. 2010. Crystal structure of salt-tolerant glutaminase from *Micrococcus luteus* K-3 in the presence and absence of its product L-glutamate and its activator Tris. FEBS J. 3: 738–748.

Yu, H.Y. and X. Li. 2012. Purification and characterization of novel organic-solvent-tolerant b-amylase and serine protease from a newly isolated *Salimicrobium halophilum* strain LY20. FEMS. Microbiol. Lett. 329: 204–211.

Zafrilla, B., R.M. Martínez-Espinosa, J. Esclapez, F. Pérez-Pomares and M.J. Bonete. 2010. SufS protein from *Haloferax volcanii* involved in Fe-S cluster assembly in haloarchaea. Biochimica. Et. Biophysica. Acta. 1804: 1476–1482.

Zhang, J., S. Lin and R. Zeng. 2007. Cloning, expression, and characterization of a cold-adapted lipase gene from an antarctic deep-sea psychrotrophic bacterium, *Psychrobacter* sp. 7195. J. Microbiol. Biotechnol. 17: 604–610.

Zhang, J.W. and R.Y. Zeng. 2008. Molecular cloning and expression of a cold-adapted lipase gene from an Antarctic deep sea psychrotrophic bacterium *Pseudomonas* sp. 7323. Mar. Biotechnol. 10: 612–621.

Zhang, Y., E.M. Zamudio Cañas, Z. Zhu, J.L. Linville, S. Chen and Q. He. 2011. Robustness of archaeal populations in anaerobic co-digestion of dairy and poultry wastes. Bioresour. Technol. 102: 779–785.

Zhao, Y.X., Z.M. Rao, Y.F. Xue, P. Gong, Y.Z. Ji and Y.H. Ma. 2015. Poly(3-hydroxybutyrate-co-3-hydroxyvalerate) production by haloarchaeon *Halogranum amylolyticum*. Appl. Microbiol. Biotechnol. 99: 7639–7649.

Zhu, C., J. Zhang, Y. Tang, X. Zhengkai and R. Song. 2011. Diversity of methanogenic archaea in a biogas reactor fed with swine feces as the mono-substrate by mcrA analysis. Microbiol. Res. 166: 27–35.

Zhu, Y., H. Li, H. Ni, A. Xiao, L. Li and H. Cai. 2015. Molecular cloning and characterization of a thermostable lipase from deep-sea thermophile *Geobacillus* sp. EPT9. World J. Microbiol. Biotechnol. 31: 295–306.

Chapter 2

Halophilic Rhizobacteria as the Acquaintance of Crop Plants Enduring Soil Salinity

Deepanwita Deka and *Dhruva Kumar Jha**

||||||||||||||||||||||||||||||||||||||||||||||||||||||||||||||||||||||||||||||||||||||||||||||||||||||||||||||||||||||||||||||||||||||||||||||||||||||||||||||||||||||||||||||||||||||||||||||||

Introduction

The beginning of the 21st century can be characterized by the universal scarcity of water resources, environmental pollution and increased salinization of soil and water (Shrivastava and Kumar 2015). Climatic change accelerates abiotic stresses such as drought, salinity besides abrupt fluctuations in temperature, etc. (Shrivastava and Kumar 2015). Soil salinity is one of the most serious and damaging environmental stresses, which results in changes of soil physico-chemical properties, degrading soil quality, causing loss of soil organic matter consequently affecting plant growth, agricultural outputs, etc. (Ahmad et al. 2016; Yamaguchi and Blumwald 2005; Shahbaz and Ashraf 2013). This also leads to increased infestation of plants by pathogens and pests (Chakraborty 2013). Salinity stress results due to many reasons such as low precipitation, high surface evaporation, weathering of native rocks, irrigation with saline water and poor cultural practices. Salinized areas are increasing at a rate of 10% annually (Shrivastava and Kumar 2015). It has been estimated that more than 50% of arable land would be salinized by the year 2050 (Jamil et al. 2011).

Soil salinization involves increase in the concentration of dissolved salts in the soil profile, which degrades soil health and consequently affects crop productivity impacting the lives of many organisms that depend on plants for food, nutrition and shelter (Shrivastava and Kumar 2015). In a saline soil the Electrical Conductivity (EC) of the saturation extract (ECe) in the root zone of a plant exceeds 4 dSm^{-1} (approximately 40 mM NaCl) at 25°C and has exchangeable sodium of 15% (Shrivastava and Kumar 2015). Under this ECe, growth and yield of most of the crop plants are reduced (Munns 2005; Jamil et al. 2011). It has been established that high salinity has afflicted 20% of total cultivated and 33% of irrigated agricultural lands worldwide (Shrivastava and Kumar 2015). More than 50% of the world energy consumption is supplied by crop plants such as wheat, rice and maize (Orhan 2016). Therefore, decrease in plant yield due to soil salinity stresses severely affects all trophic levels in the food chain directly or indirectly.

Orhan (2016) stated that universally, the total expanse of saline soils in arid and semi-arid regions is nearly 15% and almost 40% in irrigated lands. Each plant requires a measurable quantity of salt dissolved in soil for growth and development. However, high concentration of salts in soil has a negating effect on soil physical and chemical properties, thereby directly or indirectly affecting

Microbial Ecology Laboratory, Department of Botany, Guwahati University, Guwahati-781014, Assam, India.
* Corresponding author: dkjhabot07@gmail.com

the growth and diversity of plants and other organisms present in that soil. Plants growing under high soil salinity conditions (excessive sodium and chloride concentrations) for an extended period are exposed to ionic and osmotic stresses, which undesirably affect their different biochemical and physiological processes such as expansion and premature ageing of leaves, closure of stoma, photosynthesis, cell division and expansion, etc. Most cereal crops need low salinity or salt stress thresholds for high growth and productivity. For example, wheat plant can stand salinity up to a level of 6 dSm^{-1}, while the salinity threshold for maize is three times lesser than that of the wheat (approximately 2 dSm^{-1}) (Orhan 2016). A major loss of crop productivity, therefore, is the result of salt stress in salt-sensitive plants.

Increasing soil salinization as well as the challenge of feeding the world's growing population has attracted many investigators towards plant and soil productivity research (Yadav et al. 2019). This kind of problem can only be mitigated with the help of suitable biotechnological interventions like applications of appropriate soil microorganisms (Lugtenberg et al. 2002) to improve soil health and to increase crop productivity. Microorganisms such as *Pseudominas mendocina, Rhizobium,* etc. enhances uptake of essential nutrients, selective uptake of potassium ion (K+) in maize (Bano and Fatima 2009), rice plants (Jha et al. 2011) under salinity stress. Rigorous research works are going on to improve plant growth, yield and tolerance to abiotic stresses like soil salinity using Plant Growth Promoting Rhizobacteria (PGPR) (Lugtenberg and Kamilova 2009; Berg and Martinez 2015). Certain PGPR have pronounced influences on sustainable crop production even under stress conditions of salinity, high and low temperature, drought stress and pH stress (Egamberdieva et al. 2015; Egamberdieva et al. 2016; Kour et al. 2020b).

Extensive works are going on several halophilic and halotolerant rhizobacterial strains to evaluate their plant growth stimulating activities. The rhizosphere soil of halophytic plants is a reservoir for unique groups of salt-tolerant rhizobacteria (Jha et al. 2012; 2015; Shukla et al. 2012; Bharti et al. 2013). These rhizobacteria could be used as enhancers of salt-tolerance in non-halotolerant crops overcoming salinity stress and increasing the growth in plants (Ramadoss et al. 2013; Goswami et al. 2014; Sharma et al. 2016; Yuan et al. 2016). Egamberdieva et al. (2017) studied the effect of plant growth promoting rhizobacteria on plant growth and control of foot and root rot disease of tomato caused by *Fusarium solani* under different conditions of soil salinity. Orhan (2016) tested 18 halotolerant or halophilic bacteria for their plant growth promoting capabilities in *Triticum aestivum* (wheat) in a hydroponic culture. It was observed that the bacterial strains possessed different plant-growth promoting abilities *in vitro* under salt stress (200 mM NaCl). These expressively augmented the fresh weight of the plants increasing root and shoot length (Orhan 2016).

The plants growing in salt stressed soils possess suppressed growth (Paul and Roychoudhury 2012). Rhizospheric microorganisms, particularly plant growth promoting rhizobacteria (PGPR), helps the plant to withstand stress environments by fixing nitrogen, producing phytohormones, sequestering iron with the help of siderophores and phosphate solubilization (Dimkpa et al. 2009; Hayat et al. 2010; Gaba et al. 2017; Yadav 2019a; Yadav et al. 2017b). Plants which are inoculated with PGPR can produce more root hairs, procure mineral and microelements more proficiently from the soil (Berg et al. 2013; Allah et al. 2015). PGPR helps the plant in several ways, for example, in lentil (*Lens esculenta*) (Faisal 2013), pea (*Pisum sativum* L.) (Meena et al. 2015), cucumber (*Cucumis sativus*) (Egamberdieva et al. 2011), rice (*Oryza sativa*) (Yadav et al. 2014) and soybean (*Glycine max*) (Egamberdieva et al. 2015), wheat (Rana et al. 2020a), great millet (Kour et al. 2020a), Foxtail millet (Kour et al. 2020c) growth is enhanced by PGPR synthesizing active compounds such as plant growth stimulators (Parray et al. 2016; Egamberdieva et al. 2017), osmolytes (Berg et al. 2013), antimicrobial compounds (Landa et al. 2004).

PGPR also induces systemic tolerance in plants through alteration of plant physiology growing in saline soil (Wang et al. 2012; Verma et al. 2016; Yadav et al. 2015). Some recent studies revealed that a few PGPR Induces Systemic Tolerance (IST) in plants through elevation of antioxidant responses by enhancing enzyme activity and metabolite production (Hashem et al. 2015; Hashem et al. 2016; Jha et al. 2011). This antioxidant defence mechanism scavenges the Reactive Oxygen Species (ROS) which in turn plays an important role in adaptation of plant to salinity stress (Ahanger et al. 2014).

It can be said that the problem of soil salinization is a menace for agricultural throughput worldwide. Crops growing on saline soils face problems with high osmotic stress, nutritional disorders, toxicities and reduced crop productivity because of poor physico-chemical conditions of the soil (Yadav 2017a; Yadav 2017b). About 1128 m ha area is affected by salinity stresses worldwide. The maximum salt affected area is in the Middle East (189 m ha) followed by Australia (169 m ha) and North Africa (144 m ha) (Wicke et al. 2011). South Asia, including India, has about 52 m ha salt-affected area (Wicke et al. 2011). India has about 6.73 million ha salt-affected areas including the states of Gujarat (2.23 m ha), Uttar Pradesh (1.37 m ha), Maharashtra (0.61 m ha), West Bengal (0.44 m ha) and Rajasthan (0.38 m ha) constituting almost 75% of saline soils in the country (Singh, 2009). Estimates suggests that the present day 6.73 million ha area under salt-affected soils in the country would almost triple to 20 million ha by 2050 (Sharma et al. 2014). According to one estimate, in the state of Haryana approximately half a million hectare lands are affected by salinity and the potential annual monetary loss due to this problem is about Rs. 1669 million (Datta and De Jong 2002). This chapter briefly focuses on the effect of soil salinity on agriculture, augmentation and mechanism of resistance in plants against salinity stress in order to increase plant growth and productivity by the application of plant growth promoting rhizobacteria.

Impact of soil salinization on agriculture

Impact on soil

Salts in the soil occur as ions which are released from weathering minerals in the soil (Glick et al. 2007). They may also come from the water used for irrigation or sometimes from fertilizers, or may sometimes drift upward in the soil from groundwater. When precipitation is inadequate to percolate ions from the soil profile, salts hoard in the soil leading to soil salinization (Blaylock 1994). Plants grasp essential nutrients in the form of water-soluble salts present in the soil, but excessive accumulation reversibly suppresses the plant growth (Yadav and Yadav 2018a). During the last decades, physical, chemical and/or biological degradation processes of global natural resources, have resulted in serious consequences to their loss globally. Due to the introduction of irrigation in new areas each year, soil salinization is increasing gradually (Patel et al. 2011). Bano and Fatima (2009) stated that plants growing in saline soil, face significant reduction of Phosphorus (P) uptake due to phosphate ions precipitation with Ca ions. Moreover, due to high concentration under saline conditions, some elements such as sodium, chlorine and boron, have specific toxic effects on crop plants.

Impact on crop production

Arid and semi-arid regions of the world with high temperature and dry soils, are commonly saline limiting the agricultural output (Etesami and Beattie 2018). In India, approximately 7 million hectares of land is saline (Patel et al. 2011). In such areas most of the crops are grown under irrigation, however, scarce irrigation management results into secondary salinization (Patel et al. 2011). High salinity conditions of the soil, affects plant growth adversely due to low osmotic potential of soil solution (osmotic stress), specific ion effects (salt stress), nutritional imbalances or a combination of these factors (Ashraf 2004). Plant growth and development are affected by all these factors at the physiological, biochemical as well as at molecular levels (Munns and James 2003; Tester and Davenport 2003). Due to the extreme accumulation of sodium ion in cell walls rapid osmotic stress and cell death occur in crop plants (Munns et al. 2002). Plants which are sensitive to some elements like sodium, chlorine, boron, etc., may be affected adversely at relatively low salt concentrations if the soil contains enough of the toxic element (Munns et al. 2002). There are many salts in the soil required by the plants as nutrients. However, high salt levels in the soil can disturb the nutrient balance in the crop plant or obstruct with the uptake of some nutrients (Blaylock 1994).

Lower growth rate in saline conditions due to nutrient imbalances results into the reduction of photosynthesis mainly through a reduction in leaf area, chlorophyll content and stomatal conductance (Netondo et al. 2004). Plant reproductive development is also adversely effected by soil salinity by inhibiting microsporogenesis and stamen filament elongation, augmenting programmed cell death in some tissue types, ovule abortion, senescence of fertilized embryos, etc. (Netondo et al. 2004). Crops growing under salinity stress undergo complex interfaces among morphological, physiological and biochemical progressions including seed germination, plant growth, and water and nutrient uptake (Akbarimoghaddam et al. 2011; Singh and Chatrath 2001). Salinity devastates almost all phases of plant life including photosynthesis, vegetative growth, reproductive development and germination (Yadav 2019a; Yadav 2019b).

Crops growing in a saline medium face osmotic balance failure which results in loss of turgidity, cell dehydration and eventually death of cells. The supply of photosynthetic assimilates or hormones to the growing tissues are weakened by the soil salinity, which affects plant growth (Ashraf 2004). During salinity stress, ion toxicity is the result of replacement of K^+ by Na^+ in biochemical reactions and the conformational changes in proteins induced by Na^+ and Cl^-. K^+ is one of the most important ion for plant survival. K^+ acts as a cofactor for several enzyme activities which cannot be substituted by Na^+ and a high concentration of K^+ is required for binding tRNA to ribosomes during protein synthesis (Zhu 2002). Metabolic imbalance influenced by the ion toxicity and osmotic stress, leads to oxidative stress in plants growing in salinity (Chinnusamy et al. 2006).

The negative impact of salinity on plant development is more profound during the reproductive phase (Munns and Rawson 1999). Munns and Rawson (1999) reported that wheat plants growing at 100–175 mM NaCl at a stressed condition showed a significant drop in spikelets in each spike, delayed spike emergence and reduced fertility which results in poor grain yields. However, in the shoot apex of these wheat plants Na^+ and Cl^- concentrations were below 50 and 30 mM respectively which were observed to be too low to limit the metabolic reactions (Munns and Rawson 1999). Therefore the adverse effects of salinity stress may be endorsed to the failure of the cell cycle and differentiation of the plant. Salinity halts the cell cycle rapidly by reducing the expression and activity of cyclins and cyclin-dependent kinases causing lessened cell number in the meristem, thus restricting the plant growth. Recent reports have also confirmed that the activity of cyclin-dependent kinase is reduced by post-translational inhibition during salt stress effecting plant growth and development, encumbering seed-germination, seedling growth, enzyme activity (Seckin et al. 2009), DNA, RNA, protein synthesis and mitosis (Tabur and Demir 2010; Javid et al. 2011).

Assuagement of salinity stress in crops using rhizobacteria

In order to reduce the loss of productivity by plants due to salinity, certain practices are being used. Extensive research is being carried out throughout the world to develop strategies to tackle the problem of salinity stress, which include development of salt tolerant plant varieties, shifting of crop calendars, applying resource management practices, etc. (Wang et al. 2003) and by the use of Plant Growth-Promoting Bacteria (PGPB) (Dimkpa et al. 2009).

Microorganisms settling on different parts of plants play a significant role in plant growth promotion, nutrient management and disease control. In recent years, a few rhizosphere bacteria belonging to different genera like *Rhizobium, Bacillus, Pseudomonas, Pantoea, Paenibacillus, Burkholderia, Methylobacterium, Variovorax, Enterobacter, Azospirillum, Achromobacter, Microbacterium*, etc. have been reported to provide tolerance to host plants under salt stress environments (Grover et al. 2011; Yadav et al. 2020a; Yadav et al. 2020b; Yadav et al. 2020c). These beneficial microorganisms endorse plant growth through different direct and indirect mechanisms (Nia et al. 2012; Ramadoss et al. 2013). PGPR enables plant growth directly by aiding nutrient uptake through phytohormone production (e.g., auxin, cytokinin and gibberellins), by enzymatic dropping of plant ethylene levels and/or by production of siderophores (Kohler et al. 2006) and indirectly by reducing plant pathogens (Singh et al. 2020).

Cho et al. (2006) demonstrated that inoculations with AM (Arbuscular Mycorrhizal) fungi improved plant growth under salt stress. Egamberdiyeva (2007) reported three PGPR isolates *Pseudomonas alcaligenes* PsA15, *Bacillus polymyxa* BcP26 and *Mycobacterium phlei* MbP18 which could tolerate high salt concentration like calcisol and capable of surviving in saline soils. Kohler et al. (2009) examined the impact of PGPR *Pseudomonas mendocina*, inoculated alone or in combination with an AM fungus *Glomus intraradices* or *G. mosseae*, on growth, nutrient uptake and other physiological activities of *Lactuca sativa* growing in saline soil. They observed that plants inoculated with *P. mendocina* had threateningly greater shoot biomass than the controls and recommended that inoculation with selected PGPR could be an effective tool for assuaging salinity stress in salt sensitive plants (Kohler et al. 2009). It has been established that rhizosphere bacteria isolated from different stressed habitats possessed stress tolerance capacity along with the plant growth-promoting characters and consequently might be suggested as potential candidates for seed bacterization for salt stress tolerance (Kohler et al. 2009). Plants inoculated with PGPR showed heightened root and shoot biomass and biochemical levels such as chlorophyll, carotenoids and protein (Tiwari et al. 2011). However, the interaction of PGPR with other microbes and their effect on the physiological response of crop plants under different soil salinity conditions are yet to be extensively investigated. Inoculations of PGPR and other microbes to salt sensitive crop plants could serve as a potential device for lessening salinity stress in those plants (Yadav et al. 2017a; Yadav and Yadav 2019). Hence, extensive exploration is required in this area and the utilization of PGPR and other symbiotic microorganisms can be advantageous in developing stratagems to enable sustainable agriculture in saline soils.

Various studies suggest that utilization of PGPR has become a promising substitute to lessen plant stress caused by salinity (Yao et al. 2010) and its application in the management of salinity stresses is gaining importance day by day. The application of PGPR prompted tolerance to salinity stresses in crop plants (Dodd and Perez-Alfocea 2012; Yang et al. 2009). PGPRs which were capable of solubilizing phosphate, producing phytohormones and siderophores in saline condition, supported the growth of tomato plants under 2% NaCl stress (Tank and Saraf 2010). Barassi et al. (2006); Yildirim and Taylor (2005) observed that application of plant growth promoting bacteria enriched the growth of tomato, pepper, canola, bean and lettuce growing under saline conditions. Moreover, *Pseudomonas putida* inoculant stimulated cotton growth and germination under salt stress conditions (Yao et al. 2010). A few years earlier, Ramadoss et al. (2013) also evaluated the effect of five plant growth-enhancing halotolerant bacteria on wheat growth inhabiting salinized condition and observed that application of these halotolerant bacterial strains increased the root length to 71.7% in wheat seedlings as compared to positive controls growing in salt stress.

The application of *Halobacillus* sp. and *Bacillus halodenitrificans* especially exhibited more than 90% enhancement in root elongation and 17.4% increase in dry weight as compared to un-inoculated wheat seedlings at 320 mM NaCl stress, which implied a notable decrease in the lethal effects of NaCl (Ramadoss et al. 2013). It can be stated that through direct or indirect mechanisms halotolerant bacteria isolated from saline environments possess the potential to improve plant growth under salinity stress and could be applied as the most suitable bioinoculants under such stress conditions (Ramadoss et al. 2013). However, on the other hand, screening of the halotolerant bacteria for their PGP traits should be carried out. The isolation of native rhizosphere microorganisms from the stress affected soils and their screening for salinity stress tolerance and PGP characters is advantageous in the prompt selection of effective strains which might be used as bioinoculants for crops growing in stress (Kumar et al. 2019; Yadav and Saxena 2018). Thus, application of the plant growth promoting rhizobacteria can alleviate salt stresses in crop plants opting as a new and emerging tool for salt tolerance (Kour et al. 2020d; Rana et al. 2019; Yadav 2018) (Table 2.1).

Table 2.1 Summarizes some of the advances and investigations carried out to assess the role of rhizobacteria as salinity stress premeditators.

Plant growth promoting rhizobacteria (PGPR)	Crop plant	Beneficial effects	References
Arthrobacter woluwensis, Microbacterium oxydans, Arthrobacter aurescens, Bacillus megaterium and *Bacillus aryabhattai*	Soybean	Increased production of indole-3-acetic acid (IAA), gibberellin (GA), and siderophores and increased phosphate solubilization	Khan et al. (2019)
Halomonas sp. and *Bacillus* sp.	Alfalfa	-	Kearl et al. (2019)
Pseudomonas sp., *Micriobacterium natoriense, Pseudomonas brassicacearum, Curtobacterium flaccumfaciens, Pseudomonas* sp. and *Mesorhizobium*	*Glycyrrhiza uralensis*	Production of ACC deaminase and IAA, siderophores, etc., increase the absorption of the nitrogen, magnesium, phosphorus, potassium, etc. and increase in plant biomass	Egamberdieva et al. (2017)
Streptomyces sp. strain PGPA39	*Medica gotruncatula*	ACC deaminase activity and IAA production and phosphate solubilization	Yaish et al. (2016)
Enterobacter, Halomonas, Marinobacter, Pseudomonas, Pseudoxanthomonas, Rhizobium and *Thalassospira* spp.	*S. salsa, Hordeum secalinum, Plantago winteri*	Production of ACC deaminase, phytohormone IAA and siderophores, etc. Enhanced uptake of essential nutrients, enhanced germination percentage	Yuan et al. (2016), Cardinale et al. (2015)
Rhizobium sp., *Pseudomonas putida, Enterobacter cloacae, Serratiaficaria* sp. and *Pseudomonas fluorescens*	Salt sensitive rice GJ-17	Enhanced germination percentage, germination rate, and index and improved the nutrient status of the wheat plants	Jha and Subramanian (2014)
Pseudomonas pseudoalcaligenes, Bacillus pumilus	Barley and oats	Reduce lipid peroxidation and superoxide dismutase activity	Chang et al. (2014)
Acinetobacter spp. and *Pseudomonas* sp.	'Micro tom' tomato	Production of ACC deaminase and IAA	Palaniyandi et al. (2014)
Brachybacterium saurashtrense (JG06), *Brevibacterium casei* (JG-08), and *Haererohalobacter* (JG-11)	Mung bean (*V. radiata* L.)	ACC deaminase activity and increased water use efficiency	Ahmad et al. (2013)
Rhizobium phaseoli and PGPR (*Pseudomonas syringae*, Mk1; *Pseudomonas fluorescens*, Mk20 and *Pseudomonas fluorescens* Biotype G, Mk25)	Wheat	IAA production and ACC deaminase activity	Nadeem et al. (2013)
PGPR (Mk1, *Pseudomonas syringae*; Mk20, *Pseudomonas fluorescens* and Mk25, *Pseudomonas fluorescens* and *Rhizobium phaseoli* strains M1, M6, and M9	Cotton (G. hirsutum), Groundnut (*A. hypogaea* L.)	ACC deaminase activity and improvement in growth and nodulation in mung bean	Wu et al. (2012), Shukla et al. (2012)
Raoultella planticola	Mung bean (*V. radiata* L.)	ACC deaminase activity, high K$^+$/Na$^+$ ratio and higher Ca^{2+}, phosphorus, and nitrogen content	Ahmad et al. (2012)

Table 2.1 Contd. ...

...Table 2.1 Contd.

Plant growth promoting rhizobacteria (PGPR)	Crop plant	Beneficial effects	References
Pseudomonas putida	Mung bean (*Vigna radiata* L.)	Increase the absorption of the Mg^{2+}, K^+ and Ca^{2+} and decrease the uptake of the Na^{2+} from the soil	Ahmad et al. (2011)
Rhizobium, Pseudomonas	Rice (*Oryza sativa*)	Decreased electrolyte leakage and, increase in proline production, maintenance of relative water content of leaves, and selective uptake of K ion	Jha et al. (2011)
Pseudomonas pseudoalcaligenes, Bacillus pumilus	Cotton (*Gossipium hirsutum*)	Increased concentration of glycine betaine (compatible solute)	Yao et al. (2010)
Pseudomonas mendocina	Maize (Zea mays)	ACC deaminase activity and enhanced uptake of essential nutrients	Bano and Fatima (2009)
Bacillus subtilis	*Arabidopsis thaliana,* Lettuce (*L. sativa* L. cv. Tafalla)	Tissue specific regulation of sodium transporter HKT1	Kohler et al. (2009)
Pseudomonas fluorescens	Groundnut (*Arachis hypogea*)	Enhanced ACC deaminase activity	Saravanakumar and Samiyappan (2007), Zhang et al. (2008)
Pseudomonas syringae, Pseudomonas fluorescens, Enterobacter aerogenes	Maize (Zea mays)	Increased ACC deaminase activity	Nadeem et al. (2007)

Mechanism of salt tolerance in PGPR colonized plants

A variety of mechanisms have been proposed previously by different investigators based on studies done on microbes stimulated stress tolerance in plants. It has been established that application of PGPR can restrain salinity stress in plants through direct and indirect mechanisms involved in inducing systemic tolerance (Yang et al. 2009). Many PGPR have been inspected for their beneficial action in improving plant-water relations, ion homeostasis and photosynthetic efficiency in plants dwelling salt stress fields (Rana et al. 2020b; Yadav 2020). Figure 2.1 shows salt tolerance mechanisms induced by plant growth promoting rhizobacteria in plants surviving in salt-stress condition. Amelioration mechanisms of PGPR in salt-stressed plants, are complexed and often hard to understand (Yadav and Yadav 2018b; Rastegari et al. 2020; Singh and Yadav 2020). To regulate these mechanisms an intricate network of signalling actions occurs during the plant–microbe interactions and subsequently results into stress relief (Smith et al. 2017). Different techniques have been inferred in order to understand the dynamic function of PGPR in relation to ion transport, water and nutrient uptake, stomatal conductance, signal transducing proteins, phytohormonal status, carbohydrate metabolism, antioxidants, and enzymes for shaping the Induced Systemic Tolerance (IST) in plants (Fig. 2.1).

Phytohormones like auxin, indole acetic acid, gibberellins and few other determining factors by PGPR result in augmented root length, root surface area and the number of root tips, leading to an improved uptake of soil nutrients thereby enhancing plant health under stress conditions (Egamberdieva and Kucharova 2009). A PGPR strain, *Achromobacter piechaudii* ARV8 released 1-aminocyclopropane-1-carboxylate (ACC) deaminase which developed induced systemic resistance in pepper and tomato against drought and salinity stress (Mayak et al. 2004). The multifaceted and

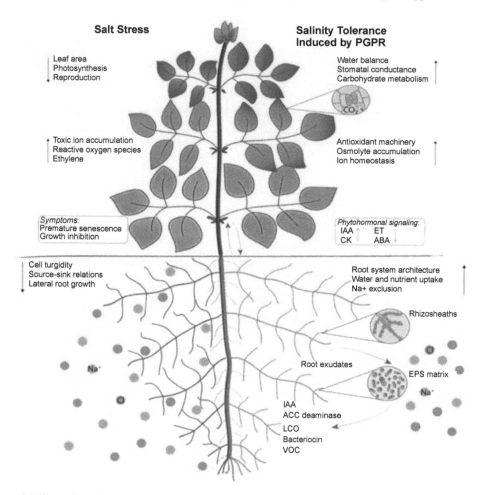

Figure 2.1 Illustration of salt tolerance mechanisms induced by plant growth promoting rhizobacteria (PGPR). Root surfaces are colonized by PGPR and extracellular polysaccharide matrix acts as a protective barrier against salt stress. Some extracellular molecules function as signaling cues that manipulate phytohormonal status in plants. Enhanced root-to-shoot communication improves water and nutritional balance, source-sink relations and stomatal conductance. Stimulating osmolyte accumulation, carbohydrate metabolism and antioxidant activity delay leaf senescence, which in turn contribute to photosynthesis. Regulation of physiological processes are indicated by black arrows and signaling pathways are indicated by purple arrows (Adapted from Illangumaran and Smith 2017).

dynamic interactions among microorganisms, roots, soil and water in the rhizosphere zone have variations in physicochemical and structural characters of the soil (Haynes and Swift 1990). The polysaccharides secreted by the rhizosphere microbes bind to the soil particles to form micro- and macro-aggregates. Plant roots and fungal hyphae occupy the inter-microaggregate spaces and leads to stabilization of macro-aggregates. Plants treated with bacteria generating Exo-Poly Saccharides (EPS), exhibited improved resistance to salinity stress due to enhanced soil structure (Sandhya et al. 2009). During saline conditions EPS bind to cations including Na^+ making it unattainable to plants (Sandhya et al. 2009). Naz et al. (2009) isolated bacterial strains from Khewra salt range of Pakistan, which exhibited salt tolerance when tested on saline media simulated by rhizosphere soil filtrate. Interestingly, the isolates produced Abscisic acid in a much higher concentration than that of normal media used.

Moreover, inoculation of those isolates in soybean plants increased the production of proline, enhanced shoot/root length, and dry weight under salt stress condition (Naz et al. 2009). Similarly, Upadhyay et al. (2011) examined the influence of PGPR inoculation on growth and the antioxidant level of wheat plant growing in saline conditions and found that co-inoculation with PGPR *Bacillus subtilis* and *Arthrobacter* sp. could lessen the negative impact of soil salinity on wheat growth resulting in the enhanced dry biomass, soluble sugars and proline content. Jha et al. (2011) described that the combination of an endophytic bacterium *Pseudomonas pseudoalcaligenes* with a rhizospheric *Bacillus pumilus* in paddy, was able to protect the plant from salt stress by stimulation of osmoprotectant and antioxidant proteins. However, the effect of the rhizospheric or endophytic bacteria was not promising when treated alone at higher salinity levels. Nia et al. (2012) also observed that the yield of wheat in saline conditions after inoculation with PGPR strains like *B. pumilus, P. pseudo alcaligenes* isolated from saline or non-saline soil, increased salinity tolerance in wheat plants. Salt-stress resistant isolates could considerably increase shoot dry weight and plant yield under extreme salinity condition (Nia et al. 2012). Moreover, they also found that plants inoculated with saline-adapted *Azospirillum* strains showed increased N concentrations at different salinity levels. The effect of a plant growth promoting the strain of *Streptomyces* which could produce auxin and siderophore was studied by Sadeghi et al. (2012) and they found that growth and development of wheat plant was increased after application of PGP *Streptomyces*. They observed a significant surge in shoot length, dry weight, germination rate, etc. as compared to the control and also an increase in the concentration of N, P, Fe and Mn in wheat shoots growing in normal as well as saline soil, concluding that *Streptomyces* isolate could be applied as bio-fertilizers under in saline conditions (Sadeghi et al. 2012).

Rhizobacteria which are bactericidal or bacteriostatic, are able to produce bacteriocins (small peptides) that may protect the plant from phytopathogens during salt-stress condition (Kirkup and Riley 2004). The application of the gene thuricin 17, extracted from *Bacillus thuriengenesis* NEB 17 inhabiting soybean plant as an endosymbiont, differentially transformed the proteome of salt-stressed (250 mM NaCl) *Arabidopsis* plants (Subramanian et al. 2016a). Bacterial signals modulate the expression of proteins intricate in carbon and energy metabolism pathways. Along with other stress-related proteins, proteins involved in photosynthesis such as, PEP carboxylase, pyruvate kinase, RuBisCo-oxygenase large subunit and proteins of photosystems I and II, were also upregulated by bacterial signals (Subramanian et al. 2016b). Subramanian et al. (2016a) observed similar kinds of changes in the proteome of soybean seeds prompted by the bacterial signal compounds at 48 hours under 100 mM NaCl. Furthermore, isocitrate-lyase and antioxidant Glutathione-S-transferase were also increased in soybean seeds (Subramanian et al. 2016a). These findings by Subramanian et al. (2016a) suggested that application of thuricin 17 positively influenced the plant proteome profile and augment physiological forbearance to salinity.

A few investigators described Volatile Organic Compounds (VOC) released by PGPR, stimulating plant growth, ensuing increased shoot dry weight and restrained salt stress. However, sensitivity of the plants towards volatiles and subsequent induction of salinity resistant mechanisms require further research (Bailly and Weisskopf 2012). VOCs secreted by *Bacillus* GB03, controlled the tissue-specific regulations of Na^+ homeostasis in salt-stressed plants (Bailly and Weisskopf 2012). When treated with VOCs, *Arabidopsis* growing in 100 mM NaCl, showed diminished accumulation Na^+ by synchronously downregulating expression of HKT1 in roots but up-regulating it in shoots (Bailly and Weisskopf 2012). Presumingly the stimulation of HKT1 dependent shoot-to-root recirculation ensued in the reduced Na^+ accumulation by about 50% throughout the plant and finally resulted in improved leaf surface area, root mass, and total K^+ content as compared to the controls (Bailly and Weisskopf 2012). Treatment with VOCs, reduced the total Na^+ content by 18% and enriched shoot and root growth of sos3 mutants growing in 30 mM NaCl (Zhang et al. 2008). A complex of VOCs, released by *Pseudomonas simiae* AU, decreased Na^+ accumulation in roots and prompted salt-tolerance, proline and chlorophyll content in soybean (*Glycine max*) inhabiting 100 mM NaCl. Chemical profiling identified the emanated compounds as 2-undecanone, 7-hexanol, 3-methylbutanol molecules (Ledger et al. 2016).

Conclusion and future prospects

The impact of salinity on agricultural crops and other plants is highly destructive which in turn affects the biodiversity and environment. Salt-stressed soils reduce the growth and development of plants growing in that area. Salinity also has negative impacts on the sustainability of beneficial microorganisms inhabiting the plant rhizosphere. Hence there is a great demand for sound, eco-compatible and eco-friendly techniques to resist the loss of plants inhabiting saline agricultural fields throughout the world. Use of plant growth promoting microorganisms (PGPR) as biofertilizers and biocontrol agents is replacing traditional agricultural techniques in organic farming practices in recent years. Plants possess multifarious mechanisms to endure salt stresses. Halo-tolerant PGPRs possess different mechanisms for osmo-tolerance and could be a valuable help to plants growing in saline soils, providing osmo-tolerance, improved growth, vigour and yield. PGPRs augment plant growth by different mechanisms such as asymbiotic N2 fixation, solubilization of mineral phosphate and other essential nutrients, release of phyto-hormones, siderophores, etc. Commercialization of the application of plant beneficial microbes to improve crop growth and yield in saline environments is an imminent approach for saline soil agriculture. A number of PGPR inoculants have currently been commercialized which endorse plant growth or inhibit pathogens or induce systemic resistance against pathogens or mitigate halo-tolerance in plants. The study of plant-microbe interactions in response to salinity, leads to new prospects of understanding the regulatory mechanisms of plant halo-tolerance mediated by rhizosphere bacteria which in turn will be beneficial for microbial inoculants for successful plant growth and enhancing tolerance to salinity. The forthcoming utilization of PGPR for the benefits of plants living in salt-stressed agricultural fields seems immensely wide, and as yet not been explored indepth.

Acknowledgement

Deepanwita Deka is thankful to the University Grant Commission, New Delhi for providing Basic Scientific Research (UGC-BSR) fellowship under Special Assistance Program (UGC-SAP, DRS-I) and Department of Botany, Gauhati University, Guwahati.

References

Ahanger, M.A., A. Hashem, E.F.A. Allah and P. Ahmad. 2014. Arbuscular mycorrhiza in crop improvement under environmental stress. pp. 69–95. *In*: Ahmad, P. (ed.). Emerging Technologies and Management of Crop Stress Tolerance. Elsevier, Oxford.

Ahmad, I., A. Mian and F.J.M. Maathuis. 2016. Overexpression of the rice AKT1 potassium channel affects potassium nutrition and rice drought tolerance. J. Exp. Bot. 67(9): 2689–2698.

Ahmad, M., A. Zahir, H.N. Asghar and M. Asghar. 2011. Inducing salt tolerance in mung bean through co-inoculation with rhizobia and plant-growth-promoting rhizobacteria containing 1-aminocyclopropane-1-carboxylate deaminase. Can. J. Microbiol. 57(7): 578–589.

Ahmad, M., Z.A Zahir, H.N. Asghar and M. Arshad. 2012. The combined application of rhizobial strains and plant growth promoting rhizobacteria improves growth and productivity of mung bean (*Vigna radiata* L.) under salt-stressed conditions. Ann. Microbiol. 62: 1321–1330.

Ahmad, M., Z.A. Zahir, F. Nazli, F. Akram and K.M.M. Arshad. 2013. Effectiveness of halo-tolerant, auxin producing *Pseudomonas* and *Rhizobium* strains to improve osmotic stress tolerance in mung bean (*Vigna radiata* L.). Braz. J. Microbiol. 44(4): 1341–1348.

Akbarimoghaddam, H., M. Galavi, A. Ghanbari and N. Panjehkeh. 2011. Salinity effects on seed germination and seedling growth of bread wheat cultivars. Trakia. J. Sci. 9(1): 43–50.

Allah, E.F.A., A. Hashem, A.A. Alqarawi, A.H. Bahkali and M.S. Alwhibi. 2015. Enhancing growth performance and systemic acquired resistance of medicinal plant *Sesbaniasesban* (L.) Merr using arbuscular mycorrhizal fungi under salt stress. Saudi. J. Biol. Sci. 22: 274–283.

Ashraf, M. 2004. Some important physiological selection criteria for salt tolerance in plants. Flora. 199: 361–376.

Bailly, A. and L. Weisskopf. 2012. The modulating effect of bacterial volatiles on plant growth: current knowledge and future challenges. Plant Sig. Behav. 7: 79–85.

Bano, A. and M. Fatima. 2009. Salt tolerance in *Zea mays* (L.) following inoculation with *Rhizobium* and *Pseudomonas*. Biol. Fert. Soils. 45: 405–413.

Barassi, C.A., G. Ayrault, C.M. Creus, R.J. Sueldo and M.T. Sobrero. 2006. Seed inoculation with *Azospirillum* mitigates NaCl effects on lettuce. Sci. Hort. 109: 8–14.

Berg, G., C. Zachow, H. Muller, J. Philipps and R. Tilcher. 2013. Next generation bio-products sowing the seeds of success for sustainable agriculture. Agronomy 3: 648–656.

Berg, G. and J.L. Martinez. 2015. Friends or foes: can we make a distinction between beneficial and harmful strains of the Stenotrophomonas maltophilia complex? Front. Microbiol. 6: 241.

Bharti, N., D. Yadav, D. Barnawal, D. Maji and A. Kalra. 2013. *Exiguobacterium oxidotolerans*, a halotolerant plant growth promoting rhizobacteria, improves yield and content of secondary metabolites in *Bacopa monnieri* (L.) Pennell under primary and secondary salt stress. World J. Microbiol. Biotechnol. 29: 379–387.

Blaylock, A.D. 1994. Soil salinity, salt tolerance and growth potential of horticultural and landscape plants. Co-operative Extension Service, University of Wyoming, Department of Plant, Soil and Insect Sciences, College of Agriculture, Laramie, Wyoming, B-988.

Cardinale, M., S. Ratering, C. Suarez, A.M.Z. Montoya, R.G. Plaum and S. Schnell. 2015. Paradox of plant growth promotion potential of rhizobacteria and their actual promotion effect on growth of barley (*Hordeum vulgare* L.) under salt stress. Microbiol. Res. 181: 22–32.

Chakraborty, S. 2013. Migrate or evolve: options for plant pathogens under climate change. Glob. Change Biol. 19(7): 1985–2000.

Chang, P., K.E. Gerhardt, X.D. Huang, X.M. Yu, B.R. Glick, P.D. Gerwing et al. 2014. Plant growth-promoting bacteria facilitate the growth of barley and oats in salt-impacted soil: implications for phytoremediation of saline soils. Int. J. Phytorem. 16(11): 1133–1147.

Chinnusamy, V., J. Zhu and J.K. Zhu. 2006. Gene regulation during cold acclimation in plants. Physiol. Plant. 126(1): 52–61.

Cho, K., H. Toler, J. Lee, B. Owenley, J.C. Stutz, J.L. Moore et al. 2006. Mycorrhizal symbiosis and response of sorghum plants to combined drought and salinity stresses. J. Plant Physiol. 163: 517–528.

Datta, K.K. and C.D. Jong. 2002. Adverse effect of waterlogging and soil salinity on crop and land productivity in northwest region of Haryana, India. Agri. Water Manag. 57(3): 223–238.

Dimkpa, C., T. Weinand and F. Ash. 2009. Plant-rhizobacteria interactions alleviate abiotic stress conditions. Plant Cell Environ. 32: 1682–1694.

Dodd, I.C. and F. Perez-Alfocea. 2012. Microbial amelioration of crop salinity stress. J. Exp. Bot. 63(9): 3415–3428.

Egamberdiyeva, D. 2007. The effect of plant growth promoting bacteria on growth and nutrient uptake of maize in two different soils. Appl. Soil Ecol. 36: 184–189.

Egamberdieva, D. and Z. Kucharova. 2009. Selection for root colonising bacteria stimulating wheat growth in saline soils. Biol. Fertil. Soils. 45. 561–573.

Egamberdieva, D., Z. Kucharova, K. Davranov, G. Berg, N. Makarova and T. Azarova. 2011. Bacteria able to control foot and root rot and to promote growth of cucumber in salinated soils. Biol. Fert. Soils. 47: 197–205.

Egamberdieva, D., D. Jabborova and G. Berg. 2015. Synergistic interactions between *Bradyrhizobium japonicum* and the endophyte *Stenotrophomonas rhizophila* and their effects on growth, nodulation and nutrition of soybean under salt stress. Plant Soil. 405: 3–5.

Egamberdieva, D., U. Behrendt, E.F.A. Allah and G. Berg. 2016. Biochar treatment resulted in a combined effect on soybean growth promotion and a shift in plant growth promoting rhizobacteria. Front. Microbiol. 7: 1–11.

Egamberdieva, D., K. Davranov, S. Wirth, A. Hashem and E.F.A. Allah. 2017. Impact of soil salinity on the plant-growth-promoting and biological control abilities of root associated bacteria. Saudi J. Biol. Sci. 24(7): 1601–1608.

Etesami, H. and G.A. Beattie. 2018. Mining halophytes for plants growth-promoting halotolerant bacteria to enhance the salinity tolerance of non-halophytic crops. Front. Microbiol. 9: 148.

Faisal, M. 2013. Inoculation of plant growth promoting bacteria *Ochrobactrum intermedium*, *Brevibacterium* sp. and *Bacillus cereus* induce plant growth parameters. J. Appl. Biotech. 1: 4–53.

Fricke, W., G. Akhiyarova, W.X. Wei, E. Alexandersson, A. Miller, P.O. Kjellbom et al. 2006. The short-term growth response to salt of the developing barley leaf. J. Exp. Bot. 57: 1079–1095.

Gaba, S., R.N. Singh, S. Abrol, A.N. Yadav, A.K. Saxena and R. Kaushik. 2017. Draft genome sequence of *Halolamina pelagica* CDK2 isolated from Natural Salterns from Rann of Kutch, Gujarat, India. Genome Announcements 5: 1–2.

Glick, B.R., Z. Cheng, J. Czarny and J. Duan. 2007. Promotion of plant growth by ACC deaminase-producing soil bacteria. Eur. J. Plant Pathol. 119: 329–339.

Goswami, D., P. Dhandhukia, P. Patel and J.N. Thakker. 2014. Screening of PGPR from saline desert of Kutch: growth promotion in Arachis hypogea by *Bacillus licheniformis* A2. Microbiol. Res. 169: 66–75.

Gray, E.J. and D.L. Smith. 2005. Intracellular and extracellular PGPR: commonalities and distinctions in the plant-bacterium signalling processes. Soil Biol. Biochem. 37: 395–412.

Grover, M., S.Z. Ali, V. Sandhya, A. Rasul and B. Venkateswarlu. 2011. Role of microorganisms in adaptation of agriculture crops to abiotic stresses. World J. Microbiol. Biotechnol. 27: 1231–1240.

Hashem, A., E.F. Allah, A.A. Alqarawi, A. Aldebasi and D. Egamberdieva. 2015. Arbuscular mycorrhizal fungi enhance salinity tolerance of Panicum turgidum Forssk by altering photosynthetic and antioxidant pathways. J. Plant Interact. 10: 230–242.

Hashem, A., E.F. Allah, A. Alqarawi, A.A. Auqail, S. Wirth and D. Egamberdieva. 2016. The interaction between arbuscular mycorrhizal fungi and endophytic bacteria enhances plant growth of Acacia gerrardii under salt stress. Front. Plant Sci. 7: 1089.

Hayat, R., S. Ali, U. Amara, R. Khalid and I. Ahmed. 2010. Soil beneficial bacteria and their role in plant growth promotion: a review. Ann. Microbiol. 60: 579–598.

Haynes, R.J. and R.S. Swift. 1990. Stability of soil aggregates in relation to organic constituents and soil water content. J. Soil Sci. 41: 73–83.

Illangumaran, G. and D.L. Smith. 2017. Plant growth promoting Rhizobacteria in amelioration of salinity stress: a systems biology perspective. Front. Plant Sci. 8: 1–14.

Jamil, A., S. Riaz, M. Ashraf and M.R. Foolad. 2011. Gene expression profiling of plants under salt stress. Plant Sci. 30(5): 435–458.

Javid, M.G., A. Sorooshzadeh, F. Moradi, A.M.M.S. Seyed and I. Allahdadi. 2011. The role of phytohormones in alleviating salt stress in crop plants. Aust. J. Crop. Sci. 5(6): 726–734.

Jha, B., I. Gontia and A. Hartmann. 2012. The roots of the halophyte *Salicornia brachiata* are a source of new halotolerant diazotrophic bacteria with plant growth-promoting potential. Plant Soil. 356: 265–277.

Jha, B., V.K. Singh, A. Weiss, A. Hartmann and M. Schmid. 2015. *Zhihengliuel lasomnathii* sp. nov., a halotolerant actinobacterium from the rhizosphere of a halophyte *Salicornia brachiata*. Int. J. Syst. Evol. Microbiol. 65: 3137–3142.

Jha, Y., R.B. Subramanian and S. Patel. 2011. Combination of endophytic and rhizospheric plant growth promoting rhizobacteria in *Oryzasativa* shows higher accumulation of osmoprotectant against saline stress. Acta Physiol. Plant 33: 797–802.

Jha, Y. and R.B. Subramanian. 2014. PGPR regulate caspase-like activity, programmed cell death, and antioxidant enzyme activity in paddy under salinity. Physiol. Mol. Biol. Plants 20(2): 201–207.

Kearl, J., C. McNary, J.S. Lowman, C. Mei, Z.T. Aanderud, S.T. Steven et al. 2019. Salt-tolerant halophyte rhizosphere bacteria stimulate growth of alfalfa in salty soil. Front. Microbiol. 10: 1849.

Khan, M.A., S. Asaf, A.T. Khan, A. Adhikari, R. Jan, S. Ali et al. 2019. Halotolerant rhizobacterial strains mitigate the adverse effects of NaCl stress in soybean seedlings. BioMed. Res. International. 2019: 1–15.

Kirkup, B.C. and M.A. Riley. 2004. Antibiotic-mediated antagonism leads to a bacterial game of rock-paper-scissors *in vivo*. Nature 428: 412–414.

Kohler, J., F. Caravaca, L. Carrasco and A. Roldan. 2006. Contribution of *Pseudomonas mendocina* and *Glomus intraradices* to aggregates stabilization and promotion of biological properties in rhizosphere soil of lettuce plants under field conditions. Soil Use Manage. 22: 298–304.

Kohler, J., J.A. Hernandez, F. Caravaca and A. Roldan. 2009. Induction of antioxidant enzymes is involved in the greater effectiveness of a PGPR versus AM fungi with respect to increasing the tolerance of lettuce to severe salt stress. Environ. Exp. Bot. 65: 245–252.

Kour, D., K.L. Rana, T. Kaur, I. Sheikh, A.N. Yadav, V. Kumar et al. 2020a. Microbe-mediated alleviation of drought stress and acquisition of phosphorus in great millet (*Sorghum bicolour* L.) by drought-adaptive and phosphorus-solubilizing microbes. Biocatal. Agric. Biotechnol. 23: 101501. https://doi.org/10.1016/j.bcab.2020.101501.

Kour, D., K.L. Rana, I. Sheikh, V. Kumar, A.N. Yadav, H.S. Dhaliwal et al. 2020b. Alleviation of drought stress and plant growth promotion by *Pseudomonas libanensis* EU-LWNA-33, a drought-adaptive phosphorus-solubilizing bacterium. Proc. Natl. Acad. Sci. India Sec. B Biol. Sci. doi:10.1007/s40011-019-01151-4.

Kour, D., K.L. Rana, A.N. Yadav, I. Sheikh, V. Kumar, H.S. Dhaliwal et al. 2020c. Amelioration of drought stress in Foxtail millet (*Setaria italica* L.) by P-solubilizing drought-tolerant microbes with multifarious plant growth promoting attributes. Environ. Sustain. 3: 23–34. doi:10.1007/s42398-020-00094-1.

Kour, D., K.L. Rana, A.N. Yadav, N. Yadav, M. Kumar, V. Kumar et al. 2020d. Microbial biofertilizers: Bioresources and eco-friendly technologies for agricultural and environmental sustainability. Biocatal. Agric. Biotechnol. 23: 101487. https://doi.org/10.1016/j.bcab.2019.101487.

Kumar, M., D. Kour, A.N. Yadav, R. Saxena, P.K. Rai, A. Jyoti et al. 2019. Biodiversity of methylotrophic microbial communities and their potential role in mitigation of abiotic stresses in plants. Biologia 74: 287–308. doi:10.2478/s11756-019-00190-6.

Landa, B.B., J.A. Navas-Cortes and R.M. Jimenez-Diaza. 2004. Influence of temperature on plant–rhizobacteria interactions related to biocontrol potential for suppression of *Fusarium* wilt of chickpea. Plant Pathol. 53: 341–352.

Ledger, T., S. Rojas, T. Timmermann, I. Pinedo, M.J. Poupina and T. Garrido. 2016. Volatile-mediated effects predominate in *Paraburkholderia phytofirmans* growth promotion and salt stress tolerance of *Arabidopsis thaliana*. Front. Microbiol. 7: 1838.

Lugtenberg, B., T. Chin-A-Woeng and G. Bloemberg. 2002. Microbe-plant interactions: principles and mechanisms. Antonie Van Leeuwenhoek. 81: 373–383.

Lutgtenberg, B. and F. Kamilova. 2009. Plant-growth-promoting rhizobacteria. Annu. Rev. Microbiol. 63: 541–556.

Mayak, S., T. Tirosh and B.R. Glick. 2004. Plant growth-promoting bacteria confer resistance in tomato plants to salt stress. Plant Physiol. Biochem. 42: 565–572.

Meena, V.S., B.R. Maurya, J.P. Verma, A. Aeron, A. Kumar, K. Kim and V.K. Bajpai. 2015. Potassium solubilizing rhizobacteria (KSR): isolation, identification, and K-release dynamics from waste mica. Ecol. Eng. 81: 340–347.

Metternicht, G.I. and J.A. Zinck. 2003. Remote sensing of soil salinity: potentials and constraints. Remote Sens. Environ. 85: 1–20.

Munns, R. and H.M. Rawson. 1999. Effect of salinity on salt accumulation and reproductive development in the apical meristem of wheat and barley. Aust. J. Plant Physiol. 26: 459–464.

Munns, R., S. Husain, A.R. Rivelli, R.A. James, A.G. Condon, M.P. Lindsay et al. 2002. Avenues for increasing salt tolerance of crops, and the role of physiologically based selection traits. Plant Soil. 247(1): 93–105.

Munns, R. and R.A. James. 2003. Screening methods for salinity tolerance: a case study with tetraploid wheat. Plant Soil. 253: 201–218.

Munns, R. 2005. Genes and salt tolerance: bringing them together. New Phytol. 167: 645–663.

Nadeem, S.M., Z.A. Zahir, M. Naveed and M. Arshad. 2007. Preliminary investigation on inducing salt tolerance in maize through inoculation with rhizobacteria containing ACC-deaminase activity. Can. J. Microbiol. 53: 1141–1149.

Nadeem, S.M., Z.A. Zahir, M. Naveed and S. Nawaz. 2013. Mitigation of salinity-induced negative impact on the growth and yield of wheat by plant growth-promoting rhizobacteria in naturally saline conditions. Ann. Microbiol. 63(1): 225–232.

Naz, I., A. Bano and T. Hassan. 2009. Isolation of phytohormones producing plant growth promoting rhizobacteria from weeds growing in Khewra salt range, Pakistan and their implication in providing salt tolerance to *Glycine max* L. Afr. J. Biotechnol. 8(21): 5762–5766.

Netondo, G.W., J.C. Onyango and E. Beck. 2004. Sorghum and salinity: II. Gas exchange and chlorophyll fluorescence of sorghum under salt stress. Crop Sci. 44: 806–811.

Nia, S.H., M.J. Zarea, F. Rejali and A. Verma. 2012. Yield and yield components of wheat as affected by salinity and inoculation with *Azospirillum* strains from saline or non-saline soil. J. Saudi Soc. Agri. Sci. 11: 113–121.

Orhan, F. 2016. Alleviation of salt stress by halotolerant and halophilic plant growth-promoting bacteria in wheat (*Triticum aestivum*). Braz. J. Microbiol. 47(3): 621–627.

Palaniyandi, S.A., K. Damodharan, S.H. Yang and J.W. Suh. 2014. *Streptomyces* sp. strain PGPA39 alleviates salt stress and promotes growth of 'Micro Tom' tomato plants. J. Appl. Microbiol. 117: 766–773.

Parray, A.P., S. Jan, A.N. Kamili, R.A. Qadri, D. Egamberdieva and P. Ahmad. 2016. Current perspectives on plant growth promoting rhizobacteria. Plant Growth Regul. 35(3): 877–902.

Patel, B.B., B.B. Patel and R.S. Dave. 2011. Studies on infiltration of saline-alkali soils of several parts of Mehsana and Patan districts of north Gujarat. J. Appl. Technol. Environ. Sanitation. 1(1): 87–92.

Paul, S. and A. Roychoudhury. 2017. Effect of seed priming with spermine/spermidine on transcriptional regulation of stress-responsive genes in salt-stressed seedlings of an aromatic rice cultivar. Plant Gene. 11: 133–142.

Perez-Alfocea, F., A. Albacete, M.E. Ghanem and I.C. Dodd. 2010. Hormonal regulation of source-sink relations to maintain crop productivity under salinity: a case study of root-to-shoot signalling in tomato. Funct. Plant Biol. 37: 592–603.

Ramadoss, D., V.K. Lakkineni, P. Bose, S. Ali and K. Annapurna. 2013. Mitigation of salt stress in wheat seedlings by halotolerant bacteria isolated from saline habitats. SpringerPlus. 2(6): 1–7.

Rana, K.L., D. Kour and A.N. Yadav. 2019. Endophytic microbiomes: biodiversity, ecological significance and biotechnological applications. Res. J. Biotechnol. 14: 142–162.

Rana, K.L., D. Kour, T. Kaur, I. Sheikh, A.N. Yadav, V. Kumar et al. 2020a. Endophytic microbes from diverse wheat genotypes and their potential biotechnological applications in plant growth promotion and nutrient uptake. Proc. Natl. Acad. Sci. India, Sect. B Biol. Sci. doi:10.1007/s40011-020-01168-0.

Rana, K.L., D. Kour, A.N. Yadav, N. Yadav and A.K. Saxena. 2020b. Agriculturally important microbial biofilms: Biodiversity, ecological significances, and biotechnological applications. pp. 221–265. *In*: Yadav, M.K. and B.P. Singh (eds.). New and Future Developments in Microbial Biotechnology and Bioengineering: Microbial Biofilms. Elsevier, USA. https://doi.org/10.1016/B978-0-444-64279-0.00016-5.

Rastegari, A.A., A.N. Yadav, A.A. Awasthi and N. Yadav. 2020. Trends of Microbial Biotechnology for Sustainable Agriculture and Biomedicine Systems: Diversity and Functional Perspectives. Elsevier, Cambridge, USA.

Sadeghi, A., E. Karimi, P.A. Dahaji, M.G. Javid, Y. Dalvand and H. Askari. 2012. Plant growth promoting activity of an auxin and siderophore producing isolate of *Streptomyces* under saline soil conditions. World J. Microbiol. Biotechnol. 28: 1503–1509.

Sandhya, V., S.Z. Ali, M. Grover, G. Reddy and B. Venkateswarlu. 2009. Alleviation of drought stress effects in sunflower seedlings by exopolysaccharides producing *Pseudomonas putida* strain P45. Biol. Fertility Soil. 46: 17–26.

Saravanakumar, D. and R. Samiyappan. 2007. ACC deaminase from *Pseudomonas fluorescens* mediated saline resistance in groundnut (*Arachis hypogea*) plants. J. Appl. Microbiol. 102: 1283–1292.

Seckin, B., A.H. Sekmen and I. Turkan. 2009. An enhancing effect of exogenous mannitol on the antioxidant enzyme activities in roots of wheat under salt stress. J. Plant Growth Regul. 28: 12–20.

Shahbaz, M. and M. Ashraf. 2013. Improving salinity tolerance in cereals. Plant Sci. 32: 237–249.

Sharma, S., J. Kulkarni and B. Jha. 2016. Halotolerant rhizobacteria promote growth and enhance salinity tolerance in peanut. Front. Microbiol. 7: 16.

Sharma, R., A. Sahoo, R. Devendran and M. Jain. 2014. Over-expression of a rice tau class glutathione S-transferase gene improves tolerance to salinity and oxidative stresses in *Arabidopsis*. PLoS ONE 9(3): e92900.

Shrivastava, P. and R. Kumar. 2015. Soil salinity: A serious environmental issue and plant growth promoting bacteria as one of the tools for its alleviation. Saudi J. Biol. Sci. 22: 123–131.

Shukla, P.S., P.K. Agarwal and B. Jha. 2012. Improved Salinity tolerance of *Arachis hypogaea* (L.) by the interaction of halotolerant plant-growth-promoting rhizobacteria. J. Plant Growth Regul. 31: 195–206.

Singh, A., R. Kumari, A.N. Yadav, S. Mishra, A. Sachan and S.G. Sachan. 2020. Tiny microbes, big yields: microorganisms for enhancing food crop production sustainable development. pp. 1–16. *In*: Rastegari, A.A., A.N. Yadav, A.K. Awasthi and N. Yadav (eds.). Trends of Microbial Biotechnology for Sustainable Agriculture and Biomedicine Systems: Diversity and Functional Perspectives. Elsevier, Amsterdam. https://doi.org/10.1016/B978-0-12-820526-6.00001-4.

Singh, G. 2009. Salinity-related desertification and management strategies: Indian experience. Land Degrad Dev 20: 367–385.

Singh, J. and A.N. Yadav. 2020. Natural Bioactive Products in Sustainable Agriculture. Springer, Singapore.

Singh, K.N. and R. Chatrath. 2001. Salinity tolerance. pp. 101–110. *In*: Reynolds, M.P., J.I.O. Monasterio and A. McNab (eds.). Application of Physiology in Wheat Breeding. CIMMYT, Mexico, DF.

Smith, D.L., V. Gravel and E. Yergeau. 2017. Signaling in the phytomicrobiome. Front. Plant Sci. 8: 6–11.

Subramanian, S., E. Ricci, A. Souleimanov and D.L. Smith. 2016a. A proteomic approach to lipo-chitooligosaccharide and thuricin 17 effects on soybean germination unstressed and salt stress. PLoS ONE 1: e0160660.

Subramanian, S., A. Souleimanov and D.L. Smith. 2016b. Proteomic studies on the effects of lipo-chitooligosaccharide and thuricin 17 under unstressed and salt stressed conditions in *Arabidopsis thaliana*. Front Plant Sci. 7: 1314.

Tabur, S. and K. Demir. 2010. Role of some growth regulators on cytogenetic activity of barley under salt stress. J. Plant Growth Regul. 60: 99–104.

Tank, N. and M. Saraf. 2010. Salinity-resistant plant growth promoting rhizobacteria ameliorates sodium chloride stress on tomato plants. J. Plant Interact. 5: 51–58.

Tester, M. 2003. Davenport R. Na$^+$ tolerance and Na$^+$ transport in higher plants. Ann. Bot. 91: 503–507.

Tiwari, S., P. Singh, R. Tiwari, K.K. Meena, M. Yandigeri, D.P. Singh et al. 2011. Salt-tolerant rhizobacteria-mediated induced tolerance in wheat (*Triticum aestivum*) and chemical diversity in rhizosphere enhance plant growth. Biol. Fertility Soils. 47: 907–916.

Upadhyay, S.K., J.S. Singh, A.K. Saxena and D.P. Singh. 2011. Impact of PGPR inoculation on growth and antioxidant status of wheat under saline conditions. Plant Biol. 14: 605–611.

Verma, P., A.N. Yadav, K.S. Khannam, S. Kumar, A.K. Saxena and A. Suman. 2016. Molecular diversity and multifarious plant growth promoting attributes of Bacilli associated with wheat (*Triticum aestivum* L.) rhizosphere from six diverse agro-ecological zones of India. J. Basic Microbiol. 56: 44–58.

Wang, C.J., W. Yang, C. Wang, C. Gu, D.D. Niu and H.X. Liu. 2012. Induction of drought tolerance in cucumber plants by a consortium of three plant growth-promoting rhizobacterium strains. PLoS ONE 7: pe52565.

Wang, W., B. Vinocur and A. Altman. 2003. Plant responses to drought, salinity and extreme temperatures: towards genetic engineering for stress tolerance. Planta. 218: 1–14.

Wicke, B., E. Smeets, V. Dornburg, B. Vashev, T. Gaiser, W. Turkenburg et al. 2011. The global technical and economic potential of bioenergy from salt-affected soils. Energy Environ. Sci. 4: 2669–2681.

Wu, Z., H. Yue, J. Lu and C. Li. 2012. Characterization of rhizobacterial strain Rs-2 with ACC deaminase activity and its performance in promoting cotton growth under salinity stress. World J. Microbiol. Biotechnol. 28: 2383–2393.

Yadav, A.N., D. Sharma, S. Gulati, S. Singh, R. Dey, K.K. Pal et al. 2015. Haloarchaea endowed with phosphorus solubilization attribute implicated in phosphorus cycle. Sci. Rep. 5: 12293.

Yadav, A.N. 2017a. Agriculturally important microbiomes: biodiversity and multifarious PGP attributes for amelioration of diverse abiotic stresses in crops for sustainable agriculture. Biomed. J. Sci. Tech. Res. 1: 1–4.

Yadav, A.N., R. Kumar, S. Kumar, V. Kumar, T. Sugitha, B. Singh et al. 2017a. Beneficial microbiomes: biodiversity and potential biotechnological applications for sustainable agriculture and human health. J. Appl. Biol. Biotechnol. 5: 45–57.

Yadav, A.N., P. Verma, R. Kaushik, H.S. Dhaliwal and A.K. Saxena. 2017b. Archaea endowed with plant growth promoting attributes. EC Microbiol. 8: 294–298.

Yadav, A.N. 2017b. Beneficial role of extremophilic microbes for plant health and soil fertility. J. Agric. Sci. 1: 1–4.

Yadav, A.N. 2018. Biodiversity and biotechnological applications of host-specific endophytic fungi for sustainable agriculture and allied sectors. Acta Sci. Microbiol. 1: 01–05.

Yadav, A.N. and A.K. Saxena. 2018. Biodiversity and biotechnological applications of halophilic microbes for sustainable agriculture. J. Appl. Biol. Biotechnol. 6: 48–55.

Yadav, A.N. and N. Yadav. 2018a. Stress-adaptive microbes for plant growth promotion and alleviation of drought stress in plants. Acta Sci. Agric. 2: 85–88.

Yadav, A.N., S. Gulati, D. Sharma, R.N. Singh, M.V.S. Rajawat, R. Kumar et al. 2019. Seasonal variations in culturable archaea and their plant growth promoting attributes to predict their role in establishment of vegetation in Rann of Kutch. Biologia 74: 1031–1043. doi:10.2478/s11756-019-00259-2.

Yadav, A.N. 2019a. Endophytic fungi for plant growth promotion and adaptation under abiotic stress conditions. Acta Sci. Agric. 3: 91–93.

Yadav, A.N. 2019b. Microbiomes of wheat (*Triticum aestivum* L.) endowed with multifunctional plant growth promoting attributes. EC Microbiol. 15: 1–6.

Yadav, A.N. 2020. Plant microbiomes for sustainable agriculture: current research and future challenges. pp. 475–482. *In*: Yadav, A.N., J. Singh, A.A. Rastegari and N. Yadav (eds.). Plant Microbiomes for Sustainable Agriculture. Springer International Publishing, Cham. doi:10.1007/978-3-030-38453-1_16.

Yadav, A.N., A.A. Rastegari, N. Yadav and D. Kour. 2020a. Advances in Plant Microbiome and Sustainable Agriculture: Diversity and Biotechnological Applications. Springer, Singapore.

Yadav, A.N., A.A. Rastegari, N. Yadav and D. Kour. 2020b. Advances in Plant Microbiome and Sustainable Agriculture: Functional Annotation and Future Challenges. Springer, Singapore.

Yadav, A.N., J. Singh, A.A. Rastegari and N. Yadav. 2020c. Plant Microbiomes for Sustainable Agriculture. Springer International Publishing, Cham.

Yadav, J., J.P. Verma, D.K. Jaiswal and A. Kumar. 2014. Evaluation of PGPR and different concentration of phosphorus level on plant growth, yield and nutrient content of rice (*Oryza sativa*). Ecol. Eng. 62: 123–128.

Yadav, N. and A. Yadav. 2018b. Biodiversity and biotechnological applications of novel plant growth promoting methylotrophs. J. Appl. Biotechnol. Bioeng. 5: 342–344.

Yadav, N. and A.N. Yadav. 2019. Actinobacteria for sustainable agriculture. J. Appl. Biotechnol. Bioeng. 6: 38–41.

Yaish, M.W., A. Al-Lawati, G.A. Jana, H.V. Patankar and B.R. Glick. 2016. Impact of soil salinity on the structure of the bacterial endophytic community identified from the roots of Caliph Medic (*Medicago truncatula*). PLoS ONE 11(7): pe0159007.

Yamaguchi, T. and E. Blumwald. 2005. Developing salt-tolerant crop plants: challenges and opportunities. Trends Plant Sci. 10(12): 615–620.

Yang, J., J.W. Kloepper and C.M. Ryu. 2009. Rhizosphere bacteria help plants tolerate abiotic stress. Trends Plant Sci. 14: 1–4.

Yao, L., Z. Wu, Y. Zheng, I. Kaleem and C. Li. 2010. Growth promotion and protection against salt stress by *Pseudomonas putida* Rs-198 on cotton. Eur. J. Soil Biol. 46: 49–54.

Yensen, N.P. 2008. Halophyte uses for the twenty-first century. pp. 367–396. *In*: Khan, M.A. and D.J. Weber (eds.). Ecophysiology of High Salinity Tolerant Plants. Springer, Dordrecht.

Yildirim, E. and A.G. Taylor. 2005. Effect of biological treatments on growth of bean plans under salt stress. Ann. Rep. Bean Improvement Cooperative 48: 176–177.

Yuan, Z., I.S. Druzhinina, J. Labbe, R. Redman, Y. Qin and R. Rodriguez. 2016. Specialized microbiome of a halophyte and its role in helping non-host plants to withstand salinity. Sci. Rep. 6(32467): 1–13.

Zhang, H., M.S. Kim, Y. Sun, S.E. Dowd, H. Shi and P.W. Pare. 2008. Soil bacteria confer plant salt tolerance by tissue-specific regulation of the sodium transporter HKT1. Mol. Plant Microbe Interact. 21: 731–744.

Zhu, J.K. 2002. Salt and drought stress signal transduction in plants. Annu. Rev. Plant Bol. 53: 247–273.

Chapter 3

Halophilic Microbiome
Biodiversity and Biotechnological Applications

Mrugesh Dhirajlal Khunt,[1] Rajesh Ramdas Waghunde,[2,]*
Chandrashekhar Uttamrao Shinde[3] and Dipak Maganlal Pathak[2]

Introduction

Microorganisms are the smartest living things on the earth; they requires different elements which they take from the earth for growth (Shelake et al. 2019a). They are an essential component of the environment and their resilience will help to understand the impact of climate change on ecosystems (Uritskiy et al. 2019). Microorganisms and plants have modified themselves as required by time and environment (Shelake et al. 2019b). Several beneficial microbes could survive and remain viable under different stress conditions (Waghunde et al. 2016; Yadav et al. 2015b). Halophiles are salt loving microorganisms encompassing all three domains of Archaea, Bacteria and Eucarya in which archaeal members have heterotrophic, chemotrophic as well as aerobic and anaerobic characteristics (Yadav et al. 2019a; Yadav et al. 2015a). These microbes have an ability to grow under diversified conditions, i.e., thalassohaline as well as at halassohaline environments (Sarwar et al. 2015; Ma et al. 2010). Halophiles were categorized based on salt concentrations by Kushner in 1985, as slight (0.2 to 0.5 M), moderate (0.5 to 2.5 M), and extreme (2.5 to 5.2 M NaCl) halophiles based on their halotolerance.

The saltern pond brines and natural salt lakes are prominent sources for evolutional studies by researchers to understand development, selection and biodiversity of these microbes. Halophiles genera, i.e., *Halobacterium*, *Haloferax*, and *Haloarcula* are models to study the archaeal domain because of their simple handling compared to other archaebacterial members such as methanogens and hyperthermophiles (Ma et al. 2010; Gaba et al. 2017). Halophiles received a great deal of attention due to their ability to survive under extreme conditions and the production of several industrial, pharmaceuticals, cosmetics and food industry related products with high biotechnological potential (Yadav 2017; Yadav et al. 2017). Besides their increasing biotechnological prospects, halophilic microorganisms are an ideal model to study the physiological and biochemical process variations under high salinity stress (Grant 2004) including molecular methods, i.e., DNA replication, gene

---
[1] Department of Plant Pathology, N. M. College of Agriculture, Navsari Agricultural University, Navsari, Gujarat, India.
[2] Department of Plant Pathology, College of Agriculture, Navsari Agricultural University, Bharuch, Gujarat, India.
[3] Department of Agricultural Entomology, N. M. College of Agriculture, Navsari Agricultural University, Navsari, Gujarat, India.
* Corresponding author: rajeshpathology191@gmail.com

expression, genomic evolution and speciation. Halophiles archaea *Haloferax volcanii* and *Haloferax mediterranei* were among first organisms to detect CRISPR/Cas system (Mojica et al. 1995).

The detailed study of halophiles and their applications resulted in a big demand in the industry because of their diversity in ecosystems and their potential to produce various metabolites. Halophiles are considered as a model of early life forms and for characteristics which may be observed in the hypothetical inhabitants of extraterrestrial, salt rich habitats (Litchfield 1998; Gómez et al. 2012). This chapter briefly describes the halophiles diversity and their applications in the different area of interest.

Biodiversity of culturable and non-culturable halophiles

Culturable halophiles

There is a vast diversity of culturable halophiles present in the environment, they can be easily identified and characterized because of their culturable trait. The list of some culturable halophiles is mentioned in Table 3.1.

Non-culturable halophiles

As per definitions, non culturable halophiles can not be cultivated on an artificial medium under laboratory conditions. Therefore, it poses a problem in characterization and identification by conventional laboratory methods. Non culturable halophiles can be identified by various advanced molecular biology tools which give only a partial idea about the identity of a particular halophilic species. Many investigators have characterized and identified extreme halophiles and moderately halophilic isolates from different saline habitats. Non culturable halophiles can be compared with its culturable partner in terms of per cent similarities. Most of the non culturable halophile genes have been sequenced and when compared with available databases, were found to share homology with culturable halophiles such as *Rhodococcus* sp., *Marinobacter* sp. and *Pseudomonas* sp. from solar salterns of India (Kabilan 2016), *Bacillus oceanisediminis* and *Acinetobacter indicus* from fossil sediments of Ariyalur (Edward 2017).

Salient features of halophiles

Halophiles survive under extreme saline conditions and this adaptation mechanism needs to be studied more for better understanding of halophiles. Various physical, biochemical, structural and genetically metabolisms can be altered by halophiles for existence under unfavorable conditions that are connected to the surrounding environment. Halophiles have developed several strategies for growth and survival under saline conditions. In particular, extreme halophiles have been related with several salient features that grant them salinity tolerance. True halophiles have definite distinctive characters over non-halophilic organisms. Some of these features for survival are described below.

Cell wall

The cell wall is the protective layer around the cell that provides protection under osmotic stress conditions. The major difference between an eubacterial and archaeal cell wall is the presence of pseudopeptidoglycan (also known as pseudomurein) in archaea over peptidoglycan (also known as murein) in the non-halophilic eubacteria. Apart from protection against osmotic stress, the cell wall also determines the shape of the cell. In archaea, different shapes of cells are observed such as pleomorphic *Heloferax,* rod shaped *Halobacterium,* spherical shaped *Halococcus, Natronococcus,* etc. (Englert et al. 1992). Additionally, a small quantity of archaea shows an unusual and unique shape which is not possessed by any other member, due to an absence of significant turgor pressure inside the cell (Walsby 1971). Few haloarchaeal members such as *Halococcus* and *Natronococcus* have a thick and rigid cell wall for structural stability against deleterious salt concentration.

Table 3.1 The list of culturable halophiles with their source of isolation.

Sr. No.	Culturable halophiles	Source of isolation	Reference
1	*Bacillus alcalophilus, B. firmus, B. macquariensis, B. brevis, B. circulans, B. coagulans, B. subtilis*	Excreta of wild Ass, Kutch desert, India	Khunt et al. 2011
2	*Halococcus salifodinae, Haloferax volcanii, Haloarcula argentinensis, Haloarcula japonica, Haloarcula hispanica, Halorubrum* sp., *Haloferax alexandrinus*	Solar Saltern, Goa, India	Mani et al. 2012
3	*Salinibacter iranicus, Salinibacter luteus*	Aran-Bidgol salt Lake, Iran	Makhdoumi-Kakhki et al. 2012
4	*Halobacillus trueperi, Bacillus licheniformis, Bacillus pumilus, Staphylococcus succinus, Bacillus atrophaeus, Bacillus subtilis, Halobacillus* sp., *Oceanobacillus* sp.	Saline soils of Coahuila State, Mexico	Delgado-García et al. 2013
5	*Halobacillus trueperi, Shewanella algae, Halomonas venusta, Marinomonas* sp.	Lunsu, Himachal Pradesh, India	Gupta et al. 2015
6	*Desulfonatronobacter acetoxydans*	Hypersaline soda Lake, Kulunda Steppe, Altai, Russia	Sorokin et al. 2015
7	*Haloferax prahovense, Haloarcula japonica, Halobellus clavatus, Halogeometricum rufum, Halorubrum chaoviator, Haloarcula argentinensisarg, Halolamin apelagica, Haloarculam arismortui, Haloferax alexandrines, Haloarcula quadrata, Haloarcula hispanica, Psuedomonas halophila, Halorhabdus tiamatea, Haloarcula amylolytica, Halobacterium salinarum, Rhodovibrio sodomensis, Haloarcul asalaria, Halorubrum lipolyticum, Halostagnicola larsenii, Salicola salis*	IncheBroun wetland, Iran	Rasooli et al. 2016
8	*Haloarcula marismortui, Haloarcula salaria, Haloarcula argentinensis, Haloarcula quadrata*	Solar Saltern Pond, Mumbai, India	Nagrale et al. 2017
9	*Natronospira proteinivora*	Kulunda Steppe soda lakes, Altai, Russia	Sorokin et al. 2017
10	*Salinifilum proteinilyticum*	Wetland in Iran	Nikou et al. 2017
11	*Marinobacter salexigens*	Marine sediments, Weihai, China	Han et al. 2017
12	*Marinobacter aquaticus*	Marine saltern located in Huelva, Spain	Leon et al. 2017
13	*Halomonas aquamarina, Sediminibacillus* sp., *Halobacillus* sp., *Halobacillus dabanensis*	Arabian Gulf water, The Dead Sea	Bin-Salman et al. 2018
14	*Halobacillus andaensis, Halobacillus dabanensis, Marinococcus luteus, Alkalibacillus salilacus, Alkalibacillus filiformis, Bacillus halochares, Aquisalibacillus elongatus*	Cuatro Cienegas, Sayula and San Marcos lakes	Delgado-García et al. 2018

Eubacterial cells possess DAP or teichoic acid that is generally not found in the *Halobacterium* (Kushner and Onishi 1968) but the negative charge protein remains stable by attracting sodium ions (Na$^+$) from saline water.

Lipid and cytoplasmic membrane

Cytoplasmic membrane of the cell participates in the electron transport chain, movement of ions across the membrane and a sensory function with the environment. The biological membrane is generally composed of three macromolecules, i.e., lipid, carbohydrates and proteins. The eubacteria contain ester linkage between fatty acids and glycerol while archaebacteria have ether linkage between fatty acids and glycerol. Substitutions of ester linkage with ether linkage in archaea provide great stability of the cell membrane against a high salt concentration. Cell membrane lipids with ether linkage in lipids were found more stable at a broad temperature range and high salt concentrations (Edwards 1990). The presence of specific phospholipid, called phosphatidylglycerol methyl phosphate along with sulfated and desulfated archaeols in the halophilic cell membrane makes them unique (Kates 1993). Additionally, halophiles also possess specific phospholipid known as Archaeol, with elongated hydrocarbon chains of C20–C25 which are found in some halobacteriales (Kamekura and Kates 1999). The negative charge phospholipid phosphatidyl choline and cardiolipin usually increases and neutral phospholipids decrease with increment in salt concentration (Russell 1993; Vreeland 1987).

Capsules

Many of the halophilic bacteria produce extracellular gummy and mucilaginous polysaccharide known as a capsule. A capsule generally protects the cell against changing pH, temperature and other environmental factors. Halophilic archaea secrete a highly glycosylated and sulfated water-rich capsule for the protection of a cell against extremely low water activity (Bolhuis et al. 2006) with an additional role of a barrier against phage attack (Zenke et al. 2015).

Biotechnological applications

Halophilic pigments

The presence of color in the salt lakes is the most common phenomenon in the halophilic inhabitant regions throughout the world, red coloration is governed by halophilic carotenoids. Carotenoid pigments usually give tolerance to halophiles in the salt lakes against intense sunlight (Shun Ichi et al. 1996). Carotenoid pigments are C_{40} lipophilic isoprenoids and possess red, orange, yellow color or may also be colorless (Yatsunami et al. 2014). Due to its color properties, it may find applications in food industries as a food colorant. Carotenoids can be synthesized by chemical methods; however, due to consumer preference for natural carotenoid products, the high cost of the chemical synthesis process, environmental-damaging effect has forced one to switch on some alternative and ecofriendly methods for carotenoid production (Naziri et al. 2014). Microbial mediated carotenoid production may compete with carotenoid production through the synthetic process (Bhosale 2004), due to the ill effects of artificial pigments used as food additives on human health and its serious toxic effect (Srivastava 2015).

Carotenoid has great potential for human health (Fiedor and Burda 2014) and is frequently used l in the food, cosmetic and pharma industries (Vilchez et al. 2011). It has potential application as a food and cosmetic colorant, feed additives for poultry, fish and livestocks, antioxidants, antitumor as well as heart disease prevention agents, vitamin A precursors, *in vitro* antibody production enhancer, etc. (Rodrigo-Banos et al. 2015). Microbial mediated carotenoids consumption has shown the positive effect on human health as it could reduce the risk of esophageal cancer (Woodside et al. 2014), prostate cancer (Jampani and Raghavarao 2015), and cardiovascular diseases (Cardoso et al. 2017).

Carotenoid can be derived from bacteria, fungi and algae. Though more research publications are available suggesting carotenoids profile from eubacteria, some researchers support the role of archaea in carotenoids production (Naziri et al. 2014). Carotenoids production from bacteria, fungi and algae may not be cost effective and economical due to high inputs in maintenance of aseptic production conditions and the high cost of downstream processes. Halophilic microbial mediated carotenoids production offers selective advantages, and there is no need to maintain sterile conditions due to higher salinity (Andrei et al. 2012) and cost saving due to easy downstream processes used for extraction of pigments by just lowering NaCl concentration to lyse cells (El-Banna et al. 2012).

The haloarchaeal family *Haloferacaceae* can synthesize C_{50} carotenoids such as, bacterioruberin and its precursors 2-isopentyl-3,4-dehydrorhodopsin, bisanhydrobacterioruberin, and monoanhydrobacterioruberin (Kelly et al. 1967; Kushwaha et al. 1975). Carotenoids from extremely halophilic archaea *Haloarcula japonica* has the highest proportion of bacterioruberin (68.1%) followed by monoanhydrobacterioruberin (22.5%), bisanhydrobacterioruberin (9.3%), isopentenyldehydrorhodopin (< 0.1%) and traces of lycopene and phytoene. Further, bacterioruberin, extracted from *Haloarcula japonica* have higher *in vitro* scavenging capacity than of β-carotene (Yatsunami et al. 2014). Apart from halophilic archaea, marine green algae *Dunaliella salina* is also explored as a major source of β-carotene (Borowitzka 1999). *Dunaliella salina* and *D. bardawil* are the halophilic algae that could accumulate β-carotene pigment, some times more than 10% of their dry weight under high intensity, high salinity and nutrient deficient environment (Oren 2010). Thus, different halophilic archaea and marine alga can be the major source of halophilic pigments for its potential biotechnological applications.

Besides carotenoids, many halophilic microbes are known to produce black or brown color pigments known as melanin. Microbes synthesize such pigments to remain protected against different environmental stresses such as heavy metals, oxidizing agents, Ultra Violet (UV) exposure, and Reactive Oxygen Species (ROS) (Allam and El-Zaher 2012). Melanins are used as a part of sunscreens, preparation of solid plastic films, paints, and lenses, surface protection against UV (Gallas and Eisner 2006). Melanin also posseses pharmacological properties such as anti-inflammatory, hepatic, gastrointestinal benefits, immunomodulatory and radioprotective action, etc. and can be the best chemical in the medical field (ElObeid et al. 2017). Halophilic black yeast *Hortaea werneckii* is a potential melanin producer, which could be cultivated on cheaper agro-waste rice bran for increased melanin production that has antibacterial activity against deadly human pathogens such as *Vibrio parahaemolyticus, Klebsiella pneumoniae* and *Salmonella typhi* (Rani et al. 2013). Kogej et al. (2004) reported that three halophilic ascomycetous black yeast, i.e., *Hortaea werneckii, Phaeotheca triangularis,* and *Trimmatostroma salinum* from solar salterns hypersaline water produced dark colored 1,8-dihydroxynaphthalene melanin under saline and non-saline conditions.

Compatible solutes

Halophilic and halotolerant microbes grow under high ionic concentration and low water activity. Additionally, there may be a more severe problem if salinity changes due to water evaporation and water dilution during rainfall. In order to cope with this osmotic stress, halophiles have developed many alternate strategies; an accumulation of osmolytes is one of them (Yancey et al. 1982). Halophiles accumulate osmolytes, also called 'compatible solutes' are molecules that usually compensate the osmotic pressure (Brown 1976) and thereby reduce the osmotic shock in a saline environment. Therefore, compatible solutes impart survival against the deleterious effect on cell growth and metabolism due to salinity. Halophiles could accumulate wide varieties of compatible solutes such as amino acids, polyols and its derivatives, different nitrogen containing compounds (Rhodes and Hanson 1993; Hagemann and Pade 2015), as well as amino acid derivatives like glycine-betaine and ectoine (Ventosa et al. 1998; Waditee et al. 2005). Another potential compatible solute N^{ε}-acetyl lysine is usually synthesized by *Haloacillus halophilus,* however, it has to still be confirmed by further studies (Saum et al. 2013). Thus, halophiles are a source of a wide range of compatible solutes, however there is need to explore them in different biotechnological applications.

Halophilic compatible solute molecules can be synthesized under stress conditions to give macromolecular stabilization under related stress conditions. A compatible solute ectoines have gained much attention in biotechnology as an enzyme protective agent and also to prevent cell damage caused by freezing, drying and heating (Shivanand and Mugeraya 2011). Ectoine and its derivative hydroxyectoine are important as a protein stabilizer and also work as membrane protectors against dissociation (Lentzen and Schwarz 2006). Ectoines have properties to counteract the deleterious UV-A mediated effect and reduce skin aging; therefore, it can be a potent biological agent as cosmetic additives in the dermatological skin cosmetics for the care of dry, irritated and aged skin (Shivanand and Mugeraya 2011). Current research has also emphasized the need of ectoine based nasal spray and eye drops in comparison of routinely applied pharmacological treatments against allergic rhinitis (Werkhauser et al. 2014). Several studies also reported the use of ectoines as a remedy agent of neurodegenerative diseases such as Alzheimer's diseases as it prevents amyloid formation (Kanapathipillai et al. 2005). Some of the major ectoine producing prokaryotes are *Brevibacterium epidermis, Chromohalobacter israelensis, Chromohalobacter salexigens, Ectothiorhodospira halochloris, Halomonas boliviensis, Halomonas elongata,* etc. (Shivanand and Mugeraya 2011).

Betaine is another compatible solute, synthesized by halophilic microbes in response of abiotic stresses, is an osmoprotectant and methyl group donor that inhibits nuclear factor-κB activity and NLRP3 inflammasome activation, regulates energy metabolism and mitigates endoplasmic reticulum stress, apoptosis, and in turn also has the beneficial effect of human diseases such as obesity, diabetes, cancer, and Alzheimer's disease (Zhao et al. 2018). Betaine has a positive effect on internal human organs as it improves vascular system and enhances its performance, while also being an important nutrient for the prevention of chronic diseases (Craig 2004). It also finds application in PCR amplification to increase yield and specificity of GC rich DNA templates (Roberts 2005). *Actinoployspora* sp., *Halorhodopira halochloris, Thioalkalivibrio versutus* have been characterized for their betaine production (Shivanand and Mugeraya 2011).

Hydrolytic enzyme

Microbial enzymes are gaining more and more attention in industrial processes due to their unique mechanism of action and being non-polluting to the environment. Halophiles secrete a variety of hydrolytic enzymes that could function in harsh conditions with high specificity and accuracy. Halophiles produce wide ranges of hydrolytic enzymes with salt and thermotolerant potency (Khunt et al. 2011; Kour et al. 2019a; Rana et al. 2019; Yadav and Saxena 2018). Hydrolytic enzymes like lipases, proteases, amylases, chitinases, xylanases, DNases, and pullulanses have wide applications in food, feed, pharmaceutical and chemical industries (Rao et al. 1998; Pandey et al. 1999; Kulkarni et al. 1999; Sanchez-Porro et al. 2003). Several halophilic enzymes may also be useful in environmental applications such as bioremediation and waste water treatment (Waditee-Sirisattha et al. 2016; Yadav et al. 2019c).

Lipases (Triacylglyceril acylhydrolase, EC 3.1.1.3) is a hydrolytic enzyme which degrade ester bond of triacylglycerol, gives glycerol and fatty acid at oil-water interphase but does not hydrolyze dissolved substrates in bulk fluid (Sharma et al. 2001). Lipases are used extensively in the food industries for flavor and aroma development during cheese ripening, in bakery products, preparation of sausages, yoghurt and beverages (Jaeger et al. 1994), also for the processing of fats, as an additive of detergents, chemical and pharmaceutical industries, paper-pulp industries, cosmetic preparations, etc. (Rubin and Dennis 1997). Halophiles are known to produce exceptional lipases as they show a high degree of tolerance against harsh conditions such as high temperature and the presence of chemicals. Lipases extracted from moderately halophilic *Salinivibrio* showed activity even at 50°C (Amoozegar et al. 2008), suggesting that halophilic lipases possess more thermo-tolerance over normal lipases. The partially purified extracellular lipases from extremely halophilic potential isolate showed the highest activity at 60°C, also tolerating the presence of protein denaturant chemicals like urea and NaCl (Khunt and Pandhi 2012). They showed great potential of halophilic lipases in different industrial processes.

Protease is another hydrolytic enzyme with great industrial and biotechnological potential. It encompasses nearly 65% of world's aggregate enzyme market and the global sale which is estimated to be US$ 2.21 billion by 2021 (Mokashe et al. 2018). Among all microbial proteases, halophiles are known to produce protein degrading enzymes with high stability at saturated salt concentrations and organic solvents (Akolkar et al. 2008). Apart from salt and solvent tolerance, halophilic proteases occasionally possess poly-extremophilic tolerance to different conditions like alkaline pH, high temperature, etc. (Mokashe et al. 2018). Studdert et al. (2001) purified and characterized 130 kDa novel protease from halophilic *Natronococcus occultus* with a high degree of stability over pH 5.5–12, 1–2 M NaCl or KCl and also showed optimal activity at 60°C temperature in 1–2 M NaCl. Another halophilic archaea *Halogeometricum borinquense* strain TSS101 known to produce extremely stable protease with the highest activity at 60°C and pH 10.0 in 20% NaCl. The enzyme exhibited high thermal stability and retained 80% of its activity after 1 hour at 90°C (Vidyasagar et al. 2006). Thus, halophiles are known to secrete extracellular protein degrading enzymes that could show their optimal activity at thermophilic temperature and saline conditions. Biotechnological applications of proteases include an additive to the detergent to remove blood and other proteinaceous stains (Sinha and Khare 2012), in different industries such as food, leather, brewing, pharmaceutical, and as a precursor of artificial sweeteners like aspartame (Rao et al. 1998).

Apart from potential industrial applications of lipases and proteases, several other halophilic enzymes have also been explored with high degrees of thermo and solvent tolerance. Moderately halophilic microbes are more useful for the industrial enzymes production due to the broad range of salt tolerance over extreme halophiles and produce salt-adapted enzymes (Rastegari et al. 2020). Sanchez-Porro et al. (2003) explored moderately halophilic microbes for their potential role in the production of amylases, lipases, proteases, inulinases, cellulases, pullulanases, xylanases, DNases and pectinases. Overall, Gram positive halophiles showed more potential in enzyme secretion over Gram negative halophiles. Many more microbes inhabitants of saline environments have been studied for their extracellular enzyme production like *Salinococcus* for the production of extracellular protease, amylase, gelatinase and inulinase (Jayachandra et al. 2012), nucleases from *Micrococcus varians* subsp. *halophiles* (Kamekura et al. 1982), cellulases from haloalkalophilic *Bacillus* sp. (Zhang et al. 2012), and xylanases from *Bacillus pumilus* (Subramaniyan 2012).

Bio-surfactants

Surfactants are chemical substances that usually reduce surface tension of liquid in which it is dissolved. Due to increase in environmental awareness, there has been more and more interest in microbial mediated Bio-Surfactants (BS) over its chemical counterparts (Sarafin et al. 2014). Ecofriendliness is governed by unique properties of BS such as high biodegradability, lower toxicity, environmental compatibility, mild production conditions, low critical micelle concentrations, better activity at the extremities of temperature, salt and pH (Das and Mukherjee 2007). Halophilic BS have been explored in different areas such as agriculture, food industry, pharmaceutical industry, petrochemical industry, paper-pulp industries, etc. (Khopade et al. 2012). Apart from these, BS can be also used in environmental applications related to emulsification, detergency, wetting, foaming, dispersion, bioremediation, microbial mediated oil recovery and hydrophobic compound solubilization (Joshi et al. 2015).

One of the potential applications of BS, with tremendous potential in the petrochemical industry is Microbial Enhanced Oil Recovery (MEOR). Bio-surfactants have many advantages in enhancement of oil displacement through the movement of oil in oil-bearing rocks by means of interfacial tension reduction, modification of porous media wettability, emulsification of crude oil, etc. (Wu et al. 2014). The BS mediated MEOR could be performed *in situ,* where BS producing microbes on the oil containing well is injected, with or without additional nutrients or could also be by *ex situ* methods (Geetha et al. 2018). An efficient BS producing *Bacillus* sp. BS2, isolated from Gang Xi and Kong Dian blocks of the Dagang oil field was found potent in terms of BS production and also possessed adaptability to temperature, pH and salinity (Wu et al. 2014). Marine actinomycetes *Nocardiopsis* B4 also reported to have BS production with surface tension reduction activity of 29 mN/m and emulsification index E24 to the tune of 80% in 6 to 9 days (Khopade et al. 2012). Another way of exploring halophilic BS is in the biodegradation of crude oil that may pollute the environment, specifically marine environments by accidental spillage of oil in them. Moderately halophilic bacteria *Bacillus cerus* ND1, isolated from the Dwarka seacoast, Gujarat, India showed 90% crude oil degradation due to BS production with very high oil displacement activity and also reduced surface tension of crude oil (Dixini and Mistry 2018). Thus, halophilic BS may find wide scale applications in the petroleum industries for MEOR and also for biodegradation of crude oil.

Most halophilic BS have successfully been explored in the pharmaceutical industry for the management of pathogenic microorganisms. Halophilic bacterium *Halomonas* sp. produce potential BS 1, 2-Ethanediamine N, N, N', N'-tetra, 8-Methyl-6-nonenamide, (Z)-9-octadecenamide and a fatty acid derivative, with efficient growth inhibition activities against human pathogenic bacteria, fungi, aquaculturally important virus WSSV, and also suppress the proliferation of mammary epithelial carcinoma cells (Donio et al. 2013). The halophilic isolate *Bacillus tequilensis* CH, isolated from Lake Chilika, Odisha, India produced potential BS lipopeptide that was found to inhibit biofilm forming *E. coli* and *Streptococcus mutans* at the minimal concentration of 50 µg/ml (Pradhan et al. 2013). Thus, halophilic mediated BS could be the cost effective and ecofriendly way of novel antimicrobial and anticancer drugs in the treatment of various diseases.

Medical applications

Extremophilic mediated novel compounds have been explored in the medical field due to their unique nature and mode of action in the treatment of diseases and disorders. Protein instability is a major concern during the administration of protein-based medicines in the form of aqueous solution (Raddadi et al. 2015). Several extremolytes, derived from the extremophilic microorganism which could stabilize protein under *in vivo* and *in vitro* conditions, and therefore, work as a protein stabilizer, for the preservation of sensitive proteins (Avanti et al. 2014). A common compatible solute ectoine possesses an outstanding water-binding activity and therefore, it can be explored in the prevention of water loss from dry atopic skin, improving skin viability and preventing skin aging (Graf et al. 2008). Several other compatible solutes such as mannosylglycerate (firoin) and ectoine from extremophiles have been investigated for their role in reducing signaling pathways triggered by carbon nanoparticles in the lung cells. Pretreatment of lung epithelial cells with both these substances decreases the particle specific activation of mitogen activated protein kinases and also endpoints increase and apoptosis (Autengruber et al. 2014). Ectoine and hydroxyectoine are compatible solutes that possess potential inhibitory activity against the aggregate formation of Alzheimer's Abeta (Kanapathipillai et al. 2005).

Microbes isolated from the marine environment produce protein, and purified extract of halophilic protein, that could be a potential source for inhibiting growth and metabolic activities of various pathogenic microorganisms (Rastegari et al. 2019b; Rastegari et al. 2019c; Yadav et al. 2019b; Todkar et al. 2012). Halophilic bacterium *Serratia marcescens* produced red pigment prodigiosin (Vora et al. 2014). Prodigiosin has a potential application in cancer treatment as it induced apoptosis of human hematopoietic cancer cell lines and no marked toxicity against the non-malignant cell line (Campas et al. 2003) besides use in the treatment of cancer, *Halobacterium halobium* derived 84 KDa proteinic molecule has been explored for the detection of antibodies against the human cmyc oncogene product in cancer patients (Ben-Mahrez et al. 1988). Thus, many halophilic mediated bioactive compounds possess the potential of effective medicine to cure different diseases and disorders in more liberal ways over different conventional options.

Food preparation

The use of halophilic microbes in food preparation and food preservation is immemorial. A variety of fermented food is usually prepared by adding a large amount of salt and mainly dominated by salt tolerant or halophilic microorganisms. Salt rich food products are generally dominated in the Far East such as fermented seafood 'jeotgal' in Korea, fermented salted puffer fish ovaries 'fugunoko nukazuke' in Japan and a fish sauce 'nam-pla' in Thailand (Aljohny 2015). Salt concentration during such fermentation preparations is too high and supports the growth of archaea of the family *Halobacteriaceae*. Salt loving microbes have been routinely used to produce a wide variety of fermented foods such as vegetable pickles, fermented sea foods, fermented meat, fruit, rice noodles and flours (Roh et al. 2007; Kivisto et al. 2010; Yadav et al. 2019d).

Biological active molecules from halophiles have been widespread in the food industries for the preparation/modification of food. Amylases are starch hydrolyzing, extracellular, hydrolytic enzymes from halophiles that are widely used in food preparation, mainly in starch liquefaction, convert starch into glucose and fructose syrup (Couto and Sanroman 2006). Amylases are also useful in food processing such as brewing, baking, digestive aids and fruit juice preparation (Souza 2010). Another potential enzyme used in food processing is lipases. A lipid degrading enzyme from halophilic bacterium *Marinobacter lipolyticus* has been explored for its excellent fish oil enrichment potential of Poly-Unsaturated Fatty Acids (PUFAs) (Perez et al. 2011).

Functional food preparation from halophilic microbes is another application in food industries, where the food prepared supply nutrition and also provides health benefits. Sumitha et al. (2018) isolated halophilic microbes from fermented food such as pickles of citron, chilly, mango, mackerel, emperor angelfish and anchovies. Halophilic *Spirulina* has an unusual high protein content and higher amino-acids spectrum than plant proteins (Ventosa and Nieto 1995); therefore, can be useful as a functional food to prevent malnutrition due to protein deficiency in the diet. Thus, many halophilic microorganisms can be explored as functional foods due to their properties that confer human health benefits.

Bio-fuel production

The current era is dominated with the use of fossil fuels; however, they are non-renewable sources of energy and have also raised environmental concerns (Stephanopoulos 2007). Alkenes are the major constituents of gasoline, diesel and jet fuel, that are naturally produced by many species (Schirmer et al. 2010). Biobutanol is another contender in biofuel and has certain advantages like easy transportation, lower vapor pressure and less absorption of moisture, lower corrosion, etc. (Karimi and Pandey 2014). Butanol is generally produced by Acetone Butanol Ethanol (ABE) fermentation by using anaerobic bacterium *Clostridium* sp. (Duran-Padilla et al. 2014). However, being an anaerobic bacteria, *Clostridium* grows very slow and genetic manipulations are also difficult in a strain improvement program, leading to inferior productivity over aerobic bacteria (Jang et al. 2012). Therefore, the use of aerobic microorganisms in ABE fermentation may solve the problems associated with *Clostridium* at an industrial scale bio-fuel production. Amiri et al. (2016) first reported that moderately halophilic bacterium *Nesterenkonia* sp. strain F produces acetone, butanol, ethanol, acetic and butyric acid in the presence and absence of oxygen. Thus, use of such salt tolerant and aerobic microorganisms in the ABE fermentation may resolve issues pertaining to the *Clostridium* at an industrial scale bio-fuel production.

The utilization of microbial hydrolytic enzymes in the conversion of biomass into simpler molecules and then into bio-fuel by fermentation is another way of producing ecofriendly biological fuel (Kour et al. 2019b; Kumar et al. 2019; Rastegari et al. 2019a). Various salt resistant hydrolytic enzymes from halophilic origin, viz. lipases, proteases, amylases, cellulases, chitinases, xylanases and mannanases have great potential in the hydrolysis of complex biomass molecules into simpler ones (Johnson et al. 2017). Lipases are the potential lipid degrading enzymes that have wide scale applications in biodiesel preparation. Immobilization of lipases in nanomaterials as a carrier can have great advantages in the biodiesel production due to high enzyme folding, multiple time uses, effective protection against denaturation, etc. (Kim et al. 2018). Additionally, use of cellulose degrading enzymes from halophilic microbes have great potential for developing biotechnologies for the production of biodiesel and a source of lignocellulolytic enzymes for treating biomass (Begemann et al. 2011). Overall, halophilic microorganisms have the potential to produce wide ranges of ecofriendly biofuel, as an alternate of fossil fuel.

Other biotechnological applications

A halophilic microorganism has wide scale novel biotechnological application in different processes. Nowadays, due to advancement in industrialization and urbanization, plastic wastes have been generated in considerable amounts. One study reported that 270 Kt of plastic residues are adrift in oceans (Eriksen et al. 2014), that may potentially harm an ecosystem. Therefore, scientists are focusing on biodegradable bio-plastics derived from living organisms. Polyhydroxyalkanoates (PHA) are renewable and microbially synthesized polymers during growth under excess carbon and phosphate depletion conditions (Poli et al. 2011). Polymer PHAs are almost similar to conventional plastics in terms of thermoplastic and elastomeric properties, but biodegradable unlike conventional plastics (Philip et al. 2006). Current PHAs production at an industrial level is carried out by a recombinant strain of *Escherichia coli*; however, halophilic archaeon *Haloferax mediterranei* and *Halomonas boliviensis* have great potential for PHAs production (Quillaguaman et al. 2010).

Halophilic mediated pigment carotenoid production absorbs light and is helpful in increasing evaporation of saltern crystallizer ponds. Another halophilic pigment bacteriorhodopsin is being used as storage material in computer memory (Kaleem et al. 2015). Halophile mediated environmental biodegradation and bioremediation is also increasing due to exploitation of such microbes in the field of environmental biotechnology. Many researchers have emphasized the need of halophilic or halotolerant microorganisms in the process of oil biodegradation (Margesin and Schinner 2001; Hao and Lu 2009) and bioremediation of heavy metals (Voica et al. 2016). Moderately halophilic bacteria may remove phosphate from the saline environment and therefore is a cheaper alternative for chemical approaches (Ramos-Cormenzana 1989), to minimize the extensive problem of eutrofication in natural habitats. Other potential applications of halophiles in environmental biotechnology include recovery of saline soil, alkaline industrial waste water treatment, and degradation of toxic compounds (Manikandan and Palani 2017).

Conclusions and future prospective

Nowadays, halophiles are the most prominent alternative for the pharmaceutical, food and petroleum industry. The demands of halophiles are increasing due to their quality to sustain at high salinity and other adverse conditions. As ecofriendly and chemical free products are in great demand, and halophiles are the best source for it. Halophiles can be used for enzyme and metabolites production and as a biodegrader in environment. There is great scope for characterization of halophiles, their diversity and production of different secondary metabolites and their purification methods. The remarkable diversity of halophiles that are present in the environment is still unknown, researchers need to study it indepth. Halophiles need to be used for new vaccine production, as an environment cleaner, food preservation and biofuel production, etc. Newer genomics and protcomics tools are needed to explore and are of demand currently. Therefore, it may be expected that new techniques and novel applications will be developed to study the hidden world of halophiles.

References

Akolkar, A.V., G.M. Deshpande, K.N. Raval, D. Durai, A.S. Nerurkar and A.J. Desai. 2008. Organic solvent tolerance of *Halobacterium* sp. SP (1) and its extracellular protease. J. Basic Microbiol. 48: 421–425.

Aljohny, B.O. 2015. Halophilic bacterium—a review of new studies. Biosci. Biotechnol. Res. Asia 12: 2061–2069.

Allam, N.G. and E.A. El-Zaher. 2012. Protective role of *Aspergillus fumigatus* melanin against ultraviolet (UV) irradiation and *Bjerkandera adusta* melanin as a candidate vaccine against systemic candidiasis. Afr. J. Biotechnol. 11: 6566–6577.

Amiri, H., R. Azarbaijani, L.P. Yeganeh, A.S. Fazeli, N.M. Tabatabaei, G.H. Salekdeh et al. 2016. *Nesterenkonia* sp. strain F, a halophilic bacterium producing acetone, butanol, and ethanol under aerobic conditions. Sci. Rep. 6: 18408–18418.

Amoozegar, M.A., E. Salehghamari, K. Khajeh, M. Kabiri and S. Naddaf. 2008. Production of an extracellular thermohalophilic lipase from a moderately halophilic bacterium, *Salinivibrio* sp. strain SA-2. J. Basic Microbiol. 48: 160–167.

Andrei, A.Ş., H.L. Banciu and A. Oren. 2012. Living with salt: metabolic and phylogenetic diversity of archaea inhabiting saline ecosystems. FEMS Microbiol. Lett. 330: 1–9.

Autengruber, A., U. Sydlik, M. Kroker, T. Hornstein, N. Ale-Agha, D. Stockmann et al. 2014. Signaling dependent adverse health effects of carbon nanoparticles are prevented by the compatible solute mannosylglycerate (firoin) *in vitro* and *in vivo*. PLoS One 9: e111485.

Avanti, C., V. Saluja, E.L.P. van Streun, H.W. Frijlink and W.L.J. Hinrichs. 2014. Stability of lysozyme in aqueous extremolyte solutions during heat shock and accelerated thermal conditions. PLoS ONE 9: e86244.

Begemann, M.B., M.R. Mormile, V.G. Paul and D.J. Vidt. 2011. Potential enhancement of biofuel production through enzymatic biomass degradation activity and biodiesel production by halophilic microorganisms. *In*: Ventosa, A., A. Oren and Y. Ma (eds.). Halophiles and Hypersaline Environments. Springer, Berlin, Heidelberg.

Ben-Mahrez, K., W. Sougakoff, M. Nakayama and M. Kohiyama. 1988. Stimulation of an alpha like DNA polymerase by v-myc related protein of Halobacterium halobium. Arch. Microbiol. 149: 175–180.

Bhosale, P. 2004. Environmental and cultural stimulants in the production of carotenoids from microorganisms. Appl. Microbiol. Biotechnol. 63: 351–361.

Bin-Salman, S.A., R.H. Amasha, S.D. Jastaniah, M.M. Aly and K. Altaif. 2018. Isolation, molecular characterization and extracellular enzymatic activity of culturable halophilic bacteria from hypersaline natural habitats. Biodiversitas. 19: 1828–1834. 10.13057/biodiv/d190533.

Bolhuis, H., P. Palm, A. Wende, M. Falb, M. Rampp, F. Rodriguez-Valera et al. 2006. The genome of the square archaeon *Haloquadratum walsbyi*: Life at the limits of water activity. BMC Genomics 7: 169. doi: 10.1186/1471-2164-7-169.

Borowitzka, M.A. 1999. Commercial production of microalgae: ponds, tanks, tubes and fermenters. J. Biotechnol. 70: 313–321.

Brown, A.D. 1976. Microbial water stress. Bacteriol. Rev. 40: 803–846.

Campas, C., M. Dalmau, B, Montaner, M. Barragan, B. Bellosillo and D. Colomer. 2003. Prodigiosin induces apoptosis of B and T cells from B-cell chronic lymphocytic leukemia. Leukemia 17: 746–750.

Cardoso, L.A.C., K.Y.F. Kanno and S.G. Karp. 2017. Microbial production of carotenoids—a review. Afr. J. Biotechnol. 16(4): 139–146.

Couto, S.R. and M.A. Sanroman. 2006. Application of solid-state fermentation to food industry—a review. J. Food Eng. 76(3): 291–302.

Craig, S.A.S. 2004. Betaine in human nutrition. Am. J. Clin. Nutr. 80(3): 539–549.

Das, K. and A.K. Mukherjee. 2007. Comparison of lipopeptide biosurfactants production by *Bacillus subtilis* strains in submerged and solid state fermentation systems using a cheap carbon source: some industrial applications of biosurfactants. Process. Biochem. 42: 1191–1199.

Delgado-García, M., I. De la Garza-Rodríguez, M.A. Cruz Hernández, N. Balagurusamy, C. Aguilar and R. Rodríguez-Herrera. 2013. Characterization and selection of halophilic microorganisms isolated from Mexican soils. ARPN J. Agric. Biol. Sci. 8(6): 457–464.

Delgado-García, M., S.M. Contreras-Ramos, J.A. Rodríguez, J.C. Mateos-Díaz, C.N. Aguilar and R.M. Camacho-Ruíz. 2018. Isolation of halophilic bacteria associated with saline and alkaline-sodic soils by culture dependent approach. Heliyon. 4(11): e00954. doi:10.1016/j.heliyon.2018.e00954.

Dixini, N. and K. Mistry. 2018. Biosurfactant assistance in crude oil degradation by halophilic *Bacillus cerus* ND1. Indian J. Mar. Sci. 47(08): 1640–1647.

Donio, M.B.S., F.A. Ronica, V.T. Viji, S. Velmurugan, J.S. Jenifer, M. Michaelbabu et al. 2013. *Halomonas* sp. BS4, A biosurfactant producing halophilic bacterium isolated from solar salt works in India and their biomedical importance. Springerplus. 2(1): 149. doi: 10.1186/2193-1801-2-149.

Durán-Padilla, V.R., G. Davila-Vazquez, N.A. Chávez-Vela, J.R. Tinoco-Valencia and J. Jauregui-Rincon. 2014. Iron effect on the fermentative metabolism of *Clostridium acetobutylicum* ATCC 824 using cheese whey as substrate. Biofuel Res. J. 1(4): 129–133.

Edward, A. 2017. Phylogenetic studies of unculturable halophilic bacteria in the fossil sediments of Ariyalur basin. Int. J. Pure App. Biosci. 5(5): 348–352.

Edwards, C. 1990. Microbiology of Extreme Environments. Milton Keynes: Open University Press.

El-Banna, A.A.E.R., A.M.A. El-Razek and A.R. El-Mahdy. 2012. Isolation, identification and screening of carotenoid-producing strains of *Rhodotorula glutinis*. Food Nutr. Sci. 3: 627–633.

ElObeid, A.S., A. Kamal-Eldin, M.A.K. Abdelhalim and A.M. Haseeb. 2017. Pharmacological properties of melanin and its function in health. Basic Clin. Pharmacol. Toxicol. 120(6): 515–522. doi: 10.1111/bcpt.12748. Epub 2017 Mar 7.

Englert, C., G. Wanner and F. Pfeifer. 1992. Functional analysis of the gas vesicle gene cluster of the halophilic archaeon *Haloferax mediterranei* defines the vac-region boundary and suggests a regulatory role for the *gvpD* gene or its product. Mol. Microbiol. 6: 3543–3550.

Eriksen, M., L.C.M. Lebreton, H.S. Carson, M. Thiel, C.J. Moore, J.C. Borerro et al. 2014. Plastic pollution in the world's oceans: more than 5 trillion plastic pieces weighing over 250,000 tons afloat at sea. PLoS ONE. doi:10.1371/journal.pone.0111913.

Fiedor, J. and K. Burda. 2014. Potential role of carotenoids as antioxidants in human health and disease. Nutrients 6: 466–488.

Gaba, S., R.N. Singh, S. Abrol, A.N. Yadav, A.K. Saxena and R. Kaushik. 2017. Draft Genome Sequence of Halolamina pelagica CDK2 Isolated from Natural Salterns from Rann of Kutch, Gujarat, India. Genome Announc. 5: 1–2.

Gallas, J. and M. Eisner. 2006. Melanin polyvinyl alcohol plastic laminates for optical applications. US Patent 7029758.

Geetha, S.J., I.M. Banat and S.J. Joshi. 2018. Biosurfactants: production and potential applications in microbial enhanced oil recovery (MEOR). Biocatal. Agric. Biotechnol. 14: 23–32.

Gómez, F., J.A. Rodríguez-Manfredi, N. Rodriguez, M. Fernández-Sampedro, F.J. Caballero-Castrejón and R. Amils. 2012. Habitability: where to look for life? Halophilic habitats: earth analogs to study Mars habitability. Planet. Space. Sci. 68: 48–55.

Graf, R., S. Anzali, J. Bunger, F. Pflucker and H. Driller. 2008. The multifunctional role of ectoine as a natural cell protectant. Clinics Dermatol. 26: 326–333.

Grant, W.D. 2004. Life at low water activity. Phil. Trans. R Soc. Lond. B. Biol. Sci. 359: 1249–1267.

Gupta, S., P. Sharma, K. Dec, M. Srivastava and A. Sourirajan. 2015. A diverse group of halophilic bacteria exist in Lunsu, a natural salt water body of Himachal Pradesh, India. Springer Plus. volume 4, Article number: 274.

Hagemann, M. and N. Pade. 2015. Heterosides—compatible solutes occurring in prokaryotic and eukaryotic phototrophs. Plant Biol. 17: 927–934.

Han, J.R., S.K. Ling, W.N. Yu, G.J. Chen and Z.J. Du. 2017. *Marinobacter salexigens* sp. nov, isolated from marine sediment. Int. J. Syst. Evol. Microbiol. 67(11): 4595–4600. DOI: 10.1099/ijsem.0.002337.

Hao, R. and A. Lu. 2009. Biodegradation of heavy oils by halophilic bacterium. Pro. Nat. Sci. 19: 997–1001. https://doi.org/10.1016/j.pnsc.2008.11.010.

Jaeger, K.E., S. Ransac, B.W. Dijkstra, C. Colson, M. Heuvel and O. Misset. 1994. Bacterial lipases. FEMS Microbiolo. Rev. 15: 29–63.

Jampani, C. and K.S.M.S. Raghavarao. 2015. Process integration for purification and concentration of red cabbage (*Brassica oleracea* L.) anthocyanins. Sep. Purif. Technol. 141: 10–16.

Jang, Y.S., J. Lee, A. Malaviya, Y. Seung do, J.H. Cho and S.Y. Lee. 2012. Butanol production from renewable biomass: rediscovery of metabolic pathways and metabolic engineering. Biotechnol. J. 7(2): 186–198.

Jayachandra, S.Y., A. Kumar, D.P. Merley and M.B. Sulochana. 2012. Isolation and characterization of extreme halophilic bacterium *Salinicoccus* sp. JAS4 producing extracellular hydrolytic enzymes. Recent Res. Sci. Technol. 4(4): 46–49.

Johnson, J., P.V.D.N. Sudheer, Y.H. Yang, Y.G. Kim and K.Y. Choi. 2017. Hydrolytic activities of hydrolase enzymes from halophilic microorganisms. Biotechnol. Bioproc. E. 22(4): 450–461. https://doi.org/10.1007/s12257-017-0113-4.

Joshi, P.A., N. Singh and D.B. Shekhawat. 2015. Effect of metal ions on growth and biosurfactant production by halophilic bacteria. Adv. Appl. Sci. Res. 6(4): 152–156.

Kabilan, M. 2016. Microbial Diversity of Halophilic Archaea and Bacteria in Solar Salterns and Studies on their Production of Antiarchaeal Substances. Ph.D. thesis submitted to Birla Institute of Technology and Science, Pilani.

Kaleem, S., I. Azam and I. Tahir. 2015. Biology and applications of halophilic bacteria and archaea: a review. Electron. J. Biol. 11: 98–103.

Kamekura, M., T. Hamakawa and H. Onishi. 1982. Application of halophilic nuclease H of *Micrococcus varians* subsp. *halophilus* to commercial production of flavoring agent 5′-GMP. Appl. Environ. Microbiol. 44: 994–995.

Kamekura, M. and M. Kates. 1999. Structural diversity of membrane lipids in members of Halobacteriaceae. Biosci. Biotechnol. Biochem. 63: 969–972.

Kanapathipillai, M., G. Lentzen, M. Sierks and C.B. Park. 2005. Ectoine and hydroxyectoine inhibit aggregation and neurotoxicity of Alzheimer's beta-amyloid. FEBS Lett. 579: 4775–4780. doi: 10.1016/j.febslet.2005.07.057.

Karimi, K. and A. Pandey. 2014. Current and future ABE processes. Biofuel Res. J. 1(3): 77.

Kates, M. 1993. Biology of halophilic bacteria, Part II. Membrane lipids of extreme halophiles: biosynthesis, function and evolutionary significance. Experientia. 49: 1027–1036.

Kelly, M., S.L. Jensen and S. Liaaen. 1967. Bacterial carotenoids. XXVI. C50-carotenoids. 2. Bacterioruberin. Acta Chem. Scand. 21: 2578–2580.

Khopade, A., R. Biao, X. Liu, K. Mahadik, L. Zhang and C. Kokare. 2012. Production and stability studies of the biosurfactant isolated from marine *Nocardiopsis* sp. B4. Desalination 3: 198–204.

Khunt, M., N. Pandhi and A. Rana. 2011. Amylase from moderate halophiles isolated from wild ass excreta. Int. J. Pharm. Biol. Sci. 1(4): 586–592.

Khunt, M. and N. Pandhi. 2012. Purification and characterization of lipase from extreme halophiles isolated from Rann of Kutch, Gujarat, India. Int. J. Life Sci. Pharma Res. 2: L55–61.

Kim, K.H., O.K. Lee and E.Y. Lee. 2018. Nano-immobilized biocatalysts for biodiesel production from renewable and sustainable resources. Catalysts. 8(2): 68. https://doi.org/10.3390/catal8020068.

Kivisto, A., V. Santala and M. Karp. 2010. Hydrogen production from glycerol using halophilic fermentative bacteria. Bioresour. Technol. 101: 8671–8677.

Kogej, T., M.H. Wheeler, T. Lanisnik Rizner and N. Gunde-Cimerman. 2004. Evidence for 1,8-dihydroxynaphthalene melanin in three halophilic black yeasts grown under saline and non-saline conditions. FEMS Microbiol. Lett. 232: 203–209.

Kour, D., K.L. Rana, T. Kaur, B. Singh, V.S. Chauhan, A. Kumar et al. 2019a. Extremophiles for hydrolytic enzymes productions: biodiversity and potential biotechnological applications. pp. 321–372. *In*: Molina, G., V.K. Gupta, B. Singh and N. Gathergood (eds.). Bioprocessing for Biomolecules Production. doi:10.1002/9781119434436.ch16.

Kour, D., K.L. Rana, N. Yadav, A.N. Yadav, A.A. Rastegari, C. Singh et al. 2019b. Technologies for biofuel production: current development, challenges, and future prospects. pp. 1–50. *In*: Rastegari, A.A., A.N. Yadav and A. Gupta (eds.). Prospects of Renewable Bioprocessing in Future Energy Systems. Springer International Publishing, Cham. doi:10.1007/978-3-030-14463-0_1.

Kulkarni, N., A. Shendye and M. Rao. 1999. Molecular and biotechnological aspects of xylanases. FEMS Microbiol. 23: 411–456.

Kumar, S., S. Sharma, S. Thakur, T. Mishra, P. Negi, S. Mishra et al. 2019. Bioprospecting of microbes for biohydrogen production: Current status and future challenges. pp. 443–471. *In*: Molina, G., V.K. Gupta, B.N. Singh and N. Gathergood (eds.). Bioprocessing for Biomolecules Production. Wiley, USA.

Kushner, D.J. and H. Onishi. 1968. Absence of normal cell wall constituents from the outer layers of *Halobacterium cutirubrum*. Can. J. Biochem. 46: 997–998.

Kushner, D.J. 1985. The Halobacteriaceae. pp. 171–214. *In*: Woese, C.R. and R.S. Wolfe (eds.). The Bacteria, vol. 8, Academic Press, London.

Kushwaha, S.C., J.K. Kramer and M. Kates. 1975. Isolation and characterization of C50-carotenoid pigments and other polar isoprenoids from *Halobacterium cutirubrum*. Biochim. Biophys. Acta. 398: 303–314.

Lentzen, G. and T. Schwarz. 2006. Extremolytes: natural compounds from extremophiles for versatile applications. Appl. Microbiol. Biotechnol. 72: 623–634.

Leon, M.J., C. Sánchez-Porro and A. Ventosa. 2017. *Marinobacter aquaticus* sp. nov., a moderately halophilic bacterium from a solar saltern. Int. J. Syst. Evol. Microbiol. 67(8): 2622–2627. DOI: 10.1099/ijsem.0.001984.

Litchfield, C.D. 1998. Survival strategies for microorganisms in hypersaline environments and their relevance to life on early Mars. Meteor. Planet. Sci. 33: 813–819.

Ma, Y., E.A. Galinski, W.D. Grant, W.D. Oren and A. Ventosa. 2010. Halophiles 2010: Life in saline environments. Appl. Environ. Microbiol. 76: 6971–6981.

Makhdoumi-Kakhki, A., M.A. Amoozegar and A. Ventosa. 2012. *Salinibacter iranicus* sp. nov and *Salinibacter luteus* sp. nov., isolated from a salt lake, and emended descriptions of the genus *Salinibacter* and of *Salinibacter ruber*. Int. J. Syst. Evol. Microbiol. 62(7): 1521–1527. DOI: 10.1099/ijs.0.031971-0.

Mani, K., B.B. Salgaonkar and J.M. Braganca. 2012. Culturable halophilic archaea at the initial and crystallization stages of salt production in a natural solar saltern of Goa, India. Aquat. Biosyst. 8(1): 15. doi:10.1186/2046-9063-8-15.

Manikandan, P. and K.S. Palani. 2017. On overview of saltpan halophilic bacterium. Int. J. Biol. Res. 2(4): 28–33.

Margesin, R. and F. Schinner. 2001. Potential of halotolerant and halophilic microorganisms for biotechnology. Extremophiles 5: 75–83.

Mojica, F.J.M., C. Ferrer, G. Juez and F. Rodriguez-Valera. 1995. Long stretches of short tandem repeats are present in the largest replicons of the Archaea *Haloferax mediterranei* and *Haloferax volcanii* and could be involved in replicon partitioning. Mol. Microbiol. 17: 85–93.

Mokashe, N., B. Chaudhari and U. Patil. 2018. Operative utility of salt-stable proteases of halophilic and halotolerant bacteria in the biotechnology sector. Int. J. Biol. Macromol. 117: 493–522. doi: 10.1016/j.ijbiomac.2018.05.217.

Nagrale, D.T., Renu and P. Das. 2017. Genetic diversity and phylogenetic analyses of culturable extremely haloarchaea isolated from marine solar saltern pond in Mumbai, India. J. App. Biol. Biotech. 5(1): 29–34. DOI: 10.7324/JABB.2017.50105.

Naziri, D., M. Hamidi, S. Hassanzadeh, H. Salar, T. Vahideh, B.M. Maleki et al. 2014. Analysis of carotenoid production by *Halorubrum* sp. TBZ126; an extremely halophilic archeon from Urmia Lake. Adv. Pharm. Bull. 4(1): 61–67. doi:10.5681/apb.2014.010.

Nikou, M.M., M. Ramezani, S. Harirchi, S. Makzoom, M.A. Amoozegar, S.A. Shahzadeh Fazeli et al. 2017. *Salinifilum* gen. nov., with description of *Salinifilum proteinilyticum* sp. nov., an extremely halophilic actinomycete isolated from Meighan wetland, Iran, and reclassification of *Saccharopolyspora aidingensis* as *Salinifilum aidingensis* comb. nov. and *Saccharopolyspora ghardaiensis* as *Salinifilum ghardaiensis* comb. Nov. Int. J. Syst. Evol. Microbiol. 67(10): 4221–4227.

Oren, A. 2010. Industrial and environmental applications of halophilic microorganisms. Environ. Technol. 31(8-9): 825–834. doi: 10.1080/09593330903370026.

Pandey, A., S. Benjamin, C.R. Soccol, P. Nigam, N. Krieger and V.T. Soccol. 1999. The realm of microbial lipases in biotechnology. Biotech. Appl. Biochem. 29: 119–131.

Perez, D., S. Martin, G. Fernandez-Lorente, M. Filice, J.M. Guisan, A. Ventosa et al. 2011. A novel halophilic lipase, LipBL, showing high efficiency in the production of eicosapentaenoic acid (EPA). PloS One 6(8): e23325.

Philip, S., T. Keshavarz and I. Roy. 2006. Polyhydroxyalkanoates: Biodegradable polymers with a range of applications. J. Chem. Technol. Biotechnol. 82: 233–247.

Poli, A., P. Di Donato, G.R. Abbamondi and B. Nicolaus. 2011. Synthesis, production, and biotechnological applications of exopolysaccharides and polyhydroxyalkanoates by archaea. Archaea. (Vancouver, B.C.). 2011. doi: 10.1155/2011/693253.

Pradhan, A.K., N. Pradhan, G. Mall, H.T. Panda, L.B. Sukla, P.K. Panda et al. 2013. Application of lipopeptide biosurfactant isolated from a halophile: *Bacillus tequilensis* CH for inhibition of biofilm. Appl. Biochem. Biotechnol. 171: 1362–1375.

Quillaguaman, J., H. Guzman, D. Van-Thuoc and R. Hatti-Kaul. 2010. Synthesis and production of polyhydroxyalkanoates by halophiles: current potential and future prospects. Appl. Microbiol. Biotechnol. 85(6): 1687–1696. doi: 10.1007/s00253-009-2397-6.

Raddadi, N., A. Cherif, D. Daffonchio, M. Neifar and F. Fava. 2015. Biotechnological applications of extremophiles, extremozymes and extremolytes. Appl. Microbiol. Biotechnol. 99: 7907–7913.

Ramos-Cormenzana, A. 1989. Ecological distribution and biotechnological potential of halophilic microorganisms. pp. 289–309. *In*: DaCosta, M.S., J.C. Duarte and R.A.D. Williams (eds.). Microbiology of Extreme Environments and its Potential for Biotechnology. London: Elsevier.

Rana, K.L., D. Kour, I. Sheikh, N. Yadav, A.N. Yadav, V. Kumar et al. 2019. Biodiversity of endophytic fungi from diverse niches and their biotechnological applications. pp. 105–144. *In*: Singh, B.P. (ed.). Advances in Endophytic Fungal Research: Present Status and Future Challenges. Springer International Publishing, Cham. doi:10.1007/978-3-030-03589-1_6.

Rani, M.H.S., T. Ramesh, J. Subramanian and M. Kalaiselvam. 2013. Production and characterization of melanin pigment from halophilic black yeast *Hortaea werneckii*. Int. J. Pharm. Res. Rev. 2(8): 9–17.

Rao, M.B., A.M. Tanksale, M.S. Ghatge and V.V Deshpande. 1998. Molecular and Biotechnological aspects of microbial proteases. Microbiol. Mol. Biol. Rev. 62: 597–635.

Rasooli, M., M.A. Amoozegar, A.A. Sepahy, H. Babavalian and H. Tebyanian. 2016. Isolation, identification and extracellular enzymatic activity of culturable extremely halophilic Archaea and Bacteria of Incheboroun wetland. Int. Lett. Nat. Sci. 56: 40. doi:10.18052/www.scipress.com/ILNS.56.40.

Rastegari, A.A., A.N. Yadav and A. Gupta. 2019a. Prospects of Renewable Bioprocessing in Future Energy Systems. Springer International Publishing, Cham.

Rastegari, A.A., A.N. Yadav and N. Yadav. 2019b. Genetic manipulation of secondary metabolites producers. pp. 13–29. *In*: Gupta, V.K. and A. Pandey (eds.). New and Future Developments in Microbial Biotechnology and Bioengineering. Elsevier, Amsterdam. doi:https://doi.org/10.1016/B978-0-444-63504-4.00002-5.

Rastegari, A.A., A.N. Yadav, N. Yadav and N. Tataei Sarshari. 2019c. Bioengineering of secondary metabolites. pp. 55–68. *In*: Gupta, V.K. and A. Pandey (eds.). New and Future Developments in Microbial Biotechnology and Bioengineering. Elsevier, Amsterdam. doi:https://doi.org/10.1016/B978-0-444-63504-4.00004-9.

Rastegari, A.A., A.N. Yadav, A.A. Awasthi and N. Yadav. 2020. Trends of Microbial Biotechnology for Sustainable Agriculture and Biomedicine Systems: Diversity and Functional Perspectives. Elsevier, Cambridge, USA.

Rhodes, D. and A.D. Hanson. 1993. Quaternary ammonium and tertiary sulfonium compounds in higher plants. Annu. Rev. Plant Physiol. Plant Mol. Biol. 44: 357–384.

Roberts, M.F. 2005. Organic compatible solutes of halotolerant and halophilic microorganisms. Saline Syst. 1: 1–30.

Rodrigo-Baños, M., I. Garbayo, C. Vílchez, M.J. Bonete and R.M. Martínez-Espinosa. 2015. Carotenoids from Haloarchaea and their potential in biotechnology. Mar Drugs. 13: 5508–5532.

Roh, S.W., Y.D. Nam, Y.D. Chang, Y. Sung, K.H. Kim, H.M. Oh et al. 2007. *Halalkalicoccus jeotgali* sp. nov., a halophilic archaeon from shrimp jeotgal, a traditional Korean fermented seafood. Int. J. Syst. Evol. Microbiol. 57: 2296–2298.

Rubin, B. and E.A. Dennis. 1997. Lipases: Part A and B. Biotechnology Methods in Enzymology. Academic Press: New York.

Russell, N.J. 1993. Lipids of Halophilic and Halotolerant Microorganisms, the Biology of Halophilic Bacteria. CRC Press, Boca Raton, pp. 163–210.

Sanchez-Porro, C., S. Martin, E. Mellado and A. Ventosa. 2003. Diversity of moderately halophilic bacteria producing extracellular hydrolytic enzymes. J. Appl. Microbiol. 94(2): 295–300.

Sarafin, Y., M.B.S. Donio, S. Velmurugan, M. Michaelbabu and T. Citarasu. 2014. *Kocuria marina* BS-15 a biosurfactant producing halophilic bacteria isolated from solar salt works in India. Saudi J. Biol. Sci. 21: 511–519. doi: 10.1016/j.sjbs.2014.01.001.

Sarwar, M.K., I. Azam and I. Tahir. 2015. Biology and applications of halophilic bacteria and archaea: a review. Elect. J. of Biol. 11: 98–103.

Saum, S., F. Pfeiffer, P. Palm, M. Rampp, S. Schuster, V. Müller et al. 2013. Chloride and organic osmolytes: A hybrid strategy to cope with elevated salinities by the moderately halophilic, chloride-dependent bacterium *Halobacillus halophilus*. Environ. Microbiol. 15(5): 1619–1633.

Schirmer, A., M.A. Rude, X. Li, E. Popova and S.B. del Cardayre. 2010. Microbial biosynthesis of alkanes. Science 329: 559–562.

Sharma, R., Y. Chisti and U.C. Banerjee. 2001. Production, purification, characterization and applications of lipases. Biotechnol. Adv. 19: 627–662.

Shelake, R.M., R.R. Waghunde and J.Y. Kim. 2019a. Plant–Microbe–Metal (PMM) interactions and strategies for remediating metal ions. *In*: Plant Metal Interactions. Springer, Singapore.

Shelake, R.M., D. Pramanik and J.Y. Kim. 2019b. Exploration of plant-microbe interactions for sustainable agriculture in CRISPR era. Microorganisms 7: 269. doi:10.3390/microorganisms7080269.

Shivanand, P. and G. Mugeraya. 2011. Halophilic bacteria and their compatible solutes—osmoregulation and potential applications. Curr. Sci. 100: 1516–1521.

Shun Ichi, S., S. Hiroyuki and T. Hiroaki. 1996. Voltage-dependent absorbance change of carotenoids in halophilic Archaebacteria. Biochim. Biophys. Acta. 1284: 79–85.

Sinha, R. and S. Khare. 2012. Characterization of detergent compatible protease of a halophilic *Bacillus* sp. EMB9: Differential role of metal ions in stability and activity. Bioresour. Technol. 145: 357–361.

Sorokin, D.Y., N.A. Chernyh and M.N. Poroshina. 2015. *Desulfonatronobacter acetoxydans* sp. nov.: A first acetate-oxidizing, extremely salt-tolerant alkaliphilic SRB from a hypersaline soda lake. Extremophiles 19(5): 899–907. DOI: 10.1007/s00792-015-0765-y.

Sorokin, D.Y., I.V. Kublanov and T.V. Khijniak. 2017. *Natronospira proteinivora* gen. nov., sp. nov., an extremely salt-tolerant, alkaliphilic gammaproteobacterium from hypersaline soda lakes. Int. J. Syst. Evol. Microbiol. 67(8): 2604–2608. DOI: 10.1099/ijsem.0.001983.

Souza, P.M.D. 2010. Application of microbial α-amylase in industry—A review. Braz. J. Microbiol. 41(4): 850–861.

Srivastava, S. 2015. Food adulteration affecting the nutrition and health of human beings. Biol. Sci. Med. 1(1): 65–70.

Stephanopoulos, G. 2007. Challenges in engineering microbes for biofuels production. Science 315: 801–804.

Studdert, C.A., M.K. Herrera Seitz, M.I. Plasencia Gil, J.J. Sanchez and R.E. De Castro. 2001. Purification, biochemical characterization of the haloalkaliphilic archeon *Natronococcus occultus* extracellular serine protease. J. Gen. Microbiol. 41: 375–383.

Subramaniyan, S. 2012. Isolation, purification and characterization of low molecular weight xylanase from *Bacillus pumilus* SSP-34. Appl. Biochem. Biotechnol. 166: 1831–1842.

Sumitha, D., D. Preetha and J.C. Daniel. 2018. Functional properties of halophilic bacteria isolated from fermented foods. Indian J. Appl. Microbiol. 21(1): 37–45.

Todkar, S., R. Todkar, L. Kowale, K. Karmarkar and A. Kulkarni. 2012. Isolation and screening of antibiotic producing halophiles from Ratnagri coastal area, State of Maharashtra. Int. J. Sci. Res. Pub. 2: 2250–3153.

Uritskiy, G., S. Getsin, A. Munn, B. Gomez-Silva, A. Davila, B. Glass et al. 2019. Halophilic microbial community compositional shift after a rare rainfall in the Atacama Desert. ISME J. 13: 2737–2749. doi: https://doi.org/10.1101/442525.

Ventosa, A., J.J. Nieto and A. Oren. 1998. Biology of moderately halophilic aerobic bacteria. Microbiol Mol. Biol. Rev. 62: 504–544.

Ventosa, J. and J.J. Nieto. 1995. Biotechnological applications and potentialities of halophilic microorganisms. World J. Microbiol. Biotechnol. 11(1): 85–94.

Vidyasagar, M., S. Prakash, C. Litchfield and K. Sreeramulu. 2006. Purification and characterization of a thermostable, haloalkaliphilic extracellular serine protease from the extreme halophilic archaeon *Halogeometricum borinquense* strain TSS101. Archaea. 2: 51–57.

Vilchez, C., E. Forjan, M. Cuaresma, F. Bedmar, I. Garbayo and J.M. Vega. 2011. Marine carotenoids: Biological functions and commercial applications. Mar. Drugs 9: 319–333.

Voica, D.M., L. Bartha, H.L. Banciu and A. Oren. 2016. Heavy metal resistance in halophilic Bacteria and Archaea. FEMS Microbiol. Lett. 363(41): https://doi.org/10.1093/femsle/fnw146.

Vora, J.U., N.K. Jain and H.A. Modi. 2014. Extraction, characterization and application studies of red pigment of halophile *Serratia marcescens* KH1R KM035849 isolated from Kharaghoda soil. Int. J. Pure Appl. Biosci. 2(6): 160–168.

Vreeland, R.H. 1987. Mechanisms of halotolerance in microorganisms. CRC Crit. Rev. Microbiol. 14: 311–356.

Waditee, R., M.N. Bhuiyan, V. Rai, K. Aoki, Y. Tanaka, T. Hibino et al. 2005. Genes for direct methylation of glycine provide high levels of glycinebetaine and abiotic-stress tolerance in *Synechococcus* and *Arabidopsis*. Proc. Natl. Acad. Sci. USA 102: 1318–1323.

Waditee-Sirisattha, R., H. Kageyama and T. Takabe. 2016. Halophilic microorganism resources and their applications in industrial and environmental biotechnology. AIMS Microbiol. 2(1): 42–54. doi:10.3934/microbiol.2016.1.42.

Waghunde, R.R., R.M. Shelake and A.N. Sabalpara. 2016. Trichoderma: a significant fungus for agriculture and environment. Afr. J. Agric. Res. 11: 1952–1965.

Walsby, A.E. 1971. The pressure relationships of gas vacuoles. Proc. R. Soc. London B. 178: 301–326.

Werkhauser, N., A. Bilstein and U. Sonnemann. 2014. Treatment of allergic rhinitis with ectoine containing nasal spray and eye drops in comparison with azelastine containing nasal spray and eye drops or with cromoglycic acid containing nasal spray. J. Allergy. Article ID 176597. doi: 10.1155/2014/176597.

Woodside, J.V., A.J. McGrath, N. Lyner and M.C. McKinley. 2014. Carotenoids and health in older people. Maturitas. 80: 63–68.

Wu, L., J. Yao, A.K. Jain, R. Chandankere, X. Duan and H.H. Richnow. 2014. An efficient thermotolerant and halophilic biosurfactant-producing bacterium isolated from Dagang oil field for MEOR application. Int. J. Curr. Microbiol. App. Sci. 3(7): 586–599.

Yadav, A.N., D. Sharma, S. Gulati, S. Singh, R. Dey, K.K. Pal et al. 2015a. Haloarchaea endowed with phosphorus solubilization attribute implicated in phosphorus cycle. Sci. Rep. 5: 12293.

Yadav, A.N., P. Verma, M. Kumar, K.K. Pal, R. Dey, A. Gupta et al. 2015b. Diversity and phylogenetic profiling of niche-specific Bacilli from extreme environments of India. Ann. Microbiol. 65: 611–629.

Yadav, A.N., R. Kumar, S. Kumar, V. Kumar, T. Sugitha, B. Singh et al. 2017. Beneficial microbiomes: biodiversity and potential biotechnological applications for sustainable agriculture and human health. J. Appl. Biol. Biotechnol. 5: 45–57.

Yadav, A.N. 2017. Beneficial role of extremophilic microbes for plant health and soil fertility. J. Agric. Sci. 1: 1–4.

Yadav, A.N. and A.K. Saxena. 2018. Biodiversity and biotechnological applications of halophilic microbes for sustainable agriculture. J. Appl. Biol. Biotechnol. 6: 48–55.

Yadav, A.N., S. Gulati, D. Sharma, R.N. Singh, M.V.S. Rajawat, R. Kumar et al. 2019a. Seasonal variations in culturable archaea and their plant growth promoting attributes to predict their role in establishment of vegetation in Rann of Kutch. Biologia 74: 1031–1043. doi:10.2478/s11756-019-00259-2.

Yadav, A.N., D. Kour, K.L. Rana, N. Yadav, B. Singh, V.S. Chauhan et al. 2019b. Metabolic engineering to synthetic biology of secondary metabolites production. pp. 279–320. *In*: Gupta, V.K. and A. Pandey (eds.). New and Future Developments in Microbial Biotechnology and Bioengineering. Elsevier, Amsterdam. doi:https://doi.org/10.1016/B978-0-444-63504-4.00020-7.

Yadav, A.N., S. Mishra, S. Singh and A. Gupta. 2019c. Recent Advancement in White Biotechnology Through Fungi. Volume 1: Diversity and Enzymes Perspectives. Springer International Publishing, Cham.

Yadav, A.N., S. Singh, S. Mishra and A. Gupta. 2019d. Recent Advancement in White Biotechnology Through Fungi. Volume 2: Perspective for Value-Added Products and Environments. Springer International Publishing, Cham.

Yancey, P.H., M.E. Clark, S.C. Hand, R.D. Bowlus and G.N. Somero. 1982. Living with water stress: evolution of osmolyte systems. Science 217: 1214–1222.

Yatsunami, R., A. Ando, Y. Yang, S. Takaichi, M. Kohno, Y. Matsumura et al. 2014. Identification of carotenoids from the extremely halophilic archaeon *Haloarcula japonica*. Front. Microbiol. 5: 100–105.

Zenke, R., S. von Gronau, H. Bolhuis, M. Gruska, P. Pfeiffer and D. Oesterhelt. 2015. Fluorescence microscopy visualization of halomucin, a secreted 927 kDa protein surrounding *Haloquadratum walsbyi* cells. Front. Microbiol. 6: 249. doi: 10.3389/fmicb.2015.00249.

Zhang, G., S. Li, Y. Xue, L. Mao and Y. Ma. 2012. Effects of salts on activity of halophilic cellulase with glucomannanase activity isolated from alkaliphilic and halophilic *Bacillus* sp. BG-CS10. Extremophiles 16(1): 35–43. doi: 10.1007/s00792-011-0403-2.

Zhao, G., F. He, C. Wu, P. Li, N. Li, J. Deng et al. 2018. Betaine in inflammation: mechanistic aspects and applications. Front Immunol. 9. https://doi.org/10.3389/fimmu.2018.01070.

Chapter 4

Use of Halo-tolerant Bacteria to Improve the Bioactive Secondary Metabolites in Medicinally Important Plants under Saline Stress

Yachana Jha[1],* and *R.B. Subramanian*[2]

Introduction

Plants produce a large pool of diverse natural chemicals (natural products/phytochemicals), i.e., secondary metabolites, which is regularly produced by plants for genetic adaptation, to protect plants from the changing environment or to develop resistant against infection. Plant secondary metabolites are one of the essential components for the fitness and survival of plants in their natural habitat. Such secondary metabolites are generally produced from different intermediates or end products of primary metabolic pathways. Such secondary metabolites contain a variety of important bioactive compounds for biochemical, physiological and defense activity. Formation of such secondary metabolites is expensive and in plants it takes place under the influence of external stimuli factor such as, relative humidity, temperature, light intensity, environmental variation, seasonal changes and other agronomical conditions (Patra et al. 2013).

The concentration and composition of secondary metabolites in plants depends on the plants genetic composition. The composition of secondary metabolites in plants also depends on agronomical condition of the field such as, soil types, salt concentration, water status and nutrients in the soil. Secondary metabolites do not have any direct role in normal plant growth, development or reproduction, but can effectively act against adverse environmental conditions (Jha 2019a). They act as natural defense molecules against diseases caused due to pathogens, parasites or metabolic activity, and also help plants to overcome inter/intra species competition for better survival of the host plant (Singh and Yadav 2020; Yadav et al. 2020). Plant growth and development is continuously affected by abiotic factors like relative humidity, temperature, light, mineral nutrients, CO_2 and availability of water, as well as pollutants, ionizing radiation, wind pressure and resources which determine plant growth or biotic factors including other symbiotic organisms, parasites, herbivores and phytopathogen.

Secondary metabolites are an unusually varied group of natural products synthesized by all living organism including bacteria, fungi, algae, plant and animals (Yadav et al. 2019b). Secondary

---

[1] N. V. Patel College of Pure and Applied Sciences, S. P. University, V V Nagar, Anand, Gujarat, India.
[2] B. R. D. School of Biosciences, Sardar Patel University, Post Box no. 39, V V Nagar-388120 (Gujarat) India.
   Email: subramanianrb@gmail.com
* Corresponding author: yachanajha@ymail.com

metabolites are classified on the basis of their biosynthetic origin, such as phenolic, alkaloids and terpenes. The constitution of the bioactive compound in many medicinal, spice, aromatic and colorant plants, are often associated to a narrow range of species within a phylogenetic group. Different classes of such compounds are generally organic molecules that are not required for normal growth and are classified as secondary metabolites. Secondary metabolites are frequently produced at the highest levels during transition from active growth to the stationary phase. The producer organism can grow in the absence of their synthesis, suggesting that secondary metabolism is not essential, at least for short term survival. The genes involved in secondary metabolism provide a 'genetic playing field' that allows mutation and natural selection to fix new beneficial traits via evolution.

Secondary metabolism as an integral part of cellular metabolism and relies primarily on metabolism to supply the required enzymes, energy, substrates and cellular machinery and contributes to the long term survival of the plant (Roze et al. 2011; Rastegari et al. 2019a; Rastegari et al. 2019b). A simple classification of secondary metabolites includes three main groups: terpenes (such as plant volatiles, cardiac glycosides, carotenoids and sterols), phenolics (such as phenolic acids, coumarins, lignans, stilbenes, flavonoids, tannins and lignin) and nitrogen containing compounds (such as alkaloids and glucosinolates). These days, medicinal and aromatic plants have undergone a transition from unknown or minor agricultural plantings to major crops that farmers may consider as alternatives to traditional food or feed crops. The steadily increasing agricultural role is driven by consumer interest in these plants for culinary, medicinal and other anthropogenic applications. The use of plants, foods and herbal products is increasing due to consumer awareness of their various health benefits. Due to the ever-increasing population, the pressure on arable lands for cultivation of food crops has amplified; therefore, utilization of degraded wastelands is a viable option for cultivation of medicinal and aromatic plants. Hence for the enhanced production of secondary metabolite halo-tolerant bacteria inoculated *Withania somnifera* can be cultivated at normal as well as stressed conditions for human welfare.

Isolation and inoculation of halo-tolerant bacteria (HTB)

Plants have always been associated with many beneficial microbes and mechanism of mutualism that are highly complex and associative. This mechanism is so complicated that it is difficult to reproduce, to take full advantage of such an association for improved growth of plants. The halo-tolerant bacterial association with plants can offer many advantages to the associated host plant, especially to develop resistance against stress, protection from phytopathogens and growth. When host plant is under stress, halo-tolerant bacteria connects better with the environment for its efficient interaction with the host plant (Coutinho et al. 2015).

Therefore to avail the benefits of such halo-tolerant bacteria, bacteria are isolated from the root of *Suaeda nudiflora* wild mosque plant from khambhat near the seashore of Gujarat as the previously described method (Jha 2017). The soil sample is also collected from the site to analyze the physio-chemical property of the soil. The soil sample was tested in SICART (Sophisticated Instrumentation Centre for Applied Research and Testing) laboratory by extracting the water sample method and the soil possessing the following physio-chemical properties; pH 6.58, electrical conductivity 1480 µS/cm, salinity 8.6%, nitrate 112.5 mg kg^{-1}, chloride 128 mg kg^{-1}, sulfate 155 mg kg^{-1}, ammonia nitrogen 23.3 mg kg^{-1}, CEC: 3 cmol, organic carbon: 5500 mg kg^{-1} confirmed that the soil was highly saline. The isolated bacteria were then purified on the JNFB to analyze its nitrogen fixation ability. The two selected bacteria were selected for molecular identification, which was done with total genomic DNA and PCR amplification with 16S rDNA specific primers 16S F: 5'AGAGTTTGATCCTGGCTCAG3' and 16S R: 5'AGGTTACCTTGTTACGACTT3' followed by sequencing as our described method (Jha and Subramanian 2014). PCR amplicons of 16S rDNA of about 1500 bp were obtained for both the isolates as discrete bands in agarose gel. The phylogenetic trees were constructed using BLAST software by the comparison of the 16S rDNA sequence of isolates and related genera from a database using the Neighbor-Joining (NJ) algorithm and Maximum Likelihood (ML) method. The two isolates were identified by nucleotides homology and phylogenetic analysis as *Pseudomonas pseudoalcaligenes* (GenBank Accession Number: EU921258) and *Pseudomonas aeruginosa* (GenBank Accession Number: JQ790515). Then this bacteria was used for *Withania somnifera* plant inoculation, to analyze its effect on plant primary and secondary metabolites. Seeds of *Withania somnifera* were obtained from the DMAPR Anand and the surface sterilized with 0.1% $HgCl_2$ solution for 4 minutes and 70% ethanol for 10 minutes, after properly washing it with distilled water. The surface sterilized seeds were then tested for possible contamination, by placing it on petri dishes containing tryptone glucose yeast extract agar medium at 30°C.

The germinated seedlings devoid of any contamination were used for inoculation experiments. To study the effect of the isolated bacteria on the physiological and biochemical parameters, 4 days old germinated seedlings devoid of any contamination were transferred to culture tubes containing 400 μl Hoagland's nutrient medium, 400 μl micronutrients and 1% agar in 40 ml distilled water. Before the transfer, bacterial inoculums of the isolated bacteria *Pseudomonas pseudoalcaligenes* and *Pseudomonas aeruginosa* were added to the medium at a concentration of 6×10^8 cfu ml^{-1}. To obtain a mixture of both bacterial cultures, an equal volume of both the cultures were mixed in the medium to give a concentration of 6×10^8 cfu ml^{-1}. The tubes were incubated at 27°C with 12 hours light–dark cycle in a growth chamber. Seven days old plants were carefully removed from different test tubes inoculated with the strain of bacterium, and planted in a pot. Similarly the control plants (uninoculated) were also transferred to a fresh pot. Seedlings were planted at the rate of four plants per pot and watered at the time of transplantation of the seedlings to analyze the effect of halo-tolerant bacteria on plant growth promotion ability. The result of the present study indicated that inoculation with halo-tolerant bacteria, viz. *P. pseudoalcaligenes* and *P. aeruginosa* both alone or in combination significantly enhanced all the growth parameter (Table 4.1).

Table 4.1 Effect of halo-tolerant bacteria on growth parameters of *Withania somnifera* under stress.

Treatments	Germination %	Survival %	Plant height (m)	Dry weight (kg)	Growth index
Normal					
Control	67.4[d]	83.2[d]	1.321[fg]	0.353[cde]	37.3
Control + *P. aeruginosa*	73.6[bc]	92.3[bc]	1.522[c]	0.367[fg]	39.6
Control + *P. pseudoalcaligenes*	82.3[b]	89.3[b]	1.497[a]	0.398[ab]	41.7
Control + *P. aeruginosa* + *P. pseudoalcaligenes*	87.1[a]	92.8[a]	1.574[ab]	0.401[a]	43.4
Stressed					
Control	22.3[d]	32.2[cd]	1.021[h]	0.378[gh]	32.7
Control + *P. aeruginosa*	24.1[bc]	32.6[bc]	1.276[def]	0.414[def]	36.4
Control + *P. pseudoalcaligenes*	25.7[b]	33.1[ab]	1.312[cd]	0.442[bc]	37.2
Control + *P. aeruginosa* + *P. pseudoalcaligenes*	27.2[a]	33.8[a]	1.297[de]	0.431[bcd]	37.9

Values are the means of replicates. Values with different letters are significantly different at $P < 0.05$ (Duncan's Test). Values in columns followed by the same letter are not significantly different at ($P \leq 0.05$).

Halo-tolerant bacteria mediated regulation of nutrients for secondary metabolites production

The production of secondary metabolites in plants, is greatly increased under stress by suppressing plant growth, due to the extra allocation of fixed carbon for secondary metabolites production. In plant allocation of biomass to shoots increases if carbon fixation is affected by above ground resources such as CO_2 and light, while plants biomass allocation to roots increase under low levels of below ground resources, such as nutrients and water, according to the functional balance theory (Hendrik et al. 2012). But under prolonged stress, the nutrient predominantly allocated for secondary metabolites production, reverts for maintenance of cell osmotic balance that is necessary for proper metabolic activity of plants. Stress adversely affects plant nutrient acquisition, especially in the root, resulting in a significant decrease in shoots dry biomass.

Autotrophic plants need minerals for their life cycle and adequate supply of mineral nutrients that are necessary for optimum plant growth. However, when adequate amounts of essential nutrients are present in the soil, plants may still show deficiencies due to the non-availability of these mineral nutrients. Microorganisms can help plants to grow under such conditions, by providing soluble mineral nutrients from insoluble mineral converted by acidification, or via mobilization of essential nutrients that are necessary for improvement in the growth of plants growth (Yadav et al. 2015). The plant nutrient status has been increased by such halo-tolerant bacteria to promote growth and yield in different non-leguminous plants; by various mechanisms such as, associative nitrogen fixation, phosphorus, potassium solubilization, siderophores production, changing the absorptive capacity of the root (Karlidag et al. 2007; Kour et al. 2019; Yadav et al. 2019a). The association of halo-tolerant bacteria and their effect on the biological growth response of *Withania somnifera* plants under stress is complex. In our study, the foliar contents of N, P, K, Fe and Zn in halo-tolerant bacteria inoculated plants were estimated by taking 1 g of plant material digested in tri-acid mixture in the ratio of 9:3:1 by using a specific filter on digital flame photometry.

The foliar Fe concentration is higher in non-inoculated control plants, while P concentration is higher in plants inoculated with the halo-tolerant bacteria under stress. The plants inoculated with halo-tolerant bacteria alone and in combination show higher levels of foliar K. Potassium is an osmotically active solute that contributes to water absorption at the cell and whole plant level (Table 4.2) and help stressed plants in maintaining central metabolic activity for its survival. In our study, inoculations of plants with halo-tolerant bacteria always have higher nitrogen and carbon concentration under normal and stress conditions. Deficiencics of important nutrients like N, P, K and S usually cause a great deal of alteration in the concentration of secondary metabolites such as phenolic compounds and abundant nitrogen generally reduces the phenolic accumulation in plants (Ghasemzadeh et al. 2010), which is easily regulated and maintained by halo-tolerant bacteria. The levels of phenolic compounds are directly related to secondary metabolism and show the sensitivity of plant response to nutrient deficiency. The inoculation of halo-tolerant bacteria in plants, alone or

Table 4.2 Effect of halo-tolerant bacteria on nutrient minerals concentration of *Withania somnifera* under stress.

Treatments	N (mg kg⁻¹)	P (mg kg⁻¹)	K (mg kg⁻¹)	Fe (mg kg⁻¹)	Zn (mg kg⁻¹)
Normal					
Control	0.876ᵈ	0.781ᵈ	0.537ᶜᵈ	0.463ᵉᶠ	0.414ᶜᵈ
Control + *P. aeruginosa*	1.123ᶜ	0.836ᶜ	0.583ᵇᶜ	0.482ᶜᵈ	0.435ᵃᵇ
Control + *P. pseudoalcaligenes*	1.321ᵇ	0.887ᵃᵇ	0.627ᵃᵇ	0.484ᵇᶜ	0.446ᵇᶜ
Control + *P. aeruginosa* + *P. pseudoalcaligenes*	1.487ᵃ	0.989ᵃ	0.713ᵃ	0.508ᵇ	0.471ᵃ
Stressed					
Control	0.765ʰ	0.613ʰ	0.466ᵍʰ	0.413ᵃ	0.378ᵈᵉ
Control + *P. aeruginosa*	0.853ᶠᵍ	0.682ᶠ	0.519ᵉᶠ	0.435ᵈᵉ	0.392ᶠ
Control + *P. pseudoalcaligenes*	0.914ᵉᶠ	0.747ᶠᵍ	0.574ᶠᵍ	0.423ᵍᶠ	0.391ᶠᵍ
Control + *P. aeruginosa* + *P. pseudoalcaligenes*	1.182ᵉ	0.821ᵉ	0.624ᵉ	0.462ʰ	0.412ʰ

Values are the means of replicates. Values with different letters are significantly different at $P < 0.05$ (Duncan's Test). Values in columns followed by the same letter are not significantly different at ($P \leq 0.05$).

in groups can lead to tolerance of plants against adverse environmental conditions and also improve other nutrient availability, helping the plant to overcome stress by regulating secondary metabolite production.

Halo-tolerant bacteria mediated regulation of concentration of photosynthetic pigments for secondary metabolites production

The multiple effects on plant biochemical, molecular and physiological processes are collectively responsible for plant growth. But plant growth is greatly hampered under stressful environments by altering the ultra-structure of the organelles as well as the concentration of photosynthetic pigments (Wu and Kubota 2008). The photosynthetic pigments absorb energy from light, which is necessary for photosynthesis. Soil salinity is responsible for osmotic and ionic stresses in plants because of the generation of reactive oxygen results in oxidative stress and nutritional imbalances (Yang and Guo 2018). Such stress adversely affects the biochemical, molecular and ultimately physiological processes including transpiration, water relations, cellular homeostasis, enzymatic activities and photosynthesis, as well as gene expression patterns in plants (Deinlein et al. 2014).

Concentration of photosynthetic pigments is a useful indicator of plant responses to salinity stress. Fresh leaves are used for chlorophyll measurements, by placing fresh leaf samples (0.5 g) in a shaker with 80% acetone until the leaves are completely bleached. The extract is centrifuged at 13,000 rpm for 10 minutes, and the supernatant is used to measure chlorophyll a (Chl a), chlorophyll b (Chl b), and carotenoid by taking absorbance at 663, 645, and 470 nm respectively, using a spectrophotometer. In the present study halo-tolerant bacteria inoculated plants show higher concentration of chlorophyll a, b and carotenoids in comparison to control plants (Table 4.3). Chlorophyll content is influenced by the nitrogen concentration, as the levels of N_2 increased, chlorophyll a, b and total chlorophyll are also enhanced. Halo-tolerant bacteria having a nitrogen-fixing ability must enhance the concentration of N_2 in plants and results in high level of chlorophylls and also carotenoids. The increase in chlorophyll content with increasing nitrogen has been reported by Suza and Valio (2003). Enhanced chlorophyll concentrations might result in increased photosynthetic rate and consequently, enhanced plant growth under saline conditions (Kang et al. 2014).

Table 4.3 Effect of halo-tolerant bacteria on the photosynthetic pigment, total sugar and photosynthesis in *Withania somnifera* under stress.

Treatments	Chl a (mg g^{-1} FW)	Chl b (mg g^{-1} FW)	Carotenoid (mg g^{-1} FW)	Total Sugar (%wt/wt)	Photosynthesis (μmol m^{-2}s^{-1})
Normal					
Control	0.666[d]	0.381[d]	0.537[cd]	22.3[d]	32[d]
Control + *P. aeruginosa*	0.715[c]	0.422[c]	0.564[bc]	26.7[bc]	37[bc]
Control + *P. pseudoalcaligenes*	0.823[b]	0.432[ab]	0.523[ab]	28.3[b]	42[b]
Control + *P. aeruginosa* + *P. pseudoalcaligenes*	0.821[a]	0.473[a]	0.575[a]	32.2[a]	45[a]
Stressed					
Control	0.562[h]	0.326[h]	0.436[gh]	17.3[d]	35[cd]
Control + *P. aeruginosa*	0.583[g]	0.373[f]	0.412[ef]	22.4[bc]	43[bc]
Control + *P. pseudoalcaligenes*	0.568[ef]	0.376[fg]	0.424[fg]	26.7[b]	48[ab]
Control + *P. aeruginosa* + *P. pseudoalcaligenes*	0.611[e]	0.412[e]	0.465[e]	31.1[a]	53[a]

Values are the means of replicates. Values with different letters are significantly different at $P < 0.05$ (Duncan's Test). Values in columns followed by the same letter are not significantly different at ($P \leq 0.05$).

Carotenoids is another important pigment for photosynthesis and has a main role in photoprotection, it also act as a signaling precursor to protect the development of plants from advert environmental stress. It is responsible for improving nutritional status and yield of the plant under stress (Jha and Subramanian 2018a). There are negative correlations between the concentration of secondary metabolites and concentration of chlorophyll a, b and total chlorophyll. According to the prediction of the protein competition model, there is contest between chlorophyll contents and concentration of secondary metabolites, in which the concentration of secondary metabolites in the plant is a product of competition among secondary metabolites biosynthesis and protein. This negative correlation between secondary metabolites and concentration of chlorophyll indicates the steady shift of investment from protein to polyphenolics production (Meyer et al. 2006). The inoculation with halo-tolerant bacteria helps in the enhanced production of chlorophyll to compete with the production of secondary metabolites, but under stress it promotes production of secondary metabolites for the survival of the host plant. Affendy et al. (2010), reported, an increase in the production of secondary metabolites of *O. stimaneus* under low irradiance due to increase in availability of phenylalanine, a precursor for secondary metabolites and protein production. The production of secondary metabolites is prioritized under low nitrogen levels due to the restriction of protein production as exhibited by reduced chlorophyll production.

Halo-tolerant bacteria mediated regulation of photosynthesis for secondary metabolites production

Photosynthesis is the process by which plants, algae and certain types of bacteria-synthesized carbohydrates from carbon dioxide and water, use light as an energy source. The process can be broken down into two stages—the light and dark reactions. NADPH and ATP generated during the light reaction is fuel for the formation of carbohydrates in dark reactions. The most substantial physiological process for the plant is photosynthesis and in all its stages, is influenced by environmental stress. Photosynthesis involves various cell components, such as photosystems, photosynthetic pigments, CO_2 reduction pathways and electron transport system and any alteration at any phase of photosynthesis due to stress results in reduction of the general photosynthetic capacity of a green plant (Jha and Subramanian 2018b). In plant photosynthesis one of the most important metabolic process, responsible for the synthesizing photo-assimilates, can act as precursor for the many other metabolic pathways. A second important metabolic process is respiration, which uses photo-assimilates as substrate to generate energy (ATP), carbon/sugar intermediates and reductants (NADH and NADPH), as products.

The energy generated by respiration will be the source energy for the production of secondary metabolites and the products of respiration act as the precursors of secondary metabolites (Li et al. 2017). The rate of photosynthesis is strictly under the control of two of the most important metabolic processes, i.e., RuBisCO (ribulose-1,5-bis-phosphate) carboxylation and RuBP (ribulose-1,5-bis-phosphate) regeneration. Environmental stress is responsible for enhanced activity of RuBisCO to increase the rate of photosynthesis, while halo-tolerant bacteria tends to alter partitioning of the photo-assimilates in the plant for the production of secondary metabolite (Table 4.3). Thus the altered environmental condition increases the rate of respiration and photosynthesis, for enhanced accumulation of biomass, and enhances the rate of respiration and photosynthesis to modify the allocation of photo-assimilates towards secondary metabolite production, resulting in enhanced secondary metabolites production (Jha 2019b).

Research reveals that the availability of plant nutrients is one of the most important factors in defining secondary metabolite within plants (Strik 2008). The concentration of a mineral nutrient is a necessary component in regulating the yield as well as the quality of the plant's secondary metabolite (Cakmak 2010), and halo-tolerant bacteria are an integral component of saline soil microbial community and play a major role in the mineral cycle in soil rendering the unavailable minerals like P, K, Fe to the plants (Jha 2019c). These halo-tolerant bacteria have great potential to use fixed minerals and it very slowly releases minerals under soil systems with low minerals availability. Moreover, the stability of the halo-tolerant bacteria after inoculation in soil is also important for solubilization of minerals to benefit plant growth and development. Thus plants inoculated with halo-tolerant mineral mobilizing bacteria, can enhance photosynthesis under stress and is correlated with changes in plant physiology. Halo-tolerant bacteria regulate nutrient status, chlorophyll content, and carbohydrate/sugar concentration of the plant and modulate secondary metabolite production under stress to help in the survival of the plant. In this sense, halo-tolerant mineral mobilizing bacteria can be used to improve photosynthesis, plant health and growth rate, not only by induction as well as by modifying secondary metabolite for better survival of plants, without contaminating the environment under stress.

Halo-tolerant bacteria mediated regulation of essential oil production

Medicinal plants that have valuable aromatic compounds have been used by humans from ancient times due to the presence of important secondary metabolites (i.e., essential oils). Such oils are special organic compounds synthesized by specialized cells of medicinal plants by natural mechanism (Adorjan and Buchbauer 2010). Most secondary metabolites, having a variety of bioactive compounds are biosynthesized by the number of metabolic pathways in plants and assigned to interact with the environment. When plants encounter environmental stress like salinity; production, accumulation and secretion of such secondary metabolites including essential oils are accordingly in response to abiotic stresses. The chemical composition of essential oils shows that these oils are comprised of monoterpenoids, sesquiterpenoids, phenylpropanoids and the production of essential oils is primarily associated with primary metabolism and availability of soil nutrients (Dhifi et al. 2016). A variety of essential oils has been synthesized by plants for many different purposes like repellent against insects, pollinators' attraction, allelopathic activity, protection against phytopathogen and dispersal agents to favor the dispersion of seeds and pollens.

Essential oils could act as sedatives, antimicrobial, anti-inflammatory, antiviral, bactericidal, antifungal and as food preservatives. Essential oils present in the genus *Withania somnifera* are an important source of bioactive constituents, especially because of their biological properties such as, cytotoxic, insecticide and antimicrobial. Essential oils are volatile compounds that can be synthesized by all plant organs and are stored in epidermis cells, cell cavities, secretary cells and cell canals. In this study, Essential Oil (EO) was extracted from fresh herbage by hydro-distillation in Clevanger's apparatus for 1 hour 30 minutes, followed by estimation of oil content (w/v) and total oil yield (ml). The essential oil was analyzed on an Agilent 4890D Gas Chromatograph fitted with a column (30 m × 0.25 mm, film thickness 0.25 μm, Supelco Wax-10). The leaves were the source of the essential oil and are thus the most economically viable parts of the *Withania somnifera* and the oil yield is hence directly proportional to the number of leaves.

In the present study, the essential oil yield in plants inoculated with halo-tolerant bacteria is higher at normal conditions, but under stress inoculation with halo-tolerant bacteria do not have a significant effect on oil yield as shown in Table 4.4. Inoculation with root associated halo-tolerant bacteria under stress reduces the yield of essential oil, and was also reported by Arvin et al. (2012), but as it enhances the overall plant growth, results in enhanced overall essential oil yield per plant. The decreased oil yield at a specific location, due to inoculation with halo-tolerant bacteria may be due to the reduced effect of stress on the plant, which ultimately reduces the essential oil content. In the present study, the Gas Chromatographic (GC) analysis of the *Withania somnifera* essential oil enabled us to compare seven major compounds. The important secondary metabolites of *Withania somnifera* include alkaloids (isopelletierine, anaferine, cuseohygrine, anahygrine, etc.), steroidal lactones (withanolides, withaferins) and saponins (Singh et al. 2011). Among these bioactive compounds sitoindosides and acylsteryl glucosides have remarkable anti-stress activity as well as induce defense mechanism against pathogens. The production of 5-dehydroxy withanolide-R and withasomniferin-A compounds takes place in the aerial parts of *Withania somnifera*.

Table 4.4 Effect of halo-tolerant bacteria on the total steroid, alkaloid, Withanolide A, Withaferin A and oil content in *Withania somnifera* under stress.

Treatments	Total steroid content (mg CHOL/g)	Total alkaloid content (mg CF/g)	Withanolide A (µg/g DW)	Withaferin A (µg/g DW)	Oil content %
Normal					
Control	32.3[d]	13.2[d]	72.1[fg]	353.1[cde]	0.684[ef]
Control + *P. aeruginosa*	36.7[bc]	15.4[bc]	85.2[c]	367.3[fg]	0.721[cd]
Control + *P. pseudoalcaligenes*	38.3[b]	19.2[b]	94.7[a]	398.5[ab]	0.726[bc]
Control + *P. aeruginosa* + *P. pseudoalcaligenes*	37.2[a]	20.5[a]	105.4[ab]	401.7[a]	0.791[b]
Stressed					
Control	35.3[d]	15.2[cd]	83.4[h]	378.2[gh]	0.736[a]
Control + *P. aeruginosa*	32.4[bc]	17.3[bc]	97.6[def]	414.2[def]	0.743[de]
Control + *P. pseudoalcaligenes*	32.7[b]	18.7[ab]	131.2[cd]	442.7[bc]	0.711[gf]
Control + *P. aeruginosa* + *P. pseudoalcaligenes*	34.1[a]	28.9[a]	149.7[de]	531.3[bcd]	0.724[h]

Values are the means of replicates. Values with different letters are significantly different at $P < 0.05$ (Duncan's Test). Values in columns followed by the same letter are not significantly different at ($P \leq 0.05$).

The composition of essential oil in a plant mostly depends on the metabolic and physiological state of the plant. Plant inoculated with halo-tolerant bacteria has enhanced production of secondary metabolites including essential oil, which directly depends on primary metabolites and nutrient status of the plant (Jha and Subramanian 2016). Analysis of other important secondary metabolites like total steroid, alkaloid withanolide A and withaferin A has also been estimated and results have shown that both inoculation with halo-tolerant bacteria as well as stress, were able to enhance the production of these secondary metabolites (Table 4.4). So for the biosynthesis of a large variety of secondary metabolites, plants act as a chemical factory in which the root-associated bacteria act as a catalyst. These secondary metabolites are used as pesticides, scents, medicines, dyes and has many other commercial uses Secondary metabolites are not essential for plant growth, but are produced by the plant for its survival under stress. Adverse environmental signals like carbon-nutrition imbalance, ontogenesis, abiotic and biotic stimuli, are usually responsible for the production of secondary metabolites in plants (Mary Ann Lila 2006). Co-evolution between plants and their microbial partners is a mediator for plant chemical defense and for the protection of plants.

Halo-tolerant bacteria mediated regulation of secondary metabolites production under biotic stress

Plants being static, encounter many stresses and adverse environmental conditions. And the response of plants towards such stress takes place via several mechanisms as modification of biochemical pathways or molecular mechanisms or development of physical barrier or interaction among specific signaling pathways (Nejat and Mantri 2017). Among various stresses, biotic stress in plant are due to pathogens, pests or parasite interaction with plants, which results in plant infection (Yadav et al. 2017; Yadav and Saxena 2018). Plants have developed complex chemical defense mechanisms, for metabolic adaptation, to protect themselves from various types of pathogenic attacks. To counteract defense responses, plant pathogen induce modification of plant metabolites to acquire nutrients for their (pathogen) establishment and growth. At the same time plants also stimulate a complex defense mechanism to combat against the phytopathogen and acquire the potential to develop resistance against many pathogens. During evolution plants develop a multilayered defense system including pre-formed barriers such as a rigid cell wall, chemical barriers as secondary metabolite

(González-Lamothe et al. 2009) and induce various other defense mechanisms. Such chemical barriers are products of modified metabolism of plants under stress that will function as a first line defense, which act as repellents or antimicrobial compounds to prevent pathogen entry in plants. Biocontrol using halo-tolerant bacteria having the potential to modify the chemical or secondary metabolites profile under stress may be the best alternative to protect plants from pathogen infection, to maintain its growth pace under stress (Jha 2018a). The induced systemic resistant activated by the beneficial halo-tolerant bacteria on interaction with the plant, will act as an elicitor to activate plant defense mechanism and can protect plants from a wide range of pathogens, by modifying secondary metabolites profiles, such as β-1, 3-glucanases, phenolic and PAL activity of the host plant (Kour et al. 2020; Rastegari et al. 2020).

The present study showed that, root association of halo-tolerant bacteria can cause efficient induction of pathogenesis related protein like β-1, 3-glucanases, phenolic, and PAL in the inoculated plant compared to non-inoculated plants, may be due to its elicitation effect prior to infection (Table 4.5). Elicitation is the induced or enhanced biosynthesis of metabolites due to the addition of trace amounts of elicitors. Chamam et al. (2013) reported that root inoculation of rice with root-associated bacteria under stress, changes the metabolite profile of rice, with enhanced concentration of phenolic compounds such as flavonoids and hydroxyl cinnamic derivatives. However the change in composition of secondary metabolites takes place in shoots indicating the systemic effect of root-associated bacteria in plants for its survival under stress. There is thus a direct relation between plant growth and the production of secondary metabolites under stress conditions. Root associated halo-tolerant bacteria manage plants growth to acquire the maximum yield of biomass and hence also the phyto-medicinal compounds (Jha et al. 2014). As it is well known that plant secondary metabolites are produced by the plant have no essential role in maintaining the life processes of the plants, but have a significant role in plants interaction with adverse environmental conditions for its adaptation and develop resistant against a wide range of pathogens. Secondary metabolites have significant practical applications in medicinal, nutritive and cosmetic purposes, besides, being of importance in plant stress physiology for adaptation.

Table 4.5 Effect of halo-tolerant bacteria Proline, Glycine, PAL, β-1, 3-glucanases, Phenolic and Flavonoid content of *Withania somnifera* under stress.

Treatment	Proline (mMol min^{-1} g^{-1})	Glycine betaine (mMol min^{-1} g^{-1})	PAL (nmol of trans cinnamic acid min^{-1} g^{-1})	β-1, 3-glucanases (nmol of Glucose min^{-1} g^{-1})	Total phenolics (mg GAE/g)	Total flavonoids (mg CE/g)
Normal						
Control	2.12[cd]	0.98[d]	0.34[d]	2.15[d]	71.24[cd]	14.2[d]
Control + *P. aeruginosa*	19.6[c]	1.21[c]	0.38[bc]	2.23[c]	82.42[bc]	19.3[c]
Control + *P. pseudoalcaligenes*	2.13[b]	1.31[b]	0.36[b]	2.31[b]	87.21[ab]	22.6[ab]
Control + *P. pseudoalcaligenes* + *P. aeruginosa*	2.24[a]	1.32[a]	0.41[a]	2.41[a]	89.19[a]	24.7[a]
Stressed						
Control	1.91[d]	1.12[cd]	0.31[d]	2.19[d]	63.45[cd]	12.3[d]
Control + *P. aeruginosa*	1.86[bc]	1.27[c]	0.35[ab]	2.24[b]	71.06[c]	16.8[bc]
Control + *P. pseudoalcaligenes*	2.17[b]	1.29[ab]	0.39[a]	2.31[a]	77.13[b]	19.7[b]
Control + *P. pseudoalcaligenes* + *P. aeruginosa*	2.31[a]	1.33[a]	0.42[c]	2.47[bc]	83.26[a]	21.1[a]

For each parameter, values in columns followed by the same letter are not significantly different at (P ≤ 0.05).

Halo-tolerant bacteria mediated regulation of secondary metabolites production under abiotic stress

Plant growth and production of secondary metabolites are badly affected by the change in environmental conditions. Dehydration is the most common effect of many abiotic stresses, resulting in reduced availability of water and subsequent cause of osmotic and oxidative stress in plants, and is responsible for drastic changes in the plant physiology. Such stress significantly affects the metabolic activity and secondary metabolites production of the plant, which have a significant role in stress tolerance. Normally plants under stress have an increased level of secondary metabolites production, which may be because growth is often inhibited more than photosynthesis, and the carbon fixed is not allocated to growth, but instead to secondary metabolites production. Tripathi et al. (2018), reported enhanced production of withaferin A and withanone in *Withania somnifera*, that play an important role in combating oxidative stress. Enhanced production of such secondary metabolites (withanolides) under abiotic stress, can contribute in relocation of carbon for production of secondary metabolites in place of biomass production in plants.

Abiotic stress causes reduction in plant growth with a resulting increase in production of secondary metabolites like phenolic compounds as a defense mechanism (Arbona et al. 2013). Phenolic compounds contain one or more aromatic rings with added hydroxyl groups. They are broadly distributed in the plant kingdom and are the most abundant secondary metabolites of plants. In the present study, the leaf extract is used for the determination of the total phenol content by spectrophotometer (absorbance 735 nm) using gallic acid as standard. It was seen that inoculation of halo-tolerant bacteria alone is sufficient to increase the phenolic content in the *Withania somnifera* plant in the normal state and stress does not show any further increase in it (Table 4.5).

Plants produce various types of secondary plant products in response to a wide range of abiotic factors, to manage environmental changes, but inoculation with halo-tolerant bacteria modulates both primary and secondary metabolites for better survival of plants under stress (Jha 2018b). Osmotic stress caused by abiotic stress induces dehydration, which stimulates production of osmo-protectant for combating the osmotic stress. So the secondary metabolite as Glycine Betaine (GB) equivalents and proline production has been induced as a major osmo-protectant. In this study, accumulation of glycine betaine-like quaternary compounds and proline are significantly higher in plants leaves inoculated with both *P. pseudoalcaligenes* and *P. aeruginosa* (Table 4.5). Accumulation of glycine betaine-like quaternary compounds enhanced in the *Withania somnifera* leaves inoculated with both *P. pseudoalcaligenes* and *P. aeruginosa* compared to *Withania somnifera* plant treated with either of the *P. pseudoalcaligenes* and *P. aeruginosa* alone under stress. Many plant species naturally accumulate glycine betaine-like quaternary compounds and proline as major organic osmolytes when subjected to different abiotic stresses. These compounds are thought to play an adaptive role in mediating osmotic adjustment and protecting subcellular structures in stressed plants (Chen and Jiang 2010).

Global changes in environmental conditions due to human activities appear to influence endogenous plant metabolites for adaptation. Moreover, plants have adapted to produce several metabolites that are species-specific and depend on environmental factors. Various plant metabolites, such as polyamines, flavonoids, jasmonic acid, methyl-jasmonate, glycine betaine, have a protective role under abiotic stress. These phytochemical derivatives of secondary metabolism confer a multitude of adaptive and evolutionary advantages to the producing plants (Bilgin et al. 2010).

When the environment is adverse, plant metabolism is profoundly involved in signaling, physiological regulation, defense responses and affects plant growth. At the same time, in feedback, abiotic stresses affect the biosynthesis, concentration, transport and storage of primary and secondary metabolites. Halo-tolerant bacteria induced different small protein molecules in plants under stress as well as control conditions to establish itself in the host plant and to protect the host plant under stress. Mechanisms of halo-tolerant bacteria-mediated phytostimulation would help us to find more capable strains having the ability to function efficiently for sustainable production of important secondary metabolites under different agro-ecological conditions (Jha 2019d).

Conclusion and future prospects

Secondary metabolism comprises a coordinated series of coupled enzymatic conversions that utilizes limited products of primary metabolism as substrates. Secondary metabolism uses highly organized systematic mechanisms that integrate into developmental, morphological and biochemical regulatory patterns of the entire plant metabolic network. Elicitation or compounds produced by the halo-tolerant bacteria and stress is responsible for triggering the formation of secondary metabolites. Halo-tolerant bacteria have the ability to modulate the primary effect for the production of desired secondary metabolites for the survival of the host plant in normal as well as under stress conditions.

References

Adorjan, B. and G. Buchbauer. 2010. Biological properties of essential oils: an updated review. Flavour. Fragr. J. 25: 407–426.

Affendy, H., M. Aminuddin, M. Azmy, M.A. Amini, K. Assis and A.T. Tamir. 2010. Effects of light intensity on *Orthosiphon stamineus* Benth treated with different organic fertilizers. Int. J. Agric. Res. 5: 201–207.

Arbona, V., M. Manzi, C.D. Ollas and A. Gómez-Cadenas. 2013. Metabolomics as a tool to investigate abiotic stress tolerance in plants. Int. J. Mol. Sci. 14: 4885–4911.

Arvin, P., Vafabakhsh, J., Mazaheri, D., Noormohamadi, G. and Azizi, M. (2012). Study of Drought Stress and Plant Growth Promoting Rhizobacteria (PGPR) on yield, yield components and seed oil content of different cultivars and species of brassica oilseed rape. Ann Biol Res 3: 4444–4451.

Bilgin, D.D., J.A. Zavala, J. Zhu, S.J. Clough, D.R. Ort and E.H. De Lucia. 2010. Biotic stress globally downregulates photosynthesis genes. Plant Cell Environ. 33: 1597–1613.

Cakmak, I. 2010. Potassium for better crop production and quality. Plant Soil 335: 1–2.

Chamam, A., H. Sanguin, F. Bellvert, G. Meiffren, G. Comte and F. Wisniewski-Dyé. 2013. Plant secondary metabolite profiling evidences strain-dependent effect in the *Azospirillum Oryza sativa* association. Phytochem. Lett. 87: 65–77.

Chen, H. and J.G. Jiang. 2010. Osmotic adjustment and plant adaptation to environmental changes related to drought and salinity. Environ. Rev. 18: 309–319.

Coutinho, B.G., D. Licastro, L. Mendonc Previato, M. Cámara and V. Venturi. 2015. Plant-influenced gene expression in the rice endophyte *Burkholderia kururiensis* M130. Mol. Plant-Microbe Interact. 28: 10–21.

Deinlein, U., A.B. Stephan, T. Horie, W. Luo, G. Xu and J.I. Schroeder. 2014. Plant salt-tolerance mechanisms. Trends Plant Sci. 19: 371–379.

Dhifi, W., S. Bellili, S. Jazi, N. Bahloul and W. Mnif. 2016. Essential oils' chemical characterization and investigation of some biological activities: a critical review. Medicines (Basel) 3(4): 25.

Ghasemzadeh, A., H.Z.E. Jaafar and A. Rahmat. 2010. Elevated carbon dioxide increases contents of flavonoids and phenolic compounds, and antioxidant activities in Malaysian young ginger (*Zingiber officinale* Roscoe.) varieties. Molecules 15: 7907–7922.

González-Lamothe, R., G. Mitchell, M. Gattuso, M.S. Di-arra, F. Malouin, K. Bouarab. (2009). Plant antimicrobial agents and their effects on plant and human pathogens. Int J Mol Sci. 10: 3400–3419.

Hendrik, P., J.N. Karl, B. Peter, J. Reich, P. Oleksyn and M. Poot Liesje. 2012. Biomass allocation to leaves stems and roots: meta-analyses of inter specific variation and environmental control. New Phytol. 193: 30–50.

Jha, Y. and R.B. Subramanian. 2014. Identification of plant growth promoting rhizobacteria from *Suaeda nudiflora* plant and its effect on maize. Indian J. Plant Prot. 42(4): 422–429.

Jha, Y., R.B. Subramanian and S. Sahoo. 2014. Antifungal potential of fenugreek coriander, mint, spinach herbs extracts against *Aspergillus niger* and *Pseudomonas aeruginosa* phyto-pathogenic fungi. Allelopathy J. 34: 325–334.

Jha, Y. and R.B. Subramanian. 2016. Rhizobacteria enhance oil content and physiological status of *Hyptis suaveolens* under salinity stress. Rhizosphere 1: 33–35.

Jha, Y. 2017. Potassium mobilizing bacteria: enhance potassium intake in paddy to regulate membrane permeability and accumulate carbohydrates under salinity stress. Braz. J. Biol. Sci. 4(8): 333–344.

Jha, Y. 2018a. Induction of anatomical, enzymatic, and molecular events in maize by PGPR under biotic stress. *In*: Meena, V. (ed.). Role of Rhizospheric Microbes in Soil. Springer, Singapore.

Jha, Y. and R.B. Subramanian. 2018a. From interaction to gene induction: An eco-friendly mechanism of PGPR-mediated stress management in the plant. Plant Microbiome: Stress Response. 217–232. 10.1007/978-981-10-5514-0_10.

Jha, Y. and R.B. Subramanian. 2018b. Effect of root-associated bacteria on soluble sugar metabolism in plant under environmental stress. Plant Metabolites and Regulation Under Environmental Stress. 231–240. doi.org/10.1016/B978-0-12-812689-9.00012-1.

Jha, Y. 2018b. Effects of salinity on growth physiology, accumulation of osmo-protectant and autophagy-dependent cell death of two maize varieties. Russ. Agric. Sci. 44: 124–130.

Jha, Y. 2019a. Regulation of water status, chlorophyll content, sugar, and photosynthesis in maize under salinity by mineral mobilizing bacteria. Photosynthesis, Productivity and Environmental Stress.75–93. doi.org/10.1002/9781119501800.ch5.

Jha, Y. 2019b. Endophytic bacteria as a modern tool for sustainable crop management under stress. *In*: Giri, B., R. Prasad, Q.S. Wu and A. Varma (eds.). Biofertilizers for Sustainable Agriculture and Environment. Soil Biology, 55: Springer, Cham.

Jha, Y. 2019c. The importance of zinc-mobilizing rhizosphere bacteria to the enhancement of physiology and growth parameters for paddy under salt-stress conditions. Jordan J. Biol. Sci. 12: 167–173.

Jha, Y. 2019d. Endophytic bacteria-mediated regulation of secondary metabolites for the growth induction in Hyptis suaveolens under stress. *In*: Egamberdieva, D. and A. Tiezzi (eds.). Medically Important Plant Biomes: Source of Secondary Metabolites. Microorganisms for Sustainability, vol. 15. Springer, Singapore.

Kang, S.M., A.L. Khan, M. Waqas, Y.H. You, J.H. Kim, J.G. Kim et al. 2014. Plant growth-promoting rhizobacteria reduce adverse effects of salinity and osmotic stress by regulating phytohormones and antioxidants in *Cucumis sativus*. J. Plant Interact. 9: 673–682.

Karlidag, H., A. Esitken, M. Turan and F. Sahin. 2007. Effects of root inoculation of plant growth promoting rhizobacteria (PGPR) on yield, growth and nutrient element contents of leaves of apple. Sci. Hortic. 114: 16–20.

Kour, D., K.L. Rana, N. Yadav, A.N. Yadav, A. Kumar, V.S. Meena et al. 2019. Rhizospheric microbiomes: biodiversity, mechanisms of plant growth promotion, and biotechnological applications for sustainable agriculture. pp. 19–65. *In*: Kumar, A. and V.S. Meena (eds.). Plant Growth Promoting Rhizobacteria for Agricultural Sustainability: From Theory to Practices. Springer Singapore, Singapore. doi:10.1007/978-981-13-7553-8_2.

Kour, D., K.L. Rana, A.N. Yadav, N. Yadav, M. Kumar, V. Kumar et al. 2020. Microbial biofertilizers: Bioresources and eco-friendly technologies for agricultural and environmental sustainability. Biocatal. Agric. Biotechnol. 23: 101487. doi:https://doi.org/10.1016/j.bcab.2019.101487.

Li, X., L. Zhang, G.J. Ahammed, Z.X. Li, J.P. Wei, C. Shen et al. 2017. Stimulation in primary and secondary metabolism by elevated carbon dioxide alters green tea quality in *Camellia sinensis* L. Sci. Rep. 7: 7937.

Mary, Ann. Lila. 2006. The nature-versus-nurture debate on bioactive phytochemicals: the genome versus terroir. J. Sci. Food Agric. 86: 2510–2515.

Meyer, S., Z.G. Cerovic, Y. Goulas, P. Montpied, S. Demotes, L.P.R. Bidel et al. 2006. Relationship between assessed polyphenols and chlorophyll contents and leaf mass per area ratio in woody plants. Plant Cell Environ. 29: 1338–134.

Nejat, N. and N. Mantri. 2017. Plant immune system: Crosstalk between responses to biotic and abiotic stresses the missing link in understanding plant defense. Curr. Issues Mol. Biol. 23: 1–16.

Patra, B., C. Schluttenhofer, Y. Wu, S. Pattanaik and L. Yuan. 2013. Transcriptional regulation of secondary metabolite biosynthesis in plants. Biochim. Biophys. Acta. 1829: 1236–1247.

Rastegari, A.A., A.N. Yadav and N. Yadav. 2019a. Genetic manipulation of secondary metabolites producers. pp. 13–29. *In*: Gupta, V.K. and A. Pandey (eds.). New and Future Developments in Microbial Biotechnology and Bioengineering. Elsevier, Amsterdam. doi:https://doi.org/10.1016/B978-0-444-63504-4.00002-5.

Rastegari, A.A., A.N. Yadav, N. Yadav and N. Tataei Sarshari. 2019b. Bioengineering of secondary metabolites. pp. 55–68. *In*: Gupta, V.K. and A. Pandey (eds.). New and Future Developments in Microbial Biotechnology and Bioengineering. Elsevier, Amsterdam. doi:https://doi.org/10.1016/B978-0-444-63504-4.00004-9.

Rastegari, A.A., A.N. Yadav, A.A. Awasthi and N. Yadav. 2020. Trends of Microbial Biotechnology for Sustainable Agriculture and Biomedicine Systems: Diversity and Functional Perspectives. Elsevier, Cambridge, USA.

Roze, L.V., A. Chanda and J.E. Linz. 2011. Compartmentalization and molecular traffic in secondary metabolism: a new understanding of established cellular processes. Fungal Genet. Biol. 48: 35–48.

Singh, J. and A.N. Yadav. 2020. Natural Bioactive Products in Sustainable Agriculture. Springer, Singapore.

Singh, N., M. Bhalla and P.M. de Jager. 2011. An overview on ashwagandha: a rasayana (rejuvenator) of ayurveda. Afr. J. Tradit. Complement Altern. Med. 8(5): 208–213.

Strik, B.C. 2008. A review of nitrogen nutrition of Rubus. In Proceedings of the IXth International Rubus and Ribes Symposium, Pucon, Chile, 403–410.

Suza, R. and I.F.M. Valio. 2003. Leaf optical properties as affected by shade in samplings of six tropical tree species differing in succesional status. Braz. J. Plant Physiol. 15: 49–54.

Tripathi, N., D. Shrivastava, B.A. Mir, S. Kumar, S. Govil, M. Vahedi et al. 2018. Metabolomic and biotechnological approaches to determine therapeutic potential of *Withania somnifera* (L.) Dunal: A review. Phytomedicine 50: 127–136.

Wu, M. and C. Kubota. 2008. Effects of high electrical conductivity of nutrient solution and its application timing on lycopene, chlorophyll and sugar concentrations of hydroponic tomatoes during ripening. Sci. Hortic. 116: 122–129.

Yadav, A.N., D. Sharma, S. Gulati, S. Singh, R. Dey, K.K. Pal et al. 2015. Haloarchaea endowed with phosphorus solubilization attribute implicated in phosphorus cycle. Sci. Rep. 5: 12293.

Yadav, A.N., R. Kumar, S. Kumar, V. Kumar, T. Sugitha, B. Singh et al. 2017. Beneficial microbiomes: biodiversity and potential biotechnological applications for sustainable agriculture and human health. J. Appl. Biol. Biotechnol. 5: 45–57.

Yadav, A.N. and A.K. Saxena. 2018. Biodiversity and biotechnological applications of halophilic microbes for sustainable agriculture. J. Appl. Biol. Biotechnol. 6: 48–55.

Yadav, A.N., S. Gulati, D. Sharma, R.N. Singh, M.V.S. Rajawat, R. Kumar et al. 2019a. Seasonal variations in culturable archaea and their plant growth promoting attributes to predict their role in establishment of vegetation in Rann of Kutch. Biologia 74: 1031–1043. doi:10.2478/s11756-019-00259-2.

Yadav, A.N., D. Kour, K.L. Rana, N. Yadav, B. Singh, V.S. Chauhan et al. 2019b. Metabolic engineering to synthetic biology of secondary metabolites production. pp. 279–320. *In*: Gupta, V.K. and A. Pandey (eds.). New and Future Developments in Microbial Biotechnology and Bioengineering. Elsevier, Amsterdam. doi:https://doi.org/10.1016/B978-0-444-63504-4.00020-7.

Yadav, A.N., J. Singh, A.A. Rastegari and N. Yadav. 2020. Plant Microbiomes for Sustainable Agriculture. Springer International Publishing, Cham.

Yang, Y. and Y. Guo. 2018. Elucidating the molecular mechanisms mediating plant salt-stress responses. New Phytol. 217: 523–539.

Chapter 5

Hot Springs Thermophilic Microbiomes
Biodiversity and Biotechnological Applications

Juan-José Escuder-Rodríguez, María-Eugenia DeCastro,
Esther Rodríguez-Belmonte, Manuel Becerra and *María-Isabel González-Siso**

Introduction

Hot springs (continental waters that emanate from Earth, usually as a consequence of volcanic or geothermal activity) represent a world-spread natural-occurring niche for the development of thermophile microbial communities. Such communities, and specifically their microbiomes, are highly revealing about heat-adaptation mechanisms to sustain life at very high temperatures, and constitute a most desirable source for enzymes of industrial interest (DeCastro et al. 2018; Gomri et al. 2018a; Kaushal et al. 2018). Traditionally, thermophiles were considered demanding to work with using standard laboratory procedures, and many were regarded to be unculturable or yet-to-be cultured (Hedlund et al. 2015). Despite this fact, works exist that have optimized media composition and procedures for the growth of some of these microorganisms (Vartoukian et al. 2010; Meyer-Dombard et al. 2012; Grivalský et al. 2016), and new species (Kanoksilapatham et al. 2016; Patel et al. 2017) and enzymes of interest (Mehetre et al. 2018) are still found using what can be called the most classical approach to assess microbiomes.

Development of culture-independent techniques, such as marker gene sequencing (mostly 16S RNA) (Song et al. 2013; Lin et al. 2015; Lee et al. 2018; Mardanov et al. 2018; Selvarajan et al. 2018) and metagenomics (Inskeep et al. 2013; López-López et al. 2015; Mirete et al. 2016), overcame the great plate count anomaly, although some bottlenecks are still present both in the functional metagenomic library screening (Ekkers et al. 2012) and in the sequencing approaches. Integration of multiple-omics, like metatranscriptomics (Liu et al. 2011) as well as isotope-labeled carbon and nitrogen assimilation rates (Alcamán-Arias et al. 2018) further helps understanding the composition of these communities and their active metabolism.

One of the key factors in any microbiome study is collecting enough information about the sample (metadata) (Knight et al. 2018), ranging from date of collection to geochemical composition and temperature, to analyze how they affect the community composition and its diversity and to draw valid comparisons between different hot springs communities (Poddar and Das 2018; Power et al. 2018a). In hot springs, temperature is an obvious factor to take into account (Sharp et al.

Grupo EXPRELA, Centro de Investigacións Científicas Avanzadas (CICA), Facultade de Ciencias, Universidade da Coruña, A Coruña, Spain.

* Corresponding author: isabel.gsiso@udc.es

2014; Sofía Urbieta et al. 2014; Amin et al. 2017; Alcorta et al. 2018; Tang et al. 2018), but other parameters must also be considered (Chan et al. 2017; Chiriac et al. 2017), such as pH, salinity, conductivity and trace elements like arsenic (Meyer-Dombard and Amend 2014; Jiang et al. 2016) or dissolved sulfide and/or elemental sulfur (Inskeep 2013) and sulfates (Arce-Rodríguez et al. 2019), among others.

Other driving factors of the microbiome in hot springs include intake of exogenous organic carbon (Schubotz et al. 2013) and biomineralization by microbial mats (layered communities) (Valeriani et al. 2018). The composition of the community has also been shown to change over time (Bowen De León et al. 2013) and differ between sediments, the water column (Cole et al. 2013; Chernyh et al. 2015) and biological mats (Kambura et al. 2016; Rozanov et al. 2017; Mahato et al. 2019). Most recently, the spatial distribution of the biodiversity in hot springs microbiomes on a global scale has been assessed by Diversity-Area Relationship (DAR) analysis (Li and Ma 2019).

The community interaction has been described frequently in terms of their metabolism (Badhai et al. 2015; Jiang and Takacs-Vesbach 2017; Merkel et al. 2017; Saxena et al. 2017; Omae et al. 2019), but other interactions like the production of cyanotoxins by *Cyanobacteria* (Cirés et al. 2017) and other secondary metabolites with antimicrobial activity (Liu et al. 2016) are less explored in these extreme habitats. Moreover, thermophilic microbiomes in hot springs are thought to resemble those of early life on Earth, including their nitrogen fixation mechanisms (Nishihara et al. 2018) and iron redox-based metabolism (Fortney et al. 2018), which makes them interesting from both an evolutionary point of view and for the prospect of astrobiological studies.

The most studied microorganisms in hot springs belong to the *Bacteria* and *Archaea* domains, but work has also been done to study microbial thermophilic eukaryotes (Oliverio et al. 2018), including in some cases fungi (Liu et al. 2018), although the temperature threshold at which these organisms can thrive is lower. These facts raise the question of whether these organisms are actively growing or exist in a dormant state in these environments (Salano et al. 2017). Metagenomic sequencing studies have also enabled the description of hot spring-associated virus populations, known to play key roles in the modulation and evolution of microbiomes (Guajardo-Leiva et al. 2018; Munson-McGee et al. 2018; Sharma et al. 2018; Zablocki et al. 2018). Furthermore, the microbiome not only offers information on the composition of the microbial community and its primary metabolism, but has also been exploited as a source of potentially interesting industrial enzymes (Wemheuer et al. 2013; López-López et al. 2015; Wohlgemuth et al. 2018), and enzymes whose thermophilic variants remain uncharacterized (Ferrandi et al. 2017; Zapata-Pérez et al. 2017).

Despite this, the number of studies to assess microbiomes from hot springs remains comparatively scant in comparison with other much more studied ones like the human microbiome and soil microbiomes, as can be seen from a simple metric like a keyword search from a public citations database such as Pubmed (Fig. 5.1A). Another metric worth highlighting is the number of public projects included in the JGI GOLD Genome Online Database (Mukherjee et al. 2019) with the sample appearing from a hot spring: using an ecosystem filter one can find 56 public studies that have up to 2,043 biosamples emerging from them, but these are relatively low numbers compared to other environments with a larger number of studies, like freshwater and marine ecosystems (251 and 384 respectively) or biosamples like the host-associated human digestive system (29,074) (Fig. 5.1B).

It is worth noting that most of the samples with the thermal spring 'ecosystem type' tag in this database were obtained from the USA, mainly from hot springs of Yellowstone National Park (YNP), so an additional effort must be made in order to store more varied microbiomes at a global scale in these kinds of databases. Taking this into consideration, the potential for gaining new knowledge and products from microbiomes found in this extreme habitat remains largely untapped.

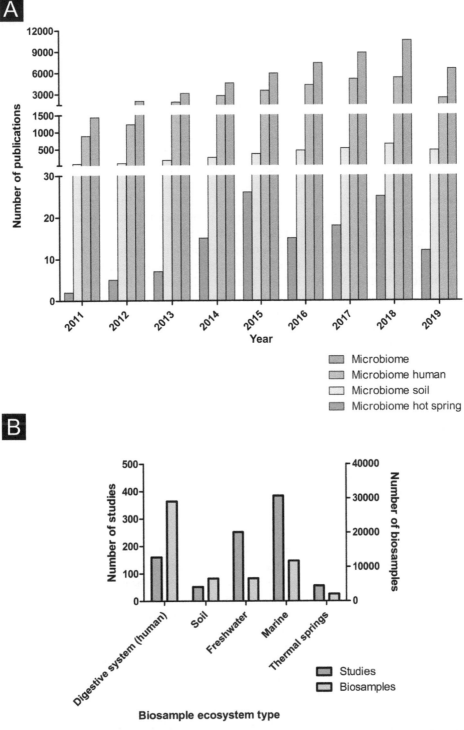

Figure 5.1 Metrics for research efforts regarding microbiomes found in hot springs. (A) Pubmed results for publications during the last few years (from 2011 to 2019) using different combinations of keywords (microbiome and a specific habitat). The Y axis is split in three segments, each with a different scale for a better visualization. Searches were performed on the 9 July 2019. (B) JGI GOLD Genome Online Database number of public projects and biosamples distribution across the five most abundant ecosystem types in the database: the host-associated human digestive system and four environmental habitats: soil, freshwater, marine and thermal springs. Searches were performed on the 23 July 2019.

Methods for assessing the microbiome

As mentioned before, many early studies tried to assess the microbiome using traditional culture techniques, but the development of culture-independent strategies has proven to be crucial to study the microbiome from extreme habitats like hot springs and similar high temperature environments. Recent reviews had already covered the methods that researchers generally use to assess thermophile microbiomes, namely marker gene sequencing and metagenomics (Urbieta et al. 2015; Knight et al. 2018), including the bioinformatics tools required to interpret the sequencing data (Roumpeka et al. 2017) and examples of good practices for proper normalization of the data (McMurdie 2018) and for sampling and data analysis (Staley and Sadowsky 2018). Some notable methods include the combination of microfluidics and metagenomics, called mini-metagenomics, allowing for the reconstruction of genomes with the shotgun sequencing strategy but with single cell-resolution (Yu et al. 2017). Researchers have begun to question some traditional methods for the classification of microbiomes, namely the effect of the database used when the classification method relies on Last Common Ancestor (LCA) (Nasko et al. 2018) or the preference for exact sequence variants instead of OTUs when resolving sequence data (Knight et al. 2018).

In addition, novel bioinformatics programs are constantly being developed, providing new tools to analyze the enormous amount of data generated by Next Generation Sequencing (NGS). Some of this tools include MetaQUBIC for gene module detection integrating both metagenomic and metatranscriptomic data, to help in understanding the active metabolism of a given microbial community (Ma et al. 2019). MNEMONIC for the comparison of microbial shifts in different conditions and comparison of the shifts between conditions, providing a useful tool for meta-analysis (Perz et al. 2019). ASAR an interactive analysis tool in R for performing simultaneous and hierarchical analysis at both the taxonomic and the functional levels of metagenomic data (Orakov et al. 2018). Metaviz for the visualization of data analysis of microbiomes derived from metagenomics or maker gene sequencing with hierarchical organization and available as a web service (Wagner et al. 2018) and PICRUSt for making inferences for functional annotation from marker gene sequencing data (Douglas et al. 2018). It is worth mentioning that most of these tools have been developed and optimized for microbiomes associated to a human host, but they could be easily adapted or readily be used for other habitats. Older algorithms are also being updated or replaced by less resource-intensive alternatives, such as Stripped UniFract (McDonald et al. 2018) to calculate the Uni Frac metric (a phylogenetic distance between microbiome profiles).

Biodiversity

The study of microbiomes from hot springs is becoming of great interest because of the specific impact in diverse industrial processes that require high temperatures or improve yields at high temperatures (Kumar et al. 2014a; Yadav et al. 2015). As stated earlier, precise information about the environmental conditions of collected samples, or metadata (Knight et al. 2018) is crucial to understand the complexity of those communities whose variability, and therefore interactions, can be affected by physicochemical parameters and the nutrient input from bordering areas (Zhang et al. 2018) especially when environmental conditions can vary seasonably. Even though a large number of hot spring studies have been published, compiled data correlating biodiversity with environmentally significant parameters used to be dispersed or scarce in the past (Kumar et al. 2014b; Saxena et al. 2016; Singh et al. 2016). However, currently there is a growing level of interest in studies that provide as much information as possible about the environmental conditions in order to feed and enrich the searching filters in databases, as JGI GOLD (Mukherjee et al. 2019) because this is especially helpful to understand these peculiar environments and to make valid comparisons between their communities.

Most of the knowledge we have about hot spring microbiomes comes from the analysis of the data obtained through culture-dependent and culture-independent techniques. Important information may be lost using any of the two approaches. Culture-based methods may affect the original community composition as some microorganisms would not be detected if they cannot be cultured in the conditions used, while others may be overestimated, especially if enriched cultures are used. The direct metagenomic analysis seems to be the most suitable technique to study the real complexity of a community in a specific geographical point and in a specific lapse of time, with the only limitation being its sensibility. Single-cell genomics have allowed to overcome this hindrance some time ago (Rinke et al. 2013) and the combination of metagenomics with microfluidics (Yu et al. 2017) highly improved this technology, which will accelerate our understanding of the actual complexity of microbiomes in extreme environmental habitats, including the so called 'microbial dark matter' not revealed so far.

Even though metagenomics has expanded as a very powerful tool to study the microbial communities inhabiting hot springs, there are still studies that keep on using traditional culture-based methods to assess the microbial diversity of these environments (Najar et al. 2018; Yasir et al. 2019) although very often they are not simply focused on taxonomic studies. Enzyme production, resistance to heavy metals, antimicrobial activities in water and other studies with possible biotechnological applications, still rely on culture-based methods with very interesting results. Culture-based methods are also being used to study changes in the microbial community composition owing to diverse natural or induced effects. Prieto-Barajas et al. (2017) explored the diversity of culturable bacterial communities inhabiting two hot springs (Tina and Bonita) located in Araró, México, and analyzed the effect of seasonality and related changes in physicochemical parameters of spring water.

They found that salt content was the principal factor that influenced the bacterial community inhabiting the Tina hot spring, while in the Bonita hot spring diversity was significantly correlated with temperature, pH, and arsenic content. Kikani et al. (2017) also studied the impact of the seasonal variation on the microbial diversity of Tulasi Shyam hot spring (India) combining culture-dependent and culture-independent methods (16s rRNA amplification followed by DGGE analysis). Nishida et al. (2018) studied the effect of light wavelength on hot spring microbial mat biodiversity showing significant variations between phototrophic and chemotrophic microbes, and also characterized their commensal relationship.

Certain diversity studies just focus on a part of the community. For example (Omae et al. 2019) evaluated the diversity of thermophilic hydrogenogenic carboxydotrophs by the analysis of the correlation between carbon monoxide dehydrogenase (CODH)–Energy-Converting Hydrogenase (ECH) gene cluster and taxonomic affiliation of 100 sediment samples from different hydrothermal environments of Japan. Industrial applications are probably one of the major drivers in most studies in which a specific part of a microbial community is the research focus, although this is not always the case. Other studies are centered on exploring special biodiversity, showing only one part—but a very informative one-, of the complexity of the hot spring environments, describing the presence of particular variability, ranging from viruses to eukaryotic microorganisms. Hot aquatic environments constitute favorable habitats for viruses (Wang et al. 2015; Gudbergsdóttir et al. 2016) therefore, some studies focus on hot spring phage diversity, such as those from (Parmar et al. 2018) and (Zablocki et al. 2017; 2018).

In India (Sharma et al. 2018) studied the viral diversity of the Manikaran hot springs using shotgun metagenomics sequencing and reconstructed 65 bacteriophage genomes and four potential phage hosts. Interestingly, they found signs of lateral gene transfer activity between these phage genomes and their hosts in this extreme environment. Guajardo-Leiva et al. (2018) used metagenomic and metatranscriptomic approaches to study the viral community composition of the Porcelana hot spring (Chile). In this study, they reported the full genome of a new thermophilic cyanopodovirus (TC-CHP58) and demonstrated the interaction between this phage and its host. Particular phyla of bacteria were detected and analyzed in some hot springs. For example (Gomri et al. 2018b) used culture-dependent and independent methods in order to assess the diversity of endospore-forming bacteria in two Algerian hot springs. Strunecký et al. (2019) focused on the cyanobacterial diversity of the Rupite hot spring (Bulgaria) with temperatures up to $71°C$ through a combination of microscopy, cultivation, single-cell PCR and 16s rRNA sequencing, showing that this hot spring has one of the richest cyanobacterial diversity reported from a site above 50°C. (Singh et al. 2018) used a culture-based method to study the cyanobacterial community of nine hot springs of North-Western Himalayas and the isolates were subjected to morphology and molecular characterization for determining their identity, diversity and phylogenetic relationship.

Even though bacteria and archaea are predominant, eukaryotic microorganisms can also be found in hot environments. Liu et al. (2018) focused on the fungal populations of Rehai thermal springs of Tengchong (China) revealing a relatively high diversity of fungi in those hot springs, dominated by genera such as *Penicillium* or *Entyloma*. Similar results were found in the hot springs of Soda lakes (Kenya) where the phylum *Ascomycota* and *Basidiomycota* were the most abundant fungi (Salano et al. 2017).

Direct metagenomic analysis of the complete community of specific environments is the best way to understand their complexity, as mentioned earlier, especially if relevant physicochemical parameters are provided, as is the case of the studies shown in Table 5.1. The studies are focused on the data published during the last three years, from 2017 to date in 2019. The different type of hot spring samples from water, soil, sediment, mats, filaments, stromatolites or biofilms seem to be influenced primarily by temperature. Bacteria are the main representative domain in these environments, but *Archaea* can also be found. At high temperatures, *Proteobacteria* seems to be the predominant phylum in all types of hot spring environments, though, at temperatures < 70°C the variability increases, probably because of the dependence in pH, a point that was already observed by (Power et al. 2018b) who studied the microbial biogeography of 925 geothermal springs in New Zealand (with a temperature range from 13.9 to100.6°C and a pH range from below 1 to 9.7) and determined that diversity is primarily influenced by pH at temperatures < 70°C; with temperature only having a significant effect for values > 70°C. They found that the phyla *Proteobacteria* and *Aquificae* dominated in the studied ecosystems (65.2% total average relative abundance across all springs).

Moreover, *Proteobacteria* was the most abundant phylum across all samples found predominantly at temperatures lower than 50°C. The abundance of phylum *Proteobacteria* in geothermal springs has been previously shown in a large number of geographically distant hot springs like Lobios in Spain (López-López et al. 2015) Deulajhari and Tattapani hot springs in India (Mohanrao et al. 2016; Singh and Subudhi 2016), Aguas Calientes in the Amazon rainforest of Perú (Paul et al. 2016) and El Coquito in the Colombian Andes (Bohorquez et al. 2012). This predominance is also reflected in Table 5.1, in which phylum *Proteobacteria* clearly dominates a high number of the hot springs studied from 2017 onwards. For example (Mahato et al. 2019) analyzed the community DNA of the 95°C water of Manikaran hot spring in India finding that microbial population is dominated by the phylum *Proteobacteria* followed by phyla *Bacteroidetes*, *Deinococcus–Thermus* and *Actinobacteria* (Table 5.1). This study also demonstrates a clear influence of temperature (at a genus level) in the composition of the microbial community, as could be seen in the sediment samples which had different temperatures. The phylum *Proteobacteria* exhibited the highest relative abundance across the four soil and water samples of the Tattapani hot spring (India) studied by (Kaushal et al. 2018) and in the sediments of geothermal springs of Pakistan (Amin et al. 2017). It is the largest phylum of *Bacteria* and comprises a vast diversity of gram-negative bacterial strains that are generally adapted to the harsh environments (Kaushal et al. 2018).

Proteobacteria seems to also dominate in moderate temperature hot springs. In this aspect (Najar et al. 2018) analyzed the microbial diversity of Polok (75°C–77°C) and Borong (50°C–52°C) hot springs in India and found that the dominance of the phylum *Proteobacteria* was more pronounced in the Borong hot spring which has a lower temperature. On the other hand (Chiriac et al. 2017), studied three moderate high-temperature hot springs from Rumania and reported that the percentage of *Proteobacteria* increased with temperature, while the proportion of *Cyanobacteria* decreased. Similar results were observed in the Ganzi Prefecture hot spring (China) in which *Cyanobacteria* become less abundant with an increase of temperature. With the rise in temperature, green microbial

Table 5.1 Examples of hot springs studied with sequence metagenomics.

Location	Type of sample/s	pH	Temp (°C)	Sequencer	Type of sequencing	Total reads	Dominant domain	Dominant phylum	References
Tato Field, TattaPani, and Murtazaabad, Pakistan	Sediment	6.2–9.4	60–95	GS-FLX-Titanium	16S rRNA	70,836	*Bacteria*	*Proteobacteria*	(Amin et al. 2017)
Ulu Slim, Malaysia	Water and sediment	7.2	90	IlluminaMiSeq	16S rRNA	1,053,625	*Bacteria*	*Proteobacteria*	(Chan et al. 2017)
SungaiKlah, Malaysia		8.2	75			480,983		*Firmicutes*	
DusunTua, Malaysia		7	70			1,028,376		*Proteobacteria*	
SungaiSerai, Malaysia		6.9	43			1,761,134		*Proteobacteria*	
Semenyih, Malaysia		6.9	43			167,238		*Firmicutes*	
Ayer Hangat, Malaysia		7.1	45			1,100,225		*Chlorobi*	
Bakreshwar, India	Water	7.8–8	54–65	IlluminaMiSeq	16S rRNA	416,074–442,064	*Bacteria*	*Firmicutes*	(Chaudhuri et al. 2017)
Chiraleu, Romania	Microbial mat	7.86	40–53	IlluminaMiSeq	16S rRNA	498,539	*Bacteria*	*Proteobacteria* and *Cyanobacteria*	(Chiriac et al. 2017)
Ciocaia, Romania		7.75	20–65					*Proteobacteria* and *Cyanobacteria*	
MihaiBravu, Romania		7.81	53–65					*Proteobacteria*	
Ankwar, Eritrea	Water, microbial mat and wet sediment	7.22	49.5	IlluminaMiSeq	16S rRNA	890,752	*Bacteria*	*Proteobacteria*	(Ghilamicael et al. 2017)
Elegedi, Eritrea		7.01	100					*Proteobacteria*	
Garbanabra, Eritrea		7.05	51.3					*Firmicutes* (microbialmat), *Proteobacteria* (sediment and water)	
Gelti, Eritrea		7.18	52.6					*Proteobacteria*	
Maiwooi, Eritrea		7.54	51.4					*Proteobacteria*	
Porcelana, Chile	Microbial mat	7.1–6.8	48–66	Illumina Hi-seq	Whole metagenome sequencing	-	*Bacteria*	*Cyanobacteria* and *Chloroflexi*	(Guajardo-Leiva et al. 2018)

Table 5.1 Contd. ...

...Table 5.1 Contd.

Location	Type of sample/s	pH	Temp (°C)	Sequencer	Type of sequencing	Total reads	Dominant domain	Dominant phylum	References
Leh (Nubravalley, Ladakh), India	Soil and water	8–8.5	60–80	GS-FLX-Titanium	16S rRNA	5,551	Bacteria	Bacteroidetes	(Gupta, et al. 2017a)
Ma'in, Jordania	Water	7.44–7.76	48–59	IlluminaMiSeq	16S rRNA	31,431	Bacteria	Proteobacteria	(Hussein et al. 2017)
Afra, Jordania		8.41	38					Proteobacteria	
Norris (YNP), USA	Sediment and biofilm	4.34	84	454 GS FLX Titanium	16S rRNA	6,432	Bacteria	Cyanobacteria	(Jiang and Takacs-Vesbach 2017)
Mary Bay Area (YNP), USA		4.32	80		16S rRNA			Cyanobacteria	
MudKettles (YNP), USA		4.35	72		16S rRNA			Aquificae	
Seven Mile Hole (YNP), USA		4.05	55		16S rRNA and shotgun metagenome sequencing	848,583		Chloroflexi	
Manikaran, Himachal Pradesh, India	Soil	8	50	ABI 3730xl	16S rRNA	-	Bacteria	Proteobacteria	(Kaur et al. 2018)
Tattapani, India	Soil and water	7.7–8.3	55–98	IlluminaHiseq 1000	Whole metagenome sequencing	49,095,827–89,832,690	Bacteria	Proteobacteria and Deinococcus-Thermus	(Kaushal et al. 2018)
Alla (Baikal Rift Zone), Russia	Microbial mat	9.7	65	454 GS-FLX-Titanium	16S rRNA	18,219	Bacteria	Deinococcus-Thermus	(Lavrentyeva et al. 2018)
Seya(Baikal Rift Zone), Russia		9.9	50					Chloroflexi	
Tsenkher, (Baikal Rift Zone), Russia		9.8	40					Cyanobacteria	
SungaiKlah, Malasya	Water	8.6	68	IlluminaMiSeq	16S rRNA	510,983	Bacteria	Proteobacteria	(Lee et al. 2018)
Manikaran, Himachal Pradesh, India	Water	-	95	Illumina GAII	Whole metagenome sequencing	52,726,234–53,144,772	Bacteria	Proteobacteria and Bacteroidetes	(Mahato et al. 2019)
Uzon Caldera, Kamchatka, Russia	Water and sediment	3.5–5.6	60–91	GS-FLX-Titanium	16S rRNA	20,000–80,000	Archaea	Chrenarchaeota	(Mardanov et al. 2018)

Location	Sample type	pH	Temperature	Platform	Gene	Reads	Domain	Phylum	Reference
Polok (Sikkim), India	Water	7.5–8.5	75–77	IlluminaMiSeq	16S rRNA	398,782	Bacteria	Proteobacteria and Bacteroidetes	(Najar et al. 2018)
Borong (Sikkim), India		5.1–5.6	50–52			372,480		Proteobacteria and Bacteroidetes	
Wall site (Nakabusa), Japan	Microbial mat	8.5–8.9	72	IlluminaMiSeq	16S rRNA	33,049	Bacteria	Aquificae	(Nishihara et al. 2018)
Streamsite (Nakabusa), Japan			75			38,824		Aquificae	
Coamo, Puerto Rico	Water	8.22	47	IlluminaMiSeq	Whole metagenome sequencing	-	Bacteria	Proteobacteria	(Padilla-Del Valle et al. 2017)
Yumthang, India	Microbial mat	8	39–41	IlluminaMiSeq	16S rRNA	1,381,343	Bacteria	Proteobacteria and Actinobacteria	(Panda et al. 2017)
Garga (Baikal rift zone), Russia	Microbial mat	-	45–74	IlluminaMiSeq	16S rRNA	232,310	Bacteria	Cyanobacteria and Proteobacteria	(Rozanov et al. 2017)
Anhoni, India	Water	7.5–7.8	43.5–55	IlluminaNextSeq 500	16S rRNA	21,881,886	Bacteria	Proteobacteria	(Saxena et al. 2017)
Tattapani, India		7–7.8	61.5–98					Proteobacteria	
Site 1, MudVolcanoArea (YNP), USA	Filaments, microbial mat and stromatolite	6.28	48.4	IlluminaMiSeq	16S rRNA	334,832	Bacteria	Cyanobacteria	(Schuler et al. 2017)
Site 2, MudVolcanoArea (YNP), USA	Microbial mat	6.07	45.1					Cyanobacteria	
Site 3, MudVolcanoArea (YNP), USA	Filaments	3.87	29.4					Sphingobacteria	
Site 4, MudVolcanoArea (YNP), USA	Filaments	5.44	68.4					Aquificae	
Brandvlei, South Africa	Microbial mat	6.2	55	454 GS-FLX-Titanium	16S rRNA	2,214	Bacteria	Proteobacteria	(Selvarajan et al. 2018)
Calitzdorp, South Africa	Microbial mat	6.3	58			2,729		Bacteroidetes	
Soldhar (Uttarakhand), India	Sediment	-	95	Ion Torrent PGM	16S rRNA	-	Bacteria	Proteobacteria	(Sharma et al. 2017)

Table 5.1 Contd. ...

...Table 5.1 Contd.

Location	Type of sample/s	pH	Temp (°C)	Sequencer	Type of sequencing	Total reads	Dominant domain	Dominant phylum	References
GanziPrefecture, China	Water, microbial mat and wet sediment	6.32– 8.84	40.8– 95	IlluminaMiSeq	16S rRNA	899,830	Bacteria	*Proteobacteria, Aquificae, Cyanobacteria*	(Tang et al. 2018)
Costa Rica	Microbial mat	6–7.5	37–63	GS- FLX- Titanium	16S rRNA	-	Bacteria	*Cyanobacteria* and *Chloroflexi*	(Uribe-Lorio et al. 2019)
Bullicame (Viterbo), Italy	Water and microbial mat	6.5	54	IlluminaMiSeq	16S rRNA	-	Bacteria	*Proteobacteria* (water), *Firmicutes* (mat)	(Valeriani et al. 2018)
Okuoku-hachikurou, Japan	Water and biofilm	6.8	44.3	Illumina MiSeq	16S rRNA	141,125	Bacteria	*Proteobacteria*	(Ward et al. 2017)

mats were absent and typical hyper thermophilic organisms such as *Aquificae* and some thermophilic members of *Proteobacteria* became more abundant (Tang et al. 2018). Not only temperature but the type of sample seems to be another important factor affecting variability, as *Cyanobacteria* can be found in sediments and biofilms, even at temperatures higher than 70°C (Jiang and Takacs-Vesbach, 2017).

Some other studies also reported the influence of temperature (< 70°C) for the presence of *Cyanobacteria* and *Chloroflexi* in hot springs. This can be clearly seen in the data reported in Table 5.1, as the phylum *Cyanobacteria* and *ChlorOflexi* predominate in sites with less temperature like some hot springs in YNP in USA (Schuler et al. 2017; Yuan et al. 2017), the Garga, Seya and Tsehker hot springs in Russia (Rozanov et al. 2017; Lavrentyeva et al. 2018) Porcelana geothermal spring in Chile (Guajardo-Leiva et al. 2018) or several Costa Rican hot springs (Uribe-Lorío et al. 2019).

Biotechnological applications

An important attraction of geothermal springs is that they constitute a potential source of microorganisms producers of thermo-tolerant enzymes that can be likely used in industrial applications. Therefore, a high number of studies have focused on a determined enzymatic activity of a microbial community. Recent reviews have been written on the various biotechnological applications of these thermozymes (DeCastro et al. 2016; Gomes et al. 2016; Pandey et al. 2016; Cabrera and Blamey 2018; Wohlgemuth et al. 2018). Although the main approaches to obtain them include functional metagenomics (Table 5.2), sequence-based screenings (Table 5.3), and isolation from thermophilic microorganisms (Table 5.4), improvements can be made to all these strategies. For example (Jardine et al. 2018) identified enzymes of interest through conventional plate assays and liquid chromatography-tandem mass spectrometry, LC-MS/MS. To identify proteins by LC-MS/MS, bacterial isolates from hot springs in South Africa were cultured and the supernatant was processed through polyacrylamide gel electrophoresis, gel was stained with Coomassie, protein bands of interest were digested with trypsin and subsequent LC-MS/MS. LC-MS/MS conceded to obtain more information from one sample than conventional plate assays which might be, in the long run, more cost-effective.

Wohlgemuth et al. (2018) studied 15 hot spring metagenomes looking for novel hydrolases in a combined sequence and function-based screening. As result of this project, novel hydrolases were studied and characterized such as carboxylesterases, enol lactonases, quorum sensing lactonases, gluconolactonases, cellulases and two epoxide hydrolases found by sequence metagenomics in the metagenomes of a Russian and a Chinese hot spring (Ferrandi et al. 2018). Gupta et al. (2017b) analyzed the bacterial diversity of hot sulfur springs of Leh (Nubra valley, Ladakh, India) looking for laccases. Lee et al. (2018) explored the bacterial diversity of Malaysian Y-shaped Sungai Klah hot spring, located in a wooded area, bioprospecting for Glycoside Hydrolases (GHs) that could be potentially used as cellulases and hemicellulases. The study revealed the huge potential of this hot spring as a source of these kinds of thermozymes. Using a similar procedure, the 140 thermophilic bacteria isolated from Manikaran and Yumthang hot springs (India) were studied for phylogenetic profiling, growth properties at varying conditions and potential sources of extracellular thermostable hydrolytic enzymes such as protease, amylase, xylanase and cellulase (Sahay et al. 2017). Kaushal et al. (2018) focused on the carbohydrate-related thermozymes of four thermal water reservoirs (55 to 98°C) located in Tattapani geothermal field of Chhattisgarh, India, using a metagenomic approach. McKay et al. (2017) centered their studies on the archaeal diversity of the sediment of Washburn and Heart Lake methanogenic hot springs (YNP, USA) in order to examine the presence and expression of novel methyl-coenzyme M reductase gene (mcrA) variants. Chuzel et al. (2018) used functional metagenomics screening of Dixie Valley hot spring mats (Nevada, USA) to seek enzymes capable of hydrolysis of sialic acids, sialidases. Other examples of different types of thermozymes obtained using classical approaches published since 2017 are listed in Tables 5.2, 5.3 and 5.4.

Table 5.2 Examples of thermozymes from hot springs obtained through functional metagenomics.

Family	Activity	Location	Type of sample	Vector	Host	Substrate	Total no. of clones	Positive clones	References
Glycoside hydrolases	Exosialidase	Dixie Valley, Nevada, USA	Microbial mat	pSMART FOS	*E. coli*	X-Neu5AC and 4MU-α-Neu5AC	616	1	(Chuzel et al. 2018)
	Endoglucanase	Puga, Ladakh, India	Sediment	pUC19	*E. coli* DH10B	Congo Red (CMC)	60,000	1	(Gupta et al. 2017a)
Lipases	Esterase	Khir Ganga, India	Microbialmat	pUC19	*E. coli* DH10B	Tributyrin	168,000	5	(Ranjan et al. 2018)
	Lipase	Taptapani, Odisha, India	Sediment	pUC19	*E. coli* DH5α	Olive oil and Rhodamine B	13,298	7	(Sahoo et al. 2017)
	Lipase/esterase	EryuanNiujie, China	Water	pCC1BAC	*E. coli* EPI300T1	Tributyrin	68,352	10	(Yan et al. 2017)

Table 5.3 Examples of thermozymes from hot springs obtained through sequence metagenomics.

Type of enzyme	Name	Optimum temperature (°C)	Optimum pH	Reference
Amine transferase	B3-TA	90	9	(Ferrandi et al. 2017)
	It6-TA	50	10	
	Is3-TA	50	9	
Epoxidehydrolase	Sibe-EH	30	-	(Ferrandi et al. 2018)
	CH65-EH	50	-	
β-xylosidase	AR19M-311-2	90	5	(Sato et al. 2017)
Nicotinamidase	-	90	9.5–10	(Zapata-Pérez et al. 2017)

Table 5.4 Examples of thermozymes from hot springs obtained by microbial isolation.

Type of enzyme	Strain	Hot spring	Reference
Catalase	*Bacillus* sp. strain KP10	TattaPani, Azad Kashmir	(Erum et al. 2017)
Protease	*Brevibacillus* sp. OA30	Ouled Ali, Algeria	(Gomri et al. 2018a)
Phytase	*Bacillus licheniformis, Bacillus coagulans Bacillus stearothermophillus*	Sulili, South Sulawesi	(Ibnu Irwan et al. 2017)
Amylase	*Bacillus subtilis* RK6 (KX247637)	Munger (Bihar, India)	(Kiran et al. 2018)
Lipase	*Bacillus* sp. HT19	Kalianda Island, Indonesia	(Li and Liu 2017)
Cellulase	*Bacillus licheniformis* NCIM 5556	Maharashtra, India	(Shajahan et al. 2017)
Amylase	*Geobacillus bacterium* (K1C)	Manikaran, India	(Sudan et al. 2018)
Amylase	*Bacillus*	Tangshan and Laoshan (China)	(Wu et al. 2018)

In addition to thermozymes, hot springs microbiomes may be sources of other interesting biotechnological products, such as antibiotics or bioactive compounds (Mahajan and Balachandran 2017). Antibiotic production of isolated strains from Ma'in hot springs (Jordan) water and antimicrobial activity of the water were tested by culture methods (Shakhatreh et al. 2017). Although in this case the bacterial isolates did not show antimicrobial activity, Ma'in hot springs water was found active against some gram-positive bacteria and this property was attributed to the chemical composition of the water. However, previously, different species of hot springs microorganisms with antibacterial activity and compounds with anti-tuberculosis activity have been isolated (Mahajan and Balachandran 2017). Moreover, compounds with potential anticancer activity were also isolated from the deep-sea hydrothermal vent mussel *Bathymodiolus thermophiles* (Adrianasolo et al. 2011). Thermophilic cyanobacterium *Leptolyngbya* sp. and microalgae *Graesiella* sp. from hot springs in the Korbous region of northern Tunisia were studied and demonstrated to possess abundant natural antioxidant products with potential therapeutic effects (Trabelsi et al. 2016a,b).

Not only are the compounds from thermophilic microorganisms are of potential interest but the microorganisms themselves may also have biotechnological applications of advantage. Ratnawati et al. (2015) selected thermophilic microbes resistant to ethanol from Ciater hot-springs (Subang District, West Java, Indonesia). Jardine et al. (2018) through traditional culture methods found a thermophilic strain of *Anoxybacillusrupiensis* in hot springs in South Africa which degrades starch, protein and phenol and produces several important enzymes for water bioremediation. Representative isolates from Unkeshwar hot springs, India, were found to tolerate heavy metals (chromium and arsenic) with potential environmental applications (Mehetre et al. 2018) in addition to being sources of important enzymes (cellulase, xylanase, amylase and protease).

Concluding remarks

Hot springs microbiomes are, till date and in spite of the abundant research performed, a mostly untapped source of new microbial species and metabolites with biotechnological potential. In the last few years, metagenomics techniques have been increasingly used to study biodiversity, both from a taxonomic and functional perspective, complementing but not completely replacing culture-based techniques. A new promising approach called mini-metagenomics allows dissecting the microbiomes with single cell resolution and covering low abundant species. Using these methods, the deciphering of metabolic networks in complex populations is advancing. We now regard communities as the result of a dynamic process where multiple physicochemical parameters also affect their composition and species abundance, highlighting the importance of recording significant metadata, which would in turn improve our understanding of the processes underlying changes in the microbiome and allow in drawing valid comparisons between the various conditions considered. Lastly a substantial number of heat-tolerant enzymes have been discovered and recovered from hot springs inhabitants, which are valuable for industrial needs, and that inevitably will be increased in the near future owing to the development of these novel approaches to assess microbiomes.

Acknowledgements

General funding to the EXPRELA Group by the Xunta de Galicia (Consolidación Grupos Referencia Competitiva contract number ED431C2016-012), co-financed by FEDER (EEC).

References

Adrianasolo, E., L. Haramaty, K. McPhail, E. White, C. Vetriani, P. Falkowski et al. 2011. Bathymodiolamides A and B, ceramide derivatives from a deep-sea hydrothermal vent invertebrate mussel, Bathymodiolus thermophiles. J. Nat. Prod. 74: 842–846.

Alcamán-Arias, M.E., C. Pedrós-Alió, J. Tamames, C. Fernández, D. Pérez-Pantoja, M. Vásquez et al. 2018. Diurnal changes in active carbon and nitrogen pathways along the temperature gradient in porcelana hot spring. Microbial Mat. Front. Microbiol. 9: 2353.

Alcorta, J., S. Espinoza, T. Viver, M.E. Alcamán-Arias, N. Trefault, R. Rosselló-Móra et al. 2018. Temperature modulates Fischerella thermalis ecotypes in Porcelana hot spring. Syst. Appl. Microbiol. 41: 531–543.

Amin, A., I. Ahmed, N. Salam, B.Y. Kim, D. Singh, X.Y. Zhi et al. 2017. Diversity and distribution of thermophilic bacteria in hot springs of Pakistan. Microbial Ecol. 74: 116–127.

Arce-Rodríguez, A., F. Puente-Sánchez, R. Avendaño, M. Martínez-Cruz, J.M. de Moor, D.H. Pieper et al. 2019. Thermoplasmatales and sulfur-oxidizing bacteria dominate the microbial community at the surface water of a CO_2-rich hydrothermal spring located in Tenorio Volcano National Park, Costa Rica. Extremophiles 23: 177–187.

Badhai, J., T.S. Ghosh and S.K. Das. 2015. Taxonomic and functional characteristics of microbial communities and their correlation with physicochemical properties of four geothermal springs in Odisha, India. Front. Microbiol. 6: 1166.

Bohorquez, L.C., L. Delgado-Serrano, G. López, C. Osorio-Forero, V. Klepac-Ceraj, R. Kolter et al. 2012. In-depth characterization via complementing culture-independent approaches of the microbial community in an acidic hot spring of the Colombian Andes. Microbial Ecol. 63: 103–115.

Bowen De León, K., R. Gerlach, B.M. Peyton and M.W. Fields. 2013. Archaeal and bacterial communities in three alkaline hot springs in Heart Lake Geyser Basin, Yellowstone National Park. Front. Microbiol. 4: 1–10.

Cabrera, M.Á and J.M. Blamey. 2018. Biotechnological applications of archaeal enzymes from extreme environments. Biol. Res. 1–15.

Chan, C.S., K.G. Chan, R. Ee, K.W. Hong, M.S. Urbieta, E.R. Donati et al. 2017. Effects of physiochemical factors on prokaryotic biodiversity in Malaysian circumneutral hot springs. Front. Microbiol. 8: 1252.

Chaudhuri, B., T. Chowdhury and B. Chattopadhyay. 2017. Comparative analysis of microbial diversity in two hot springs of Bakreshwar, West Bengal, India. Genomics Data 12: 122–129.

Chernyh, N.A., A.V. Mardanov, V.M. Gumerov, M.L. Miroshnichenko, A.V. Lebedinsky, A.Y. Merkel et al. 2015. Microbial life in Bourlyashchy, the hottest thermal pool of Uzon Caldera, Kamchatka. Extremophiles 19: 1157–1171.

Chiriac, C.M., E. Szekeres, K. Rudi, A. Baricz, A. Hegedus and N. Dragoş et al. 2017. Differences in temperature and water chemistry shape distinct diversity patterns in thermophilic microbial communities. Appl. Environ. Microbiol. 83: 1–20.

Chuzel, L., M.B. Ganatra, E. Rapp, B. Henrissat and C.H. Taron. 2018. Functional metagenomics identifies an exosialidase with an inverting catalytic mechanism that defines a new glycoside hydrolase family (GH156). J. Biol. Chem. 293: 18138–18150.

Cirés, S., M. Casero and A. Quesada. 2017. Toxicity at the edge of life: a review on cyanobacterial toxins from extreme environments. Mar. Drugs 15: 233.

Cole, J.K., J.P. Peacock, J.A. Dodsworth, A.J. Williams, D.B. Thompson, H. Dong et al. 2013. Sediment microbial communities in Great Boiling Spring are controlled by temperature and distinct from water communities. ISME J. 7: 718–729.

DeCastro, M.E., E. Rodríguez-Belmonte and M.I. González-Siso. 2016. Metagenomics of thermophiles with a focus on discovery of novel thermozymes. Front. Microbiol. 7: 1–21.

DeCastro, M.E., J.J. Escuder-Rodriguez, M.E. Cerdan, M. Becerra, E. Rodriguez-Belmonte and M.I. Gonzalez-Siso. 2018. Heat-loving β-galactosidases from cultured and uncultured microorganisms. Curr. Protein Pept. Sci. 19: 1224–1234.

Douglas, G.M., R.G. Beiko and M.G.I. Langille. 2018. Predicting the functional potential of the microbiome from marker genes using PICRUSt. Method Mol. Biol. 1849: 169–177.

Ekkers, D.M., M.S. Cretoiu, A.M. Kielak and J.D. van Elsas. 2012. The great screen anomaly—a new frontier in product discovery through functional metagenomics. Appl. Microbiol. Biotechnol. 93: 1005–1020.

Erum, N., Z. Mushtaq and A. Jamil. 2017. Isolation and identification of a catalase producing thermoduric alkalotolerant *Bacillus* sp. strain KP10 from hot springs of Tatta Pani, Azad Kashmir. J. Anim. Plant. Sci. 27: 2056–2062.

Ferrandi, E.E., A. Previdi, I. Bassanini, S. Riva, X. Peng and D. Monti. 2017. Novel thermostable amine transferases from hot spring metagenomes. Appl. Microbiol. Biotechnol. 101: 4963–4979.

Ferrandi, E.E., C. Sayer, S.A. De Rose, E. Guazzelli, C. Marchesi, V. Saneei et al. 2018. New thermophilic α/β class epoxide hydrolases found in metagenomes from hot environments. Front. Bioeng. Biotechnol. 6: 144.

Fortney, N.W., S. He, B.J. Converse, E.S. Boyd and E.E. Roden. 2018. Investigating the composition and metabolic potential of microbial communities in chocolate pots hot springs. Front. Microbiol. 9: 2075.

Ghilamicael, A.M., N.L.M. Budambula, S.E. Anami, T. Mehari and H.I. Boga. 2017. Evaluation of prokaryotic diversity of five hot springs in Eritrea. BMC Microbiol. 17: 203.

Gomes, E., A.R. de Souza, G.L. Orjuela, R. Da Silva, T.B. de Oliveira and A. Rodrigues. 2016. Applications and benefits of thermophilic microorganisms and their enzymes for industrial biotechnology. pp. 459–492. *In*: Schmoll, M. and C. Dattenböck (eds.). Gene Expression Systems in Fungi: Advancements and Applications. Fungal Biology. Springer, Cham.

Gomri, M., A. Rico-Díaz, J.J. Escuder-Rodríguez, T. El Moulouk Khaldi, M.I. González-Siso and K. Kharroub. 2018a. Production and characterization of an extracellular acid protease from thermophilic *Brevibacillus* sp. OA30 isolated from an Algerian hot spring. Microorganisms 6: 31.

Gomri, M.A., T. El Moulouk Khaldi and K. Kharroub. 2018b. Analysis of the diversity of aerobic, thermophilic endospore-forming bacteria in two Algerian hot springs using cultural and non-cultural methods. Ann. Microbiol. 68: 915–929.

Grivalský, T., M. Bučková, A. Puškárová, L. Kraková and D. Pangallo. 2016. Water-related environments: a multistep procedure to assess the diversity and enzymatic properties of cultivable bacteria. World J. Microbiol. Biotechnol. 32: 42.

Guajardo-Leiva, S., C. Pedrós-Alió, O. Salgado, F. Pinto and B. Díez. 2018. Active crossfire between cyanobacteria and cyanophages in phototrophic mat communities within hot springs. Front. Microbiol. 9: 2039.

Gudbergsdóttir, S.R., P. Menzel, A. Krogh, M. Young and X. Peng. 2016. Novel viral genomes identified from six metagenomes reveal wide distribution of archaeal viruses and high viral diversity in terrestrial hot springs. Environ. Microbiol. 18: 863–874.

Gupta, P., A.K. Mishra and J. Vakhlu. 2017a. Cloning and characterization of thermo-alkalistable and surfactant stable endoglucanase from Puga hot spring metagenome of Ladakh (J&K). Int. J. Biol. Macromol. 103: 870–877.

Gupta, V., N. Gupta, N. Capalash and P. Sharma. 2017b. Bio-prospecting bacterial diversity of hot springs in Northern Himalayan region of India for Laccases. Indian J. Microbiol. 57: 285–291.

Hedlund, B.P., S.K. Murugapiran, T.W. Alba, A. Levy, J.A. Dodsworth, G.B. Goertz et al. 2015. Uncultivated thermophiles: current status and spotlight on 'Aigarchaeota'. Curr. Opi. Microbiol. 25: 136–145.

Hussein, E.I., J.H. Jacob, M.A.K. Shakhatreh, M.A. Abd Al-razaq, A.F. Juhmani and C.T. Cornelison. 2017. Exploring the microbial diversity in Jordanian hot springs by comparative metagenomic analysis. Microbiologyopen. 6(6): e00521.

Ibnu Irwan, I., L. Agustina, A. Natsir and A. Ahmad. 2017. Isolation and characterization of phytase-producing thermophilic bacteria from Sulili hot springs in South Sulawesi. Sci. Res. J. (SCIRJ) 5: 16.

Inskeep, W. 2013. The YNP metagenome project: environmental parameters responsible for microbial distribution in the Yellowstone geothermal ecosystem. Front. Microbiol. 4: 1–15.

Inskeep, W.P., Z.J. Jay, M.J. Herrgard, M.A. Kozubal, D.B. Rusch and S.G. Tringe. 2013. Phylogenetic and functional analysis of metagenome sequence from high-temperature archaeal habitats demonstrate linkages between metabolic potential and geochemistry. Front. Microbiol. 4: 95.

Jardine, J.L., S. Stoychev, V. Mavumengwana and E. Ubomba-Jaswa. 2018. Screening of potential bioremediation enzymes from hot spring bacteria using conventional plate assays and liquid chromatography–Tandem mass spectrometry (Lc-Ms/Ms). J. Environ. Manag. 223: 787–796.

Jiang, X. and C.D. Takacs-Vesbach. 2017. Microbial community analysis of pH 4 thermal springs in Yellowstone National Park. Extremophiles 21: 135–152.

Jiang, Z., P. Li, D. Jiang, X. Dai, R. Zhang, Y. Wang et al. 2016. Microbial community structure and arsenic biogeochemistry in an acid vapor-formed spring in tengchong geothermal area, China. PLOS ONE 11: e0146331.

Kambura, A.K., R.K. Mwirichia, R.W. Kasili, E.N. Karanja, H.M. Makonde and H.I. Boga. 2016. Bacteria and archaea diversity within the hot springs of Lake Magadi and Little Magadi in Kenya. BMC Microbiol. 16: 136.

Kanoksilapatham, W., P. Pasomsup, P. Keawram, A. Cuecas, M.C. Portillo and J.M. Gonzalez. 2016. *Fervidobacterium thailandense* sp. nov., an extremely thermophilic bacterium isolated from a hot spring. Int. J. Syst. Evol. Microbiol. 66: 5023–5027.

Kaur, R., C. Rajesh, R. Sharma, J.K. Boparai and P.K. Sharma. 2018. Metagenomic investigation of bacterial diversity of hot spring soil from Manikaran, Himachal Pradesh, India. Ecol. Gene. Genom. 6: 16–21.

Kaushal, G., J. Kumar, R.S. Sangwan and S.P. Singh. 2018. Metagenomic analysis of geothermal water reservoir sites exploring carbohydrate-related thermozymes. Intl. J. Biol. Macromol. 119: 882–895.

Kikani, B.A., A.K. Sharma and S.P. Singh. 2017. Metagenomic and culture-dependent analysis of the bacterial diversity of a hot spring reservoir as a function of the seasonal variation. Intl. J. Environ. Res. 11: 25–38.

Kiran, S., A. Singh, C. Prabha, S. Kumari and S. Kumari. 2018. Isolation and characterization of thermostable amylase producing bacteria from hot springs of Bihar, India. Intl. J. Pharma Med. Biol. Sci. 7: 28–34.

Knight, R., A. Vrbanac, B.C. Taylor, A. Aksenov, C. Callewaert, J. Debelius et al. 2018. Best practices for analysing microbiomes. Natur. Rev. Microbiol. 16: 410–422.

Kumar, M., A.N. Yadav, R. Tiwari, R. Prasanna and A.K. Saxena. 2014a. Deciphering the diversity of culturable thermotolerant bacteria from Manikaran hot springs. Ann. Microbiol. 64: 741–751.

Kumar, M., A.N. Yadav, R. Tiwari, R. Prasanna and A.K. Saxena. 2014b. Evaluating the diversity of culturable thermotolerant bacteria from four hot springs of India. J. Biodivers Biopros. Dev. 1: 1–9.

Lavrentyeva, E.V., A.A. Radnagurueva, D.D. Barkhutova, N.L. Belkova, S.V. Zaitseva, Z.B. Namsaraev et al. 2018. Bacterial diversity and functional activity of microbial communities in hot springs of the baikal rift zone. Microbiology 87: 272–281.

Lee, L.S., K.M. Goh, C.S. Chan, G.Y. Annie Tan, W. Yin, C.S. Chong et al. 2018. Microbial diversity of thermophiles with biomass deconstruction potential in a foliage-rich hot spring. Microbiologyopen. 7: e00615.

Li, J. and X. Liu. 2017. Identification and characterization of a novel thermophilic, organic solvent stable lipase of *Bacillus* from a hot spring. Lipids 52: 619–627.

Li, L. and Z. Ma. 2019. Global microbiome diversity scaling in hot springs with DAR (diversity-area relationship) profiles. Front. Microbiol. 10: 118.

Lin, K.H., B.Y. Liao, H.W. Chang, S.W. Huang, T.Y. Chang, C.Y. Yang et al. 2015. Metabolic characteristics of dominant microbes and key rare species from an acidic hot spring in Taiwan revealed by metagenomics. BMC Genom. 16: 1029.

Liu, K.H., X.W. Ding, N. Salam, B. Zhang, X.F. Tang, B. Deng et al. 2018. Unexpected fungal communities in the Rehai thermal springs of Tengchong influenced by abiotic factors. Extremophiles 22: 525–535.

Liu, L., N. Salam, J.Y. Jiao, H.C. Jiang, E.M. Zhou, Y.R. Yin et al. 2016. Diversity of culturable thermophilic actinobacteria in hot springs in Tengchong, China and studies of their biosynthetic gene profiles. Microbial. Ecol. 72: 150–162.

Liu, Z., C.G. Klatt, J.M. Wood, D.B. Rusch, M. Ludwig, N. Wittekindt et al. 2011. Metatranscriptomic analyses of chlorophototrophs of a hot-spring microbial mat. ISME J. 5: 1279–1290.

López-López, O., K. Knapik, M. Cerdán and M.I. González-Siso. 2015. Metagenomics of an alkaline hot spring in Galicia (Spain): Microbial diversity analysis and screening for novel lipolytic enzymes. Front. Microbiol. 6: 1291.

Ma, A., M. Sun, A. McDermaid, B. Liu and Q. Ma. 2019. MetaQUBIC: a computational pipeline for gene-level functional profiling of metagenome and metatranscriptome. Bioinformatics 1–4.

Mahajan, G.B. and L. Balachandran. 2017. Sources of antibiotics: Hot springs. Biochem. Pharmacol. 134: 35–41.

Mahato, N.K., A. Sharma, Y. Singh and R. Lal. 2019. Comparative metagenomic analyses of a high-altitude Himalayan geothermal spring revealed temperature-constrained habitat-specific microbial community and metabolic dynamics. Arch. Microbiol. 201: 377–388.

Mardanov, A.V., V.M. Gumerov, A.V. Beletsky and N.V. Ravin. 2018. Microbial diversity in acidic thermal pools in the Uzon Caldera, Kamchatka. Antonie van Leeuwenhoek 111: 35–43.

McDonald, D., Y. Vázquez-Baeza, D. Koslicki, J. McClelland, N. Reeve, Z. Xu et al. 2018. Striped UniFrac: enabling microbiome analysis at unprecedented scale. Natur. Method 15: 847–848.

McKay, L.J., R. Hatzenpichler, W.P. Inskeep and M.W. Fields. 2017. Occurrence and expression of novel methyl-coenzyme M reductase gene (mcrA) variants in hot spring sediments. Sci. Rep. 7: 7252.

McMurdie, P.J. 2018. Normalization of microbiome profiling data. Method Mol. Biol. 1849: 143–168.

Mehetre, G., M. Shah, S.G. Dastager and M.S. Dharne. 2018. Untapped bacterial diversity and metabolic potential within Unkeshwar hot springs, India. Arch. Microbiol. 200: 753–770.

Merkel, A.Y., N.V. Pimenov, I.I. Rusanov, A.I. Slobodkin, G.B. Slobodkina, I.Y. Tarnovetckii et al. 2017. Microbial diversity and autotrophic activity in Kamchatka hot springs. Extremophiles 21: 307–317.

Meyer-Dombard, D.R., E.L. Shock and J.P. Amend. 2012. Effects of trace element concentrations on culturing thermophiles. Extremophiles 16: 317–331.

Meyer-Dombard, D.R. and J.P. Amend. 2014. Geochemistry and microbial ecology in alkaline hot springs of Ambitle Island, Papua New Guinea. Extremophiles 18: 763–778.

Mirete, S., V. Morgante and J.E. González-Pastor. 2016. Functional metagenomics of extreme environments. Curr. Opin. Biotechnol. 38: 143–149.

Mohanrao, M.M., D.P. Singh, K. Kanika, E. Goyal and A.K. Singh. 2016. Deciphering the microbial diversity of Tattapani hot water spring using metagenomic approach. Intl. J. Agric. Sci. Res. 6: 371–382.

Mukherjee, S., D. Stamatis, J. Bertsch, G. Ovchinnikova, H.Y. Katta, A. Mojica et al. 2019. Genomes OnLine database (GOLD) v.7: updates and new features. Nucleic Acids Res. 47: D649–D659.

Munson-McGee, J., J. Snyder and M. Young. 2018. Archaeal viruses from high-temperature environments. Genes 9: 128.

Najar, I.N., M.T. Sherpa, Sayak Das, Saurav Das and N. Thakur. 2018. Microbial ecology of two hot springs of Sikkim: Predominate population and geochemistry. Sci. Total Environ. 637: 730–745.

Nasko, D.J., S. Koren, A.M. Phillippy and T.J. Treangen. 2018. Ref Seq database growth influences the accuracy of k-mer-based lowest common ancestor species identification. Genom. Biol. 19: 165.

Nishida, A., V. Thiel, M. Nakagawa, S. Ayukawa and M. Yamamura. 2018. Effect of light wavelength on hot spring microbial mat biodiversity. PLOS ONE 13: e0191650.

Nishihara, A., S. Haruta, S.E. McGlynn, V. Thiel and K. Matsuura. 2018. Nitrogen fixation in thermophilic chemosynthetic microbial communities depending on hydrogen, sulfate, and carbon dioxide. Microb. Environ. 33: 10–18.

Oliverio, A.M., J.F. Power, A. Washburne, S.C. Cary, M.B. Stott and N. Fierer. 2018. The ecology and diversity of microbial eukaryotes in geothermal springs. ISME J. 12: 1918–1928.

Omae, K., Y. Fukuyama, H. Yasuda, K. Mise, T. Yoshida and Y. Sako. 2019. Diversity and distribution of thermophilic hydrogenogenic carboxydotrophs revealed by microbial community analysis in sediments from multiple hydrothermal environments in Japan. Arch. Microbiol. 1–14.

Orakov, A.N., N.K. Sakenova, A. Sorokin and I.I. Goryanin. 2018. ASAR: Visual analysis of metagenomes in R. Bioinformatics 34: 1404–1405.

Padilla-Del Valle, R., L.R. Morales-Vale and C. Ríos-Velázquez. 2017. Unraveling the microbial and functional diversity of Coamo thermal spring in Puerto Rico using metagenomic library generation and shotgun sequencing. Genomics Data 11: 98–101.

Panda, A.K., S.S. Bisht, B.R. Kaushal, S. De Mandal, N.S. Kumar and B.C. Basistha. 2017. Bacterial diversity analysis of Yumthang hot spring, North Sikkim, India by Illumina sequencing. Big Data Anal. 2: 7.

Pandey, R.K., A. Barh, D. Chandra, S. Chandra, V. Pandey Pankaj and L. Tewari. 2016. Biotechnological applications of hyperthermophilic enzymes. Intl. J. Curr. Res. Acad. Rev. 5(3): 39–47.

Parmar, K., N. Dafale, R. Pal, H. Tikariha and H. Purohit. 2018. An insight into phage diversity at environmental habitats using comparative metagenomics approach. Curr. Microbiol. 75: 132–141.

Patel, K.S., J.H. Naik, S. Chaudhari and N. Amaresan. 2017. Characterization of culturable bacteria isolated from hot springs for plant growth promoting traits and effect on tomato (*Lycopersicon esculentum*) seedling. C. R. Biol. 340: 244–249.

Paul, S., Y. Cortez, N. Vera, G.K. Villena and M. Gutiérrez-Correa. 2016. Metagenomic analysis of microbial community of an Amazonian geothermal spring in Peru. Genom. Data 9: 63–66.

Perz, A.I., C.B. Giles, C.A. Brown, H. Porter, X. Roopnarinesingh and J.D. Wren. 2019. MNEMONIC: Metagenomic experiment mining to create an OTU network of inhabitant correlations. BMC Bioinformatics 20: 96.

Poddar, A. and S.K. Das. 2018. Microbiological studies of hot springs in India: a review. Arch. Microbiol. 200: 1–18.

Power, J.F., C.R. Carere, C.K. Lee, G.L.J. Wakerley, D.W. Evans, M. Button et al. 2018a. Microbial biogeography of 925 geothermal springs in New Zealand. Natur. Communicat. 9: 2876.

Power, J.F., C.R. Carere, C.K. Lee, G.L.J. Wakerley, D.W. Evans, M. Button et al. 2018b. Microbial biogeography of 1,000 geothermal springs in New Zealand. bioRxiv 247759.

Prieto-Barajas, C.M., R. Alfaro-Cuevas, E. Valencia-Cantero and G. Santoyo. 2017. Effect of seasonality and physicochemical parameters on bacterial communities in two hot spring microbial mats from Araró, Mexico. Rev. Mex. Biodivers. 88: 616–624.

Ranjan, R., M.K. Yadav, G. Suneja and R. Sharma. 2018. Discovery of a diverse set of esterases from hot spring microbial mat and sea sediment metagenomes. Int. J. Biol. Macromol. 119: 572–581.

Ratnawati, L., T. Salim, W. Agustina and Sriharti. 2015. Isolation of thermophilic microbes resistant to ethanol from Ciater hot springs, Subang-West Java. Proced. Chemi. 16: 548–554.

Rinke, C., P. Schwientek, A. Sczyrba, N.N. Ivanova, I.J. Anderson, J.F. Cheng et al. 2013. Insights into the phylogeny and coding potential of microbial dark matter. Nature 499: 431–437.

Roumpeka, D.D., R.J. Wallace, F. Escalettes, I. Fotheringham and M. Watson. 2017. A review of bioinformatics tools for bio-prospecting from metagenomic sequence data. Front. Gen. 8: 1–10.

Rozanov, A.S., A.V. Bryanskaya, T.V. Ivanisenko, T.K. Malup and S.E. Peltek. 2017. Biodiversity of the microbial mat of the Garga hot spring. BMC Evolution. Biol. 17: 254.

Sahay, H., A.N. Yadav, A.K. Singh, S. Singh, R. Kaushik and A.K. Saxena. 2017. Hot springs of Indian Himalayas: potential sources of microbial diversity and thermostable hydrolytic enzymes. 3 Biotech 7: 118.

Sahoo, R.K., M. Kumar, L.B. Sukla and E. Subudhi. 2017. Bioprospecting hot spring metagenome: lipase for the production of biodiesel. Environ. Sci. Pollut. Res. 24: 3802–3809.

Salano, O.A., H.M. Makonde, R.W. Kasili, L.N. Wangai, M.P. Nawiri and H.I. Boga. 2017. Diversity and distribution of fungal communities within the hot springs of soda lakes in the Kenyan rift valley. Afri. J. Microbiol. Res. 11: 764–775.

Sato, M., M. Suda, J. Okuma, T. Kato, Y. Hirose, A. Nishimura et al. 2017. Isolation of highly thermostable β-xylosidases from a hot spring soil microbial community using a metagenomic approach. DNA Res. 24: 649–656.

Saxena, A.K., A.N. Yadav, M. Rajawat, R. Kaushik, R. Kumar, M. Kumar et al. 2016. Microbial diversity of extreme regions: An unseen heritage and wealth. Indian J. Plant Genet. Resour. 29: 246–248.

Saxena, R., D.B. Dhakan, P. Mittal, P. Waiker, A. Chowdhury, A. Ghatak et al. 2017. Metagenomic analysis of hot springs in central india reveals hydrocarbon degrading thermophiles and pathways essential for survival in extreme environments. Front. Microbiol. 7: 2123.

Schubotz, F., D.R. Meyer-Dombard, A.S. Bradley, H.F. Fredricks, K.U. Hinrichs, E.L. Shock et al. 2013. Spatial and temporal variability of biomarkers and microbial diversity reveal metabolic and community flexibility in streamer biofilm communities in the lower Geyser basin, Yellowstone National Park. Geobiology 11: 549–569.

Schuler, C.G., J.R. Havig and T.L. Hamilton. 2017. Hot spring microbial community composition, morphology, and carbon fixation: implications for interpreting the ancient rock record. Front. Earth Sci. 5: 97.

Selvarajan, R., T. Sibanda and M. Tekere. 2018. Thermophilic bacterial communities inhabiting the microbial mats of "indifferent" and chalybeate (iron-rich) thermal springs: Diversity and biotechnological analysis. Microbiologyopen. 7: e00560.

Shajahan, S., I.G. Moorthy, N. Sivakumar and G. Selvakumar. 2017. Statistical modeling and optimization of cellulase production by *Bacillus licheniformis* NCIM 5556 isolated from the hot spring, Maharashtra, India. J. King Saud Univ. Sci. 29: 302–310.

Shakhatreh, M.A.K., J.H. Jacob, E.I. Hussein, M.M. Masadeh, S.M. Obeidat and A.F. Juhmani. 2017. Microbiological analysis, antimicrobial activity, and heavy-metals content of Jordanian Ma'in hot-springs water. J. Infect. Public Heal. 10: 789–793.

Sharma, A., D. Paul, D. Dhotre, K. Jani, A. Pandey and Y.S. Shouche. 2017. Deep sequencing analysis of bacterial community structure of Soldhar hot spring, India. Microbiology 86: 136–142.

Sharma, A., M. Schmidt, B. Kiesel, N.K. Mahato, L. Cralle and Y. Singh. 2018. Bacterial and archaeal viruses of Himalayan hot springs at Manikaran modulate host genomes. Front. Microbiol. 9: 3095.

Sharp, C.E., A.L. Brady, G.H. Sharp, S.E. Grasby, M.B. Stott and P.F. Dunfield. 2014. Humboldt's spa: microbial diversity is controlled by temperature in geothermal environments. ISME J. 8: 1166–1174.

Singh, A. and E. Subudhi. 2016. Profiling of microbial community of Odisha hot spring based on metagenomic sequencing. Genom. data 7: 187–188.

Singh, R.N., S. Gaba, A.N. Yadav, P. Gaur, S. Gulati, R. Kaushik et al. 2016. First, High quality draft genome sequence of a plant growth promoting and cold active enzymes producing psychrotrophic *Arthrobacter agilis* strain L77. Stand Genomic Sci. 11: 54. doi:10.1186/s40793-016-0176-4.

Singh, Y., A. Gulati, D.P. Singh and J.I.S. Khattar. 2018. Cyanobacterial community structure in hot water springs of Indian North-Western Himalayas: A morphological, molecular and ecological approach. Algal Res. 29: 179–192.

Sofia Urbieta, M., E.G. Toril, M. Alejandra Giaveno, Á.A. Bazán and E.R. Donati. 2014. Archaeal and bacterial diversity in five different hydrothermal ponds in the Copahue region in Argentina. Syst. Appl. Microbiol. 37: 429–441.

Song, Z.Q., FP. Wang, X.Y. Zhi, J.Q. Chen, E.M. Zhou, F. Liang et al. 2013. Bacterial and archaeal diversities in Yunnan and Tibetan hot springs, China. Environ. Microbiol. 15: 1160–1175.

Staley, C. and M.J. Sadowsky. 2018. Practical considerations for sampling and data analysis in contemporary metagenomics-based environmental studies. J. Microbiol. Method 154: 14–18.

Strunecký, O., K. Kopejtka, F. Goecke, J. Tomasch, J. Lukavský, A. Neori et al. 2019. High diversity of thermophilic cyanobacteria in Rupite hot spring identified by microscopy, cultivation, single-cell PCR and amplicon sequencing. Extremophiles 23: 35–48.

Sudan, S.K., N. Kumar, I. Kaur and G. Sahni. 2018. Production, purification and characterization of raw starch hydrolyzing thermostable acidic α-amylase from hot springs, India. Intl. J. Biol. Macromol. 117: 831–839.

Tang, J., Y. Liang, D. Jiang, L. Li, Y. Luo, M.M.R. Shah et al. 2018. Temperature-controlled thermophilic bacterial communities in hot springs of western Sichuan, China. BMC Microbiol. 18: 134.

Trabelsi, L., A. Mnari, M.M. Abdel-Daim, S. Abid-Essafi and L. Aleya. 2016a. Therapeutic properties in Tunisian hot springs: First evidence of phenolic compounds in the cyanobacterium Leptolyngbya sp. biomass, capsular polysaccharides and releasing polysaccharides. BMC Complementary Altern. Med. 16: 1–10.

Trabelsi, L., O. Chaieb, A. Mnari, S. Abid-Essafi and L. Aley. 2016b. Partial characterization and antioxidant and antiproliferative activities of the aqueous extracellular polysaccharides from the thermophilic microalgae *Graesiella* sp. BMC Complementary Altern. Med. 16: 1–10.

Urbieta, M.S., E.R. Donati, K.G. Chan, S. Shahar, L.L. Sin and K.M. Goh. 2015. Thermophiles in the genomic era: Biodiversity, science, and applications. Biotechnol. Adv. 33: 633–647.

Uribe-Lorío, L., L. Brenes-Guillén, W. Hernández-Ascencio, R. Mora-Amador, G. González, C.J. Ramírez-Umaña et al. 2019. The influence of temperature and pH on bacterial community composition of microbial mats in hot springs from Costa Rica. Microbiologyopen. 8: e893.

Valeriani, F., S. Crognale, C. Protano, G. Gianfranceschi, M. Orsini, M. Vitali et al. 2018. Metagenomic analysis of bacterial community in a travertine depositing hot spring. New Microbiol. 41: 126–135.

Vartoukian, S.R., R.M. Palmer and W.G. Wade. 2010. Strategies for culture of 'unculturable' bacteria. FEMS Microbiol. Lett. 309: 1–7.

Wagner, J., F. Chelaru, J. Kancherla, J.N. Paulson, A. Zhang, V. Felix et al. 2018. Metaviz: interactive statistical and visual analysis of metagenomic data. Nucleic Acids Res. 46: 2777–2787.

Wang, H., Y. Yu, T. Liu, Y. Pan, S. Yan, Y. Wang et al. 2015. Diversity of putative archaeal RNA viruses in metagenomic datasets of a yellowstone acidic hot spring. Springer Plus 4: 189.

Ward, L.M., A. Idei, S. Terajima, T. Kakegawa, W.W. Fischer and S.E. McGlynn. 2017. Microbial diversity and iron oxidation at Okuoku-hachikurou Onsen, a Japanese hot spring analog of Precambrian iron formations. Geobiology 15: 817–835.

Wemheuer, B., R. Taube, P. Akyol, F. Wemheuer and R. Daniel. 2013. Microbial diversity and biochemical potential encoded by thermal spring metagenomes derived from the Kamchatka Peninsula. Archaea 2013: 1–13.

Wohlgemuth, R., J. Littlechild, D. Monti, K. Schnorr, T. van Rossum, B. Siebers et al. 2018. Discovering novel hydrolases from hot environments. Biotechnol. Adv. 36: 2077–2100.

Wu, X., Y. Wang, B. Tong, X. Chen and J. Chen. 2018. Purification and biochemical characterization of a thermostable and acid-stable alpha-amylase from *Bacillus licheniformis* B4-423. Intl. J. Biol. Macromol. 109: 329–337.

Yadav, A.N., P. Verma, M. Kumar, K.K. Pal, R. Dey, A. Gupta et al. 2015. Diversity and phylogenetic profiling of niche-specific Bacilli from extreme environments of India. Ann. Microbiol. 65: 611–629.

Yan, W., F. Li, L. Wang, Y. Zhu, Z. Dong and L. Bai. 2017. Discovery and characterizaton of a novel lipase with transesterification activity from hot spring metagenomic library. Biotechnol. Rep. 14: 27–33.

Yasir, M., A.K. Qureshi, I. Khan, F. Bibi, M. Rehan and S.B. Khan. 2019. Culturomics-based taxonomic diversity of bacterial communities in the hot springs of Saudi Arabia. OMICS: A J. Int. Biol. 23: 17–27.

Yu, F.B., P.C. Blainey, F. Schulz, T. Woyke, M.A. Horowitz and S.R. Quake. 2017. Microfluidic-based mini-metagenomics enables discovery of novel microbial lineages from complex environmental samples. eLife 6: e26580.

Yuan, C.G., X. Chen, Z. Jiang, W. Chen, L. Liu and W.D. Xian. 2017. *Altererythrobacter lauratis* sp. nov. and *Altererythrobacter palmitatis* sp. nov., isolated from a Tibetan hot spring. Antonie van Leeuwenhoek 110: 1077–1086.

Zablocki, O., L.J. Van Zyl, B. Kirby and M. Trindade. 2017. Diversity of dsDNA viruses in a south african hot spring assessed by metagenomics and microscopy. Viruses 9: 348.

Zablocki, O., L. van Zyl and M. Trindade. 2018. Biogeography and taxonomic overview of terrestrial hot spring thermophilic phages. Extremophiles 22: 827–837.

Zapata-Pérez, R., A.B. Martínez-Moñino, A.G. García-Saura, J. Cabanes, H. Takami and Á. Sánchez-Ferrer. 2017. Biochemical characterization of a new nicotinamidase from an unclassified bacterium thriving in a geothermal water stream microbial mat community. PLOS ONE 12: e0181561.

Zhang, Y., G. Wu, H. Jiang, J. Yang, W. She and I. Khan. 2018. Abundant and rare microbial biospheres respond differently to environmental and spatial factors in tibetan hot springs. Front. Microbiol. 9: 2096.

Chapter 6

Molecular Biology of Thermophilic and Psychrophilic Archaea

Chaitali Ghosh[1] and *Jitendra Singh Rathore*[2],*

Introduction

Till the 19th century there were only two domain systems which included bacteria and eukarya. Later, due to the differences at the molecular level, a third domain was successfully proposed, called 'archaea' (Woese and Fox 1977). The archaeal domain is distinguished from bacteria and eukarya, because the primary difference occurs in the membrane architecture, as well as small subunit rRNA sequences (Woese et al. 1990). The taxonomy of microorganisms was established by comparing nucleotide sequences of ribosomal RNA. It has been observed that bacteria are different from eukarya and archaea, because there is a difference in the rRNA nucleotide sequence between the position 500 and 545. Whereas eukarya have a difference in the nucleotide sequence between the region 585 and 655, as compared to both bacteria and archaea (Woese et al. 1990). Finally, archaea are distinguished by the unique structure observed in regions between positions 180 and 197 or 405 and 498 (Woese et al. 1990).

Even though the difference in the molecular composition of all the three domains, i.e., prokarya, eukarya and archaea is extreme, but still there is evidence of phenotypic resemblance of archaea with the other two, which was recognized only of late. As stated earlier, archaea has several characteristics which match with eukarya and bacteria such as, both archaea and eukarya have similar DNA binding proteins like homologues of glycosylation, gene organization, transcription, DNA replication, ribosomes and histones (White 2000). Unlike the initiator tRNA in bacteria which is a modified methioinine called formylmethionine, both archaea and eukarya have methionine. Moreover, eukaryotic RNA polymerase II and III sequences are comparable to archaeal sequences of RNA polymerase (Woese 1993). There are differences in the cell wall composition of single celled archaea and bacteria, both harbouring circular DNA and the cell wall. Additionally, archaeal and bacterial ribosomes both have the same sediment rate of 70 svedberg units.

Archaeal cell walls have pseudomurein (or pseudopeptidoglycan) and lack peptidoglycan which is universally present in all bacterial cell walls (White 2000). The membrane structure of archaea plays a significant role in their survival in hostile environments. Archaeal membranes are composed of isoprenoid alcohols and glycerol linked by ether bond, instead of ester linkage between fatty acid and glycerol as usually observed in eukaryotes and prokaryotes (White 2000). Due to the isoprenoid chain branching and resultant reduced tertiary carbon mobility, ether lipids are more

[1] Department of Zoology, Gargi College, University of Delhi, New Delhi, India.
[2] Gautam Buddha University, School of Biotechnology, Greater Noida, Yamuna Expressway, Uttar Pradesh, India.
* Corresponding author: jitendra@gbu.ac.in

stable than ester lipids, and therefore are less easily degraded, having high salt tolerance, and are able to withstand thermal and mechanical challenges (Konings et al. 2002; van de Vossenberg et al. 1998). Structural modifications such as, having cyclopentane rings that decrease membrane fluidity, also used within the archaeal membranes thrive at high temperatures. However, those that survive at low temperatures have higher numbers of double bonds in their lipids to increase membrane fluidity.

To elucidate the true mechanisms of thermophilic survival in extreme conditions, rational prediction on the basis of genomic data alone is insufficient. Therefore, other molecular approaches like transcriptomic sequencing (RNA Sequencing) have also been used. Based on the different transcriptional analyses it has been observed that the methanarchaeon *Methanococcus jannaschii*, has an optimal growth temperature of 85°C, but when they are grown at 95°C, there is upregulation of various genes. These include genes encoding prefoldin alpha subunit, small HSP, thermosome subunit, protease regulatory subunit, CRISPR-associated proteins, and a hypothetical protein (Boonyaratanakornkit et al. 2005). Several other transcriptome analyses on different archaea have been performed, which are summarized in detail elsewhere (Walther et al. 2011). Some experimental results identified that HSP60 and elongation factor Tu are also responsible for thermostability of *Thermotoga maritima* (Wang et al. 2012). As more temperature-related RNA sequence data will be available in future, it may be more beneficial for identification of specific signature sequences that are essential for the survival of thermophiles in extreme conditions.

Apart from the above mentioned methods, Horizontal Gene Transfer (HGT) is also an important process for thermophiles. For example the genome of *Thermomicrobium roseum* DSM 5159 have a circular chromosome (2.0 Mbp) and one megaplasmid (919,596 bp) (Wu et al. 2009). The gene encoding flagellar system is generally encoded by the genome. Interestingly in this bacterium, it is located in the megaplasmid and not in the bacterial chromosome. In another example, *Thermotoga maritima* are able to adapt to high temperatures because 24% of its genes are acquired from archaea (Nelson et al. 1999; Zhaxybayeva et al. 2009). Therefore, from the above findings it has been established that horizontal gene transfer work as a survival mechanism under extreme conditions (Goh et al. 2014).

To survive in extreme conditions, thermophiles have adopted various molecular strategies to stabilize their proteins. For example thermophilic proteins reduced their number of flexible regions in the native protein structure, and hence have shorter amino acid lengths than their homologoues, in non-thermophilic organisms. Comparative proteomic analysis of more than 204 bacteria and archaea, have shown that the amino acids Glu, Leu, Val, Tyr, Arg, Trp and Ile have been preferred in many optimum growth temperatures (Zeldovich et al. 2007). Specific patterns of codon usage, amino acid composition, nucleotide content and solutes used to stabilize cell component in thermophiles has also been studied (Singer and Hickey 2003; Borges et al. 2010; Empadinhas and da Costa 2011; Faria et al. 2008; Santos and Da Costa 2002).

Comparative genomic analysis by researchers revealed important ORFs which are essential for the survival of thermophiles in extreme conditions. These includes important ORFs encoding various proteins such as Heat Shock Proteins (HSPs); chaperones; chaperonins (required for folding of macromolecules); agmatine (a role in stabilization of DNA and RNA); spermidines (preserves membrane potentials); polyamines (functions as membrane stabilizers necessary for growth); α- and β-subunit prefoldins (required for protein folding), SOS regulons (stress condition as DNA damage responses); and various DNA repair systems. In genome thermostability, reverse gyrase (a heat-protective DNA chaperone) also plays a crucial role (Heine and Chandra 2009).

Although archaea were thought to inhabit only extreme environments, various adaptations are essential for their survival in thermophilic and psychrophilic conditions. This chapter explores two types of archaea, i.e., thermophilic, and psychrophilic archaea that have caught the interest of researchers in the recent years, due to their cellular and molecular adaptations and prospective applications. Well defined adaptations are cultivated by each type of archaea for their survival in respective environments at the DNA as well as protein level.

Cell membrane architecture of archaea

The cell membrane is a major barrier of the cell, which separates the vital cytoplasm from the surrounding environment. The major component of the cell membrane is phospholipids, which is involved in a wide variety of cellular processes. They work like a matrix to support integral membrane proteins having diverse functions. Other important functions governed by lipid are protein translocation, signal transduction, DNA replication and cell division, transport and many other cellular mechanisms (Dowhan 1997; Cronan 1978). Phospholipids consist of hydrophobic hydrocarbon tails as well as a polar head. Interaction between hydrophobic tails makes a bi-layer structure. In such an orientation, the hydrophobic region is oriented towards the inside of the membrane, whereas the polar head is linked by the glycerolphosphate backbone to the hydrophobic tail, facing the external aqueous face. The bilayer structure of the membrane is embedded with membrane-integral proteins and is semi permeable in nature. Therefore they allow an exchange of limited cellular constituents, including nutrients and ions (Raetz and Dowhan 1990). Composition of membrane lipid represents a taxonomic signature, used to differentiate the various kingdoms of life.

Archaeal phospholipids structure is chemically different from bacterial and eukaryotic lipids. Archaeal phospholipids are made up of highly methylated isoprenoid chains linked by an ether bond to a glycerophosphate backbone, Glycerol-1-Phosphate (G1P). On the other hand in bacteria and eukarya, straight fatty acids are linked by ester bonds to the enantiomeric form of the glycerophosphate backbone, glycerol-3-phosphate. The ether linked lipids in archaeal membrane is a key feature responsible for their survival in extreme environment (Gambacorta et al. 1995; Zhang and Rock 2008). For example hyperthermophile archaea, have an optimal growth temperature of above 85°C to 121°C having tetraether lipids in their membrane, with one or more cyclic structures in their isoprenoid side chains. Due to such a membrane architecture which provides stability to membranes, they are able to survive at high temperatures (Wu et al. 2013; Yang et al. 2007).

Structural diversity of archaeal phospholipids

Bacterial membrane lipids are composed of two Fatty Acid Residues (FARs) that are ester-linked to hydroxyls group of a glycerol derived from glycerol 3-phosphate (Raetz and Dowhan 1990) (Fig. 6.1). Generally, the hydrocarbon chain of fatty acid molecules has an even number of carbon atoms and one or more double bonds between carbon atoms, in a *cis* configuration. In the case of bacteria, phospholipid species are composed of 16 or 18 carbon fatty acids, with one unsaturation per single chain. On the other hand, archaeal membrane lipids have a different chemical structure. The hydrocarbon chain is composed of a repetition of five-carbon unit having a methyl group at every fourth carbon of a saturated isoprene unit which results in the formation of isoprenoid chains of different lengths (De Rosa and Gambacorta 1988; Kates 1992). These branched chains with a carbon number of 20 make ether bonds to an enantiomeric form of glycerol 1-phosphate (Fig. 6.2).

Apart from the general structure mentioned above there are lots of variations observed on archaeal lipids in nature, including composition and configuration as well as various modifications at the polar head groups, difference in phytanyl chain length, etc. In archaea lipids, two main lipid cores are reported. The first one is a C20 *sn*-2,3-diacylglycerol diether lipid which is termed as archaeol. The second is a C40 glyceroldialkylglycerol tetraether lipid lipid, which is commonly known as caldarchaeol. Archaeol based lipids in some thermophilic archaea have further modifications, including condensation in a macrocyclic glycerol diether, aberrant isoprenoid chain length (C25), or the presence of a tetritol diether (Comita et al. 1984; De Rosa et al. 1986a; De Rosa and Gambacorta 1988; De Rosa et al. 1986b). Such variations are present at even higher degrees in tetraether structures.

In tetraether lipids, two isoprenoid chains having a 40 carbon number are linked either to two identical or different polar groups. Therefore, due to a long chain of 40 carbon numbers, these lipids are reported to span the entire length of the membrane, and hence they form a single monolayer rather than a bilayer. The first tetraether reported is caldarchaeol, which has an antiparallel arrangement of

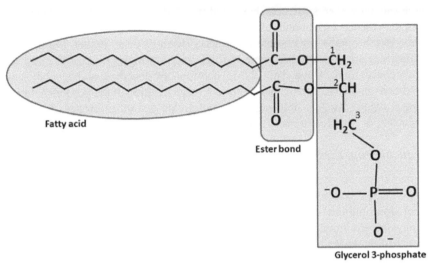

Figure 6.1 In bacteria fatty acid residues are ester-linked to hydroxyls group of a glycerol derived from glycerol 3-phosphate.

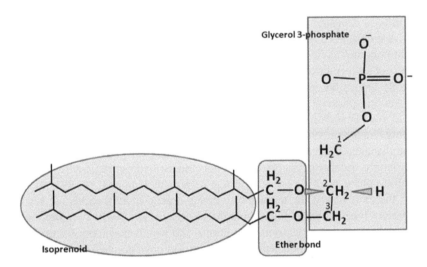

Figure 6.2 In arechaea isoprenoid are ether-linked to hydroxyls group of a glycerol derived from glycerol 3-phosphate.

two diether molecules connected via their isoprenoid chains (Langworthy 1977). Later, the parallel isomer of caldarchaeol was also reported from *Sulfolobus* and *Thermoplasma* species and they were called isocaldarchaeol (Koga and Morii 2005). A particular tetraether structure which was observed in hyper-thermophilic organisms and termed as the H-shaped caldarchaeol, have a covalent bond between two carbon atoms in the isoprenoid chain (Koga and Morii 2005).

In *Sulfolobales,* a specific structure named nonitol, have C40 diphytanyl chains linked to C6 polyol at both ends. When the attached polyol group is in an isomeric orientation, the nonitol structure is termed as calditol (Untersteller et al. 1999; DE Rosa et al. 1980). In addition to this, tetraethers of hyperthermophile archaea have an additional feature, which includes the presence of cyclopentane rings in the hydrophobic chain region. The number of rings varies with growth temperature, and therefore a maximum of 8 rings per molecule is observed at the highest growth temperatures of around 121°C. A new set of tetraether structures named crenarchaeol is found in some *Sulfolobus* species having cyclohexane rings in their lipids (De Rosa et al. 1986b; Sinninghe Damsté et al. 2002).

The archaeal tetraether structures have additional diversity of hydrocarbon chain lengths such as C30, C31 or C35 (Schouten et al. 2000) as well as the presence of unsaturated diether lipids in some psychrophilic (or cryophilic) archaea (Nichols et al. 2004). Moreover, polar head group modifications are also observed with multiple sugars which give rise to more diversity and variability to archaeal lipids (Koga et al. 1993). Therefore, the described archaeal lipids diversity can be considered as a unique taxonomic marker for our understanding of the phylogenetic relationships among archaeal organisms. It is believed that 'lipid divide' led to differences in the bacterial and archaeal organism, as depicted in many evolutionary theories (Caforio and Driessen 2017).

Physiochemical properties of archaeal lipids

Biological membranes are typically a bilayer structure of phospholipids which interact with each other via non-covalent bonds including Van der Waals bond as well as electrostatic interactions. To ensure proper lipid mobility and maintain constant fluidity, by adapting the lipid composition in response to physical environments such as temperature and pressure are the most important properties of membrane. Lipid bilayers have a complicated melting behaviour of undergoing a transition from a crystal or gel phase to a liquid state. At the optimum growth temperature, membranes are present in a liquid crystalline state. Such a state has a high degree of lipid movement which govern proper functioning of membrane proteins as well as maintenance of the barrier function (Melchior 1982). Membrane fluidity is dependent on the chemical structure of its lipids and by changing the lipid composition in response to environmental factors; the organisms can control membrane fluidity.

Bacterial membranes having 16 or 18 carbon Fatty Acid Residues (FARs), have a melting point in a narrow range of temperature (40° to 50°C), however, the introduction of one or more unsaturated bonds remarkably lowers the melting point to even below 0°C. Therefore, to keep the membrane in a liquid-crystalline phase at variable temperatures, organisms adopted different mechanisms, including the degree of saturation, change in the lipid acyl chain length, as well as the ratio of iso/anteiso branching (Gaughran 1947; Russell and Fukunaga 1990; Svobodová and Svoboda 1988). On the one hand, archaeal membranes do not prefer extreme changes in lipid composition, according to temperature variations because of the presence of the unique phytanyl chain and it ability to maintain a liquid-crystalline phase over a very wide range of growth temperatures (10° to 100°C) (Koga 2012). Only at the extremes of the temperature range, drastic modifications are seen, as outlined below. The different types of lipids found in archaeal organisms support two important membrane properties required for life, under extreme conditions, and which are thermal stability and low ion/proton permeability. Because of the chemical structure and physico-chemical characteristics of the lipid bilayer like fluidity, stability, surface charge and stability, archaeosomes are used to deliver vaccines and other skin permeation requirements (Jia et al. 2018).

Thermal stability

The archaeal membranes are stable due to the presence of the isoprenoid chain having branched methyl groups, tetraether as well as macrocyclic rings. All this together form an extensive network of hydrogen bonds between the polar head groups and is entirely responsible for a tight membrane packing (Gliozzi et al. 2002). Other factors responsible for stability of archaeal membranes at higher temperatures include the covalent linkage between the hydrocarbon chains of the tetraether lipids. Moreover, monolayer architecture of tetraether lipids gives more stability, due to intact hydrophobic core and tighter packing in archaeal membranes as compared to diether. The ether bond in archaeal lipids are more stable at high temperatures then ester bond of bacterial lipids, because the former is less prone to hydrolysis (Jacquemet et al. 2009). Further, in the case of phospholipase treatment, bacterial lipids are found to be completely degraded, where as ether lipids are resistant to such treatments.

Molecular Dynamics (MD) simulation studies were performed to determine the flexibility of tetraether lipid monolayer membrane in comparison to that of diether lipid bilayers. Here, when the membranes composed of diether, acyclic tetraether and macrocyclic tetraether lipids were compared, results supported the relevance of monolayer structure for membrane stability (Shinoda et al. 2005). The simulation studies showed that, the presence of cyclic rings which decrease mobility of the hydrophobic region, also increases membrane rigidity as well as resistance against external mechanical forces (Benvegnu et al. 2004). This is supported by the observation where hyper-thermophilic archaea showed the increment in the degree of cyclization as growth temperature increased (De Rosa and Gambacorta 1988).

The MD simulations showed that the presence cyclopentane rings to the diphytanyl chain maximum up to eight per chain, reduces the rotational freedom of the chain and hence is responsible for the maintenance of membrane fluidity and dynamics, at higher temperatures similar to archaeal-like and bacterial-like lipids having cyclopropane rings (Benvegnu et al. 2008; Gliozzi et al. 1983; Grogan and Cronan 1997). Membranes having tetraether lipids with cyclopentane rings are tightly packed and have stable lipid-lipid interaction (Gabriel and Lee Gau Chong 2000). Tight membrane packing due to the presence of cyclic rings brings the phosphate moieties of neighbouring polar head groups closer, which results in more favourable electrostatic interactions. MD simulation studies also showed that membrane stability is gained by hydrogen bonds formed by the interactions between the polar head groups. The ratio between tetraethers and diethers with and without macrocyclic structures is another mechanism by which archaea can maintain membrane fluidity and stability at different temperatures (Sprott et al. 1991).

In psychrophiles at lower temperatures, lipids will crystallize and the membranes may adopt a gel-like structure. This architectural change leads to a drastic decrease in membrane stability and on other hand it increases its permeability. Therefore, psychrophilic organisms increase the percentage of unsaturated isoprenoid chains in their lipids as the temperature decreases, which is considered as one of the mechanisms to maintain membrane fluidity at low temperatures (Russell and Nichols 1999; Oger and Cario 2013; Sakamoto and Murata 2002).

Ion permeability

The microorganism membrane works like a barrier and has selective permeability for a few ions and solutes. This property is essential for cellular homeostasis, including maintaining a constant intracellular pH and ion strength. Selective permeability is also responsible to maintain high electrochemical gradient of protons or sodium ions which is essential to drive energy-requiring processes. The composition of archaeal lipids, particularly at high temperatures also influences the membrane permeability. Due to the presence of bulky branched methyl groups in the phytanyl chains, they reduce the degree of movement of these lipids (Shinoda et al. 2004). As a result, membranes containing such lipids show a low permeability for small ions or solute molecules. These results were further supported by *in vitro* studies, where tetraether-based liposomes showed much lower ion-permeability as compared to diether or diester based liposomes.

Another study related to the permeability of liposome was performed, where liposomes were composed of total lipid, extracted from *Thermoplasma acidophilum* or *Sulfolobus acidocaldarius* (Mathai et al. 2001; Elferink et al. 1994; Komatsu and Chong 1998). Since, in both case lipids mostly contain tetraether, they showed very low permeability towards protons at high temperatures. Further, the presence of cyclopentane rings in lipids and the presence of two or more sugar moieties on a polar head group are responsible for low proton permeability (Chang 1994; Vossenberg et al. 1995; Dannenmuller et al. 2000; Shimada et al. 2008). Therefore, it has been observed that, there is a correlation between the effective proton permeability and membrane stability, where both factors are influenced by lipid packing. Thus, the chemical composition of archaeal lipids and their derived modifications are responsible for a proper balance between membrane stability and fluidity as well as for low proton permeability, even in extreme environments.

Applications of archaeal lipids

The properties, such as extreme high thermal stability as well as low permeability of liposomes derived from archaeal lipids have attracted the attention for biotechnological and biomedical applications. The advanced archaeal lipid-based liposomes formulation and their application in drug delivery is one of the promising applications of archaeal lipids. Ester-bonded phospholipid liposomes have been extensively used as carrier systems for deliver drugs (Ambrosch et al. 1997). These liposomes are quickly phagocytozed by macrophages and thus readily cleared from the blood stream. On other hand, archaeal-lipid based liposomes termed as archaeosomes, due to their intrinsic chemical properties and biocompatibility with human tissue, appear to be more suitable for drug delivery (Omri et al. 2003). They also have adjuvant like properties, and studies have shown that they are capable to evoke a stronger immune response against cancer cells or in infectious diseases, in comparison to the ester-based phospholipid liposomes (Patel and Sprott 1999; Krishnan and Sprott 2003). Such enhanced adjuvant activity of archaeosome is due to their higher stability as well as size, and therefore they are easily phagocytized by antigen-presenting cells.

Antigens delivered via such a mechanism evoke a full humoral response via improved antigen delivery to the antigen processing compartment (Krishnan and Sprott 2008; Sprott et al. 2003). Archaeosomes are extremely stable as they withstand several cycles of autoclaving at a wide range of pH without any content release (Brown et al. 2009). Because of such an outstanding property, archaeosomes are in demand for pharmaceutical applications. Moreover, as archaeosomes are also highly stable at very low pH, encapsulated drugs may be given orally and the given drug reaches its target after being absorbed into the human gastro-intestinal tract (Patel et al. 2000). Therefore, this makes archaeosomes more attractive for their biomedical applications when targets are difficult to reach and at the same time allow the drug to release at a slow pace (Barbeau et al. 2011). One major drawback of the high stability of the archaeosomes is that they release drugs at an extremely slow rate, and therefore for effective drug release further avenues need to be explored. One such possibility is the amalgamation of switchable channels that may assist in drug release under predetermined conditions.

Genomic adaptations in archaea

Genomic size

It has been observed that thermophiles have less intergenic DNA, which makes their genomes compact (Sabath et al. 2013). Usually, on an average, a gene is represented by every 1 kb of a bacterial genome, but this is not true in the case of thermophilic archaea, as their coding length is slightly shorter. On the basis of comparisons of 1553 prokaryotes, it was observed that cells that grow below 45°C have genomes sizes larger than 6 Mbp, however the average genome size for thermophiles is less than 4 Mbp (Sabath et al. 2013). Further, it has been reported that if the optimum growth temperature is higher, then the genome size is smaller than 2 Mbp, as in the case of Hyperthermus spp., having average optimum growth temperature of 100.5°C and their mean genome size of only 1.67 Mbp.

Nucleotide arrangement

The arrangement of nucleotide in genomes also provides thermostability, as higher frequencies of purinepurine and pyrimidine-pyrimidine composition was observed in hyperthermophiles which provide less flexibility to DNA and also help in supercoiling of DNA affecting thermostability of DNA. This kind of nucleotide arrangement is reported to rise with increasing optimal growth temperature above 60°C (Kawashima et al. 2000; Nakashima et al. 2003).

GC content

Due to the advances in the DNA sequencing technology, a total of 53 complete genomes of Crenarchaeota (19 genera with mean genome size of 1.75 Mbp) and 44 complete genomes of Euryarchaeota (14 genera with mean genome size of 1.75 Mbp) has been sequenced and submitted in the NCBI database. These genomes could be accessed from NCBI web page (http://www.ncbi.nlm.nih.gov/genome/browse/). It has been observed that life under extreme conditions is made possible by special genome features. The general assumption is that high GC content stabilizes the genome at a high temperature but statistics say that a high GC content does not really represent a good indicator of thermophilicity (Hickey and Singer 2004). In the case of *Aquifex aeolicus*, it has a high optimum growth temperature of 95°C but surprisingly has a low GC content of 43.4% (Deckert et al. 1998). Interestingly, many hyper-thermophilic archaea do not have high GC contents. There are various examples present in nature, such as mesophilic *E. coli* has a GC content of 50.7% whereas the *Methanotorris* spp. and *Caldisphaera* spp., have average GC contents of only 32.3 and 30%, respectively. Further research proved that thermophilicity was better symbolized by the ratio of GC content of rRNA and tRNA, as compared to the GC content of genomic DNA (Trivedi et al. 2005). Therefore, it was concluded that there is no simple relationship between the optimum growth temperature, thermophilicity and genomic features, and it may also be due to the combinations of other factors.

Methylation of nucleotides

There are a few reports which support that methylation of the nucleotides may also provide thermo protection in extremophiles. It has been observed that thermophiles generally have N4 methylcytosine (m4C) instead of N5 methylcytosine (m5C) and N6 methyladenine (m6A) (Ehrlich et al. 1985). This modification protects DNA from the mutation which occurs at higher temperatures. Methylation at a different position is also governed by different DNA methylase, which is important for thermal stability (Bujinicki and Radlinsks 1999; Gorgan 1998). A recent DNA methylome study in Crenarchaeon *Sulfolobus acidocaldarius* also showed two types of menthylation including N5 methylcytosine (m5C) and N6 methyladenine (m6A) having important roles in different biological pathways (Couturier and Lindås 2018).

Histone and non-histone proteins

It has been reported that transient binding/association of histone and non-histone proteins with DNA may provide a role in thermostability (Peak et al. 1995; Gorgan 1998; White 2003). Archaeal histones HMf and HTz are known to wrap DNA in positive supercoil at high salt concentrations, but at lower salt concentrations they induce negative supercoiling. On the other hand, MkaH induces negative supercoiling under variable salt concentrations. Alternate packing by such proteins may prove to be advantageous to archaea under variable salt concentrations (Sandman et al. 1994; Musgrave et al. 2000). This transient type of supercoiling may probably vary at the time of gene activity. Only essential genes are partially exposed for their activity and rest of the DNA are protected at high temperature. Therefore, irrespective of variable ion concentrations, histones or histone like proteins may play an essential role in folding and unfolding as well as thermal protection of DNA (Bailey et al. 2002; White 2003; Weng et al. 2004).

SSB protein

Besides histone proteins, Single Strand Binding proteins (SSB) have also been found to be responsible in DNA stabilization at higher temperatures. The roles of DNA stabilizing proteins have been extensively reviewed (White 2003). These reports support that histones or histone like proteins are also essential factors that provide protection to DNA at elevated temperatures, apart from the GC content of genome. The biochemical and biophysical studies demonstrated that the binding properties of *S. solfataricus* SSB are essentially identical for ssDNA and ssRNA. These features may represent an adaptation to a hyperthermophilic lifestyle, where DNA and RNA damage is a more frequent event (Morten et al. 2017).

DNA repair system

DNA in thermophiles are more prone to damages and such DNA damage may be caused at a higher rate in comparison to others. At high temperatures, hydrolytic deamination of adenosines and cytosines, hydrolytic depurination or oxidation of guanine, methylation of bases and strand break have been observed (Gorgan 2000; Yang et al. 2000). Therefore, repair systems should be taking care of these damages and several proteins are present in thermophiles to repair such damages (with some exceptions). Repair Associated Mysterious Proteins (RAMPs) are believed to be proteins that repair DNA damage found in only thermophiles (Makarova et al. 2002). However, it is not clear whether such unique repair proteins perform DNA repair only at elevated temperatures.

It is believed that the thermophiles have multiple activities in one enzyme and they make efficient use of its genetic material (Sandigursky and Franklin 1999; Belova et al. 2001). Thermophilic archaea has accretion of PIN-domain and more numbers of the exonuclease family. These accumulations are believed to be helpful in DNA and RNA damage repair, which occurs at elevated temperatures. Increase in PIN domains in thermophiles generally compensate and substitute lack of repairs systems in them. Recombination repair systems are also reported in *Pyrococcus furiosus* even at 95°C to protect DNA at high temperatures, as recA homologous to radA is responsible to repair double stranded breaks (Di Ruggiero 1997; White 2003; Gorgan 1998; 2004). It has been proposed that hyperthermophiles do not need efficient repair systems as the mutations which occurs due to replication or inefficient repair system may be beneficial to these organisms living under stressful conditions (Gorgan 1998). Therefore, the common underlying feature(s) believed to provide thermostability to DNA in thermophilic archaea and eubacteria thus appear to be of high GC content, difference in methylation of nucleotides, cations, histones/histone like proteins, SSB and flexibility of DNA affected by nucleotide arrangement. Recent advances in understanding of archaeal DNA repair processes such as base excision repair, nucleotide excision repair, mismatch repair and double-strand break repair is reviewed in detail elsewhere (White and Allers 2018).

Reveres gyrase

The presence of reverse gyrase is another important feature that is common in hyperthermophile archaea and eubacteria which provide relaxation to the positive supercoiling of DNA. Its key role is to initiate replication and gene regulation at higher temperatures (Napoli et al. 2001; Bao et al. 2002). A Smj12, a nonspecific DNA-binding protein has also been reported to have a key role in positive supercoiling of DNA (Napoli et al. 2001). But the presence of reverse gyrase is not always essential for all archaea as it has been observed that in the case *Picrophilus torridus* and *T. volcanium* reverse gyrase gene was absent (Futterer et al. 2004; Kawashima et al. 2000). Although *T. volcanium* has genes for gyrase and topoisomerase I, it lacks chaperones including DnaK, DnaJ and GrpE. The absence of the above proteins in *T. volcanium* is an exceptional case and may adopt alternate DNA folding strategies and therefore both *T. volcanium* and *Picrophilus torridus* has the least optimal growth rate of 60°C, among Crenarchaeota. Therefore, one cannot generally assume the role of reverse gyrase and chaperone proteins in thermal protection of DNA. However, deletion of the gene encoding the reverse gyrase of hyperthermophilic archaeon *Pyrococcus furiosus* having an optimal growth temperature of 100°C, revealed that the gene is essential for growth at 95°C, as well as at 100°C. The activity of reverse gyrase is absolutely necessary to maintain a correct DNA twist for any organism growing at such extreme temperatures (Lipscomb et al. 2017).

Cations and polycations

Hence if it is not the GC content (at least not in all cases), there should be other features for thermal protection of the genetic material. In order to maintain the double stranded structure and provide thermal stability irrespective of the nucleotide ratio particularly in archaea which grow at 60°C optimal growth temperature or above, have cations and polycations such as potassium ion in cytoplasm to neutralize the negatively charged nucleic acid (Gorgan 1998; Kawashima et al. 2000).

RNA stability

Adaptations at the transcriptional level are also essential in order to survive at high temperature conditions for thermophiles (Lobry and Chessel 2003; Singer and Hickey 2003). Generally thermophiles have purine rich genome as the purine load is preferred for efficient transcription of mRNA as well as to avoid double strand interaction. It has been proved in a study that thermophiles which have low GC content also prefer purine loading (Lao and Forsdyke 2000). These studies are further supported by findings which showed purine richness in genomes of thermophiles particularly in most of the mRNA, except for short mRNAs (Lobry and Chessel 2003; Paz et al. 2004). As purine richness is not confined to the coding region alone, therefore it could not be considered as a significant adaptive value (Fitz-Gibbon et al. 2002). Differing from this, purine rich RNA is preferred in thermophiles and in particular adenine over guanine in tRNA and rRNA besides m RNA. This preference may be in order to avoid undesirable bonds with other RNA and especially to avoid any competition with the tRNA anticodons. Purine richness could be helpful in high transcription rate of many mRNAs from pyrimidine rich template. Moreover they may be more stable to spontaneous hydrolysis at high temperature (Paz et al. 2004).

Apart from the nucleotide content of genome, another important factor is the optimal growth temperature, which governs the preference for codon usage. It has been observed that thermophiles have a different preference of codon usage for example arginine (AGR and CGN) and isoleucine (ATH), etc. (Lobry and Chessel 2003; Lynn et al. 2002; Singer and Hickey 2003). This could be beneficial for stable pre-mRNA transcript and mRNAs, as well as translation at elevated temperatures.

Thermal stability of tRNA and specifically aminoacyl tRNA is essential, which is otherwise unstable at high temperatures but not at lower temperatures (Stepanov and Nyborg 2002). There are several factors responsible for stabilizing RNA in thermophiles. These factors include increase in hydrogen bonds in helices, short length of RNA, additional Watson-Crick base pairs at the base of stem loop and shortened connections between helices, high CG content, minimizing alternative folding (Brown et al. 1993). Post transcriptional modification of nucleotide also plays a key role, for example 2 thiolation of T (54) and 5 methylcytidine is directly responsible for the thermostability of tRNA species (Edmonds et al. 1991; Yokoyama et al. 1987; Lobry and Chessel 2003). This type of post-transcriptional modification may be universally employed by thermophilic archaea and eubacteria. Apart from this, a derivative of 7-deazaguanosine, archaeosine is present exclusively in the D-loop of archaea t-RNA among other unique derivatives (Watanabe et al. 1997; Edmonds et al. 1991).

Additionally, the presence of Mg^+ ions and a high number of ribose methylated nucleotides also contribute to the stabilization of tRNA (Edmonds et al. 1991; Kowalak et al. 1994). Like tRNA, thermostability in rRNA and other non-coding RNA (ncRNA), may be achieved by increased hydrogen bonding which is contributed by double stranded duplex, post-transcriptional modifications and higher GC content (Bao et al. 2002; Klein et al. 2002; Nakashima et al. 2003; Lobry and Chessel 2003). As reported there are significant differences in the GC content in tRNA of cold-adapted archaea and thermophiles, therefore, it may be possible that additional factors provide thermal stability to tRNA at high temperature (Saunders et al. 2003). Recently, random mutagenesis studies in the hyperthermophile showed crucial roles of tRNA modifications in their cellular thermotolerance (Orita et al. 2019).

Proteomics adaption's in thermophilic archaea

There are certain exceptional areas on Earth which have extreme environments, and no living organisms can survive. Thermophiles and hyperthermophiles are a few organisms which can survive. Both of them have a different range for optimum growth in spite of similar adaptations.

One hundred and five degrees is the upper limit for the hyperthermophiles whereas 50–70°C is the range for thermophiles. Proteins undergo aggregation at such high temperatures because those proteins which are not adapted undergo irreversible unfolding (Tomazic and Klibanov 1988). There are several adaptations as mentioned below in thermophilic proteins, which help them survive at such high temperatures by retaining their structure as well as function.

Increased surface charges

As the charge on the surface of protein increases, it makes it more thermostable (Fukuchi and Nishikawa 2001). Substitution of polar uncharged surface amino acids with polar charged amino-acids leads to an increased stability. At higher temperatures, deamination of polar amino-acids such as asparagine and glutamine would reduce stability (Fukuchi and Nishikawa 2001). Therefore, substitution of such amino acids and other thermolabile amino acids with charged amino acid from the surface of protein increases both short and long-range charge interactions. This in general, helps to protect the protein from thermal denaturation (Lee et al. 2005). To further support these findings, besides from those involved in salt bridges, a site-directed mutagenesis approach was used. Multiple single point mutations were made, where surface charged amino acids were replaced with alanine in the ribosomal protein, L30e, from *Thermococcus celer* (Lee et al. 2005). It has been observed that favourable mutation to charged amino acid enhanced the thermal capacity of thermophilic protein. On the other hand thermal capacity was decreased when alanine replaced surface charged amino acids. Long range ions interactions were also essential for the stability of L30e, as removal of such electrostatic interaction made it labile to chemical and thermal denaturation.

In the case of putative DNA binding protein from *Methanothermobacter thermautotrophicus*, MTH10b which is heavily charged showed that it has an unknown activity (Liu et al. 2012). Salt deficiency led to the decrease in its thermal capacity. In crystallographic studies, its structure revealed the presence of highly charged amino acids on its surface, which governs its salt dependent stability. The studies with MTH10b proved that repulsive forces due to charged amino acids on the surface may destabilize protein, and are countered by the presence of salt and provide thermal stability (Liu et al. 2012).

Higher content of surface charged amino acids is responsible to stabilize proteins by preventing aggregate formation at elevated temperatures, but sometimes it may lead to protein structure destabilization (Mamat et al. 2002). In *M. kandleri* there are a few enzymes whose stability and protein functionality require the presence of inorganic salts (Breitung et al. 1992).

Increased salt-bridging

Salt-bridging is also an important feature of most thermophilic enzymes (Karshikoff and Ladenstein 2001). In the case of mesophiles, salt-bridging may destabilize proteins and hence hydrophobic interactions are more favourable (Hendsch and Tidor 1994). The entropic cost and desolvation penalty are generally associated with ion pairing found in salt bridges and are more easily overcome at higher temperatures (Chan et al. 2011). When these thermodynamic considerations are ignored, salt bridges are considered as a structurally stabilizing element and hence increase the thermal capacity of proteins using favourable charge-charge interactions. Biophysical studies of L30e showed that thermophilic ribosomal protein from *Thermococcus celer* produced a remarkable change in thermal capacity without causing major structural changes (Chan et al. 2011). In the site-directed mutagenesis approach where charged residues involved in salt-bridging are replaced with hydrophobic residues, results showed an increase in the heat capacity change of unfolding, ΔCp. One approach used by thermophiles to enhance the thermostability of their proteins is by lowering the ΔCp and uphold the natively folded structure over that of the unfolded one. This proves that favourable interactions of charged residues (salt bridges) are essential and in return they improve the thermal stability of proteins (Chan et al. 2011).

Increased number of disulfide bonds

The tertiary structure of a protein is maintained by the disulphide bridging between cysteine residues which are present in the primary sequence of the protein and hence important for the overall structure. For the functional activity of a protein, proper bridging is very essential, as random bridging changes the conformation of a protein and the function may be lost completely. In the case of thermostable enzymes these structural elements are essential as they increase the stability and prevent any quaternary structural alteration (Cacciapuoti et al. 2012; Cacciapuoti et al. 1994; Boutz et al. 2007).

It has been considered that disulphide bridging provide thermostability in 5'-deoxy-5'methylthioadenosine phosphorylase II, while studying the role of its CXC motif and intrasubunit disulphide bonds (Cacciapuoti et al. 2012; Cacciapuoti et al. 1994). By using a biophysical approach like circular dichroism spectroscopy it was observed that hexameric protein was disassociated into its monomeric form under reduction conditions, in a reversible manner. However, chemical and thermal treatment resulted in irreversible degradation of the overall structure. By the site-directed mutagenises approach, single and double mutants of key cysteine residues resulted in a remarkable change in its thermostability. The native protein was completely denatured at 108°C, whereas the single mutant (C262S) denatured at 102°C and CXC double mutant (C259S/C261S) was completely denatured at 99°C. These results indicated that mutations in key cysteine residues decreased thermal stability of the mutant and thus disulphide bridging in native protein is an essential structural adaptation (Cacciapuoti et al. 2012). Disulphide bonds have also shown its importance in oligomerization. In the case of citrate synthetase from *Pyrobaculum aerophilum,* disulphide bonds interlink two monomeric subunits to create cyclized protein chains (Boutz et al. 2007). The individual subunits are held together and become very stable due to the unique interaction of disulphide bonds between two monomeric units. These two examples supports the role of disulphide bridging in thermostability, either by increasing rigidity or by the interlocking the two monomeric subunits.

Oligomerization and large hydrophobic core

Standard quaternary structure which are present in mesophilic proteins are absent in many thermostable proteins. This deviation is responsible for tight packing of individual subunits as well as it promotes tighter packing of the hydrophobic core. It also hides the exposure of hydrophobic residues from the solvent (Vieille and Zeikus 2001). For example one amylase and three acetyl-CoA synthetases from thermophilic archaea do not have a quaternary structure, which also governs their thermostable nature.

Ignicoccus hospitalis and *Pyrobaculum aerophilum* has two Acetyl-CoA Synthetases (ACS) which are oligomerized (Mayer et al. 2012; Brasen et al. 2005). Such oligomerization states are absent in mesophilic variants, which generally form either monomers or homodimers. However, it has been observed that in a hyper-thermophile like *Archaeoglobus fulgidus* the ACS form a trimer, when grown at a lower temperature (Ingram-Smith and Smith 2007).

The hydrophobic interaction in protein also provides stability and folding at a higher temperature. Due to the favourable hydrophobic interactions at the dimer interface, phosphotriesterase form tighter packing in *Sulfolobus solfataricus* (Del Vecchio et al. 2009). This approach is beneficial for protein stability as it reduces the overall ratio of surface area to the volume of individual subunits as well as solvent-exposed hydrophobic regions. As a result this leads to tighter packing of the hydrophobic core and provides stability to proteins at a higher temperature.

Oligomerization of proteins is not always beneficial for thermophiles. It has been observed that the *Pyrococcus furiosus*'s (Pf) thermostable amylase lacks the oligomerization state and resembles the mesophilic homologues (Park et al. 2013). This is the first report where cyclodextrin hydrolyzing enzyme is functional as monomer. Although bacterial homologues require dimerization for its activity, but a novel N-terminus (N') domain in the Pf amylase allows it to be active as a monomer (Park et al. 2000). The bacterial N-domain has a loop-like structure which extends over the active site and functions like a structural 'lid'. Its key role is to stabilize certain substrates including maltose, whereas Pf amylase does not have any such mechanism.

With reference to thermostable ACS, it is a higher oligomeric state which is favourable. Whereas on the other hand, Pf amylase is governed by the opposite strategy such as creating structural rigidity, tighter packing of the hydrophobic core and all the necessary components into a single subunit. In both cases, it supports the hypothesis that changes in quaternary structure could be beneficial for protein stability at higher temperatures.

Industrial applications

The thermophilic enzymes are more stable and have optimal activity at high temperatures, and because of these properties they showed a high potential for industrial and biotechnological application (Unsworth et al. 2007; Kour et al. 2019; Yadav et al. 2016). The kinetics and thermodynamics of a catalyzed reaction is more favourable at high temperatures, therefore this allows higher product yield due to a more competent reaction. Other benefits are also associated with their thermostability, which include reduction in operating cost because the least involvement of no cold chain management lowers chances of bacterial contamination in food and drug applications (Unsworth et al. 2007; Champdore et al. 2007). The first application of thermophilic enzymes in molecular biology was Taq DNA polymerase, isolated from the bacterium *Thermus aquaticus* (Unsworth et al. 2007; Champdore et al. 2007). Today, apart from *T. aquaticus,* varieties of other thermophilic.

DNA polymerases are isolated from archaeal species including DeepVentR, *pfu*Turbo, Therminator (Stratagene Inc., and New England BioLabs Inc.). Another example of potential application of archaeal thermophilic enzymes is thermostable amylase, isolated from *Pyrococcus furiosus*. Mutational studies were performed in Pf amylase, and results showed that there was an increase in the production of maltoheptaose using β-cyclodextrin as substrate. Maltoheptaose and other linear malto-oligosaccharides have a large demand in the pharmaceutical, cosmetic and food industries because it can be used as a carrier (Park et al. 2013). At ambient temperatures, thermophilic enzymes lack catalytic activity and thus are of immense use to the industry. This property of thermophilic enzymes can be exploited in optical nanosensors which could bind a substrate without converting it to a product (Champdore et al. 2007). By just monitoring the enzyme fluorescence variation, substrate-enzyme complex can be easily measured. And thus it is used to quantitate the concentration of substrate present in any sample. The application of such tools has great potential in biotechnology, medical testing, and drug discovery (Champdore et al. 2007).

Proteomics adaption's in psychrophilic archaea

A class of extremophiles having optimal growth tempearture below 20°C are known as psychrophiles (Luisa Tutino et al. 2009). A lot of research has been undertaken on archaeal organisms thriving in extremely cold environments, especially on methanogens (psychrophiles) breeding in Alaska and the Antarctic (Dong and Chen 2012). Normally, at temperatures below 20°C, most of the proteins have extremely low activity; hence such conditions are not suitable for a growing cell (Feller 2010). Due to a lower mean kinetic energy at lower temperatures, enzyme activity also decreases. The conformational movements of a protein structure become slower at low temperature, therefore proteins having enzymatic activity are less efficient in such a condition (Cavicchioli et al. 2000). Moreover, at low temperatures, the energy barrier for the conversion of the substrate into a product becomes too high for an enzymatic protein, further reducing the enzyme's activity (Feller 2010). Psychrophilic organisms can survive in an extremely cold environment because they have undergone several adaptations in their proteins. The optimal activity of such proteins has been reported at a temperature which is higher than their physiological temperature (Feller 2010). Psychrophilic proteins are adapted to move and change conformation due to a structure flexibility, which governs high activity even at low temperatures (Smalas 2000).

Increased specific activity

Due to the more flexible structure of psychrophilic enzyme, their catalytic activity is much higher at low temperatures as compared to the same mesophilic enzyme. While compromising with slow reaction rates at lower temperatures, it compensates the same with high specific activity (*k*cat) which is typically 10 times greater than that of a mesophilic enzyme (Feller 2010; Georlette et al. 2003). The reason behind such greater *k*cat in psychrophilic enzyme is the increment in the size of the binding site (Smalas et al. 2000). There are many mechanisms by which the substrate binding area in psychrophilic enzyme is enlarged, while the active site remains unchanged (Feller 2010). A few mechanisms which are involved in the enlargement of substrate-binding area includes (i) removal of loops near the binding site (Russell et al. 1998), (ii) presence of glycine amino acids near the functional sites (Feller 2010), (iii) increasing the substrate accessibility by pulling the protein backbone out from binding site (Aghajari et al. 2003). Therefore, substrates are not able to interact so strongly with the psychrophilic enzyme, and ultimately Michaelis-Menten constant (*Km*) of such an enzyme becomes high (Feller 2010; D'Amico et al. 2006). Because of the reduction in the enzyme's activation energy, the enzymatic activity elevates at low temperatures due to poor substrate affinity (Feller 2010).

Weak protein interactions

The reason for this is the depletion in the energy barrier between various conformations, psychrophilic proteins have greater flexibility (Feller 2010). The main reason behind this flexibility is the amino acid composition. Cold adaptive proteins are generally compromised with a stabilizing interaction present within proteins. There are a number of adaptations in psychrophilic proteins which are essential for survival in cold environment as follows: (i) For greater conformational mobility, the number of glycine residues have been increased, (ii) Conformational rigidity has been reduced in loop regions by reducing the number of proline residues, (iii) Number of arginine residues responsible for forming the hydrogen bond are reduced, as well as the number of salt bridges are reduced, (iv) hydrophobic interactions in the protein core become weaker by reducing the number of non polar residues (Feller 2010).

In support of the afore mentioned features, when proteins were isolated from the archaeal cold-adapted halophile *Halorubum lacusprofundi,* it showed a lesser number of tryptophan residue, having large hydrophobic side chain as well as glutamic acid, having the capabilities to form hydrogen bonds. In the case of β-galactosidase isolated from *H. lacusprofundi*, protein surface showed elevated hydrophobicity, while replacing anionic electrostatic interactions (Dassarma et al. 2013; Karan et al. 2013). Such a type of amino acid substitution has also been observed in the elongation factor 2 proteins from psychrophilic methanogens (Thomas and Cavicchioli 1998).

Genome sequences for two psychrophilic methanogens, *Methanococcoides burtonii* and *Methanogenium frigidum* were available. They were studied and three dimensional models of proteins were constructed. Comparative analysis of such modelled proteins was done with other proteins designed from thermophilic and mesophilic (Saunders et al. 2003). The results showed that there was a decrease in the charged amino acids on the surface of cold adopted proteins. However, there was an increase in glutamine and threonine residues. This was may be to counter the reduction of charges on the protein surface and inhibit the chance aggregate formation due to too much hydrophobicity on the surface (Saunders et al. 2003).

Lower thermal stability

Sometime weaker interactions between amino acid residues in protein are also beneficial, as in case of psychrophilic proteins. Such weak interactions prevent them from being 'frozen' in a specific conformation and assist in molecular dynamics required for catalysis. As a result of these weaker interactions, which make them less stable proteins, is one of the key factors responsible for cold-adapted proteins to get unfolded at lower temperatures (Feller 2010; D´Amico et al. 2001; Georlette et al. 2003). It has been observed that thermal unfolding of psychrophilic proteins have occurred via a single transition. The reason behind such a transition is weaker interactions present in cold-adapted proteins that have a greater role in the overall stability as well as local unfolding. Due to the lesser stabilizing interactions the protein structure gets destabilized (Feller 2010). Archaeal cold-shock protein isolated from *Methanogenium frigidum* showed such characteristics, because of their lesser stability at optimal temperatures as compared to its mesophilic homologue present in *E. coli* (Giaquinto et al. 2007).

Industrial applications

There are many psychrophilic enzymes that have been isolated and have useful applications in the biotechnological industry. Cold-adapted lipases isolated from bacterial psychrophiles are used in commercial detergents due to their higher activity at low temperatures (Luisa Tutino et al. 2009). On the other hand, cellulases are also useful due to their reduced thermal stability, because of their inactivation. This is important in the stone-washing step in the textile industry, where prolonged exposure of cellulases is detrimental to the resistance of cottons (Luisa Tutino et al. 2009; Hasan et al. 2006). Enzymes from other bacteria are widely used, but applications of archaeal cold-adapted enzymes are not liable as such. But due to their adaptations, they may soon have possible applications in the industry.

Conclusion

Archaea survive under different extreme conditions such as high and low temperatures, acidic and alkaline conditions, high pressure, etc. wherein mesophilic bacteria are unable to survive. The survival of extremophiles in an inhospitable environment is governed by their membrane architecture, DNA stability, composition and conformation of their proteins and enzymes. Archaea's are considered to have prospective applications in the field of biotechnology because of their ability to adjust in extreme environments. There has been an escalation of research and modelling of the applications of archaea so that economical and environment friendly methods can be built up.

References

Aghajari, N., F. Van Petegem, V. Villeret, J.P. Chessa, C. Gerday, R. Haser et al. 2003. Crystal structures of a psychrophilic metalloprotease reveal new insights into catalysis by cold-adapted proteases. Proteins 50: 636–647.
Ambrosch, F., G. Wiedermann, S. Jonas, B. Althaus, B. Finkel, R. Glück et al. 1997. Immunogenicity and protectivity of a new liposomal hepatitis A vaccine. Vaccine 15: 1209–1213.
Bailey, K.A., F. Marc, K. Sandman and J.N. Reeve. 2002. Both DNA and histone fold sequences contribute to archaeal nucleosome stability. J. Biol. Chem. 277: 9293–9301.
Bao, Q., Y. Tian, W. Li, Z. Xu, Z. Xuan, S. Hu et al. 2002. A complete sequence of the *T. tengcongensis* genome. Genome Res. 12: 689–700.
Barbeau, J., S. Cammas-Marion, P. Auvray and T. Benvegnu. 2011. Preparation and characterization of stealth archaeosomes based on a synthetic PEGylated archaeal tetraether lipid. J. Drug Deliv. 396068.
Belova, G.I., R. Prasad, S.A. Kozyavkin, J.A. Lake, S.H. Wilson and A.I. Slesarev. 2001. A type IB topoisomerase with DNA repair activities. Proc. Natl. Acad. Sci. (USA) 98: 6015–6020.
Benvegnu, T., M. Brard and D. Plusquellec. 2004. Archaeabacteria bipolar lipid analogues: structure, synthesis, and lyotropic properties. Curr. Opin. Colloid Interface Sci. 8: 469–479.
Benvegnu, T., L. Lemiègre and S. Cammas-Marion. 2008. Archaeal lipids: innovative materials for biotechnological applications. European J. Org. Chem. 4725–4744.
Boonyaratanakornkit, B.B., A.J. Simpson, T.A. Whitehead, C.M. Fraser, N. El-Sayed and D.S. Clark. 2005. Transcriptional profiling of the hyperthermophilic methanarchaeon *Methanococcus jannaschii* in response to lethal heat and non-lethal cold shock. Environ. Microbiol. 7: 789–97.
Borges, N., R. Matsumi, T. Imanaka, H. Atomi and H. Santos. 2010. *Thermococcus kodakarensis* mutants deficient in di-myo-inositol phosphate use aspartate to cope with heat stress. J. Bacteriol. 192: 191–7.
Boutz, D.R., D. Cascio, J. Whitelegge, L.J. Perry and T.O. Yeates. 2007. Discovery of a thermophilic protein complex stabilized by topologically interlinked chains. J. Mol. Biol. 368(5): 1332–1344.

Bl asen, C., C. Urbanke and P. Sch onheit. 2005. A novel octameric AMP-forming acetyl-CoA synthetase from the hyperthermophilic crenarchaeon *Pyrobaculum aerophilum*. FEBS Letters 579(2): 477–482.

Breitung, J., G. Borner, S. Scholz, D. Linder, K.O. Stetter and R.K. Thauer. 1992. Salt dependence, kinetic properties and catalytic mechanism of *N*-formylmethanofuran: tetrahydromethanopterin formyltransferase from the extreme thermophile *Methanopyrus kandleri*. Eur. J. Biochem. 210(3): 71–981.

Brown, D.A., B. Venegas, P.H. Cooke, V. English and P.L.G. Chong. 2009. Bipolar tetraether archaeosomes exhibit unusual stability against autoclaving as studied by dynamic light scattering and electron microscopy. Chem. Phys. Lipids. 159: 95–103.

Brown, J.W., E.S. Haas and N.R. Pace. 1993. Characterization of ribonuclease P RNAs from thermophilic bacteria. Nucleic Acids Res. 3: 671–679.

Bujinicki, J.M. and M. Radlinsks. 1999. Molecular evolution of DNA (cytosine-N4) methyltransferases: evidence for their polyphyletic origin. Nucleic Acids Res. 22: 4501–4509.

Cacciapuoti, G., M. Porcelli, C. Bertoldo, M. De Rosa and V. Zappia. 1994. Purification and characterization of extremely thermophilic and thermostable 5'-methylthioadenosine phosphorylase from the archaeon *Sulfolobus solfataricus*. Purine nucleoside phosphorylase activity and evidence for intersubunit disulfide bonds. Int. J. Biol. Chem. 269(40): 24762–24769.

Cacciapuoti, G., F. Fuccio, L. Petraccone, P. Del Vecchio and M. Porcelli. 2012. Role of disulfide bonds in conformational stability and folding of 5'-deoxy-5'-methylthioadenosine phosphorylase II from the hyperthermophilic archaeon *Sulfolobus solfataricus*. Biochimica et Biophysica Acta. 1824(10): 1136–1143.

Caforio, A. and A.J.M. Driessen. 2017. Archaeal phospholipids: Structural properties and biosynthesis. Biochim. Biophys. Acta Mol. Cell Biol. Lipids 1862(11): 1325–1339.

Cavicchioli, R., T. Thomas and P.M.G. Curmi. 2000. Cold stress response in Archaea. Extremophiles 4(6): 321–331.

Champdoré, M. de, M. Staiano, M. Rossi and S. D'Auria. 2007. Proteins from extremophiles as stable tools for advanced biotechnological applications of high social interest. J. R. Soc. Interface. 4: 183–191.

Chan, C.H., T.H. Yu and K.B. Wong. 2011. Stabilizing salt-bridge enhances protein thermostability by reducing the heat capacity change of unfolding. PLoS One 6(6): e21624.

Chang, E.L. 1994. Unusual thermal stability of liposomes made from bipolar tetraether lipids. Biochem. Biophys. Res. Commun. 202: 673–9.

Comita, P.B., R.B. Gagosian, H. Pang and C.E. Costello. 1984. Structural elucidation of a unique macrocyclic membrane lipid from a new, extremely thermophilic, deep-sea hydrothermal vent archaebacterium, *Methanococcus jannaschii*. J. Biol. Chem. 259: 15234–41.

Couturier, M. and A.C. Lindås. 2018. The DNA methylome of the hyper-thermoacidophilic crenarchaeon *Sulfolobus acidocaldarius*. Front Microbiol. 9: 137.

Cronan, J.E. 1978. Molecular biology of bacterial membrane lipids. Annu. Rev. Biochem. 47: 163–89.

D'Amico, S., C. Gerday and G. Feller. 2001. Structural determinants of cold adaptation and stability in a large protein. J. Biol. Chem. 276(28): 25791–25796.

D'Amico, S., J.S. Sohier and G. Feller. 2006. Kinetics and energetics of ligand binding determined by microcalorimetry: insights into active site mobility in a psychrophilic α-amylase. J. Mol. Biol. 358(5): 1296–1304.

Dannenmuller, O., K. Arakawa, T. Eguchi, K. Kakinuma, S. Blanc, A.M. Albrecht et al. 2000. Membrane properties of archaeal macrocyclic diether phospholipids. Chemistry 6: 645–54.

Dassarma, S., M.D. Capes, R. Karan and P. Dassarma. 2013. Amino acid substitutions in cold-adapted proteins from *Halorubrum lacusprofundi*, an extremely halophilic microbe from antarctica. PLoS One 8(3): e585878.

De Champdor'e, M., M. Staiano, M. Rossi and S. D'Auria. 2007. Proteins from extremophiles as stable tools for advanced biotechnological applications of high social interest. J. R. Soc. Interface. 4(13): 183–191.

DE Rosa, M., S.D.E. Rosa and D. Bu. John. 1980. Structure of calditol, a new branched-chain nonitol, and of the derived tetraether lipids in thermoacidophile archaebacteria of the caldariella group. Phytochemistry. 19: 249–254.

De Rosa, M., A. Gambacorta and A. Gliozzi. 1986. Structure, biosynthesis, and physicochemical properties of archaebacterial lipids. Microbiol. Rev. 50: 70.

De Rosa, M., A. Gambacorta, V. Lanzotti, A. Trincone, J.E. Harris and W.D. Grant. 1986. A range of ether core lipids from the methanogenic archaebacterium *Methanosarcina barkeri*. Biochim. Biophys. Acta - Lipids Lipid Metab. 875: 487–492.

De Rosa, M. and A. Gambacorta. 1988. The lipids of archaebacteria. LipM Res. 27: 53–175.

Deckert, G., P.V. Warren, T. Gaasterland, W.G. Young, A.L. Lenox, D.E. Graham et al. 1998. The complete genome of the hyperthermophilic bacterium *Aquifex aeolicus*. Nature 392(6674): 353–8.

Del Vecchio, P., M. Elias, L. Merone, G. Graziano, J. Dupuy, L. Mandrich et al. 2009. Structural determinants of the high thermal stability of SsoPox from the hyperthermophilic archaeon *Sulfolobus solfataricus*. Extremophiles 13(3): 461–470.

Di Ruggiero, J., N. Santangelo, Z. Nackerdien, J. Ravel and F.T. Robb. 1997. Repair of extensive ionizing-radiation DNA damage at 95°C in the hyperthermophilic archaeon *Pyrococcus furiosus*. J. Bacteriol. 179: 4643–4645.

Dong, X. and Z. Chen. 2012. Psychrotolerant methanogenic archaea: diversity and cold adaptation mechanisms. Science China Life Sciences 55(5): 415–421.

Dowhan, W. 1997. Molecular basis for membrane phospholipid diversity: why are there so many lipids? Annu. Rev. Biochem. 66: 199–232.

Edmonds, C.G., P.F. Crain, R. Gupta, T. Hashizume, C.H. Hocart, J.A. Kowalak et al. 1991. Posttranscriptional modification of tRNA in thermophilic archaea (Archaebacteria). J. Bacteriol. 173: 3138–48.

Ehrlich, M., M.A. Gama-Sosa, L.H. Carreira, L.G. Ljungdahl, K.C. Kuo and C.W. Gehrke. 1985. DNA methylation in thermophilic bacteria: N4-methylcytosine, 5-methylcytosine, and N6-methyladenine. Nucleic Acids Res. 13: 1399–1412.

Elferink, M.G.L., J.G. de Wit, A.J.M. Driessen and W.N. Konings. 1994. Stability and proton-permeability of liposomes composed of archaeal tetraether lipids. Biochim. Biophys. Acta - Biomembr. 119: 247–254.

Empadinhas, N. and M.S. da Costa. 2011. Diversity, biological roles and biosynthetic pathways for sugar-glycerate containing compatible solutes in bacteria and archaea. Environ. Microbiol. 13: 2056–77.

Faria, T.Q., A. Mingote, F. Siopa, R. Ventura, C. Maycock and H. Santos. 2008. Design of new enzyme stabilizers inspired by glycosides of hyperthermophilic microorganisms. Carbohydr. Res. 343: 3025–33.

Fitz-Gibbon, S.T., H. Ladner, U. Kim, K.O. Stetter, M.I. Simon and J.H. Miller. 2002. Genome sequence of the hyperthermophilic crenarchaeon *Pyrobaculum aerophilum*. Proc. Natl. Acad. Sci. (USA) 99: 984–989.

Fukuchi, S. and K. Nishikawa. 2001. Protein surface amino acid compositions distinctively differ between thermophilic and mesophilic bacteria. J. Mol. Biol. 309(4): 835–843.

Futterer, O., A. Angelov, H. Liesegang, G. Gottschalk, C. Schleper, B. Schepers et al. 2004. Genome sequence of *Picrophilus torridus* and its implications for life around pH 0. Proc. Natl. Acad. Sci. (USA) 101: 9091–9096.

Feller, G. 2010. Protein stability and enzyme activity at extreme biological temperatures. J. Physics 22(32). Article ID 323101.

Gabriel, J.L. and P. Lee Gau Chong. 2000. Molecular modeling of archaebacterial bipolar tetraether lipid membranes. Chem. Phys. Lipids 105: 193–200.

Gambacorta, A., A. Gliozzi and M. De Rosa. 1995. Archaeal lipids and their biotechnological applications. World J. Microbiol. Biotechnol. 11: 115–131.

Gaughran, E.R.L. 1947. The saturation of bacterial lipids as a function of temperature. J. Bacteriol. 53(4): 506.

Georlette, D., B. Damien, V. Blaise, E. Depiereux, V.N. Uversky, C. Gerday et al. 2003. Structural and functional adaptations to extreme temperatures in psychrophilic, mesophilic, and thermophilic DNA ligases. J. Biol. Chem. 278(39): 37015–37023.

Giaquinto, L., P.M.G. Curmi, K.S. Siddiqui, A. Poljak, Ed DeLong, S. DasSarma et al. 2007. Structure and function of cold shock proteins in Archaea. J. Bacteriol. 189(15): 5738–5748.

Gliozzi, A., G. Paoli, M. De Rosa and A. Gambacorta. 1983. Effect of isoprenoid cyclization on the transition temperature of lipids in thermophilic archaebacteria. Biochim. Biophys. Acta - Biomembr. 735234–242.

Gliozzi, A., A. Relini and P.L.-G. Chong. 2002. Structure and permeability properties of biomimetic membranes of bolaform archaeal tetraether lipids. J. Memb. Sci. 206: 131–147.

Goh, K.M., H.M. Gan, K.G. Chan, G.F. Chan, S. Shahar, C.S. Chong et al. 2014. Analysis of *Anoxybacillus* genomes from the aspects of lifestyle adaptations, prophage diversity, and carbohydrate metabolism. PLoS One 9: e90549.

Grogan, D.W. and J.E. Cronan Jr. 1997. Cyclopropane ring formation in membrane lipids of bacteria. Microbiol. Mol. Biol. Rev. 6: 429–41.

Gorgan, D.W. 1998. Hyperthermophiles and the problem of DNA instability. Molec. Microbiol. 28: 1043–1049.

Gorgan, D.W. 2000. The question of DNA repair in hyperthermophilic Archaea. Trends Microbiol. 8: 179–184.

Gorgan, D.W. 2004. Stability and repair of DNA in hyperthermophilic Archaea. Curr. Issues Mol. Biol. 6(2): 137–44.

Hasan, A., A.A. Shah and A. Hameed. 2006. Industrial applications of microbial lipases. Enzyme Microb. Technol. 39(2): 235–251.

Heine, M. and S.B. Chandra. 2009. The linkage between reverse gyrase and hyperthermophiles: a review of their invariable association. J. Microbiol. 47: 229–34.

Hendsch, Z.S. and B. Tidor. 1994. Do salt bridges stabilize proteins? A continuum electrostatic analysis. Protein Science 3(2): 211–226.

Hickey, D.A. and G. Singer. 2004. Genomic and proteomic adaptations to growth at high temperature. Genome Biol. 5(10): 117.

Ingram-Smith, A. and K.S. Smith. 2007. AMP-forming acetyl-CoA synthetases in Archaea show unexpected diversity in substrate utilization. Archaea. 2(2): 95–107.

Jacquemet, A., J. Barbeau, L. Lemiègre and T. Benvegnu. 2009. Archaeal tetraether bipolar lipids: Structures, functions and applications. Biochimie. 91: 711–717.

Jia, Y., M.J. McCluskie, D. Zhang, R. Monette, U. Iqbal, M. Moreno et al. 2018. *In vitro* evaluation of archaeosome vehicles for transdermal vaccine delivery. J. Liposome Res. 28(4): 305–314.

Karan, R., M.D. Capes, P. DasSarma and S. DasSarma. 2013. Cloning, overexpression, purification, and characterization of a polyextremophilic β-galactosidase from the Antarctic haloarchaeon *Halorubrum lacusprofundi*. BMC Biotechnology 13(3).

Karshikoff, A. and R. Ladenstein. 2001. Ion pairs and the thermo-tolerance of proteins from hyperthermophiles: a "traffic rule" for hot roads. Trends Biochem. Sci. 26(9): 550–556.

Kates, M. 1992. Archaebacterial lipids: structure, biosynthesis and function. Biochem. Soc. Symp. 58: 51–72. http://www.ncbi.nlm.nih.gov/pubmed/1445410 (accessed August 29, 2014).

Kates, M., D.J. Kushner and A.T. Matheson. 1993. The Biochemistry of Archaea. Amsterdam: Elsevier Science Publishers B.V. pp. 582.

Kawashima, T., A. Amano, A. Koike, S. Makino, S. Higuchi, Y. Kawashima-Ohya et al. 2000. Archaeal adaptation to higher temperatures revealed by genomic sequence of *Thermoplasma canium*. Proc. Natl. Acad. Sci. (USA) 97: 14257–14262.

Klein, R.J., Z. Misulovin and S.R. Eddy. 2002. Noncoding RNA genes identified in AT-rich hyperthermophiles. Proc. Natl. Acad. Sci. (USA) 99: 7542–7547.

Koga, Y., M. Nishihara, H. Morii and M. Akagawa-Matsushita. 1993. Ether polar lipids of methanogenic bacteria: structures, comparative aspects, and biosyntheses. Microbiol. Rev. 57: 164–82.

Koga, Y. and H. Morii. 2005. Recent advances in structural research on ether lipids from archaea including comparative and physiological aspects. Biosci. Biotechnol. Biochem. 69: 2019–34.

Koga, Y. 2012. Thermal adaptation of the archaeal and bacterial lipid membranes. Archaea. 789652.

Komatsu, H. and P.L. Chong. 1998. Low permeability of liposomal membranes composed of bipolar tetraether lipids from thermoacidophilic archaebacterium *Sulfolobus acidocaldarius*. Biochemistry 37: 107–15.

Konings, W.N., S.V. Albers, S. Koning and A.J. Driessen. 2002. Cell membrane plays a crucial role in survival of Bacteria and Archaea in extreme environments. Antonie van Leeuwenhock. 81: 61–72.

Kour, D., K.L. Rana, T. Kaur, B. Singh, V.S. Chauhan, A. Kumar et al. 2019. Extremophiles for hydrolytic enzymes productions: biodiversity and potential biotechnological applications. pp. 321–372. *In*: Molina, G., V.K. Gupta, B. Singh and N. Gathergood (eds.). Bioprocessing for Biomolecules Production. doi:10.1002/9781119434436.ch16.

Kowalak, J.A., J.J. Dalluge, J.A. McCloskey and K.O. Stetter. 1994. The role of posttranscriptional modification in stabilization of transfer RNA from hyperthermophiles. Biochemistry 33: 7869–7876.

Krishnan, L. and G.D. Sprott. 2003. Archaeosomes as self-adjuvanting delivery systems for cancer vaccines. J. Drug Target. 11: 515–24.

Krishnan, L. and G.D. Sprott. 2008. Archaeosome adjuvants: immunological capabilities and mechanism(s) of action. Vaccine. 26: 2043–55.

Lachman, L.B., B. Ozpolat and X.M. Rao. 1996. Cytokine-containing liposomes as vaccine adjuvants. Eur. Cytokine Netw. 7: 693–8.

Langworthy, T.A. 1977. Long chain diglycerol tetraethers from *Thermoplasma acidophilum*. Biochim. Biophys. Acta. 487: 37–50.

Lao, P.J. and D.R. Forsdyke. 2000. Thermophilic bacteria strictly obey Szybalski's transcription direction rule and politely purine-load RNAs with both adenine and guanine. Genome Res. 10: 228–236.

Lee, C.F., G.I. Makhatadze and K.B. Wong. 2005. Effects of charge-to-alanine substitutions on the stability of ribosomal protein L30e from *Thermococcus celer*. Biochemistry 44(51): 16817–16825.

Lipscomb, G.L., E.M. Hahn, A.T. Crowley and M.W.W. Adams. 2017. Reverse gyrase is essential for microbial growth at 95°C. Extremophiles 21(3): 603–608.

Liu, Y.F., N. Zhang, X. Liu, X. Wang, Z.X. Wang, Y. Chen et al. 2012. Molecular mechanism underlying the interaction of typical Sac10b family proteins with DNA. PLoS One, 7(4): ID e34986.

Lobry, L.R. and D. Chessel. 2003. Internal correspondence analysis of codon and amino-acid usage in thermophilic bacteria. J. Appl. Genet. 44: 235–261.

Luisa Tutino, M., G. Di Prisco, G. Marino and D. De Pascale. 2009. Cold-adapted esterases and lipases: from fundamentals to application. Protein Peptide Lett. 16(10): 1172–1180.

Lynn, D.J., G.A. Singer and D.A. Hickey. 2002. Synonymous codon usage is subject to selection in thermophilic bacteria. Nucleic Acids Res. 30: 4272–4277.

Makarova, K.S., L. AravindL, N.V. Grishin, I.B. Rogozin and E.V. Koonin. 2002. A DNA repair system specific for thermophilic archaea and bacteria predicted by genomic context analysis. Nucleic Acids Res. 30: 482–496.

Mamat, B., A. Roth, C. Grimm, U. Ermler, C. Tziatzios, D. Schubert et al. 2002. Crystal structures and enzymatic properties of three formyltransferases from archaea: environmental adaptation and evolutionary relationship. Protein Sci. 11(9): 2168–2178.

Mathai, J.C., G.D. Sprott and M.L. Zeidel. 2001. Molecular mechanisms of water and solute transport across archaebacterial lipid membranes. J. Biol. Chem. 276: 27266–27271.

Mayer, F., U. Küper, C. Meyer, S. Daxer, V. Müller, R. Rachel et al. 2012. AMP-forming acetyl coenzyme a synthetase in the outermost membrane of the hyperthermophilic crenarchaeon *Ignicoccus hospitalis*. J. Bacteriol. 194(6): 1572–1581.

Melchior, D.L. 1982. Lipid phase transitions and regulation of membrane fluidity in prokaryotes. Curr. Top. Membr. Transp. 17: 263–316.

Morten, M.J., R. Gamsjaeger, L. Cubeddu, R. Kariawasam, J. Peregrina, J.C. Penedo et al. 2017. High-affinity RNA binding by a hyperthermophilic single-stranded DNA-binding protein. Extremophiles 21(2): 369–379.

Musgrave, D., P. Forterre and A. Slesarev. 2000. Negative constrained DNA supercoiling in archaeal nucleosomes. Mol. Microbiol. 35: 341–349.

Nakashima, H., S. Fukuchi and K. Nishikawa. 2003. Compositional changes in RNA, DNA and proteins for bacterial adaptation to higher and lower temperatures. J. Biol. Chem. 133: 507–513.

Napoli, A., M. Kvaratskelia, M.F. White, M. Rossi and M. Ciaramella. 2001. A novel member of the bacterial-archaeal regulator family is a nonspecific DNA-binding protein and induces positive supercoiling. J. Biol. Chem. 27614: 10745–10752.

Nelson, K.E., R.A. Clayton, S.R. Gill, M.L. Gwinn, R.J. Dodson, D.H. Haft et al. 1999. Evidence for lateral gene transfer between archaea and bacteria from genome sequence of *Thermotoga maritima*. Nature 399: 323–9.

Nichols, D.S., M.R. Miller, N.W. Davies, A. Goodchild, M. Raftery and R. Cavicchioli. 2004. Cold adaptation in the antarctic archaeon *Methanococcoides burtonii* involves membrane lipid unsaturation. J. Bacteriol. 186: 8508–8515.

Omri, A., B.J. Agnew and G.B. Patel. 2003. Short-term repeated-dose toxicity profile of archaeosomes administered to mice via intravenous and oral routes. Int. J. Toxicol. Jan-Feb 22(1): 9–23.

Orita, I., R. Futatsuishi, K. Adachi, T. Ohira, A. Kaneko, K. Minowa et al. 2019. Random mutagenesis of a hyperthermophilic archaeon identified tRNA modifications associated with cellular hyperthermotolerance. Nucleic Acids Res. 47(4): 1964–1976.

Oger, P.M. and A. Cario. 2013. Adaptation of the membrane in Archaea. Biophys. Chem. 183: 42–56.

Park, J.T., H.N. Song, T.Y. Jung, M.H. Lee, S.G. Park, E.J. Woo et al. 2013. A novel domain arrangement in amonomeric cyclodextrin-hydrolyzing enzyme from the hyperthermophile *Pyrococcus furiosus*. Biochimica et Biophysica Acta. 1834(1): 380–386.

Park, K.H., T.J. Kim, T.K. Cheong, J.W. Kim, B.H. Oh and B. Svensson. 2000. Structure, specificity and function of cyclomaltodextrinase, a multispecific enzyme of the α-amylase family. Biochimica et Biophysica Acta. 1478(2): 165–185.

Patel, G.B. and G.D. Sprott. 1999. Archaeobacterial ether lipid liposomes (archaeosomes) as novel vaccine and drug delivery systems. Crit. Rev. Biotechnol. 19: 317–57.

Patel, G.B., B.J. Agnew, L. Deschatelets, L.P. Fleming and G.D. Sprott. 2000. *In vitro* assessment of archaeosome stability for developing oral delivery systems. Int. J. Pharm. 194: 39–49.

Paz, A., D. Mester, I. Baca, E. Nevo and A. Korol. 2004. Adaptive role of increased frequency of polypurine tracts in mRNA sequences of thermophilic prokaryotes. Proc. Natl. Acad. Sci. (USA) 101: 2951–6.

Peak, M.J., F.T. Robb and J.G. Peak. 1995. Extreme resistance to thermally induced DNA backbone breaks in the hyperthermophilic archaeon *Pyrococcus furiosus*. J. Bacteriol. 177: 6316–6318.

Raetz, C.R. and W. Dowhan. 1990. Biosynthesis and function of phospholipids in *Escherichia coli*. J. Biol. Chem. 265: 1235–8.

Russell, N. and N. Fukunaga. 1990. A comparison of thermal adaptation of membrane lipids in psychrophilic and thermophilic bacteria. FEMS Microbiol. Lett. 75: 171–182.

Russell, N.J. and D.S. Nichols. 1999. Polyunsaturated fatty acids in marine bacteria—a dogma rewritten. Microbiology 145(Pt 4): 767–79.

Russell, R.J.M., U. Gerike, M.J. Danson, D.W. Hough and G.L. Taylor. 1998. Structural adaptations of the cold-active citrate synthase from an Antarctic bacterium. Structure 6(3): 351–362.

Sabath, N., E. Ferrada, A. Barve and A. Wagner. 2013. Growth temperature and genome size in bacteria are negatively correlated, suggesting genomic streamlining during thermal adaptation. Genome Biol. Evol. 5: 966–77.

Sakamoto, T. and N. Murata. 2002. Regulation of the desaturation of fatty acids and its role in tolerance to cold and salt stress. Curr. Opin. Microbiol. 5(2): 208–10.

Sandigursky, M. and W.A. Franklin. 1999. Thermostable uracil-DNA glycosylase from *Thermostoga maritima* a member of a novel class of DNA repair enzymes. Curr. Biol. 9: 531–534.

Sandman, K., R.A. Grayling, B. Dobrinski, R. Lurz and J.N. Reeve. 1994. Growth-phase-dependent synthesis of histones in the archaeon *Methanothermus fervidus*. Proc. Natl. Acad. Sci. (USA) 91: 12624–12628.

Santos, H. and M.S. Da Costa. 2002. Compatible solutes of organisms that live in hot saline environments. Environ. Microbiol. 4: 501–9.

Saunders, N.F.W., T. Thomas, P.M.G. Curmi, J.S. Mattick, E. Kuczek, R. Slade et al. 2003. Mechanisms of thermal adaptation revealed from genomes of the Antarctic *Archaea Methanogenium frigidum* and *Methanacoccoides burtonii*. Genome Research 13(7): 1580–1588.

Schouten, S., E.C. Hopmans, R.D. Pancost and J.S.S. Damsté. 2000. Widespread occurrence of structurally diverse tetraether membrane lipids: Evidence for the ubiquitous presence of low-temperature relatives of hyperthermophiles. Proc. Natl. Acad. Sci. 97: 14421–14426. doi:10.1073/pnas.97.26.14421.

Shimada, H., N. Nemoto, Y. Shida, T. Oshima and A. Yamagishi. 2008. Effects of pH and temperature on the composition of polar lipids in *Thermoplasma acidophilum* HO-62. J. Bacteriol. 190: 5404–11.

Shinoda, W., K. Shinoda, T. Baba and M. Mikami. 2005. Molecular dynamics study of bipolar tetraether lipid membranes. Biophys. J. 89: 3195–202.

Shinoda, W., M. Mikami, T. Baba and M. Hato. 2004. Molecular dynamics study on the effect of chain branching on the physical properties of lipid bilayers: Structural stability. J. Phys. Chem. B. 108(26): 9346–9356.

Singer, G.A. and D.A. Hickey. 2003. Thermophilic prokaryotes have characteristic patterns of codon usage, amino acid composition and nucleotide content. Gene. 317: 39–47.

Sinninghe Damsté, J.S., S. Schouten, E.C. Hopmans, A.C.T. Van Duin and J.A.J. Geenevasen. 2002. Crenarchaeol: the characteristic core glycerol dibiphytanyl glycerol tetraether membrane lipid of cosmopolitan pelagic crenarchaeota. J. Lipid Res. 43.

Smalas, A.O., H.K. Leiros, V. Os and N.P. Willassen. 2000. Cold adapted enzymes. Biotechnology Annual Review 6: 1–57.

Sprott, G.D., M. Meloche and J.C. Richards. 1991. Proportions of diether, macrocyclic diether, and tetraether lipids in *Methanococcus jannaschii* grown at different temperatures. J. Bacteriol. 173: 3907–10.

Sprott, G.D., S. Sad, L.P. Fleming, C.J. Dicaire, G.B. Patel and L. Krishnan. 2003. Archaeosomes varying in lipid composition differ in receptor-mediated endocytosis and differentially adjuvant immune responses to entrapped antigen. Archaea. 1: 151–64.

Stepanov, V.G. and J. Nyborg. 2002. Thermal stability of aminoacyl tRNAs in aqueous solutions. Extremophiles 6: 485–490.

Svobodová, J. and P. Svoboda. 1988. Membrane fluidity in *Bacillus subtilis*. Physical change and biological adaptation. Folia Microbiol. (Praha) 33: 161–169.

Thomas, T. and R. Cavicchioli. 1998. Archaeal cold-adapted proteins: structural and evolutionary analysis of the elongation factor 2 proteins from psychrophilic, mesophilic and thermophilic methanogens. FEBS Letters 439(3): 281–286.

Tomazic, S.J. and A.M. Klibanov. 1988. Mechanisms of irreversible thermal inactivation of *Bacillus* α-amylases. J. Biol. Chem. 263(7): 3086–3091.

Trivedi, S., S.R. Rao and H.S. Gehlot. 2005. Nucleic acid stability in thermophilic prokaryotes: a review. J. Mol. Cell Biol. 4: 61–9.

Unsworth, L.D., J. Van Der Oost and S. Koutsopoulos. 2007. Hyperthermophilic enzymes stability, activity and implementation strategies for high temperature applications. FEBS Journal 274(16): 4044–4056.

Untersteller, É., B. Fritz, Y. Bliériot and P. Sinaÿ. 1999. The structure of calditol isolated from the thermoacidophilic archaebacterium *Sulfolobus acidocaldarius*, Comptes Rendus l'Académie Des Sci. - Ser. IIC - Chem. 2: 429–433.

van de Vossenberg, J.L., T. Ubbink-Kok, M.G. Elferink, A.J. Driessen and W.N. Konings. 1995. Ion permeability of the cytoplasmic membrane limits the maximum growth temperature of bacteria and archaea. Mol. Microbiol. 18: 925–32.

van de Vossenberg, J.L., A.J. Driessen and W.N. Konings. 1998. The essence of being extremophilic: the role of the unique archaeal membrane lipids. Extremophiles 2: 163–170.

Vieille, A. and G.J. Zeikus. 2001. Hyperthermophilic enzymes: sources, uses, and molecular mechanisms for thermostability. Microbiol. Mol. Biol. R.x 65(1): 1–43.

Walther, J., P. Sierocinski and J. van der Oost. 2011. Hot transcriptomics. Archaea. 2010: Article ID 897585.

Wang, Z., W. Tong, Q. Wang, X. Bai, Z. Chen, J. Zhao et al. 2012. The temperature dependent proteomic analysis of *Thermotoga maritima*. PLoS One 7(10): e46463.

Watanabe, M., M. Matsuo, S. Tanaka, H. Akimoto, S. Asahi, S. Nishimura et al. 1997. Biosynthesis of archaeosine, a novel derivative of 7-deazaguanosine specific to archaeal tRNA, proceeds via a pathway involving base replacement on the tRNA polynucleotide chain. J. Biol. Chem. 272: 20146–20151.

Weng, L., Y. Feng, X. Ji, S. Cao, Y. Kosugi and I. Matsuil. 2004. Recombinant expression and characterization of an extremely hyperthermophilic archaeal histone from *Pyrococcus horikoshii* OT3. Protein Expr. Purif. 33: 145–152.

White, D. 2000. The Physiology and Biochemistry of Prokaryotes. New York: Oxford University Press. 1–36

White, M.F. 2003. Archaeal DNA repair: paradigms and puzzles. Biochem. Soc. Trans. 31: 690–3.

White, M.F. and T. Allers. 2018. DNA repair in the archaea—an emerging picture. FEMS Microbiol Rev. 42(4): 514–526.

Woese, C.R. and G.E. Fox. 1977. Phylogenetic structure of the prokaryotic domain: the primary kingdoms. Proceedings of the National Academy of Sciences of the United States of America 74: 5088–5090.

Woese, C.R., O. Kandler and M.L. Wheelis. 1990. Towards a natural system of organisms: proposal for the domains archaea, bacteria, and eucarya. Proceedings of the National Academy of Sciences of the United States of America 87: 4576–4579.

Woese, C.R. 1993. The Archaea: their history and significance. *In*: Kates M., Kushner D.J. and Matheson, A.T. (eds.). The Biochemistry of Archaea. Amsterdam: Elsevier Science Publishers B.V., pp. vii–xxix.

Wu, D., J. Raymond, M. Wu, S. Chatterji, Q. Ren, J.E. Graham et al. 2009. Complete genome sequence of the aerobic CO-oxidizing thermophile *Thermomicrobium roseum*. PLoS One 4: 1300 e4207.

Wu, W., C.L. Zhang, H. Wang, L. He, W. Li and H. Dong. 2013. Impacts of temperature and pH on the distribution of archaeal lipids in Yunnan hot springs, China. Front. Microbiol. 4: 312.

Yadav, A.N., S.G. Sachan, P. Verma, R. Kaushik and A.K. Saxena. 2016. Cold active hydrolytic enzymes production by psychrotrophic Bacilli isolated from three sub-glacial lakes of NW Indian Himalayas. J. Basic Microbiol. 56: 294–307.

Yang, H., S. Fitz-Gibbon, E.M. Marcotte, J.H. Tai, E.C. Hyman and J.H. Miller. 2000. Characterization of a thermostable DNA glycosylase specific for U/G and T/G mismatches from the hyperthermophilic archaeon *Pyrobaculum aerophilum*. J. Bacteriol. 82: 1272–9.

Yang, Y., D.T. Levick and C.K. Just. 2007. Halophilic, thermophilic, and psychrophilic archaea: cellular and molecular adaptations and potential applications. J. Young Investig. Oct.

Yokoyama, S., K. Watanabe and T. Miyazawa. 1987. Dynamic structures and functions of transfer ribonucleic acids from extreme thermophiles. Adv. Biophys. 23: 115–147.

Zeldovich, K.B., I.N. Berezovsky and E.I. Shakhnovich. 2007. Protein and DNA sequence determinants of thermophilic adaptation. PLoS Comput. Biol. 3: e5.

Zhang, Y.M. and C.O. Rock. 2008. Membrane lipid homeostasis in bacteria. Nat. Rev. Microbiol. 6: 222–33.

Zhaxybayeva, O., K.S. Swithers, P. Lapierre, G.P. Fournier, D.M. Bickhart, R.T. DeBoy et al. 2009. On the chimeric nature, thermophilic origin, and phylogenetic placement of the Thermotogales. Proc. Natl. Acad. Sci. 106: 5865–70.

Chapter 7

Microbes from Cold Deserts and Their Applications in Mitigation of Cold Stress in Plants

Murat Dikilitas,[1,*] *Sema Karakas,*[2] *Eray Simsek*[1] and *Ajar Nath Yadav*[3]

Introduction

Abiotic stresses like high salt, extreme drought, heavy metal, water stress, environmental pollution, low temperature or cold stress, etc. have all been receiving increasing trends as climatic condition changes. Their impact would further increase as they interact with each other or with biotic stressors. These stresses could affect the quality and quantity of crop plants significantly. One factor that differs from the others is its effects could last longer periods of time while its striking effect is comparatively shorter when compared to those of other stress factors; the cold stress. Even 1 or 2 hours effects of cold or low-temperature stress on crop plants may have drastic consequences when the time of stress coincides with the stages of germination or flowering in plants. Its effect at these periods could affect the entire quality and quantity of crop production in the future. Therefore, low-temperature stress plays a more significant role than those of other stresses in reducing agricultural crop production. Tolerant plants to cold stress could be a promising solution, however, generating these types of crop plants or increasing their tolerance to cold stress via biochemical approaches have not always been successful due to the complexity and duration of cold stress on crop plants. However, inoculation with efficient microorganisms exhibiting Plant Growth-Promoting (PGP) traits at cold stress or low-temperature conditions could be a logical solution and a promising technological approach to enhance crop production for the future of agriculture.

As in the other abiotic and biotic interactions such as drought stress and plant pathogens (Dikilitas et al. 2016); salinity stress and plant pathogens (Dikilitas et al. 2019a); heavy metal stress and plant pathogens (Taiti et al. 2016), etc. there is also an interaction between cold stress and plant pathogens. These two stress factors might negatively regulate the defence responses and production of crop plants. For example, the subtropical races (STR4) of *Fusarium oxysporum* f. sp. *cubense* (*Foc*), the causal disease agent for Fusarium wilt in bananas (*Musa* sp.) worldwide, cause significant loss in subtropical countries usually after a winter regime (Sutherland et al. 2013). The authors stated

[1] Department of Plant Protection, Faculty of Agriculture, Harran University, 63210, S. Urfa, Turkey.
[2] Department of Soil Science and Plant Nutrition, Faculty of Agriculture, Harran University, S. Urfa, Turkey.
[3] Department of Biotechnology, Dr. KSG Akal College of Agriculture, Eternal University, Baru Sahib, Sirmour-173101, India.
Email: skarakas@harran.edu.tr
* Corresponding author: m.dikilitas@gmail.com

that cold stress predisposed Cavendish banana plants to infection and made the plants susceptible to the effect of the disease. In this respect, plant microbiome such as epiphytic, endophytic and rhizospheric organisms could play significant roles in defence responses and tolerance mechanisms in plant development and growth mechanisms under cold stress.

The microbiome associated with plants could be classified according to the place they colonize. For example, the rhizosphere, the root zone of plants, has a significant influence on the microbial activity (Yadav et al. 2020b; Yadav et al. 2020c; Yadav et al. 2020d). They could attach to the root surfaces and benefit from root exudates. Several factors such as type of soil, soil moisture content, soil pH and temperature as well as age, conditions and types of plants could determine the efficacy and types of rhizospheric microbes (Rastegari et al. 2020). Several microbial species such as *Acinetobacter, Arthrobacter, Aspergillus, Azospirillum, Bacillus, Erwinia, Penicillium, Pseudomonas* and *Rhizobium* could be counted in this category (Barea et al. 2005; Verma et al. 2015a; Yadav et al. 2017a). The phyllosphere, on the other hand, describes the leaf surface in which the phyllospheric microorganisms may survive and proliferates by obtaining principal nutrients such as amino acids, glucose, fructose and sucrose and some other substances and uses them for the benefit of crop plants (Yadav et al. 2020a). The phyllospheric microorganisms may control the dissemination and infection of air-borne pathogens on leaf surfaces. They could tolerate a vast range of temperatures (5–55°C) and UV radiation. Quite a few microorganisms like *Agrobacterium, Azotobacter,* and *Xanthomonas,* etc. could be counted in this category (Hornschuh et al. 2002; Verma et al. 2015b; Yadav et al. 2017a). The endophytic microorganisms, on the other hand, colonize the interior parts of plants, viz: root, stem or other interior parts of the host plants without causing any harmful effects. Endophytic microorganisms are able to enter the host plants through epidermal openings, root hairs, wounds or naturally occurring gaps as a result of plant growth and development. Endophytic microorganisms can be transmitted either from the parent to offspring or among individual crop plants (Kour et al. 2020; Rana et al. 2019; Suman et al. 2016). As a summary, the main colonization route of the microorganisms seems to be the rhizosphere. Here, amino acids, sugars and the lipo- and exopolysaccharides excreted from the roots of host plants play significant roles in the attachment of microorganisms to plant tissues (Vaishnav et al. 2016).

These beneficial microorganisms have emerged as an important and promising tool for sustainable agriculture (Yadav et al. 2017b). They could promote the growth of crop plants by either facilitating nutrient uptake or regulating hormone levels in plants, or they may decrease the pathogenic characteristics of the invading microorganisms. Although the mode of actions of these microorganisms may vary, they, in general, may release plant growth regulators or phytohormones, solubilize phosphorus, potassium and zinc and fix nitrogen and produce siderophores, ammonia, HCN and other secondary metabolites which could be antagonistic against pathogenic microorganisms for the benefit of crop plants (Yadav et al. 2017b). For example, auxins can promote the growth of plants in terms of the root system and stems by increasing cell elongation or cell division as well as differentiation (Dugasa et al. 2019). They can also regulate physiological processes (Zhang et al. 2019a). In this way, they absorb more water and nutrients and improve survival of plants under harsh conditions. It would be a very logical approach to select cold-tolerant microorganisms having antipathogenic properties for the control of disease agents under cold stress in crop plants. By this approach, both cold stress and pathogenicity under cold stress could be controlled or at least one of the stress agents could be suppressed.

Along with the other methods such as a chemical application or molecular improvements such as gene modification or gene transfer of Clustered Regularly Interspaced Short Palindromic Repeats (CRISPR) technology, the application of microorganisms appears as a promising approach and has the ability to control the plant pathogens as well. Therefore, it might have more positive sites when compared to those of other approaches when cold tolerance is considered.

Cold stress on crop plants

The Plasma Membrane (PM) is regarded as the primary site of injury and pathogenic attacks, the lipids here are the first affected components especially when the plant is subjected to low-temperature stress (Uemura et al. 2006). Tolerance of PM to cold stress largely depends on the concentration of unsaturated fatty acids. The size of water molecules tends to increase at or below temperatures of freezing points in plants. At this point, the water potential tends to decrease and water moves to the intercellular spaces by creating water deficiency in the cell. After the establishment of cold temperature, aquaporins of the PM are downregulated as illustrated in cold-tolerant *Arabidopsis*. This plays a significant role in modulating homeostasis and hydraulic conductivity (Jang et al. 2004; Dong et al. 2019). As the temperature drops, calcium-permeable channels act as sensors for low-temperature stress and are responsible for Ca^{2+} influx (Wang et al. 2019). A cytosolic Ca^{2+} rise is a good criterion for cold sensing. This is a common response in chilling-sensitive and cold-tolerant plants. For example, a linear relationship was established between the severity of cold temperature and the concentration of calcium influx. Higher calcium influx was evident with more intense cold temperatures (Knight 2002; Wang et al. 2019).

Ca^{2+} increases in plants as a result of Ca^{2+} influx across the PMs may be released from vacuoles. It is possible that cold stress may open Ca^{2+} channels so that Ca^{2+} could enter the cells (Wang et al. 2019). In the cell, calcium sensors such as calcium-dependent protein kinases, calmodulin, calmodulin-like proteins sense intracellular Ca^{2+} are present (Solanke and Sharma 2008). Although it is thought that these sensory proteins work with stress-responsive genes or with the transcription factors, the exact mechanism is still to be elucidated. It is hypothesized that the products of these stress-related genes eventually lead to plant adaptation and help the plant to survive under adverse conditions. For example, application of hot water as dipping the banana fruit for 3 minutes at 52°C led to increasing in Ca^{2+}-ATPase synthesis and increased cold stress tolerance at 7°C (Wang et al. 2008a). Therefore, the production of sensory proteins through stress-responsive genes play significant roles in cold adaptation periods.

After cold stress, visual symptoms can be detected within a few hours or in a few days depending on the severity and duration of the stress and plant species, but in reality, the plant cell through signal transduction pathway responds to cold stress within minutes if not in seconds. Cold stress results in many physiological and biochemical modifications in crop plants such as germination, photosynthesis, cell division, flowering, fruiting and seed quality. Prolongation of cold stress significantly deteriorates the physiological and biochemical parameters even it is at a low grade. If crop plants are already susceptible to cold stress, other defensive responses of crop plants would be quite insufficient.

Cold stress, one of the least studied abiotic stresses, is one of the most devastating stress factors when its final effect on crop plants is considered (Yadav et al. 2019; Yadav et al. 2018). Since plants are sessile, their survival strategy largely depends on the efficient activation of tolerance and defence mechanisms to cold stress. In a very short time, its effect could deteriorate the physiological, biochemical and molecular functions of crop plants even if the following stage was optimal for plant growth. For example, Jain et al. (2007) reported that poor sprouting of stubble buds at 15- and 6°C was found to be associated with a low level of reducing sugars, reduced acid invertase activity and high accumulation of Indole Acetic Acid (IAA) as well as total phenolic compounds. It is probable that low temperatures interfered with the biochemical metabolism and metabolites essential for sprouting. The authors suggested that *in situ* accumulation of IAA and total phenols synthesized at low temperatures might be responsible for the maintenance of dormancy in stubble buds.

The effect of cold stress on crop plants could show variation and can be influenced by the species of plants, intensity and duration of the stress along with the developmental stages of crop plants. Although symptoms of cold stress have different effects on different crop plants, the most common symptoms include wilting, necrosis, chlorosis, reduced growth and development, leaf rolling, fall of flowers and leaves, etc. (Mahajan and Tuteja 2005; Joshi et al. 2018). One of the mechanisms of wilting could be explained with the reduced uptake of water through the roots and the reduced flow of water in phloem while stomata are continuously open (Bloom et al. 2004; Agurla et al. 2018). Chlorosis, on the other hand, is another symptom of cold stress in which low temperatures lead to damage to photosynthetic organelles by inhibiting ATP and NADPH synthesis (Campbell et al. 2007; Hajihashemi et al. 2018). As the cold stress prolongs and intensifies, ice crystals quickly form in the intercellular spaces and results in a reduced rate of photosynthesis and lead to necrotic symptoms.

Low-temperature stress along with the other stress factors extends the time period between flowering and the harvest, fruit formation and maturation as well as quality and quantity of crop yield and the combination of stresses intensifies and increases the severity of symptoms caused by either cold stress or other individual stresses. For example, the intensified light exposed on leaves or fruits results in further deterioration of physiological and biochemical functions of crop plants under stress conditions. This could severely affect the food market and agricultural economy (Mittler 2006; Subramanian et al. 2011).

Cold stress initially disrupts the membrane integrity of the cell wall and leads to cellular leakage and dehydration. On exposure to cold stress, unsaturated fatty acid contents significantly increase and this leads to increased viscosity of the PM, which protect the plant against cold damage (De Palma et al. 2008; Promyou et al. 2008; Barrero-Sicilia et al. 2017). Some enzyme activities also increase breaking down the starch molecules that result in the accumulation of sugar molecules. For example, Maul et al. (2008) stated that the increased β-amylase activity led to the accumulation of maltose and sucrose sugars in grapefruits which helped to protect the photosynthetic electron transport chain and proteins in chloroplast stroma and flavedo (coloured outer peel layer) during low temperatures. Some other enzymes have been associated with the accumulation of soluble sugars. Another osmoprotectant, glycine betaine also plays significant roles in the protection of cell membranes from cold stress. It leads to an increase in unsaturated fatty acids in the thylakoid membranes and helps to protect complexes and enzyme structures.

Cold tolerant plants acclimatize their metabolisms during autumn in which the cryoprotectant molecules such as soluble sugars (saccharose, raffinose, stachyose, trehalose), sugar alcohols (sorbitol, ribitol, inositol) and low-molecular-weight nitrogenous compounds (proline, glycine betaine) are accumulated (Janska et al. 2010). These molecules interact with dehydrin proteins, Cold-regulated proteins (Crp), and Heat-shock proteins (Hsp). They work together to stabilize membrane phospholipids, cytoplasmic proteins and other structural proteins as well as maintaining ion balance while scavenging Reactive Oxygen Species (ROS) during cold stress (Gusta et al. 2004; Chen and Murata 2008).

Apart from the accumulation of cryoprotectant molecules, the increased activity of the antioxidative enzymes such as superoxide dismutase (SOD), glutathione peroxidase (GPX), Glutathione Reductase (GR), ascorbate peroxidase (APX) and catalase (CAT) as well as the accumulation of non-enzymatic antioxidants such as tripeptide thiol, glutathione, ascorbic acid (vitamin C) and α-tocopherol (vitamin E) also play important roles in cold tolerance (Chen and Paul 2002; Janska et al. 2010; Phornvillay et al. 2019).

Some plants synthesize antifreeze proteins (AFPs) to respond to cold stress by inhibiting the activity of ice nucleators. The structure of these proteins is quite similar to plant Pathogen-Related (PR) proteins in terms of the sequence (Moffatt et al. 2006; Zur et al. 2013). On the other hand, some PR proteins are also synthesized against cold stress. These plants have antifreeze activity in the apoplastic space, therefore, they are able to inhibit the recrystallization of intercellular ice within the cell, and even prevent the intracellular ice formation (Griffith and Yaish 2004; Renaut et al. 2006; Dhume et al. 2019). Induced biosynthesis of flavonoids, anthocyanins, terpenoids and phenylpropanoids are also evident in plants exposed to cold stress (Kaplan et al. 2007). Not only higher levels of anthocyanin and the blue light-absorbing flavonols were expressed on exposure to cold stress in the leaf (Hannah et al. 2006; Korn et al. 2008) but also Salicylic Acid (SA), which has a significant role in plant defence against pathogens, increases on exposure to cold stress along with the increase of a range of phenylpropanoids (Janska et al. 2010; Miura and Tada 2014; Wu et al. 2019). However, it was reported that some secondary metabolites such as terpenoid indole alkaloids were suppressed at low temperature as shown in *Catharanthusroseus* (Dutta et al. 2007). Therefore, increases of biomolecules under cold and pathogen stress conditions could be a good criterion when generating cold-tolerant plants under pathological stress conditions.

Cold stress also results in ultrastructural changes in chloroplasts and thylakoid membranes and reduces the number of starch granules (Kratsch and Wise 2000). Although, most plant species show cold tolerance up to some degree, however, their tolerance depend on the temperature and the length of cold stress exposure (Janska et al. 2010). In general, the cold-tolerant plant species have quite a small leaf surface area and a high root/shoot ratio and slow-growing characteristics to reduce heat emission and chilling (Nilsen and Orcutt 1996; Kaminska et al. 2018). Cold tolerant plants store sugars in their underground tissues with generally C3 mode of photosynthesis mechanisms. Therefore, they rapidly mobilize stored products during the short growing season by a very efficient respiration system.

As in the other types of stress mechanisms, two distinct defence strategies such as stress avoidance and stress tolerance can also be observed in plants under cold stress conditions. Stress avoidance is associated with preventing sensitive tissues from freezing stress. For example, some succulent species with thick cell walls containing abundant water are able to accumulate residual heat during the daytime and dissipate it slowly during the cold night (Nilsen and Orcutt 1996). Upregulation of Late Embryogenesis Abundant (LEA) proteins in plants enables cold acclimation. It is, therefore, hypothesized that LEA proteins stabilize other proteins and membranes by preventing aggregation of proteins under stress conditions (Tunnacliffe and Wise 2007; Banerjee and Roychoudhury 2016).

Different classes of genes activated in response to cold stress are also linked to other abiotic stressors. In general, cold-responsive genes are induced right after cold stress, however, genes induced at later stages often maintain their expression levels. Functional proteins such as LEA proteins, Membrane Stabilizing (MS) proteins and Osmoprotectant Synthesis-Related (OSR) proteins are accumulated by cold-responsive (*COR*), Low-Temperature Induced (*LTI*), cold-induced (*KIN*) and early dehydration-induced (*ERD*) genes (Talanova et al. 2013). Most of these genes are up- or downregulated following cold stress.

As shown in other stress conditions, low temperatures can also lead to ROS production during cold stress. At high concentrations, these molecules can induce oxidative damage to cellular structures by oxidizing proteins, lipids and nucleic acids (Spano et al. 2017) while acting signalling molecules at very low concentrations (Dikilitas et al. 2019a). ROS cause damage to DNA and its components by oxidative stress, they therefore, need to be to detoxified to protect the cell metabolism.

Apart from the production of ROS, Reactive Nitrogen Species (RNS) are also produced in response to cold stress. For example, pea (*P. sativum* cv. Lincoln) plants exposed to various stresses such as chilling, heat and high light intensity exhibited a high level of Nitric Oxide (NO) and its derived molecule RSNO (S-nitrosothiol). These molecules were also verified in pea plants subjected to 8°C of low-temperature stress for a period of 48 hours (Corpas et al. 2008). There were about 1,000 cold-induced genes that were expressed and among them, 170 genes encoded transcription (Thomashow 2010). The accumulation of ROS and NO are able to induce *COR* genes and result in increases in the concentration of compatible osmolytes and activate tolerance mechanisms against the cold (Gupta et al. 2011; Cattivelli et al. 2002). Among ROS, H_2O_2 was responsible for the induction of nearly 60% of the tolerance genes for chilling. Most of the genes expressed within 24 hours after exposure to 10°C in Japonica rice (Yun et al. 2010).

Several enzymes such as SOD, GPX, APX and GR play significant roles in detoxifying the ROS when ROS act as signalling molecules (Dikilitas et al. 2019a). In this way, genes and proteins are activated in repairing and protection of cells against abiotic or biotic stresses. Therefore, quick response to initial cold stress, or eliciting crop plants with a low level of cold stress may elicit the defence responses. For example, *CAT2* gene was upregulated to catalyze the toxic hydrogen peroxide in *Arabidopsis* at low temperature as an early response of the plant (Du et al. 2008; Filiz et al. 2019). Figueroa-Yanez et al. (2012) also stated that the upregulation of putative *MaCAT2* in the peel of the banana fruit was evident during low temperature (10°C).

In terms of photosynthesis and respiration processes, there are clear differences between cold-tolerant and cold-sensitive plants in response to low temperatures. For example, Yamori et al. (2006) stated that cold-tolerant plants performed better than cold-sensitive plants in terms of photosynthesis and respiration and maintained ion homeostasis after low-temperature stress. Cold tolerant plants have a much greater ability to close their stomatal apertures when compared to those of non-cold tolerant plants to reduce water loss and wilting. It is suggested that the increase of osmoprotectants restrict the movement of water to intercellular spaces and are able to protect the macromolecules and cellular structures or cell walls/membranes under cold stress conditions by increasing the osmotic potential (Sharma et al. 2019).

Cold avoidance strategy also involves supercooling, in which endogenous ice nucleation is prevented even where the temperature falls as low as –40°C. For example, winter-hardy species can generate 'liquid glass' within their cells, a highly viscous solution that prevents ice nucleation. Such cells become protected to the presence of external ice when osmotic, thermal and mechanical protection mechanisms are enabled (Wisniewski and Fuller 1999; Janska et al. 2010).

Plants, in general, have three distinct stages in response to cold stress; cold acclimation (pre-hardening), which occurs at just above zero temperatures, the second stage (hardening), which occurs at sub-zero temperatures and the final stage is plant recovery (Li et al. 2008). These stages could play significant roles under pathogen involvement.

Combined effects of cold stress and plant pathogens on crop plants

As described earlier, cold stress or low-temperature stress impacts the growth and development of plants and significantly decreases the yield of agricultural crops. Since temperature is counted as one of the most important predisposing factors and decreases plant defence barrier due to the morphological, phenotypic, physiological, biochemical and molecular effects on crop plants, it has significant effects on plant-pathogen interactions by positively affecting the disease progress and severity as well. It is evident that cold-tolerant plants and microorganisms are able to adapt to seasonal temperature changes by adjusting their metabolism during cold periods such as autumn and increase the concentrations of cryoprotective chemical compounds to tolerate cold stress. Some of these compounds are synthesized *de novo*. Low temperatures not only predispose the host defence but has significant effects on pathogen growth and dissemination. One of the mechanisms was explained by the status of Resistance (R) genes. The resistance genes are regulated by temperature. For example, *Yr36* gene of *Puccinia striiformis* f. sp. *tritici* Erikss., the causal agent of wheat stripe rust, was induced at relatively higher temperatures (25–35°C) as compared to those of low temperatures (15°C) (Fu et al. 2009). Often the R gene *Xa7* in *Xanthomonas oryzae* pv. *oryzae*, was more effective at higher temperatures whereas genes *Xa3*, *Xa4*, and *Xa5* were more effective at lower temperatures (Webb et al. 2010).

When a pathogen is recognized by a host plant, signalling cascades are activated and defence-related enzymes and hormones are synthesized and defence-related genes are induced. The major plant hormones such as SA, Jasmonic Acid (JA), ethylene (ET) and abscisic acid (ABA) are also activated on response to cold stress. Most of these hormones are regulated by temperature. For example,crop plants may become susceptible at low temperatures if the expressions of ABA or other hormones are low (Huang et al. 2017).

The change in temperature has also significant effects on plant pathogens. Walker and Van West (2007) reported that oomycete pathogens produced zoospores greatly at low temperatures. In fact, a drop in temperature resulted in an increased chance for the infection as more zoospores were produced. For example, Kilinc (2018) stated that *Zymoseptoriatritici*, a causal agent for leaf spot disease in cereals, was more effective and produced more protein-degrading enzymes such as protease and laccase at relatively low temperatures (16–20°C) and this resulted in more symptoms and devastating effects on wheat plants at low temperatures compared to those growing at 23°C.

Apart from the sporulation and dissemination of the pathogens, temperature has also a significant effect on the virulence of pathogens. For example, Fang et al. (2011) stated that *Fusarium oxysporum* Schltdl. was more virulent at 27°C as compared to those growing at 17°C on strawberry plants. In contrast, Ullrich et al. (1995) stated that *Pseudomonas syringae* increased its virulence and toxicity when temperature decreased.

Small RNAs including microRNAs (miRNAs) and small interfering RNAs (siRNA) also play a crucial role in the plant defence system by silencing targeted genes required by the pathogen for infection. However, reduced expression of siRNAs, as a result of cold stress, negatively regulates the defence mechanism. For example, tobacco plants exhibiting disease symptoms resulting from Cymbidium ringspot virus showed that many severe symptoms at low temperatures were due to suppression of siRNA-mediated RNA molecules (Szittya et al. 2003; Abla et al. 2019). These cold-responsive miRNAs and siRNAs probably mediate the response to cold stress by modulating growth and development, hormone signalling, defence enzymes and secondary metabolites. If these miRNAs or siRNAs are identified clearly, they could be used for molecular breeding studies for improving cold tolerance in crop plants.

Following temperature stress, chaperones-like heat shock proteins are also produced and these proteins play a significant role for the protection of the plants. For example, Hsp70, induced following heat or cold stress, is an important protein in virulence of *P. syringae* to *Arabidopsis* (Jelenska et al. 2010). HopI1, a virulence effect or binding to Hsp70 protein was induced following heat stress and resulted in an increase in disease severity. In general, genes upregulated due to temperature stress including cold or heat produce heat or cold shock proteins, however, these proteins can be metabolized by the pathogen to increase its virulence. As a result of this, disease severity in plants could be inevitable (Lo Presti et al. 2015). For example, Lo Presti et al. (2015) stated that Dnj1-like proteins were required for successful disease development in *Ustilagomaydis* and *Fusarium oxysporum.*

The pathogen population could need time to build up at warmer temperatures. Therefore, as soon as there is an increase in temperature after a cold stress period, a significant increase in disease progress at the start of the warm period is unavoidable. Although some evidence showed that cold temperature prevents the occurrence of the diseases or reduces the severity of symptoms caused by plant pathogens, most of the findings showed that cold stress leads to increased disease severity and reduction in host resistance. For example, *Alternaria alternata* exhibited very mild symptoms on cotton plants at the beginning of spring while no visible symptoms were evident during cold seasons (Moyer et al. 2010). However, most of the findings were related to the susceptibility of crop plants under cold stress. For example, soybean seed diseases caused by necrotroph pathogens like *Fusarium graminearum* Schwabe and *Alternaria alternate* (Fr.) Keissl. increased following frost damage at –4–5°C (Osorio and McGee 1992). Recently, Velasquez et al. (2018) and Dikilitas et al. (2019b) stated that plant resistance pathways could be influenced by the environmental stress while the virulence mechanisms involving toxin production and virulence protein synthesis are encouraged in attacking pathogens thus putting resistant crop plants in great danger.

When chemicals are applied to crop plants to increase the tolerance of cold stress, every aspect that deals with plant defence responses should be considered. For example, the application of ABA on tomato fruit resulted in an increase in cold tolerance, however, susceptibility to infection by opportunistic pathogens increased as well (Ding et al. 2002). A vice versa case is also valid as PR genes are downregulated at low temperatures which can finally result in susceptibility to cold stress.

Cold stress not only predisposes the crop plants for pathogen infection but also enhances symptom occurrences and severity. For example, barley yellow dwarf virus in maize (*Zeamays* L.) exhibited severe symptoms at temperatures between 18–25°C than those between 25–30°C (Brown et al. 1984). Also, aggressiveness and symptoms of Rhizoctonia root rot caused by *Rhizoctoniasolani* Kühn increased in winter wheat at low temperatures between 6–19°C when compared to plants growing at higher temperatures between 16–27°C (Smiley and Uddin 1993). Increased development of Fusarium head blight and higher levels of deoxynivalenol (DON) production in winter wheat were also evident with the increase of rainfall associated with low temperatures (Tamburic-llincic et al. 2007). Serrano and Robertson (2018) reported that cold stress 2 to 4 days after planting increased soybean susceptibility to damping-off caused by *Pythium sylvaticum*. They suggested that increased seed exudation and delayed seedling growth resulted in seedling susceptibility. Again, Serrano et al. (2018) reported that soybean damping-off caused by *Pythium* spp. was found to be associated with the cold weather just after spring planting.

Kemal et al. (2017) stated that the resistant genotype ICC-12004 of chickpea (*Cicerarietinum* L.), which is a highly nutritious food for human consumption and rotation crop in various parts of the world showed low disease severity under Ascochyta blight (*Didymella rabiei*) at all temperatures. The authors reported that the highly aggressive isolate AR-04 of *D. rabiei* caused significant disease symptoms and produced ample amount of pycnidia in a very short period and high disease severity in all temperatures, which clearly exhibited that the fungal race was adapted to a wide range of temperatures. Genotypes exposed to chilling stress, on the other hand, showed high disease severity. They stated that chilling temperature predisposed chickpea genotypes to *D. rabiei* infection, possibly the fungus was in a latent period at warmer conditions and acted as a severe pathogenic agent at colder temperatures. Therefore, developing multiple stress resistance in winter-sown chickpea is a prerequisite for the future of agricultural studies. Emphasis should, therefore, be given to developing germplasms with high levels of cold tolerance along with the disease resistance.

Similarly, Moyer et al. (2016) stated that the development of *Erysiphenecator* during acute cold stress (1 hour at 4°C) resulted in the death of hyphal segments and increased its latent period. On the other hand, acute cold stress resulted in the host susceptibility of grapevine prior to the arrival of the pathogen. They also stated that the effects of repeated cold events had profound effects on the epidemic progress that made it difficult for the management of grape powdery mildew. On the other hand, the authors reported that acute low-temperature events before inoculation by *E. necator* induced a transient resistance response (Moyer et al. 2010; Moyer et al. 2016). They termed this phenomenon as cold Stress-Induced Disease Resistance (cold SIDR). They indicated that cold SIDR reduced infection success rates and slowed down mycelial growth and colonization and prolonged latent periods. But, in most of the cases, cold stress with pathogen combinations had devastating effects.

New races of the pathogen under cold stress could also develop, for example, Pivonia et al. (2012) stated that the young pepper plants artificially inoculated with *Pythium* sp., maintained at 20-, 14-, 10.5- and 8.6°C, showed wilting just after 2 weeks in plants growing at 8.6°C. At 10.5°C, wilting developed more slowly, and inoculated plants maintained at 14°C and 20°C did not exhibit any wilting symptoms. They claimed that the unique variation in sporangium morphology and the sequence of the ribosomal Internal Transcribed Spacer (ITS) suggested that a new species of *Pythium* were involved. The relationship between cold stress and high disease severity established that the high disease incidence in the Arava Valley of Israel during the cold winters of 1999–2000, 2004–2005 and 2006–2007 was severe (Pivonia et al. 2012).

Effects of cold stress on cell metabolism (Fig. 7.1) and cold and pathogen stress on cell metabolism (Fig. 7.2) and possible behaviours of plant pathogens under cold stress conditions (Fig. 7.3) and the possible interactions between plant pathogens and abiotic stress factors (Fig. 7.4) were briefly illustrated in the following figures and their mode of actions were verified in figure captions.

Microbes from cold stress and their applications on crop plants

Despite the prediction in global warming scenario for the coming years, modelling evaluations for the climate framework reveal that cold occurrences will be prevalent throughout the 21st century and may become more severe than the previous century (Kodra et al. 2011). These cold events

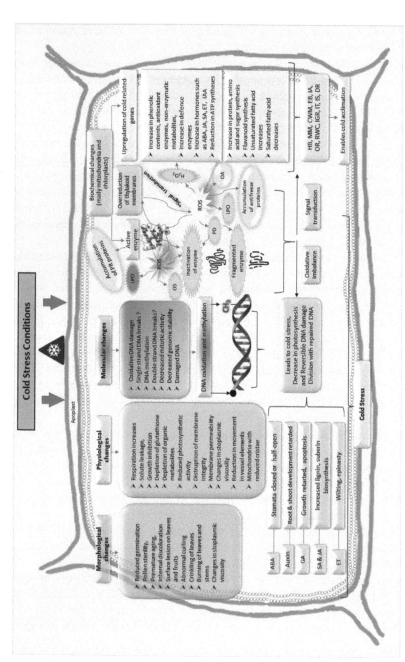

Figure 7.1 Responses of crop plants under cold stress conditions and overview of the effects of cold stress on cell metabolism. Mitochondria and chloroplasts are the major sites of ROS production. Accumulation of ROS leads to the degradation of proteins, enzymes and even DNA molecules. Major enzymatic activities that control the production of ROS also take place in mitochondria and chloroplasts. However, cold stress can reduce the production of enzymatic activities due to reduction of electron transport by reduced thylakoid membranes and cristae in mitochondria, therefore, an imbalance would occur in oxidative status. ROS are not simply toxic by-products of metabolism. At low concentration, ROS act as signalling molecules by regulating the expression of various genes, including those encoding antioxidant enzymes and H_2O_2 production. * Signal transduction pathway occurs through membrane-specific proteins, secondary molecular sensing pathway through Ca^{2+} and ROS. The signal transduction starts with transcription factors and induced gene expressions, and damage control is then enabled with the repair process through the restructuring of the plasma membrane along with the accumulation of osmolytes. This could enable molecular cold adaptation. Therefore, priming studies, as well as the application of microorganisms, would be of great importance to reduce the impact of cold stress. Accumulation of AFPs in cold-adapted plants may also possess antifungal activity. Accumulation of PR proteins may enable plants some degree of cold tolerance. However, these two issues need to be elucidated in detail since there is not much comprehensive work on them. In the illustration, the same colours indicate the pathway occurring with the same group of mechanisms.

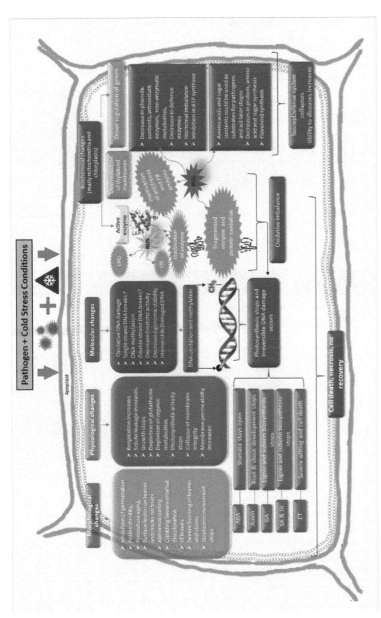

Figure 7.2 Responses of crop plants under cold and pathogen stress conditions. Cold stress disrupts the cell membrane and affects the fatty acid composition that could be a good substrate for invading pathogens. Cold stress also disrupts the structure and properties of cell wall materials; therefore, it predisposes crop plants to pathogenic infection. Although ABA plays an important role for the closure of stomata, however, overproduction of ABA leads to increased sporulation and dissemination of pathogens. ABA has a negative correlation with other hormones such as SA, JA, ET, etc., therefore, the suppression of such hormones results in decreases in defence mechanisms leading to increased attacks caused by plant pathogens. The combination of cold and pathogen stresses on crop plants could be additive, synergistic or antagonistic. However, most of the interactions studied so far have involved the additive and synergistic mechanisms. Cold and biotic stress factors can affect the ion homeostasis, and most of the defence-related genes are either downregulated or not expressed at all in cold stress and pathogen invading stress conditions. Therefore, stresses from both stressors should be well evaluated individually and in combinations to reveal the mechanisms in detail. The intensity of colours shows the degree of severity when compared to that of the previous figure. The outline of this mechanism could be affected by the severity, duration, number of exposures and combination of stress. Organs or tissues in question, stage of plant development and genotype could also affect the severity of cold stress. ROS, reactive oxygen species; ABA, abscisic acid; GA, Gibberellic Acid; SA, Salicylic Acid; JA, Jasmonic Acid; ET, ethylene; IAA, indole acetic acid; OS, osmotic stress; OA, osmolyte accumulation; LPO, lipid peroxidation; PD, protein degradation; HB, hormone balance; MM, membrane modification; CWM, Cell Wall Modification; EB, Energy Balance; IA, Increased Antioxidants; OR, Osmotic Regulation; RWC, Reduced Water Content; IGR, Increased Gene Regulation; IT, Increased Transcription; IS, Increased Stability; DR, Downregulation. Intensity of colors shows the degree of severity of stress. Similar mechanisms were expressed in a particular color.

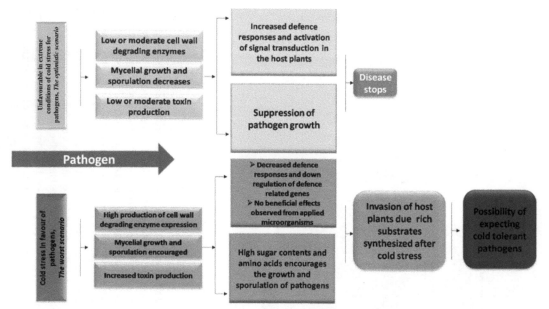

Figure 7.3 Possible pathogen behaviours under cold or chilling stress conditions. Pathogens might be freely sporulated and disseminated under mild cold stress and cause severe infections along with the negative effects of cold stress on crop plants. In contrast, cold stress might suppress pathogen dissemination and reduces virulence and pathogenicity.

	Drought	Salinity	Heat	Freezing	Cold	Chilling	Pathogen 1	Pathogen 2*	Pollution
Drought									
Salinity									
Heat									
Freezing									
Cold									
Chilling									
Pathogen 1									
Pathogen 2*									
Pollution									

Figure 7.4 Possible interactions between different stress types. Positive interactions were indicated with green colours, which show the additive effects of stressors; negative interactions were indicated with red colours, which show the possible antagonistic effects. Unknown interactions or no interactions were indicated with a yellow colour in which a possible interaction in the future might be released. *Pathogens might be suppressed depending on the pathogen species and propagules. However, occurrences of antagonistic effects between pathogens and cold stress are highly unlikely in nature.

would probably occur in both tropical and temperate regions and though they are even predicted to have a short duration from 24 hours to a few days, their effects would be devastating on crop plants.

Microorganisms thriving in harsh cold environments should have specific properties. They should produce AFPs, high levels of polyunsaturated fatty acids, ergosterol, proline, sugars, etc. Not only should these chemicals be produced in harsh conditions but also be renewed and topped up when they are consumed. The freeze-thaw cycle might deteriorate the physiological, biochemical and molecular functions. Under these circumstances, the antioxidant capacity and antimicrobial properties and even sporulation and mycelial developmental characteristics might be deteriorated (Dikilitas et al. 2011). Recently, Villarreal et al. (2018) reported that the yeasts *Vishniacozymavictoriae*, *G. gastrica* and *Leucosporidium creatinivorum*, isolated from Antarctica, were found to have cold-tolerant properties such as AFPs, fatty acids and ergosterol which could be used for industrial purposes. The North and South Poles represent the most inhospitable habitats in terms of cold and dry weather conditions, cold-adapted microorganisms in these regions could be important resources to study novel secondary metabolites and enzymes. There are quite a few microorganisms that have been successfully isolated and cultured from cold environments (Wang et al. 2017). These microorganisms have PGP properties, and could play vital roles in studies regarding plant-growth-promoting biocontrol, nitrogen fixation, antagonistic and alleviation of cold stress studies in plants. Although most of the recent studies conducted so far have largely focused on rhizobia, another interesting area that has come up recently concentrating on cold-active microorganisms that are also able to reduce the metal toxicity and harmful waste (Mishra et al. 2012).

About more than 80% of the Earth's surface is below 15°C of temperature at various times of the year (Kawahara 2002). Agricultural soils in these areas are seriously exposed to a seasonal drop in terms of temperature and rainfall for a great part of the year. During this period, any biological activity is low or negligible (Robertson and Grandy 2006; Zhang et al. 2019a). Under cold climatic conditions, crop plants do not perform efficiently due to lack of nitrogen, water and light. This, of course, has led to short cropping seasons for agricultural work in the temperate regions (Mishra et al. 2010). Naturally, cold-adapted microorganisms could be used for the benefit of crop development under cold climatic conditions (Yadav et al. 2015a; Yadav et al. 2015b).

Although most wild plants are able to adapt to harsh conditions such as salinity, drought, cold, heavy metal stresses very efficiently, however, most of the crop plants such as banana, tomato, maize, cucumber, pepper, are unlikely to acclimatize to harsh conditions and need help in tolerating adverse conditions. Basically, these plants are quite sensitive to cold stress and cannot function properly when compared to those of other crop plants. Improving plants through breeding or genetic transformations is not always as successful as the cold tolerance of plants that cannot be cold-tolerant with expressing a single gene. Instead, complex mechanisms involving multiple genes are needed to protect the plant against cold stress (Subramanian et al. 2011). However, we can benefit from microorganisms thriving under cold climates.

Many fungal or bacterial phytopathogens can be quite destructive and virulent when the soil temperature becomes low, hence it is quite appropriate to expect that biocontrol agents are also cold-tolerant and cold-adapted (Mishra et al. 2010; Kour et al. 2019a; Kour et al. 2019b). Cold-adapted microorganisms have a great capacity to survive in extremely cold temperature conditions through the synthesis of cryoprotectants, AFPs, cold shock proteins and ice nucleators. They could even live at low temperatures almost close to freezing point of water (Shivaji et al. 2011). Due to the slow growth rate and difficulty of culturing or handling, so far little attention has been paid to cold-adapted microorganisms. These microorganisms survive and retain their physiological and biochemical activities at very low temperatures and could grow much better at warmer temperatures (Mishra et al. 2010; Kumar et al. 2019; Yadav 2019). To date, very little information has been available on the use of these organisms in agriculture. Understanding the mechanisms of these microorganisms would help one carry out agriculture in cold areas and most importantly one could be able to protect the crop plants and trees from freezing temperatures.

The cold-adapted microorganisms are good sources of pigments, cold-active enzymes and antifreeze compounds, which can be used efficiently in agricultural studies as inoculants or biocontrol agents where cold stress deteriorates the conditions such as flowering or fruiting of crop plants (Singh et al. 2020; Subrahmanyam et al. 2020; Yadav 2020). For example, Sati et al. (2013) isolated 25 morphologically distinct bacteria belong to *Bacillus* and *Pseudomonas* species, and fungi including yeast from the high altitudes of the Indian Himalayan region approximately 3300 m above sea level. When the soil was used as 'inoculum' in plant-based bioassay studies from this region, it improved the plant growth-related parameters. These cold-adapted microorganisms facilitated the uptake of nutrients from the environment (Yadav et al. 2017b). These microorganisms were also able to sustain environmental health and soil productivity. For example, Subramanian et al. (2011) stated that several cold-adapted bacteria species including *Pseudomonas, Acinetobacter, Pantoea, Rhodococcus* and *Serratia* were used for the promotion of crop plants through the synthesis of IAA, HCN and ACC deaminase and for the biocontrol studies against plant pathogens such as *Alternaria, Fusarium, Sclerotium, Rhizoctonia* and *Pythium* species. For example, *Pseudomonas fluorescens* and *Pantoea agglomerans* were reported to have a significant effect on plant growth of winter wheat crops in loamy soils (Egamberdiyeva and Hoflich 2003).

Several other species of *Pseudomonas* were also found to possess PGP characteristics. For example, *Pseudomonas putida* UW4 was found to promote plant growth and development in canola at cold temperature stress (Cheng et al. 2007). Similar findings were also made by Chatterjee et al. (2017) on pepper plants. Again, Yadav et al. (2015a; 2015b) reported the microbial species such as *Alishewanella* sp., *Aurantimonas altamirensis, Bacillus baekryungensis, B. marisflavi, Desemzia incerta, Paenibacillus xylanexedens, Pontibacillus* sp., *Providencia* sp., *P. frederiksbergensis, Sinobaca beijingensis* and *Vibrio metschnikovii* from the high altitude and low-temperature environments of the Indian Himalayas. Similarly, *Sphingomonas glacialis* (Zhang et al. 2011), and *Psychrobacter pocilloporae* (Zachariah et al. 2016) were isolated from cold environments. Pandey et al. (2006) also stated that *Pseudomonas putida* (B0) growing at 4°C, isolated from a sub-Alpine site in the central Himalayas, was able to solubilize phosphate and antagonize the fungal pathogens such as *Alternaria alternata* and *Fusarium oxysporum* and induced growth of maize under greenhouse conditions. *Pseudomonas lurida* (M2RH3) growing in the high altitude regions of the Himalayas, isolated from the rhizosphere of radish plants (*Raphanus sativus*), was also able to solubilize phosphate even under longer cold incubation periods and promoted the growth and development of wheat (Selvakumar et al. 2011). *Pseudomonas* sp. (PGERs17) isolated from the garlic root grown in sub-Alpine regions of the North-Western Himalayas was found to have antagonistic properties against phytopathogens such as *Sclerotium rolfsii, Fusarium oxysporum, Rhizoctonia solani* and *Pythium* spp. Moreover, the strain was found to help the growth of wheat plants (Mishra et al. 2008). These strains are able to improve and alleviate cold stress by enhancing protein, phenolics and anthocyanin contents in leaves of crop plants. In addition, they could decrease the electrolyte leakage as well, thus reducing water loss in plants (Mishra et al. 2011; Subramanian et al. 2016).

Yadav et al. (2015c) isolated cold-tolerant pseudomonads from the Indian Himalayas screened for plant growth-promoting purposes as well as biochemical studies against *Rhizoctonia solani* and *Macrophomina phaseolina* at a low temperature (4°C). Out of 128 isolates, 17 isolates exhibited more than six plant growth-promoting attributes in *Pseudomonas cedrina, Pseudomonas deceptionensis, Pseudomonas extremaustralis, Pseudomonas fluorescens*, etc. These strains significantly improved the level of cellular metabolites like amino acids, anthocyanin, carotenoids, chlorophyll, free proline, physiologically available iron, proteins, starch content and total phenolics in wheat plants. Increased relative water content, reduced membrane injury and NPK uptake were also recorded in bacterized wheat plants. The high altitude of mountains limits biodiversity (Xu et al. 2009). However, Rehakova et al. (2017) stated that not only bacteria and mycorrhiza have been observed but also cyanobacteria could be found in cold climates. They stated that although plant richness, diversity and abundance decreased significantly with elevation, however, soil cyanobacterial communities in vegetated soil did not change across the elevational transect, in fact, bare soil cyanobacterial diversity, abundance and richness were reported to increase with the elevation.

When cold-adapted organisms were defined, two types of terms have been used. These are psychrophiles and psychrotrophs. The term psychrophiles were defined as the microorganisms growing around 15°C or less (Moyer and Morita 2007). They include gram-negative bacteria such as *Pseudoalteromonas, Moraxella, Psychrobacter, Flavobacterium Polaromonas, Psychroflexus, Polaribacter, Moritella, Vibrio* and *Pseudomonas*; Gram-positive bacteria such as *Arthrobacter, Bacillus* and *Micrococcus* species and several yeasts, fungi and microalgae species (Feller and

Gerday 2003; Berg et al. 2019). Psychrophiles grow at or below zero (0°C) and have an optimum growth temperature at 15°C and an upper limit of 20°C. In contrast, psychrotolerant microorganisms are able to grow between 0–30°C, and are also considered as cold-tolerant mesophiles (Morita 1975). Cold-tolerant microorganisms have strong metabolisms to reduce the negative effects of cold. Low temperatures primarily affect the lipid bi-layer of the bacterial cell making it impermeable to the diffusion of solutes, which helps the cell function properly. This is achieved by an increase in the amount of branched fatty acid and a decrease in cyclic fatty acids with the growth of monounsaturated straight-chain fatty acids. Cold acclimation proteins are produced by cold-tolerant bacteria in which a large amount of proteins are produced during continuous growth at low temperatures. On the other hand, Cold-shock proteins (Csp) are initiated by a sudden decrease in temperature (Piette et al. 2011). Ice nucleators are proteins which either limit supercooling or induce freezing at temperatures below 0°C by mimicking the structure of an ice crystal surface. They impose an ice crystal-like arrangement on the water molecule with their surface and reduce the necessity of energy for the initiation of ice formation (Zachariassen and Kristiansen 2000). These proteins are involved in maintaining some metabolic functions at low temperatures by replacing cold-denatured peptides. Psychrophiles also produce cold-adapted enzymes that have high specific activities at low temperatures (Mishra et al. 2010).

Cold-adapted microorganisms have unique enzymatic systems that work under cold stress conditions. These enzymes from cold microorganisms are active at cold temperatures and have the potential to be used in industrial applications to decrease energy consumption. They provide opportunities to study the adaptation of life at a low temperature and the potential for biotechnological applications (Yadav et al. 2016). These enzymes could synthesize amylase, protease, lipase, pectinase, xylanase, cellulase, β-glucosidase, β-galactosidase and chitinase (Yadav et al. 2016; Singh et al. 2016) and glucanase (Dikilitas 2003). These microorganisms can also be applied for biodegradation of agrowastes at low temperatures. For example, Shukla et al. (2016) stated that cold-tolerant microorganisms such as *Eupenicillium crustaceum*, *Paceliomyces* sp., *Bacillus atropheus* and *Bacillus* sp. were used to degrade agricultural residues and decreased the environmental pollution caused by burning of agricultural residues. For example, cold tolerance in fishes has been achieved by direct injection of anti-freezing proteins. AFPs via anal injection or by feeding was able to enhance the tolerance of tilapia juveniles (*Oreochromismossambicus* Peters) to cold stress (Wu et al. 1998). AFPs work at subzero temperatures and are found in many organisms such as fish, plants and insects. They are able to prevent the formation of ice crystals (Wu et al. 1998). Cold-active enzymes have also found application in areas of manufacturing of cleaning agents in food processing such as bakery, cheese manufacture and fermentation and molecular biology studies (Hamdan 2018). By the use of AFPs or cold-active enzymes, the cryopreservation of animals and plants are allowed and cold-tolerant organisms could be developed.

Subramanian et al. (2016) reported that tomato plants treated with the selected isolates (*Arthrobacter*, *Flavimonas*, *Flavobacterium*, *Massilia*, *Pedobacter* and *Pseudomonas*) exhibited significant tolerance to cold stress through reduction in membrane damage and activation of antioxidant enzymes along with proline synthesis in the leaves when exposed to chilling temperature conditions (15°C). Yadav et al. (2015a) stated that cold-adapted *Arthrobacternicotianae*, *Brevundimonasterrae*, *Paenibacillustylopili* and *Pseudomonas cedrina* in cold deserts exhibit multifunctional PGP attributes at low temperatures. Subramanian et al. (2016) stated that tomato (*Solanumlycopersicum* cv Mill) plants treated with psychrotolerant bacteria isolated from agricultural soils during winter improved the chilling stress of tomato plants. The isolates *Pseudomonas frederiksbergensis* OS211, *Flavobacterium glaciei* OB146, *P. vancouverensis* OB155, and *P. frederiksbergensis* OS261 consistently improved germination and plant growth at 15°C. Yarzabal et al. (2018) stated that cold-active PGP *Pseudomonas* spp. from Antarctic soils significantly enhanced root elongation of wheat (*Triticum aestivum*) plants and the bacterization of *T. aestivum* seeds notably increased germination. Moreover, *T. aestivum* seedlings showed a considerable increase in their root- and shoot-lengths compared to untreated controls when grown in sterile soil at $14 \pm 1°C$. They stated that cold-tolerant *Pseudomonas* spp. could act as cold-active biofertilizers.

Lecanicillium muscarium CCFEE 5003, isolated from Antarctica, is a powerful producer of extracellular chitin-hydrolysing cold-tolerant enzymes, was able to exert a strong mycoparasitic action against various other fungi and oomycetes at low temperatures (Fenice 2016).

Gomez-Munoz et al. (2018) stated that the effect of cold stress was not alleviated by improving soil fertility through the fertilization of phosphor or plant-available phosphor in the form of triple superphosphate. Cold stress was also not able to be alleviated by the treatment of seeds with SA. However, the addition of Mn/Zn and inoculation with *Penicillium* sp. increased biomass production of maize plants at 51 DAS (Days After Sowing) grown in soil characterized with high phosphor level as compared to that of low phosphor-containing soil. They interpreted that the addition of Mn/Zn and inoculation with *Penicillium* sp. reduced the effects of cold stress in maize plants grown in fertile soils.

Biochemical and molecular approaches to increase the potential of cold tolerance of crop plants

Since plants are sessile, the only way to survive under harsh environmental conditions is to adapt by changing their metabolisms quickly and efficiently. Cold stress is observed through membranes in the plant cell. Rigidification of the plasma membrane following cold stress has led to induce cold-responsive genes that are indicative of cold acclimation as shown in alfalfa and *Brassicanapus* (Orvar et al. 2000; Sangwan et al. 2001). Developing transgenic plants for cold tolerance is a very effective biotechnological method to increase agricultural crop production. Various agricultural crops have been generated via transgenic approaches to improve cold tolerance. Diverse studies have indicated that screening for genes regarding cold tolerance is a crucial initial step for crop improvement strategy (Ohnishi et al. 2005; Ito et al. 2006; Bhatnagar-Mathur et al. 2008). Genes encoding transcription factors are very crucial for improving cold stress tolerance in plants. Genes concerning cold tolerance in plant transgenic studies are mainly associated with LEA proteins, membrane stabilizing proteins and osmoprotectant synthesis-related proteins. They are prepared to protect the cell against cold stress conditions. However, cold resistance or tolerance is characterized by polygenic traits, it is, therefore, necessary to introduce multiple genes, signalling molecules and/or transcription factors in cold-sensitive plants. Such stress-responsive genes either protect crop plants from stress by producing important metabolic proteins such as LEA, AFPs, mRNA binding proteins, amino acids, sugars, fatty acids, proteinase inhibitors, etc. or regulates signal transduction and gene expression functioning under stress response (Yadav 2010).

Developing crop plants resistant to abiotic stress is a major challenge as well as developing crop plants to pathogen stress. However, crop plants to both abiotic stress and plant diseases are of utmost importance (Singh and Yadav 2020). Chemical control strategies have not been found successful for cold stress as combating other abiotic stress factors such as drought, salinity, heavy metal stress, etc. The most affordable and environmentally-friendly way for the management of cold tolerance would be cultivating of cold-tolerant crop plants. Conventional breeding methods can be used to improve the conditions of crop plants for cold stress, however, it is considered that it would take a great deal of time, and would be more appropriate to apply cold-tolerant microorganisms to plants under cold stress. So far, various approaches have been made to increase the cold tolerance of crop plants. For example, overexpression of *CBF3* gene increased the proline levels upto 15 times in transgenic rice (*Oryza sativa* L.) (Ito et al. 2006). A mutation in the *esk1* gene in *Arabidopsis* sp. (L.) Heynh. was also able to increase the concentration of free proline 30-fold and led to freezing tolerance (Xin and Browse 1998).

Reyes-Diaz et al. (2006) stated cold-tolerant plants increased their total soluble carbohydrates in their leaves during cold acclimation. They claimed that an increase in sugar accumulation in the apoplast could be related to freezing tolerance.Cold tolerant genes could be inserted into cold-susceptible crop plants to make them resistant. Transformation of plants with transcription factors should not only lead to enhanced cold tolerance but show tolerance to other abiotic stress factors such as salt, water stress, environmental pollution, drought, etc. For example, the transformation of sweet potato with a gene overexpressing betaine aldehyde dehydrogenase led to increased cold and salt tolerances (Fan et al. 2012). Seong et al. (2007) also showed that a zinc finger transcription factor, *CaPIF1*, increased tolerance to both cold and the bacterial pathogen *P. syringae* when overexpressed in tomato.

Although the application of chemical compounds has not been considered as successful as microbial applications, however, this might also help increasing cold tolerance in crop plants. For example, exogenous application of glycine betaine might result in increases of cold tolerance in many plants. The application of glycine betaine (5 mol L^{-1}) or chitosan (0.3%) for the cold tolerance of banana seedlings increased the cold tolerance of the banana fruits (Li et al. 2007; Li et al. 2008). Researchers stated that glycine betaine and chitosan increased the cold tolerance of banana seedlings by increasing SOD activity and preventing cellular oxidation and increasing malondialdehyde contents. Further the cold stress of young banana plants was enhanced with the application of exogenous SA. For example, Kang et al. (2003) stated that pretreatment of leaves and roots with SA one day before cold stress (5°C) increased SOD, CAT, and APX and inhibited H_2O_2 content in banana plants and protected the banana plants against ROS during cold stress conditions. Similar findings were also made by Wu et al. (2019) on *Arabidopsis thaliana* plants.

Not only have chemicals been applied to increase cold tolerance of crop plants but also physical approaches could make crop plants tolerant to cold stress. For example, cold stress tolerance at 7°C was enabled by dipping banana fruit in hot water (52°C) for 3 minutes (Wang et al. 2008b). The authors were able to succeed cold tolerance by maintaining the low level of H_2O_2 and increased levels in CAT and APX content. As in the hot water treatment, preliminary cold treatment might also increase the cold tolerance of crop plants. Wu et al. (2019) reported that exposure to cold stress for 10 hours was sufficient to activate immunity as well as H_2O_2 accumulation and callose deposition in *Arabidopsis thaliana* plants. They stated that cold-activated immunity was dependent on SA concentration.

Even correct fertilization of crop plants can sometimes make a difference. For example, Wang et al. (2013) stated that exogenous K^+ application increased the synthesis of defensive compounds, regulated stomatal conductance and increased the strength of cell membranes. These positive effects played significant roles in protecting plants from cold stress.

Not always transferring stress tolerant-related genes into crop plants make them tolerant to cold stress. Plants transformed with cold tolerant-related genes may show reduced growth under normal conditions. For example, Kasuga et al. (2004) showed that the overexpression of the *DREB1A* transcription factor in tobacco led to a slight growth retardation while cold tolerance was enabled. Also signalling pathways in some cases overlap when resistance mechanisms are developed in plants exposed to both biotic and abiotic stresses.

The development of transgenic plants with resistance against abiotic stresses still faces some challenges. Most of the studies so far have been done on *Arabidopsis* and tobacco, therefore, more crop plants have to be studied to elucidate the mechanism in detail. We have summarized physical, biochemical and molecular approaches to date with regards to cold tolerance of crop plants (Table 7.1). As shown in Table 7.1, no publications for the treatment of cold stressed plants under the pathological impact were seen. This area needs to be carefully evaluated in future studies.

The ornamental plant *Hosta capitata* was applied with AFPs (100 µg L^{-1}) obtained from fish. They regulated the expression of important stress response genes. As stated earlier, a possible way to protect plants is to treat them with AFPs. Pe et al. (2019) stated that fish AFPs can shield the rare ornamental species *Hosta capitata* from chilling stress. They elucidated the expression patterns of the cold-inducible genes C-repeat Binding Factor 1 (CBF1) and dehydrin 1 (DHN1) as well as the antioxidant genes SOD and CAT. All were upregulated at chilling stress (4°C). Similarly, Yildiztugay et al. (2017) reported that gallic acid (GLA), a naturally occurring plant phenol, was

Table 7.1 Improvement of cold or/and disease tolerance through physiological, biological, biochemical and molecular approaches.

Crop species	Cold stress	Application of other agents	Mechanisms/effect	Reference
Brassica oleracea L.	-		Induce systemic resistance.	Ghazalibiglar et al. (2016)
Cucumis sativus L.	-		Higher peroxidase and glucanase activities along with the PR proteins were expressed.	El-Borollosy and Oraby (2012)
Capsicum annuum L.	-		Increased expression of PR proteins.	Son et al. (2014)
Gossypium hirsutum L.	18°C for 7 days	Potassium nitrate (KNO_3) and distilled water seed priming	Increased seed performance under low temperature.	Cokkizgin et al. (2013)
Oryza sativa L.	5°C for 24 hours	Chitooligosaccharide treatment as an elicitor of plant immunity	Increased seedling growth and cold tolerance by the osmotic regulation through the accumulations of osmolytes.	Zhang et al. (2019b)
Glycine max L.	5 and 10°C for 72 hours	Gallic acid root priming	Improved plant tolerance by ensuring efficient water use and enhancing antioxidant enzyme systems.	Ozfidan-Konakci et al. (2019)
Vitis vinifera L.	4°C for 6 hours to 72 hours	24-Epibrassinolide treatment	Chilling-stressed injury is alleviated by 24-Epibrassinolide treatment.	Chen et al. (2019)
Camellia sinensis L.	4°C for 0, 4 and 7 days	Exogenous γ-Aminobutyric Acid (GABA) application	GABA altered levels of stress related substances such as polyamines and anthocyanins.	Zhu et al. (2019)
Hordeum vulgare L.	+5 and −5°C for	Silicon treatment as soil nutrient	Silicon has been found as an ameliorative agent against both chilling and freezing stresses.	Joudmand and Hajiboland (2019)
Triticum aestivum L.	4°C for 7 days	Exogenous salicylic acid treatment	SA improved cold tolerance by upregulating the activity of antioxidant enzymes and proline accumulation.	Ignatenko et al. (2019)
Abelmoschus esculentus L.	4°C for 12 days	Plants were dipped in Putrescine with concentrations at different levels for 10 minutes at 24°C	Putrescine efficiently reduced chilling injury by reduced seed browning by retarding the activity of antioxidative enzymes.	Phornvillay et al. (2019)
Pittocaulon praecox	5°C, 0°C, and −5°C	Silicon fertilization application	Si fertilizer enhanced bamboo growth and the tolerance of bamboo plants to cold stress by increased antioxidant enzymes.	Qian et al. (2019)
Camellia sinensis L.	4°C for 0 to 48 hours	Exogenous melatonin treatment	Melatonin protected tea plants against abiotic stress-induced damages through detoxifying ROS and modulating antioxidant systems.	Li et al. (2019)

Table 7.1 Contd. ...

...Table 7.1 Contd.

Crop species	Cold stress	Application of other agents	Mechanisms/effect	Reference
Medicago truncatula	4°C for 24 hours	Brassinolide pretreatment	Increased cold stress tolerance by regulating the expression of several cold-related genes and antioxidant enzymes.	Arfan et al. (2019)
Ocimum basilicum L.	postharvest storage at 3.5°C and at 7°C	Foliar abscisic acid treatment	Foliar preharvest applications of abscisic acid in combination with afternoon harvest were an effective strategy to alleviate chilling injury damage.	Satpute et al. (2019)
Pinus resinosa	–8°C for one hour, and then warmed to 12°C	Methyl jasmonate treatment	Methyl jasmonate induction altered tree seedling response to cold.	Connolly and Orrock (2018)
Zea mays L.	10 and 8°C	Application of exogenous salicylic acid and fertilization with manganese and zinc (Mn/Zn)	Addition of Mn/Zn and inoculation with *Penicillium* sp. reduced the effects of cold stress in maize plants grown in fertile soil.	Gomez-Munoz et al. (2018)
Hordeum vulgare L.	+5 to –5°C	Inoculation with arbuscular mycorrhizal fungi	The right combination of fungus species is important for success against cold conditions.	Hajiboland et al. (2019)
Glycine max L.	5°C for 0 to 24 hours	Paclobutrazol priming	Reduction in the shoot's dry matter and change in physiological characteristics by pre-treatment of paclobutrazol induced tolerance to cold stress.	Attarzadeh et al. (2018)
Gossypium hirsutum L.	5°C	Seed treatment with glycine betaine	Improved chilling resistance in the seedling stage by soaking the seeds in glycine betaine.	Cheng et al. (2018)
Triticum aestivum L.	10/4°C	Melatonin application	Foliar melatonin application enhanced the cold tolerance to subsequent low-temperature stress.	Sun et al. (2018)
Phaseolus vulgaris L.	4°C for 2 or 4 days	Exogenous applications of acetylsalicylic acid	Acetylsalicylic acid applications substantially improved growth and photosynthetic parameters.	Soliman et al. (2018)

(–) indicates no particular of that stress while (+) indicates the occurrence of that stress.
We express our apology from the ones whose works were not cited who carried out similar works due to constraints of time and space availability.

also able to activate the plant defence system under cold stress. For this purpose, soybean (*Glycine max*) plants were treated with gallic acid (GLA; 1 and 2 mmol L^{-1}) and cold stress (5°C and 10°C) and GLA and stress combination for 72 hours. Phornvillay et al. (2019) reported that polyamine, particularly putrescine (Put) has also been proposed to improve the cold tolerance of okra plants. The authors showed that 2 mmol L^{-1} Putrescine effectively reduced chilling injury symptoms of okra. Put treatment significantly reduced seed browning by retarding the activity polyphenol oxidase (PPO) and peroxidase (POD) enzymes. Additionally, Putrescine treatment was able to increase total phenolic contents and total antioxidant capacity and reduced H_2O_2 and malondialdehyde contents.

Enabling cold tolerance could not only be made possible before harvest but could also be succeeded during post harvest stages. For example, mango (*Mangiferaindica* L.) fruit treated with 10 µmol L^{-1} brassinolides (BL) resulted in higher cold tolerance at 5°C (Li et al. 2012). The authors compared the changes in expression profiles of PM proteins and the corresponding gene expressions between BL-treated and control fruit. Among them, four proteins remain (abscisic stress ripening-like protein, type II SK2 dehydrin, and temperature-induced lipocalin) and genes encoding these proteins were upregulated in BL-treated plants under cold stress. These findings showed that PM proteins and lipids were involved in BL-mediated responses to cold stress in the mango fruit, and provided novel evidence that BL played an important role in modulating cold stress tolerance in fruit. The procedures and possible approaches for the improvement of crop plants under cold and/or pathogen threats are summarized in Fig. 7.5.

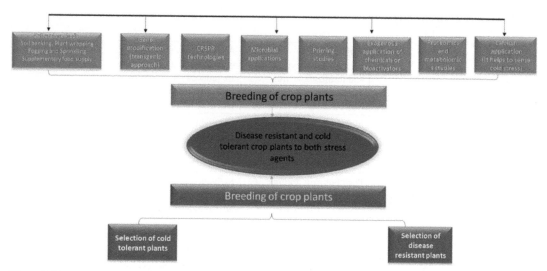

Figure 7.5 Preventing cold stress and improvement of crop plants for cold tolerance. Application of cold-tolerant PGPRs increase the physiological and biochemical conditions of crop plants by enhanced photosynthetic rate, phenolic contents, reduced ion leakage, enhanced unsaturated fatty acids, amino acid and protein synthesis, root proliferation, increased metabolite regulation, and enhanced chlorophyll and anthocyanin concentrations. PGPRs also enable plants to synthesize more CAT, POD, SOD, APX and GR to detoxify ROS.

Conclusions and future prospects

It is important to recognize that the simultaneous occurrence of several abiotic stresses is the most lethal stress to crop plants even if they are at low levels. The combination of both cold stress and pathogens impact crop plants when compared to the effects of individual stresses. Recent studies have shown that the response of plants to a combination of two different abiotic stresses is unique and cannot be directly extrapolated from the response of plants to each of the different stresses applied individually (Mittler 2006). When developing transgenic crops, the possible combination of stresses should be considered both in abiotic and abiotic; abiotic and biotic stress combinations. Simultaneous or sequential occurrences of combined stresses should not be neglected since these two combinations have a different mode of actions and should be well evaluated while generating new stress-tolerant and disease-resistant crop plants. Since breeding work in field conditions is highly influenced by other environmental features and biotic stresses, generation of cold-tolerant plants is extremely difficult. Therefore, more elaborated work is needs to be undertaken.

Cold stress not only has phenotypic effects on plants but also causes molecular changes. Following the recognition of cold stress, signalling cascades are activated in which genes are enhanced or suppressed to protect the plant cells against cold stress. Therefore, signalling pathways should be evaluated in detail to study gene expressions since the pathway for stress response is through the recognition of stressors. As it is said 'first comes, first served', which means the stress recognized first, gets treated first. Therefore, our priorities should involve evaluating the characteristics of each stress factor to apply proper chemicals or for the choice of right microorganisms to remediate the conditions of crop plants. As cold stress or freezing stress with plant pathogens not only cause severe disease symptoms and quality loss but also leads to increased susceptibility to other stresses. For example, saprophyte microorganisms such as bacteria and fungi, which are not normally found on plants, could act as pathogenic agents on crop plants under cold stress conditions. Therefore, generating crop plants to cold stress or applying microorganisms that work under cold stress is of great importance for the future of agriculture which would be carried out under adverse conditions.

Application of microorganisms could be very crucial during flowering stages. At this stage, the minimum temperature for flowering should be around 17–18°C. Therefore, any improvement or protection of crop plants at this stage is very crucial. However other growth stages such as germination and other vegetative stages are affected negatively from the effects of cold stress. The flowering and fruiting stages could be very sensitive to cold stress. For example, most of the crop plants require 6–10°C minimum at germination or maturing stages, however, other stages are far more sensitive to cold stress, therefore, any applications of microorganisms at high temperatures would be more beneficial.

It is noteworthy that global climate change including extreme degrees on both sides is emphatically predicted to have a negative impact on biotic stress resistance in plants. Thus, it is very crucial to conduct plant transcriptome analysis in response to combined stresses. As the two different stress factors act differently on plants. Recently, Wang et al. (2019) stated that the Ca^{2+} permeable channels activated by NaCl and cold stress were distinct from and independent of each other. Therefore, they recommended that future research studies should aim to elucidate the characteristics of putative Ca^{2+} ion channels, which should be verified to generate tolerant plants against multiple stresses.

Although most of the cold tolerance mechanisms have been published, however, there are some mechanisms yet to be completely understood and further studies are, therefore, needed. Until a properly integrated management system is applied, the application of cold-tolerant microorganisms seems to be the most optimum solution for crop plants under cold stress. The pathological threats would also be minimized with the application of correct microorganism under cold stress.

Dedication

We dedicate this chapter to the farmers doing their jobs at extremely adverse conditions.

References

Abla, M., H. Sun, Z. Li, C. Wei, F. Gao, Y. Zhou et al. 2019. Identification of miRNAs and their response to cold stress in Astragalus Membranaceus. Biomolecules 9(5): 182.

Agurla, S., S. Gahir, S. Munemasa,, Y. Murata and A.S. Raghavendra. 2018. Mechanism of stomatal closure in plants exposed to drought and cold stress. pp. 215–232. *In*: Mari, I., S. Minoru and U. Matsuo (eds.). Survival Strategies in Extreme Cold and Desiccation. Springer, Singapore.

Arfan, M., D. Zhang, W. Zou, S. Luo, W.R. Tan, T. Zhu et al. 2019. Hydrogen peroxide and nitric oxide crosstalk mediates brassinosteroids induced cold stress tolerance in *Medicago truncatula*. Int. J. Mol. Sci. 20: 144.

Attarzadeh, M., H. Balouchi and M.R. Baziar. 2018. Effects of paclobutrazol's pre-treatment on cold tolerance induction in soybean seedling (*Glycine max* L.). Ital. J. Agron. 13: 155–162.

Banerjee, A. and A. Roychoudhury. 2016. Group II late embryogenesis abundant (LEA) proteins: structural and functional aspects in plant abiotic stress. Plant Growth Regul. 79: 1–17.

Barea, J.M., M.J. Pozo, R. Azcon and C. Azcon-Aguilar. 2005. Microbial co-operation in the rhizosphere. J. Exp. Bot. 56: 1761–1778.

Barrero-Sicilia, C., S. Silvestre, R.P. Haslam and L.V. Michaelson. 2017. Lipid remodelling: Unravelling the response to cold stress in *Arabidopsis* and its extremophile relative *Eutremasalsugineum*. Plant Sci. 263: 194–200.

Berg, K., I. Leiros and A. Williamson. 2019. Temperature adaptation of DNA ligases from psychrophilic organisms. Extremophiles 23: 305–317.

Bhatnagar-Mathur, P., V. Vadez and K.K. Sharma. 2008. Transgenic approaches for abiotic stress tolerance in plants: retrospect and prospects. Plant Cell Rep. 27: 411–424.

Bloom, A.J., M.A. Zwieniecki, J.B. Passioura, L.B. Randall, N.M. Holbrook and D.A. St. Clair. 2004. Water relations under root chilling in a sensitive and tolerant tomato species. Plant, Cell Environ. 27: 971–979.

Brown, J.K., S.D. Wyatt and D. Hazelwood. 1984. Irrigated corn as a source of barley yellow dwarf virus and vector in eastern Washington. Phytopathology 74: 46–49.

Campbell, C., L. Atkinson, J. Zaragoza-Castells, M. Lundmark, O. Atkin and V. Hurry. 2007. Acclimation of photosynthesis and respiration is asynchronous in response to changes in temperature regardless of plant functional group. New Phytologist 176: 375–389.

Cattivelli, L., C. Crosatti, C. Marè, M. Grossi, A.M. Mastrangelo, E. Mazzucotelli et al. 2002. Expression of cold-regulated (cor) genes in Barley. pp. 121–137. *In*: Li, P.H. and E.T. Palva (eds.). Plant Cold Hardiness. Springer, Boston, MA.

Chatterjee, P., S. Samaddar, R. Anandham, Y. Kang, K. Kim, G. Selvakumar et al. 2017. Beneficial soil bacterium *Pseudomonas frederiksbergensis* OS261 augments salt tolerance and promotes red pepper plant growth. Front. Plant Sci. 8: 705.

Chen, T.H and N. Murata. 2008. Glycine betaine: an effective protectant against abiotic stress in plants. Trend. Plant Sci. 13: 499–505.

Chen, W.P. and H.L. Paul. 2002. Attenuation of reactive oxygen production during chilling in ABA-treated maize cultured cells. pp. 223–233. *In*: Paul, H.L. and E.T. Palva (eds.). Plant Cold Hardiness. Springer, Boston, MA.

Chen, Z.Y., Y.T. Wang, X.B. Pan and Z.M. Xi. 2019. Amelioration of cold-induced oxidative stress by exogenous 24-epibrassinolide treatment in grapevine seedlings: Toward regulating the ascorbate–glutathione cycle. Sci. Hortic. 244: 379–387.

Cheng, C., L.M. Pei, T.T. Yin and K.W. Zhang. 2018. Seed treatment with glycine betaine enhances tolerance of cotton to chilling stress. J. Agric. Sci. 156: 323–332.

Cheng, Z., E. Park and B.R. Glick. 2007. 1-Aminocyclopropane-1-carboxylate deaminase from *Pseudomonas putida* UW4 facilitates the growth of canola in the presence of salt. Can. J. Microbiol. 53: 912–918.

Cokkizgin, H., Y. Bolek and A. Cokkizgin. 2013. Improvement of cold tolerance in cotton seed by priming. J. Food Agric. Environ. 11: 560–563.

Connolly, B.M. and J.L. Orrock. 2018. Exogenous application of methyl jasmonate alters *Pinus resinosa* seedling response to simulated frost. Botany 96: 705–710.

Corpas, F.J., M. Chaki, A. Fernandez-Ocana, R. Valderrama, J.M. Palma, A. Carreras et al. 2008. Metabolism of reactive nitrogen species in pea plants under abiotic stress conditions. Plant Cell Physiol. 49: 1711–1722.

De Palma, M., S. Grillo, I. Massarelli, A. Costa, G. Balogh, L. Vigh et al. 2008. Regulation of desaturase gene expression, changes in membrane lipid composition and freezing tolerance in potato plants. Mol. Breeding. 21: 15–26.

Dhume, G.M., A.K. Maharana, M. Tsuji, A.K. Srivastava and S.M. Singh. 2019. Cold tolerant endoglucanase producing ability of Mrakiarobertii A2-3 isolated from cryoconites, Hamtha glacier, Himalaya. J. Basic Microbiol. 59: 667–679.

Dikilitas, M. 2003. Effect of salinity & its interactions with Verticilliumalbo-atrum on the disease development in tomato (*Lycopersiconesculentum* Mill.) and lucerne (*Medicago sativa* & M. media) plants. Ph.D. Thesis, University of Wales, Swansea.

Dikilitas, M., Y.Z. Katırcıoğlu and H. Altınok. 2011. Latest developments and methods on long term storage, protection, and recycle of fungi and fungal Material. J. Agric. Faculty Harran University 15: 55–69.

Dikilitas, M., S. Karakas, A. Hashem, E.F. Abd Allah and P. Ahmad. 2016. Oxidative stress and plant responses to pathogens under drought conditions. pp. 102–123. *In*: Ahmad, P. (ed.). Water Stress and Crop Plants: A Sustainable Approach Volume 1, First Edition. John Wiley & Sons, Ltd. New York, NY, USA.

Dikilitas, M., E. Simsek and S. Karakas. 2019a. Stress responsive signaling molecules and genes under stressful environments in plants. pp. 19–42. *In*: Khan, M.I.R., P.S. Reddy, A. Ferrante and N.A. Khan (eds.). Plant Signaling Molecules, Role and Regulation Under Stressful Environments. Woodhead Publishing, Elsevier Inc. Sawston, Cambridge.

Dikilitas, M., E. Simsek, S. Karakas and P. Ahmad. 2019b. High-temperature stress and photosynthesis under pathological impact. pp. 39–64. *In*: Parvaiz, A., A.A. Mohammad, N.A. Mohammed and A. Pravej (eds.). Photosynthesis, Productivity and Environmental Stress. John Wiley & Sons Ltd., Hoboken, New Jersey.

Ding, C.K., C. Wang, K.C. Gross and D.L. Smith. 2002. Jasmonate and salicylate induce the expression of pathogenesis-related-protein genes and increase resistance to chilling injury in tomato fruit. Planta 214: 895–901.

Dong, J., J. Zhao, S. Zhang, T. Yang, Q. Liu, X. Mao et al. 2019. Physiological and genome-wide gene expression analyses of cold-induced leaf rolling at the seedling stage in rice (*Oryza sativa* L.). The Crop J. 7: 431–443.

Du, Y.Y., P.C. Wang, J. Chen and C.P. Song. 2008. Comprehensive functional analysis of the catalase gene family in *Arabidopsis thaliana*. J. Integrative Plant Biol. 50: 1318–1326.

Dugasa, M.T., I.G. Chala and F. Wu. 2019. Genotypic difference in secondary metabolism-related enzyme activities and their relative gene expression patterns, osmolyte and plant hormones. Physiol. Plant. (in press).

Dutta, A., J. Sen and R. Deswal. 2007. Downregulation of terpenoid indole alkaloid biosynthetic pathway by low temperature and cloning of a AP2 type C-repeat binding factor (CBF) from *Catharanthusroseus* (L). G. Don. Plant Cell Rep. 26: 1869–1878.

Egamberdiyeva, D. and G. Höflich. 2003. Influence of growth-promoting bacteria on the growth of wheat in different soils and temperatures. Soil Biol. Biochem. 35: 973–978.

El-Borollosy, A.M. and M.M. Oraby. 2012. Induced systemic resistance against Cucumber mosaic cucumovirus and promotion of cucumber growth by some plant growth-promoting rhizobacteria. Ann. Agric. Sci. 57: 91–97.

Fan, W., M. Zhang, H. Zhang and P. Zhang. 2012. Improved tolerance to various abiotic stresses in transgenic sweet potato (*Ipomoea batatas*) expressing spinach betaine aldehyde dehydrogenase. PLoS One 7: e37344.

Fang, X., D. Phillips, H. Li, K. Sivasithamparam and M.J. Barbetti. 2011. Comparisons of virulence of pathogens associated with crown and root diseases of strawberry in Western Australia with special reference to the effect of temperature. Sci. Hortic. 131: 39–48.

Feller, G. and C. Gerday. 2003. Psychrophilic enzymes: hot topics in cold adaptation. Natur. Rev. Microbiol. 1: 200–208.

Fenice, M. 2016. The psychrotolerant Antarctic fungus *Lecanicilliummuscarium* CCFEE 5003: A powerful producer of cold-tolerant chitinolytic enzymes. Molecules 21: 447.

Figueroa-Yanez, L., J. Cano-Sosa, E. Castano, A.L. Arroyo-Herrera, J.H. Caamal-Velazquez, F. Sanchez-Teyer et al. 2012. Phylogenetic relationships and expression in response to low temperature of a catalase gene in banana (*Musa acuminata* cv. "Grand Nain") fruit. Plant Cell, Tissue Organ Cult. (PCTOC) 109: 429–438.

Filiz, E., I.I. Ozyigit, I.A. Saracoglu, M.E. Uras, U. Sen and B. Yalcin. 2019. Abiotic stress-induced regulation of antioxidant genes in different *Arabidopsis* ecotypes: microarray data evaluation. Biotechnol. Biotechnol. Equip. 33: 128–143.

Fu, D., C. Uauy, A. Distelfeld, A. Blechl, L. Epstein, X. Chen et al. 2009. A kinase-START gene confers temperature-dependent resistance to wheat stripe rust. Science 323: 1357–1360.

Ghazalibiglar, H., J.G. Hampton, E.V.Z. de Jong and A. Holyoake. 2016. Is induced systemic resistance the mechanism for control of black rot in *Brassica oleracea* by a *Paenibacillus* sp.? Biol. Control 92: 195–201.

Gomez-Munoz, B., J.D. Lekfeldt, J. Magid, L.S. Jensen and A. de Neergaard. 2018. Seed treatment with *Penicillium* sp. or Mn/Zn can alleviate the negative effects of cold stress in maize grown in soils dependent on soil fertility. J. Agron. Crop Sci. 204: 603–612.

Griffith, M. and M.W. Yaish. 2004. Antifreeze proteins in overwintering plants: a tale of two activities. Trend Plant Sci. 9: 399–405.

Gupta, K.J., D.K. Hincha and L.A. Mur. 2011. NO way to treat a cold. New phytol. 189: 360–363.

Gusta, L.V., M. Wisniewski, N.T. Nesbitt and M.L. Gusta. 2004. The effect of water, sugars, and proteins on the pattern of ice nucleation and propagation in acclimated and nonacclimated canola leaves. Plant Physiol. 135: 1642–1653.

Hajiboland, R., A. Joudmand, N. Aliasgharzad, R. Tolrá and C. Poschenrieder. 2019. Arbuscular mycorrhizal fungi alleviate low-temperature stress and increase freezing resistance as a substitute for acclimation treatment in barley. Crop Pasture Sci. 70: 218–233.

Hajihashemi, S., F. Noedoost, J.M. Geuns, I. Djalovic and K.H. Siddique. 2018. Effect of cold stress on photosynthetic traits, carbohydrates, morphology, and anatomy in nine cultivars of *Stevia rebaudiana*. Front. Plant Sci. 9: 1430.

Hamdan, A. 2018. Psychrophiles: Ecological significance and potential industrial application. South Afri. J. Sci. 114: 1–6.

Hannah, M.A., D. Wiese, S. Freund, O. Fiehn, A.G. Heyer and D.K. Hincha. 2006. Natural genetic variation of freezing tolerance in Arabidopsis. Plant Physiol. 142: 98–112.

Hornschuh, M., R. Grotha and U. Kutschera. 2002. Epiphytic bacteria associated with the bryophyte *Funaria hygrometrica*: effects of Methylobacterium strains on protonema development. Plant Biol. 4: 682–687.

Huang, H., B. Liu, L. Liu and S. Song. 2017. Jasmonate action in plant growth and development. J. Exp. Bot. 68: 1349–1359.

Ignatenko, A., V. Talanova, N. Repkina and A. Titov. 2019. Exogenous salicylic acid treatment induces cold tolerance in wheat through promotion of antioxidant enzyme activity and proline accumulation. Acta Physiol. Plant. 41: 80.

Ito, Y., K. Katsura, K. Maruyama, T. Taji, M. Kobayashi, M. Seki et al. 2006. Functional analysis of rice DREB1/CBF-type transcription factors involved in cold-responsive gene expression in transgenic rice. Plant Cell Physiol. 47: 141–153.

Jain, R., A.K. Shrivastava, S. Solomon and R.L. Yadav. 2007. Low temperature stress-induced biochemical changes affect stubble bud sprouting in sugarcane (*Saccharum* spp. hybrid). Plant Growth Regul. 53: 17–23.

Jang, J.Y., D.G. Kim, Y.O. Kim, J.S. Kim and H. Kang. 2004. An expression analysis of a gene family encoding plasma membrane aquaporins in response to abiotic stresses in *Arabidopsis thaliana*. Plant Mol. Biol. 54: 713–725.

Janska, A., P. Maršík, S. Zelenková and J. aOvesná. 2010. Cold stress and acclimation—what is important for metabolic adjustment? Plant Biol. 12: 395–405.

Jelenska, J., J.A. Van Hal and J.T. Greenberg. 2010. Pseudomonas syringae hijacks plant stress chaperone machinery for virulence. Proc. Natl. Acad. Sci. 107: 13177–13182.

Joshi, R., B. Singh and V. Chinnusamy. 2018. Genetically engineering cold stress-tolerant crops: approaches and challenges. pp. 179–195. *In*: Shabir, H.W. and H. Venura (eds.). Cold Tolerance in Plants. Springer Nature, Switzerland.

Joudmand, A. and R. Hajiboland. 2019. Silicon mitigates cold stress in barley plants via modifying the activity of apoplasmic enzymes and concentration of metabolites. Acta Physiol. Plant. 41: 29.

Kaminska, M., M. Gołębiewski, A. Tretyn and A. Trejgell. 2018. Efficient long-term conservation of Taraxacumpieninicum synthetic seeds in slow growth conditions. Plant Cell, Tissue Organ Cult. (PCTOC) 132: 469–478.

Kang, G., C. Wang, G. Sun and Z. Wang. 2003. Salicylic acid changes activities of H_2O_2-metabolizing enzymes and increases the chilling tolerance of banana seedlings. Environ. Exp. Bot. 50: 9–15.

Kaplan, F., J. Kopka, D.Y. Sung, W. Zhao, M. Popp, R. Porat and C.L. Guy. 2007. Transcript and metabolite profiling during cold acclimation of Arabidopsis reveals an intricate relationship of cold-regulated gene expression with modifications in metabolite content. Plant J. 50: 967–981.

Kasuga, M., S. Miura, K. Shinozaki and K. Yamaguchi-Shinozaki. 2004. A combination of the Arabidopsis DREB1A gene and stress-inducible rd29A promoter improved drought-and low-temperature stress tolerance in tobacco by gene transfer. Plant Cell Physiol. 45: 346–350.

Kawahara, H. 2002. The structures and functions of ice crystal-controlling proteins from bacteria. J. Biosci. Bioeng. 94: 492–496.

Kemal, S.A., S. KrimiBencheqroun, A. Hamwieh and M. Imtiaz. 2017. Effects of temperature stresses on the resistance of chickpea genotypes and aggressiveness of didymellarabiei isolates. Front. Plant Sci. 8: 1607.

Kilinc, N. 2018. Determination of biochemical properties of different isolates of wheat blotch disease [*Zymoseptoriatritici* (Desm. Quaedvlieg & Crous)] MSc Thesis, Harran University, Sanliurfa, Turkey.

Knight, M.R. 2002. Signal transduction leading to low–temperature tolerance in *Arabidopsis thaliana*. Philosophical Transactions of the Royal Society of London. Series B: Biol. Sci. 357: 871–875.

Kodra, E., K. Steinhaeuser and A.R. Ganguly. 2011. Persisting cold extremes under 21st-century warming scenarios. Geophys. Res. Lett. 38: 1–5.

Korn, M., S. Peterek, H.P. Mock, A.G. Heyer and D.K. Hincha. 2008. Heterosis in the freezing tolerance, and sugar and flavonoid contents of crosses between *Arabidopsis thaliana* accessions of widely varying freezing tolerance. Plant Cell Environ. 31: 813–827.

Kour, D., K.L. Rana, N. Yadav, A.N. Yadav, A. Kumar, V.S. Meena et al. 2019a. Rhizospheric microbiomes: biodiversity, mechanisms of plant growth promotion, and biotechnological applications for sustainable agriculture. pp. 19–65. *In*: Kumar, A. and V.S. Meena (eds.). Plant Growth Promoting Rhizobacteria for Agricultural Sustainability: From Theory to Practices. Springer Singapore, Singapore. doi:10.1007/978-981-13-7553-8_2.

Kour, D., K.L. Rana, N. Yadav, A.N. Yadav, J. Singh, A.A. Rastegari et al. 2019b. Agriculturally and industrially important fungi: current developments and potential biotechnological applications. pp. 1–64. *In*: Yadav, A.N., S. Singh, S. Mishra and A. Gupta (eds.). Recent Advancement in White Biotechnology through Fungi, Volume 2: Perspective for Value-Added Products and Environments. Springer International Publishing, Cham, doi:https://doi.org/10.1007/978-3-030-14846-1_1.

Kour, D., K.L. Rana, A.N. Yadav, N. Yadav, M. Kumar, V. Kumar et al. 2020. Microbial biofertilizers: bioresources and eco-friendly technologies for agricultural and environmental sustainability. Biocatal. Agric. Biotechnol. 23: 101487. doi:https://doi.org/10.1016/j.bcab.2019.101487.

Kratsch, H.A. and R.R. Wise. 2000. The ultrastructure of chilling stress. Plant Cell Environ. 23: 337–350.

Kumar, V., S. Joshi, N.C. Pant, P. Sangwan, A.N. Yadav, A. Saxena et al. 2019. Molecular approaches for combating multiple abiotic stresses in crops of arid and semi-arid region. pp. 149–170. *In*: Singh, S.P., S.K. Upadhyay, A. Pandey and S. Kumar (eds.). Molecular Approaches in Plant Biology and Environmental Challenges. Springer, Singapore. doi:10.1007/978-981-15-0690-1_8.

Li, B., C. Zhang, B. Cao, G. Qin, W. Wang and S. Tian. 2012. Brassinolide enhances cold stress tolerance of fruit by regulating plasma membrane proteins and lipids. J. Amino Acids 43: 2469–2480.

Li, J., Y. Yang, K. Sun, Y. Chen, X. Chen and X. Li. 2019. Exogenous melatonin enhances cold, salt and drought stress tolerance by improving antioxidant defense in tea plant (*Camellia sinensis* (L.) O. Kuntze). Molecules 24: 1826.

Li, M.F., S.P. Li and Y.L. Lu. 2007. Synergistic effect of betaine and chitosan on banana seedling's cold resistance [J]. Guangdong Agric. Sci. 9.

Li, W., R. Wang, M. Li, L. Li, C. Wang, R. Welti et al. 2008. Differential degradation of extraplastidic and plastidic lipids during freezing and post-freezing recovery in *Arabidopsis thaliana*. J. Biol. Chem. 283: 461–468.

Lo Presti, L., D. Lanver, G. Schweizer, S. Tanaka, L. Liang, M. Tollot et al. 2015. Fungal effectors and plant susceptibility. Annu. Rev. Plant Biol. 66: 513–545.

Mahajan, S. and N. Tuteja. 2005. Cold, salinity and drought stresses: an overview. Arch. Biochem. Biophys. 444: 139–158.

Maul, P., G.T. McCollum, M. Popp, C.L. Guy and R.O.N. Porat. 2008. Transcriptome profiling of grapefruit flavedo following exposure to low temperature and conditioning treatments uncovers principal molecular components involved in chilling tolerance and susceptibility. Plant, Cell Environ. 31: 752–768.

Mishra, P.K., S. Mishra, G. Selvakumar, S.C. Bisht, J.K. Bisht, S. Kundu et al. 2008. Characterisation of a psychrotolerant plant growth promoting *Pseudomonas* sp. strain PGERs17 (MTCC 9000) isolated from North Western Indian Himalayas. Ann. Microbiol. 58: 561–568.

Mishra, P.K., P. Joshi, S.C. Bisht, J.K. Bisht and G. Selvakumar. 2010. Cold-tolerant agriculturally important microorganisms. pp. 273–296. *In*: Dinesh, K.M. (ed.). Plant Growth and Health Promoting Bacteria. Springer, Berlin, Heidelberg.

Mishra, P.K., S.C. Bisht, P. Ruwari, G. Selvakumar, G.K. Joshi, J.K. Bisht et al. 2011. Alleviation of cold stress in inoculated wheat (*Triticum aestivum* L.) seedlings with psychrotolerant Pseudomonads from NW Himalayas. Arch. Microbiol. 193: 497–513.

Mishra, P.K., S.C. Bisht, J.K. Bisht and J.C. Bhatt. 2012. Cold-tolerant PGPRs as bioinoculants for stress management. pp. 95–118. *In*: Dinesh, K.M. (ed.). Bacteria in Agrobiology: Stress Management. Springer, Berlin, Heidelberg.

Mittler, R. 2006. Abiotic stress, the field environment and stress combination. Trends in Plant Sci. 11: 15–19.

Miura, K. and Y. Tada. 2014. Regulation of water, salinity, and cold stress responses by salicylic acid. Front Plant Sci. 5: 4.

Moffatt, B., V. Ewart and A. Eastman. 2006. Cold comfort: plant antifreeze proteins. Physiol. Plant. 126: 5–16.

Morita, R.Y. 1975. Psychrophilic bacteria. Bacteriol. Rev. 39: 144–167.

Moyer, C.L. and R.Y. Morita. 2007. Psychrophiles and psychrotrophs. eLS.

Moyer, M.M., D.M. Gadoury, L. Cadle-Davidson, I.B. Dry, P.A. Magarey, W.F. Wilcox et al. 2010. Effects of acute low-temperature events on development of *Erysiphenecator* and susceptibility of *Vitis vinifera*. Phytopathology 100: 1240–1249.

Moyer, M.M., J. Londo, D.M. Gadoury and L. Cadle-Davidson. 2016. Cold Stress-Induced Disease Resistance (SIDR): indirect effects of low temperatures on host-pathogen interactions and disease progress in the grapevine powdery mildew pathosystem. European J. Plant Pathol. 144: 695–705.

Nilsen, E.T. and D.M. Orcutt. 1996. Physiology of plants under stress. Abiotic Factors. Physiology of Plants Under Stress. Abiotic factors. John Wiley and Sons, New York, USA.

Ohnishi, T., S. Sugahara, T. Yamada, K. Kikuchi, Y. Yoshiba, H.Y. Hirano et al. 2005. OsNAC6, a member of the NAC gene family, is induced by various stresses in rice. Genes & Genetic Syst. 80: 135–139.

Orvar, B.L., V. Sangwan, F. Omann and R.S. Dhindsa. 2000. Early steps in cold sensing by plant cells: the role of actin cytoskeleton and membrane fluidity. Plant J. 23: 785–794.

Osorio, J.A. and D.C. McGee. 1992. Effects of freeze damage on soybean seed mycoflora and germination. Plant Dis. 76(9): 879–882.

Ozfidan-Konakci, C., E. Yildiztugay, A. Yildiztugay and M. Kucukoduk. 2019. Cold stress in soybean (*Glycine max* L.) roots: exogenous gallic acid promotes water status and increases antioxidant activities. Botanica Serbica 43: 59–71.

Pandey, A., P. Trivedi, B. Kumar and L.M.S. Palni. 2006. Characterization of a phosphate solubilizing and antagonistic strain of *Pseudomonas putida* (B0) isolated from a sub-alpine location in the Indian Central Himalaya. Curr. Microbiol. 53: 102–107.

Pe, P.P.W., A.H. Naing, M.Y. Chung, K.I. Park and C.K. Kim. 2019. The role of antifreeze proteins in the regulation of genes involved in the response of *Hostacapitata* to cold. 3 Biotech 9: 1–17.

Phornvillay, S., N. Pongprasert, C. Wongs-Aree, A. Uthairatanakij and V. Srilaong. 2019. Exogenous putrescine treatment delays chilling injury in okra pod (*Abelmoschusesculentus*) stored at low storage temperature. Sci. Hortic. 256: 108550.

Piette, F., S. D'Amico, G. Mazzucchelli, A. Danchin, P. Leprince and G. Feller. 2011. Life in the cold: a proteomic study of cold-repressed proteins in the Antarctic bacterium *Pseudoalteromonashaloplanktis* TAC125. Appl. Environ. Microbiol. 77: 3881–3883.

Pivonia, S., A.W.A.M. de Cock, R. Levita, E. Etiel and R. Cohen. 2012. Low temperatures enhance winter wilt of pepper plants caused by *Pythium* sp. Phytoparasitica 40: 525–531.

Promyou, S., S. Ketsa and W.G. van Doorn. 2008. Hot water treatments delay cold-induced banana peel blackening. Postharvest Biol. Technol. 48: 132–138.

Qian, Z.Z., S.Y. Zhuang, Q. Li and R.Y. Gui. 2019. Soil silicon amendment increases Phyllostachys praecox cold tolerance in a pot experiment. Forests 10: 405.

Rana, K.L., D. Kour and A.N. Yadav. 2019. Endophytic microbiomes: biodiversity, ecological significance and biotechnological applications. Res. J. Biotechnol. 14: 142–162.

Rastegari, A.A., A.N. Yadav, A.A. Awasthi and N. Yadav. 2020. Trends of Microbial Biotechnology for Sustainable Agriculture and Biomedicine Systems: Diversity and Functional Perspectives. Elsevier, Cambridge, USA.

Rehakova, K., K. Čapková, M. Dvorský, M. Kopecký, J. Altman, P. Šmilauer et al. 2017. Interactions between soil phototrophs and vascular plants in Himalayan cold deserts. Soil Biol. Biochem. 115: 568–578.

Renaut, J., J.F. Hausman and M.E. Wisniewski. 2006. Proteomics and low-temperature studies: bridging the gap between gene expression and metabolism. Physiol. Plant. 126: 97–109.

Reyes-Diaz, M., N. Ulloa, A. Zuniga-Feest, A. Gutiérrez, M. Gidekel, M. Alberdi et al. 2006. *Arabidopsis thaliana* avoids freezing by supercooling. J. Exp. Bot. 57: 3687–3696.

Robertson, G.P. and A.S. Grandy. 2006. Soil system management in temperate regions. pp. 27–39. *In*: Norman, U., S.B. Andrew, P. Cheryl, F. Erick, P. Jules, H. Hans et al. (eds.). Biological Approaches to Sustainable Soil Systems. CRC press, Boca Raton, Florida, USA.

Sangwan, V., I. Foulds, J. Singh and R.S. Dhindsa. 2001. Cold-activation of *Brassica napus* BN115 promoter is mediated by structural changes in membranes and cytoskeleton, and requires Ca^{2+} influx. Plant J. 27: 1–12.

Sati, P., K. Dhakar and A. Pandey. 2013. Microbial diversity in soil under potato cultivation from cold desert Himalaya, India. ISRN Biodiversity, vol. 2013, Article ID 767453.

Satpute, A., B. Meyering and U. Albrecht. 2019. Preharvest abscisic acid application to alleviate chilling injury of sweet basil (*Ocimumbasilicum* L.) during cold storage. Hort. Sci. 54: 155–161.

Selvakumar, G., P. Joshi, P. Suyal, P.K. Mishra, G.K. Joshi, J.K. Bisht et al. 2011. *Pseudomonas lurida* M2RH3 (MTCC 9245), a psychrotolerant bacterium from the Uttarakhand Himalayas, solubilizes phosphate and promotes wheat seedling growth. World J. Microbiol. Biotechnol. 27: 1129–1135.

Seong, E.S., K.H. Baek, S.K. Oh, S.H. Jo, S.Y. Yi, J.M. Park et al. 2007. Induction of enhanced tolerance to cold stress and disease by overexpression of the pepper CaPIF1 gene in tomato. Physiol. Plant. 129: 555–566.

Serrano, M. and A.E. Robertson. 2018. The effect of cold stress on damping-off of soybean caused by *Pythium sylvaticum*. Plant Dis. 102: 2194–2200.

Serrano, M., D. McDuffee and A.E. Robertson. 2018. Damping-off caused by *Pythium sylvaticum* on soybeans subjected to periods of cold stress is reduced by seed treatments. Canadian J. Plant Pathol. 40: 571–579.

Sharma, A., B. Shahzad, V. Kumar, S.K. Kohli, G.P.S. Sidhu, A.S. Bali et al. 2019. Phytohormones regulate accumulation of osmolytes under abiotic stress. Biomolecules 9: 285.

Shivaji, S., M.S. Pratibha, B. Sailaja, K.H. Kishore, A.K. Singh, Z. Begum et al. 2011. Bacterial diversity of soil in the vicinity of Pindari glacier, Himalayan mountain ranges, India, using culturable bacteria and soil 16S rRNA gene clones. Extremophiles 15: 1–22.

Shukla, L., A. Suman, A.N. Yadav, P. Verma and A.K. Saxena. 2016. Syntrophic microbial system for ex-situ degradation of paddy straw at low temperature under controlled and natural environment. J. Appl. Biol. Biotechnol. 4: 30–37.

Singh, A., R. Kumari, A.N. Yadav, S. Mishra, A. Sachan and S.G. Sachan. 2020. Tiny microbes, big yields: microorganisms for enhancing food crop production sustainable development. pp. 1–16. *In*: Rastegari, A.A., A.N. Yadav, A.K. Awasthi and N. Yadav (eds.). Trends of Microbial Biotechnology for Sustainable Agriculture and Biomedicine Systems: Diversity and Functional Perspectives. Elsevier, Amsterdam. doi:https://doi.org/10.1016/B978-0-12-820526-6.00001-4.

Singh, J. and A.N. Yadav. 2020. Natural Bioactive Products in Sustainable Agriculture. Springer, Singapore.

Singh, R.N., S. Gaba, A.N. Yadav, P. Gaur, S. Gulati, R. Kaushik et al. 2016. First high quality draft genome sequence of a plant growth promoting and cold active enzyme producing psychrotrophic *Arthrobacteragilis* strain L77. Standards in Genomic Sci. 11: 54.

Smiley, R.W. and W. Uddin. 1993. Influence of soil temperature on Rhizoctonia root rot (*R. solani* AG-8 and *R. oryzae*) of winter wheat. Phytopathology 83: 777–785.

Solanke, A.U. and A.K. Sharma. 2008. Signal transduction during cold stress in plants. Physiol. Mol. Biol. Plants 14: 69–79.

Soliman, M.H., A.A. Alayafi, A.A. El Kelish and A.M. Abu-Elsaoud. 2018. Acetylsalicylic acid enhance tolerance of *Phaseolus vulgaris* L. to chilling stress, improving photosynthesis, antioxidants and expression of cold stress responsive genes. Bot. Stud. 59: 1–17.

Son, J.S., M. Sumayo, Y.J. Hwang, B.S. Kim and S.Y. Ghim. 2014. Screening of plant growth-promoting rhizobacteria as elicitor of systemic resistance against gray leaf spot disease in pepper. Appl. Soil Ecol. 73: 1–8.

Spano, C., S. Bottega, M.R. Castiglione and H.E. Pedranzani. 2017. Antioxidant response to cold stress in two oil plants of the genus Jatropha. Plant, Soil Environ. 63: 271–276.

Subrahmanyam, G., A. Kumar, S.P. Sandilya, M. Chutia and A.N. Yadav. 2020. Diversity, plant growth promoting attributes, and agricultural applications of rhizospheric microbes. pp. 1–52. *In*: Yadav, A.N., J. Singh, A.A. Rastegari and N. Yadav (eds.). Plant Microbiomes for Sustainable Agriculture. Springer International Publishing, Cham. doi:10.1007/978-3-030-38453-1_1.

Subramanian, P., M.M. Joe, W.J. Yim, B.H. Hong, S.C. Tipayno, V.S. Saravanan et al. 2011. Psychrotolerance mechanisms in cold-adapted bacteria and their perspectives as plant growth-promoting bacteria in temperate agriculture. Korean J. Soil Sci. Fertilizer 44: 625–636.

Subramanian, P., K. Kim, R. Krishnamoorthy, A. Mageswari, G. Selvakumar and T. Sa. 2016. Cold stress tolerance in psychrotolerant soil bacteria and their conferred chilling resistance in tomato (*Solanum lycopersicum* Mill.) under low temperatures. PLoS One 11: e0161592.

Suman, A., A.N. Yadav and P. Verma. 2016. Endophytic microbes in crops: diversity and beneficial impact for sustainable agriculture. pp. 117–143. *In*: Singh, D., P. Abhilash and R. Prabha (eds.). Microbial Inoculants in Sustainable Agricultural Productivity, Research Perspectives. Springer-Verlag, India. doi:10.1007/978-81-322-2647-5_7.

Sun, L., X. Li, Z. Wang, Z. Sun, X. Zhu, S. Liu et al. 2018. Cold priming induced tolerance to subsequent low temperature stress is enhanced by melatonin application during recovery in wheat. Molecules 23: 1091.

Sutherland, R., A. Viljoen, A.A. Myburg and N. Van den Berg. 2013. Pathogenicity associated genes in *Fusarium oxysporum* f. sp. cubense race 4. South Afri. J. Sci. 109: 01–10.

Szittya, G., D. Silhavy, A. Molnár, Z. Havelda, Á. Lovas, L. Lakatos et al. 2003. Low temperature inhibits RNA silencing-mediated defence by the control of siRNA generation. EMBO J. 22: 633–640.

Taiti, C., E. Giorni, I. Colzi, S. Pignattelli, N. Bazihizina, A. Buccianti et al. 2016. Under fungal attack on a metalliferous soil: ROS or not ROS? Insights from Sileneparadoxa L. growing under copper stress. Environ. Pollut. 210: 282–292.

Talanova, V.V., A.F. Titov, N.S. Repkina and L.V. Topchieva. 2013. Cold-responsive COR/LEA genes participate in the response of wheat plants to heavy metals stress. *In*: Doklady Biological Sciences: Proceedings of the Academy of Sciences of the Ussr, Biological Sciences Sections, 448: 28–31.

Tamburic-Ilincic, L., A.W. Schaafsma and D.E. Falk. 2007. Indirect selection for lower deoxynivalenol (DON) content in grain in a winter wheat population. Canadian J. Plant Sci. 87: 931–936.

Thomashow, M.F. 2010. Molecular basis of plant cold acclimation: insights gained from studying the CBF cold response pathway. Plant Physiol. 154: 571–577.

Tunnacliffe, A. and M.J. Wise. 2007. The continuing conundrum of the LEA proteins. Naturwissenschaften 94: 791–812.

Uemura, M., Y. Tominaga, C. Nakagawara, S. Shigematsu, A. Minami and Y. Kawamura. 2006. Responses of the plasma membrane to low temperatures. Physiol. Plant. 126: 81–89.

Ullrich, M., A. Peñaloza-Vázquez, A.M. Bailey and C.L. Bender. 1995. A modified two-component regulatory system is involved in temperature-dependent biosynthesis of the *Pseudomonas syringae* phytotoxincoronatine. J. Bacteriol. 177: 6160–6169.

Vaishnav, A., K. Upadhyay, D. Tipre and S. Dave. 2016. Characterization of potent exopolysaccharide producing bacteria isolated from fruit pulp and potato peels and enhancement in their exopolysaccharide production potential. J. Microbiol. Biotechnol. Food Sci. 6: 874–877.

Velasquez, A.C., C.D.M. Castroverde and S.Y. He. 2018. Plant–pathogen warfare under changing climate conditions. Curr. Biol. 28: R619–R634.

Verma, P., A.N. Yadav, K.S. Khannam, N. Panjiar, S. Kumar, A.K. Saxena et al. 2015a. Assessment of genetic diversity and plant growth promoting attributes of psychrotolerant bacteria allied with wheat (*Triticum aestivum*) from the northern hills zone of India. Ann. Microbiol. 65: 1885–1899.

Verma, P., A.N. Yadav, L. Shukla, A.K. Saxena and A. Suman. 2015b. Alleviation of cold stress in wheat seedlings by *Bacillus amyloliquefaciens*. IARI-HHS2-30, an endophytic psychrotolerant K-solubilizing bacterium from NW Indian Himalayas. Natl. J. Life Sci. 12: 105–110.

Villarreal, P., M. Carrasco, S. Barahona, J. Alcaíno, V. Cifuentes and M. Baeza. 2018. Antarctic yeasts: analysis of their freeze-thaw tolerance and production of antifreeze proteins, fatty acids and ergosterol. BMC Microbiol. 18: 66.

Walker, C.A. and P. van West. 2007. Zoospore development in the oomycetes. Fungal Biol. Rev. 21: 10–18.

Wang, C., X.L. Ma, Z. Hui and W. Wang. 2008a. Glycine betaine improves thylakoid membrane function of tobacco leaves under low-temperature stress. Photosynthetica 4: 400–409.

Wang, C., Y. Teng, S. Zhu, L. Zhang and X. Liu. 2019. NaCl- and cold-induced stress activate different Ca^{2+}-permeable channels in *Arabidopsis thaliana*. Plant Growth Regul. 87: 217–225.

Wang, H.B., Z.Q. Zhang, X.M. Huang, Y.M. Jiang and X.Q. Pang. 2008b. Hot water dipping induced chilling resistance of harvested banana fruit. Acta Hort. 804: 513–522.

Wang, M., Q. Zheng, Q. Shen and S. Guo. 2013. The critical role of potassium in plant stress response. Int. J. Mol. Sci. 14: 7370–7390.

Wang, M., J. Tian, M. Xiang and X. Liu. 2017. Living strategy of cold-adapted fungi with the reference to several representative species. Mycology 8: 178–188.

Webb, K.M., I. Ona, J. Bai, K.A. Garrett, T. Mew, C.M. Vera Cruz et al. 2010. A benefit of high temperature: increased effectiveness of a rice bacterial blight disease resistance gene. New Phytol. 185: 568–576.

Wisniewski, M. and M. Fuller. 1999. Ice nucleation and deep supercooling in plants: new insights using infrared thermography. pp. 105–118. *In*: Rosa, M. and S. Franz (eds.). Cold-adapted Organisms. Springer, Berlin, Heidelberg.

Wu, S.M., P.P. Hwang, C.L. Hew and J.L. Wu. 1998. Effect of antifreeze protein on cold tolerance in juvenile tilapia (*Oreochromismossambicus* Peters) and milkfish (*Chanoschanos* Forsskal). Zool. Stud. 37: 39–44.

Wu, Z., S. Han, H. Zhou, Z.K. Tuang, Y. Wang, Y. Jin et al. 2019. Cold stress activates disease resistance in *Arabidopsis thaliana* through a salicylic acid dependent pathway. Plant Cell Environ. 42: 2645–2663.

Xin, Z. and J. Browse. 1998. Eskimo1 mutants of Arabidopsis are constitutively freezing-tolerant. Proc. Natl. Acad. Sci. 95: 7799–7804.

Xu, J., R.E. Grumbine, A. Shrestha, M. Eriksson, X. Yang, Y.U.N. Wang et al. 2009. The melting Himalayas: cascading effects of climate change on water, biodiversity, and livelihoods. Conserv. Biol. 23: 520–530.

Yadav, A.N., S.G. Sachan, P. Verma and A.K. Saxena. 2015a. Prospecting cold deserts of north western Himalayas for microbial diversity and plant growth promoting attributes. J. Biosci. Eng. 119: 683–693.

Yadav, A.N., S.G. Sachan, P. Verma, S.P. Tyagi, R. Kaushik and A.K. Saxena. 2015b. Culturable diversity and functional annotation of psychrotrophic bacteria from cold desert of Leh Ladakh (India). World J. Microbiol. Biotechnol. 31: 95–108.

Yadav, A.N., P. Verma, S.G. Sachan, R. Kaushik and A.K. Saxena. 2015c. Mitigation of cold stress for growth and yield of wheat (*Triticum aestivum* L.) by psychrotrophic pseudomonads from cold deserts of Indian Himalayas. International Symposium on "Emerging Discoveries in Microbiology", At JNU, New Delhi.

Yadav, A.N., S.G. Sachan, P. Verma, R. Kaushik and A.K. Saxena. 2016. Cold active hydrolytic enzymes production by psychrotrophic Bacilli isolated from three sub-glacial lakes of NW Indian Himalayas. J. Basic Microbiol. 56: 294–307.

Yadav, A.N., P. Verma, V. Kumar, S.G. Sachan and A.K. Saxena. 2017a. Extreme cold environments: a suitable niche for selection of novel psychrotrophic microbes for biotechnological applications. Adv. Biotechnol. Microbiol. 2: 1–4.

Yadav, A.N., P. Verma, D. Kour, K.L. Rana, V. Kumar, B. Singh et al. 2017b. Plant microbiomes and its beneficial multifunctional plant growth promoting attributes. Int. J. Environ. Sci. Nat. Res. 3: 1–8.

Yadav, A.N., P. Verma, S.G. Sachan, R. Kaushik and A.K. Saxena. 2018. Psychrotrophic microbiomes: molecular diversity and beneficial role in plant growth promotion and soil health. pp. 197–240. *In*: Panpatte, D.G., Y.K. Jhala, H.N. Shelat and R.V. Vyas (eds.). Microorganisms for Green Revolution-Volume 2: Microbes for Sustainable Agro-ecosystem. Springer, Singapore. doi:10.1007/978-981-10-7146-1_11.

Yadav, A.N. 2019. Fungal white biotechnology: conclusion and future prospects. pp. 491–498. *In*: Yadav, A.N., S. Singh, S. Mishra and A. Gupta (eds.). Recent Advancement in White Biotechnology Through Fungi: Volume 3: Perspective for Sustainable Environments. Springer International Publishing, Cham. doi:10.1007/978-3-030-25506-0_20.

Yadav, A.N., D. Kour, S. Sharma, S.G. Sachan, B. Singh, V.S. Chauhan et al. 2019. Psychrotrophic microbes: biodiversity, mechanisms of adaptation, and biotechnological implications in alleviation of cold stress in plants. pp. 219–253. *In*: Sayyed, R.Z., N.K. Arora and M.S. Reddy (eds.). Plant Growth Promoting Rhizobacteria for Sustainable Stress Management: Volume 1: Rhizobacteria in Abiotic Stress Management. Springer Singapore, Singapore. doi:10.1007/978-981-13-6536-2_12.

Yadav, A.N. 2020. Plant microbiomes for sustainable agriculture: current research and future challenges. pp. 475–482. *In*: Yadav, A.N., J. Singh, A.A. Rastegari and N. Yadav (eds.). Plant Microbiomes for Sustainable Agriculture. Springer International Publishing, Cham. doi:10.1007/978-3-030-38453-1_16.

Yadav, A.N., S. Mishra, D. Kour, N. Yadav and A. Kumar. 2020a. Agriculturally Important Fungi for Sustainable Agriculture, Volume 1: Perspective for Diversity and Crop Productivity. Springer International Publishing, Cham.

Yadav, A.N., A.A. Rastegari, N. Yadav and D. Kour. 2020b. Advances in Plant Microbiome and Sustainable Agriculture: Diversity and Biotechnological Applications. Springer, Singapore.

Yadav, A.N., A.A. Rastegari, N. Yadav and D. Kour. 2020c. Advances in Plant Microbiome and Sustainable Agriculture: Functional Annotation and Future Challenges. Springer, Singapore.

Yadav, A.N., J. Singh, A.A. Rastegari and N. Yadav. 2020d. Plant Microbiomes for Sustainable Agriculture. Springer International Publishing, Cham.

Yadav, S.K. 2010. Cold stress tolerance mechanisms in plants. A review. Agron. Sustain. Dev. 30: 515–527.

Yamori, W., K. Suzuki, K.O. Noguchi, M. Nakai and I. Terashima. 2006. Effects of Rubisco kinetics and Rubisco activation state on the temperature dependence of the photosynthetic rate in spinach leaves from contrasting growth temperatures. Plant, Cell Environ. 29: 1659–1670.

Yarzabal, L.A., L. Monserrate, L. Buela and E. Chica. 2018. Antarctic *Pseudomonas* spp. promote wheat germination and growth at low temperatures. Polar Biol. 41: 2343–2354.

Yildiztugay, E., C. Ozfidan-Konakci and M. Kucukoduk. 2017. Improvement of cold stress resistance via free radical scavenging ability and promoted water status and photosynthetic capacity of gallic acid in soybean leaves. J. Soil Sci. Plant Nutr. 17: 366–384.

Yun, K.Y., M.R. Park, B. Mohanty, V. Herath, F. Xu, R. Mauleon et al. 2010. Transcriptional regulatory network triggered by oxidative signals configures the early response mechanisms of japonica rice to chilling stress. BMC Plant Biol. 10: 1–29.

Zachariah, S., P. Kumari and S.K. Das. 2016. *Psychrobacterpocilloporae* sp. nov., isolated from a coral, *Pocilloporaeydouxi*. Int. J. Syst. Evolution. Microbiol. 66: 5091–5098.

Zachariassen, K.E. and E. Kristiansen. 2000. Ice nucleation and antinucleation in nature. Cryobiology 41: 257–279.

Zhang, D.C., H.J. Busse, H.C. Liu, Y.G. Zhou, F. Schinner and R. Margesin. 2011. *Sphingomonasglacialis* sp. nov., a psychrophilic bacterium isolated from alpine glacier cryoconite. Int. J. Syst. Evolution. Microbiol. 61: 587–591.

Zhang, H., Z. Wang, Y. Feng, Q. Cui and X. Song. 2019a. Phytohormones as stimulators to improve arachidonic acid biosynthesis in *Mortierellaalpina*. Enzy. Microb. Technol. 131: 109381.

Zhang, Y., L. Fan, M. Zhao, Q. Chen, Z. Qin, Z. Feng et al. 2019b. Chitooligosaccharide plays essential roles in regulating proline metabolism and cold stress tolerance in rice seedlings. Acta Physiol. Plant. 41: 1–11.

Zhu, X., J. Liao, X. Xia, F. Xiong, Y. Li, J. Shen et al. 2019. Physiological and iTRAQ-based proteomic analyses reveal the function of exogenous γ-aminobutyric acid (GABA) in improving tea plant (*Camellia sinensis* L.) tolerance at cold temperature. BMC Plant Biol. 19: 43.

Zur, I., G. Gołębiowska, E. Dubas, E. Golemiec, I. Matušíková, J. Libantová et al. 2013. β-1, 3-glucanase and chitinase activities in winter triticales during cold hardening and subsequent infection by *Microdochiumnivale*. Biologia 68: 241–248.

Chapter 8

Cold-active Microfungi and Their Industrial Applications

Sanjay Sahay

|||||||||||||||||||||||||||||||||||||||||||||||||||||||||||||||||||||||||||||||||||||||||||||||||||||||||||||||||||||||||||||||||||||||||||||||||||||||||||||||||||||||||||||||||||||||||||||||

Introduction

Cold-active organisms are those that are adapted to a climate characterized by longer periods in the year of temperature below 15–20°C. This condition is a trait of about 85% of the Earth's surface including the polar and alpine regions and the deep sea. The cold-active organisms may be psychrophilic or psychrotolerant depending upon their inability or ability to thrive above a threshold lower temperature respectively. This threshold temperature may be 15°C–20°C (Morita 1975; Robinson 2001; Cavicchioli et al. 2002; Yadav et al. 2018b). Microfungi constitute an artificial group of filamentous fungi belonging to Basidiomycetes and Ascomycetes. The most common microfungi are slime molds, water molds, bread molds, powdery mildews and rusts. Thus the group includes both saprophytic and pathogenic (on plants and animals) species and is believed to comprise 4468 genera and 55,989 species. For a detailed classification of microfungi, readers are suggested to visit the dedicated website (https://www.microfungi.org/).

Cold-active microfungi have been reported from almost all the cold areas existing on the Earth (Hassan et al. 2016). Their extraordinary capacity enables them to occupy diverse types of niches such as the interior (endophytes) or the surface of live plants (epiphytes), the decaying plant materials (saprophytes), live animals (zoopathogen) or insects (entomopathogen), root of higher plants (ecto- or endomycorhiza), deep sea animals (e.g., sponges), sediments, soil, water, etc. in the colder areas (Verma et al. 2015; Yadav et al. 2015a; Yadav et al. 2015b). Due to the diversity of living conditions they exhibit, they are expected to have faced diverse types of challenges enabling them to evolve varied types of biomolecules to respond to them. These biomolecules are of interest to feed various industries so as to get newer products or processes. The most attractive features of all applications are the lower energy consumption either in the bioproduction (at low temperatures) of biomolecules or application of these bio-molecules (in the process at low temperatures) or both and consequent reduction in CO_2 emission. The chapter will develop on these aspects in details.

Habitats of cold active microfungi

Cold-active microfungi exist in extensively-found and diverse types of cold habitats (Hassan et al. 2016). Almost 85% of the Earth's biosphere shows cold temperature (< 5°C) for longer periods in the year. Specifically these cold areas include the deep ocean (about 90% of volume of ocean

Government Postgraduate College, Biaora, Rajgarh, Madhya Pradesh, India.
Email: ss000@rediffmail.com

shows temperature < 5°C), snow-covered land including permafrost, sea ice and glaciers (82% of land surface), cold water lakes, cold soils (especially subsoil), cold deserts and caves. The major cold land surfaces are present in the Arctic, the Antarctic and high-mountains (Margesin and Miteva 2011; Hassan et al. 2016).

Adaptations in cold-active microbes including microfungi

Microbes in extreme cold climate exhibit both morphological and physiological adaptations. Important morphological adaptations include membrane fluidization manifested in more glazed mycelia or colonies and production of pigments, e.g., melanine (Onofri et al. 2004). However, adaptations at the physiological level are expressed as the production of various types of biomolecules or exhibiting unique features that vary from species to species and include:

a) Unsaturated fatty acids—Enhanced synthesis of unsaturated fatty acids and their acquisition by the membrane is needed to maintain its fluidity.

b) Cryoprotectants—These are organic molecules (amino acids, alcohols, sugars, etc.) synthesized *de novo* or in excess amount than normally present when cells are exposed to extremely low temperatures to protect proteins from aggregation and the membrane from losing fluidity (Mancuso et al. 2005). For example, a positive effect on growth at inhibitory low temperatures of glycine betaine on *Listeria monocytogenes* has been reported (Ko et al. 1994). Trehlose has been shown to protect proteins from denaturation in psychrophiles (Phadtare 2004) and to maintain membrane fluidity in fungi (Tibett et al. 1998). Sugars (mannitol, sucrose and trehlose) have also been found to accumulate inside cells in fungi to protect them from being frozen and desiccated when exposed to temperature below 10°C (Hoshino et al. 2009);

c) Cold-shock proteins—These are a heterogenous group of proteins reported to assist the cells through various mechanisms to thrive unfavorable cold condition (Hebraud and Potier 1999). They may regulate transcription or initiation of translation in protein biosynthesis or act as chaperones to protect protein or mRNA from misfolding or unfolding and thus help maintain their active structure and also help preserving the chromosome structure (Chaikam and Karlson 2010);

d) Antifreeze proteins (AFPs)—AFPs are proteins that can lower the freezing point without altering the melting point usually by the adsorption-prevention method. Fungi produce AFPs both in intra-cellular sites (e.g., intercellular spaces and apoplast) and extracellular site (over mycelial surface), the former protect cytoplasm and membrane from growing ice crystals and the latter help keep soil solution contiguous to mycelium unfrozen facilitating uptake of nutrients by the mycelium from the solution (Hoshino et al. 2009);

e) Cold-active enzymes—Cold-active enzymes are enzymes that exhibit higher enzymatic activity at low temperature vis-a-vis their mesophilic counterparts (Samie et al. 2012). Important adaptive features of these enzymes include enhanced flexibility in structure (Struvay and Feller 2012), up to 10-fold higher specific activity (k_{cat}) (Gerday et al. 2000), thermolability (Miyazaki et al. 2000), alteration in amino acid composition usually in the active site domain and so on (Siddiqui and Cavicchioli 2006);

f) Polyol and exopolysaccharides—Production of polyol and secretion of exopolysaccharides has been implicated as physiological adaptation against extreme cold (Robinson 2001). Some fungi in the cold climate have been reported to exhibit these traits (Pascual et al. 2002; Selbmann et al. 2002; 2005);

g) Fungal species are found to exhibit sealing or other mechanisms to repair sheared or ruptured mycelium under the influence of frost heave (Timling and Taylor 2012);

h) Some of cold-active fungi show cold avoidance such as shortening of the life cycle, preferential anamorphism, production of spores before winter and germination after winter, or cyclic entry of fungal propagules from adjoining moderate climatic sites (Marshall 1998);

i) Microbes in deep sea colder areas encounter low temperature as one of the several constraints such as very high pressures (exhibiting 105 Pascal (Pa) increase for each 10 m further depth) and salinity levels (~ 3.5%), poor light, nutrient (oligotrophy) and oxygen conditions (Raghukumar 2008; Batista-Garcia et al. 2017). Fungal isolates from the deep sea generally exhibit tolerance to most of these constraints, cold-tolerance seems less important for them so they are mostly psychrotolerants (Singh et al. 2010);

j) Many cold adapted phenotypes such as cold-adapted enzymes and fluidization of membrane however are suitable for lower temperatures only (< 10°C), at higher temperature (> 20°C) they do not work (Hoshino et al. 2009). But the presence of the most dominant arctic fungi in less-cold and other regions clearly indicate their capacity to undergo large scale molecular switching under different temperature regimes. This is more studied in the case of bacteria where upregulation of a group of genes associated with cold adaptation has been reported (Shivaji and Prakash 2010);

k) Moreover, traits such as resistance towards freezing, desiccation and ultraviolet ray and bearing small sized spore are considered potential to facilitate their dispersal (Robinson 2001).

Cold-active fungi from various habitats

Cold-active microfungi have been reported from various cold habitats such as soil [Arctic (Singh et al. 2012; Hafizah et al. 2013), Antarctica (Durán et al. 2019), cold desert (Sterflinger et al. 2012), Himalayan ranges (Petrovic et al. 2000; Freeman et al. 2009; Anupama et al. 2011; Sahay et al. 2013; Sati et al. 2014), European mountains (Margesin et al. 2007; Turchetti et al. 2008; Brunner et al. 2011)], snow covered lakes (Vivian et al. 2012; Connell et al. 2018) and deep sea (Raghukumar 2008), associated with various plants as epiphytes such as on Lichen (Zhang et al. 2015) or Moss (Gawas-Sakhalkar and Singh 2011) or other plants (Widden and Parkinson 1979), or as symbionts (Sokolovski et al. 2002; Upsan et al. 2007) or as pathogen (Davey and Currah 2009) or as endophytes (Clemmensen and Michelsen 2006; Newsham et al. 2009; Timling and Taylor 2012), or endolithic (Coleine et al. 2018) or mycorhizic (Upsan et al. 2007; Sokolovski et al. 2002; Gardes and Dahlberg 1996) or associated with root (Bjorbaekmo et al. 2010; Timling and Taylor 2012; Blaalid et al. 2014; Bonten et al. 2014) or spoiling food in refrigerator (Altunatmaz et al. 2012) or as animal pathogen (Gené et al. 2003). A list of frequently isolated cold-active microfungi from major cold habitats and niches is given in Table 8.1.

Industrial application

Cold-active microfungi have been found as sources of biomolecules of tremendous industrial importance, which can be categorized as below:

Antifreeze proteins

Antifreeze proteins are proteins that can lower the freezing point, by the adsorption-prevention method. They have been reported from bacteria, fungi, polar fishes, insects and plants with their structure being different but action broadly similar. So far five types of fish AFPs and two types of insect AFPs have been described, showing their protein structure and amino acids-composition significantly different from each other (Voets 2017). AFPs are believed to bind to a growing ice surface, shape it to a specific morphology and prevent its further enlargement to assume fatal sizes. AFPs also prevent ice crystals to fuse to form large aggregates (recrystallization) injurious to tissues. AFPs reduce freezing point without affecting melting point (Thermal hysteresis or TH). AFPs from

Table 8.1 Cold-active microfungi frequently reported from major cold habitats.

Fungi	Arctic			Antarctica			Deep sea	Alpine	Refrigerator
	Soil	Moss	Lichen	Soil	Endolithic	Lake			
Acremonium sp.			+			+			+
Alternaria sp.			+			+			
A. alternate									
Antarctomyces sp.			+						
Arthrinium			+						
Aspergillus (*Emericella*)	+						+	+	+
A. aculeatus	+								
A. flavus									+
A. nidulans	+								
A. niger									
A. ochraeus									+
A. terreus							+		+
A. ustus			+				+		
A. versicolour							+		
Atradidymella muscivora			+						
Beauveria sp.							+	+	
Botrytis sp.									+
Cadophora sp.			+	+					
C. fastigiata				+					
C. finlandica	+								
C. luteo-olivacea				+					
C. malorum				+		+			
C. Thielavia			+						
Chaetomium sp.				+					
Cenococcum sp.			+						
Cenococcum geophilum	+								
Corynespora sp.		+							
Cryomyces minteri					+				
C. antarcticus					+				
Cosmospora sp.			+						
Curvularia sp.							+	+	
Cylindrocarpon sp.							+		
Davidiella tassiana						+			
Elasticomyces elasticus			+						
Emericella sp.							+		
Emericellopsis sp.							+		
Eurotium sp.							+		
Fontanospora sp.						+			
Friedmanniomyces simplex						+			
F. endolithicus					+				
Fusarium sp.				+	+		+		
Fusarium oxysporum		+							

Table 8.1 Contd. ...

...Table 8.1 Contd.

Fungi	Arctic			Antarctica			Deep sea	Alpine	Refrigerator
	Soil	Moss	Lichen	Soil	Endolithic	Lake			
Fusarium proliferatum					+				
Galactomyces candidum							+		
Geomyces sp.	+			+	+		+	+	
Geomyces pannorum		+					+		
Gliocardium sp.				+					
Graphium sp.							+		
Gymnascella sp.							+		
Herpotrichia sp.			+						
Hormonema sp.				+					
Hormonema dematioides				+					
Lecanicillium sp.			+						
Leotiomycetes sp.								+	
Leptosphaeria sp.							+		
Microdochium sp.		+							
Monodictys sp.			+						
Mortierella sp.	+	+	+	+		+	+	+	
M. alpina		+						+	
M. plumbeus									+
M. schmuckeri		+							
M. simplex		+							
Mucor sp.		+						+	+
Mucor hiemalis		+						+	
Myrothecium sp.	+								
Oidiodendron sp.			+						
Paecilomyces sp.							+		
Penicillium sp.	+	+	+	+	+		+	+	+
P. canesense								+	
P. chrysogenum								+	
P. citrinum			+						
P. commune							+		
P. dipodomyicola							+		
P. echinulatum				+					
P. expansum				+					
P. frequentans		+							
P. italicum									+
P. lagena							+		
P. mali				+					
P. roquefortii				+					
P. rugulosum			+						
Periconia sp.								+	
Pestalitiopsis sp.				+			+		
Petriella sp.				+			+		

Table 8.1 Contd. ...

...Table 8.1 Contd.

Fungi	Arctic			Antarctica			Deep sea	Alpine	Refrigerator
	Soil	Moss	Lichen	Soil	Endolithic	Lake			
Phaeosphaeria sp.			+						
Phialocephala sp.			+						
Phialocephala fortinii	+								
Phialophora sp.	+	+							
Phoma sp.							+	+	
Phoma sclerotioides								+	
Pithomyces chartarum		+							
Polyblastia terrestris			+						
Preussia sp.	+		+						
Pseudogymnoascus sp.				+			+		
Ps. pannorum								+	
Ps. roseus								+	
Pseudeurotium sp.								+	
Rhizopus sp.				+					
Rhizoscyphus sp.							+		
Rhizoscyphus ericae			+						
Sagenomella sp.							+		
Scopulariopsis sp.								+	
Spicellum sp.							+		
Sporormiaceae					+				
Talaromyces sp.				+					
Tetracladium sp.				+					
Thelebolus sp.				+			+		
Th. globosus							+		
Th. ellipsoideus							+		
Th. microspores							+	+	
Trichoderma sp.							+	+	
T. atrovirde						+			
Truncatella angustata								+	
Umbelopsis isabellina								+	
Xeromyces sp.							+		

plants exhibit mild TH (0.1–0.5°C), those from fishes moderate TH (1–5°C) and from insects show higher TH (5°C). AFPs from fishes are also found to protect membrane from cold-induced injury. These unique properties of AFPs enable them to act as green or biocryoprotectant for the foods or other biological samples preserved at very low or chilling temperatures.

Food industries

Chilled and frozen meat and fish exports form a huge international business and important item of foreign exchange earnings for some countries. Short or long term preservation is thus required to deliver high quality meat and fish. Low temperature preservation is a common method, but the method involves freezing and thawing that promotes ice formation and ice re-crystallization leading to possible membrane damage and leaking out of cytoplasmic materials. The damaging effect of bigger ice crystal is also unfavorable for normal texture, drip and sensory properties of meat and fish, which leads to lipid oxidation. These effects altogether cause much damage to the quality of the animal products. To overcome freezing damage conventional cryoprotectants are used that has the potential to alter the taste by their sweetening effect. In some studies both AFP and AFGP are found as better alternatives to the conventional cryoprotectants (Vestergaard and Parolari 1999). Indirect evidence of AFP's ability to preserve membrane integrity came from an experiment that revealed that the marker membrane bound enzyme Ca^{2+} ATPase of actomyosin during freezing and chilled storage was preserved to a much higher degree (100%) in the presence of Type III AFP (50 g/L) as compared to conventional cryoprotectant (sucrose-sorbitol mixture) (50%) (Boonsupthip and Lee 2003).

Ice cream products: Currently, the ice cream market is on the boom and demand for quality ice cream is growing. One of the major problems, ice cream manufacturers face is the generation of larger ice crystal due to fluctuation of temperature during freezing. The larger ice crystals affect the texture and organoleptic features of ice cream adversely. Addition of AFPs overcomes the problem by inhibiting recrystallization during manufacturing and storage (Regand and Goff 2006). Supplementation of AFP to the ice cream mixture during freezing inhibits the ice formation and recrystallization. Usually a rapid cooling process is adopted to avoid ice formation during manufacturing, but with AFP added to the ice cream mixture rapid cooling is not required. Rather a simple hardening by gradual reduction in temperature from –18 to –30°C is sufficient (Huang et al. 1992; Warren et al. 1992). An experiment with two sets of two solutions: A, simple sucrose solution (control); B, simple sucrose solution supplemented with 0.25% cold-acclimatized winter wheat extract containing AFPs (AWWE); C, ice cream mixture (control); D, ice cream mixture with 0.25% AWWE was carried out. All the solutions were subjected first to super cooling temperature (–50°C), the temperature then raised to –10°C and finally to –5°C (Regand and Goff 2006). It was found that the solutions containing AFPs contained higher number but smaller sized ice crystals. In another experiment the three experimental solutions containing AWWE were incubated at –35°C, –18°C and –10°C for 12 hours. The inhibition of ice growth was found in the case of solution at –10°C only, possibly because in the former two conditions diffusion of AFPs was restrained due to low temperature related higher viscosity conditions. Currently, the ice cream industries are the largest users AFP.

Frozen dough is in high demand but suffers from one problem related to cold intolerance of baker's yeast *Saccharomyces cerevisiae*. At freezing temperatures yeast cell exhibits cell shrinkage, membrane rupture and protein denaturation. Since the quality of dough products depend on the gas producing efficiency of yeasts, the destruction of yeast cells causes a negative impact on the quality of products dough is used for. To over the problem higher concentration of yeast cells (by 4–10% of normal dose) is recommended which is expensive. Since cold-tolerant baker's yeast strain is not available, transgenic yeast containing AFP gene is being used to (Panadero et al. 2005). Type 1 AFP from grubby sculpin named GS-5 has been used for baker's yeast transgenesis (Panadero et al. 2005). Experimental dough containing obtained transgenic yeasts and the control one containing wild type yeast cells when kept at –20°C for 40 days, while the former exhibited 30% more residual gassing capacity, though in short term (3 hours) cold storage at freezing temperature both types of dough lost about 30% of CO_2 production ability (Panadero et al. 2005).

Aquaculture

Fishes constitute a very important food item because of availability, taste and nutritional values. Apart from the natural supply, fishes are also reared under controlled conditions within a specified area called aquaculture. In some areas of the world for example north of the Atlantic region, the extreme winter condition (say –1.9°C) is unfavorable for marine fishes to survive and thus they migrate to warmer areas. Sea cage farming or other types of aquaculture under this condition becomes non-feasible (Hew and Ewart 2002). Availability of freeze tolerance strains of commercially desirable fishes however may make aquaculture more practical. Towards this end an attempt has been made to transfer Type 1 AFP gene from winter flounder to Atlantic salmon (Hew and Ewart 2002). Transgenic rainbow trout has also been raised by transferring the winter flounder AFP gene to wild type trout, the transgene thus found is able to impart cold tolerance to temperatures as low as $-1.4°C$ (Hew and Ewart 2002). Wild type rainbow trout cannot survive at this temperature. AFP can thus be used in the development of aquaculture.

Fruits and vegetables

Fruits and vegetables are usually frozen during long distance transport, or for long duration storage. Ice formation and recrystallization in the extracellular or intracellular spaces due to low rate freezing and temperature fluctuations may destroy their texture and cause damage to the plasma membrane (Hightower et al. 1991; Venketesh and Dayananda 2008). Leakage of nutrient rich sap from injured fruits and vegetables is an important result of freezing. This exterior and interior damage reduces the commercial values of the fruits and vegetables substantially and thus a technique to overcome these damages is desirable (Hightower et al. 1991). Rapid freezing may overcome these problems to a large extent but the technique is very costly. AFP may rescue fruits and vegetables during freezing and cold storage (Fletcher et al. 1999). Towards this end a chimeric construct *spa-afa5* containing AFP III gene from winter flounder called *afa3* AFP III and a truncated protein A from bacterium *Staphylococcus* called *spa* was transferred to a tomato plant (Hightower et al. 1991). To ensure its expression in the tomato plant, the chimeric gene was engineered to place between the plant promoter in the upstream position and polyadenylate region in the downstream position (Hightower et al. 1991). To study individual contributions, *Spa-afa5* and *afa3* were also expressed separately in tomato plants. Soluble protein extract from three types of transgenic tomatoes was tested by first freezing rapidly down to $-70°C$ and then raising temperature to $-7°C$. Monitoring of ice crystal growth in three types of extract for 1–2 hours revealed that chimeric protein was able to inhibit recrystallization three times more effectively than pure *afa5* protein.

AFP as an allergen

To test the potential allergic reaction caused by AFP in human beings, analysis of Type III AFP from *ocean pout* has been conducted. The AFP was subjected to pepsin digestion followed by amino acid sequence determination of the protein fragments. Amino acid sequence was then compared with known protein allergens and the potential IgE mediated response of protein thereby determined. It was found that tested Type III AFP could be rapidly hydrolyzed by pepsin and that the amino acid sequence of this protein had no match in any known allergen. An *in vivo* test was also conducted in selected persons with no protein sensitivity problem to see whether AFP triggers IgE-mediated reaction in them (Baderschneider et al. 2002). A negative result was obtained indicating that AFP does not trigger such allergic reaction. Both *in vitro* and *in vivo* genotoxic tests have also been found to exhibit negative results (Manning et al. 2004). Further general health and immunogenicity test for the Type III AFP in healthy human volunteers for 8 weeks revealed that the protein causes no unfavorable reaction in the tested persons (Crevel et al. 2007).

Augmenting cold tolerance in plants

Both AFP and AFGP have been used to transform various plants (e.g., tobacco, tomato and potato) for cold tolerance (Hightower et al. 1991; Fan et al. 2002; Gupta and Deswal 2014). Transgenic plants showed various cold tolerance phenotypes as in case of *DcAFP* (carrot AFP gene) carrying transgenic tobacco exhibited inhibition in ice recrystallization, a TH of 0.35 to 0.56°C, lesser ion leakage (1–30% cf wild type showing 1–80% leakage) and faster recovery following cold stress (Fan et al. 2002). Transgenic tomato plant carrying AFP gene from *Lolium* perenne has been found to exhibit three times more relative water content and 2.6 times less electrolyte leakage as compared to wild type counterpart (Balamurugan et al. 2018). Inhibition of ice recrystallization and production of TH have generally been exhibited by AFP transgenic plants (Griffith and Yaish 2004). Likewise, expression of AFP from *Lolium perenne* in *Arabidopsis* and *E. coli* enhanced cold hardiness in the sensitive wild types (Zhang et al. 2010). Chimeric construct (e.g., spa-afa-5) has also been tested in tomato and was found to exhibit 10 times more activity than pure afa-3 (Hightower et al. 1991). Protein-A has been speculated to impart protection of small afp5 against proteolysis.

Medical applications

Preservation of cells, tissues or organs for research and application forms an important part of medical applications. AFPs have been reported to exhibit both cryoprotective and cytotoxic actions depending on type and dose of AFPs, type of biological materials, storage protocol and the composition of cryoprotectant (Wang 2000). Application of AFPs in the preservation of cells has been reported both in reproductive biology and other medical research.

In reproductive biology, there is need to preserve gametes and embryos for reproductive management of animals, assisted reproduction in human beings and gene banking. AFPs have been applied to protect from cold injury and reduce the use of conventional cytotoxic Cryo Protection Agents (CPAs) during cryopreservation of sperms (Prathalingam et al. 2006; Kim et al. 2011; Qadeer et al. 2016) and oocytes (Rubinsky et al. 1991; Jo et al. 2012; Chaves et al. 2016; Wen et al. 2014). AFPs have also been tested to supplement CPAs in cryopreservation of embryos from horses (Lagneaux et al. 1997), cows (Ideta et al. 2015), sheep (Baguisi et al. 1997), mice (Shaw et al. 1995), rabbits (Nishijima et al. 2014) and fish (Martínez-Páramo et al. 2009). Type I AFP has been found effective in improving survival after freezing at 4°C or 10°C of embryos of fish.

Other medical research animals including human being cells or organ physiology or pathogenesis related research requires cells and organs preservation usually cryopreservation. To protect them from cold injury conventionally CPAs are used many of which are cytotoxic. AFPs have been tested to reduce the use of CPAs as in case of human liver cells (Hirano et al. 2008), RIN-5F insulin tumor cells (Kamijima et al. 2013), diatoms (Koh et al. 2015), red blood cells (Chao et al. 1997; Lee et al. 2012), muscle cells (Mugnano et al. 1995), gut cell (Tursman and Duman 1995), islet cells (Mayordomo et al. 2000), and human cell lines (Kim et al. 2015) including HeLa cells, NIH/3T3 cells, preosteoblasts (MC3T3-E1cells), and human ketatinocytes (HaCaT cells).

In some cases AFPs have also been found to have an adverse effect on the survival of the cell/organ/gametes (Carpenter and Hansen 1992; Payne et al. 1994; Wang et al. 1994) which was on analysis found to be due to higher AFP concentration leading to AFPs forming needle-like ice that can damage preserved material during freezing (Carpenter and Hansen 1992; Lee et al. 2012; Kim et al. 2015; Mayordomo et al. 2006). It was also found that the favorable concentrations of AFPs vary with the type of AFPs and type of biological materials used. Thus the suitable LeIBP concentration is 0.1–0.8 mg/mL for red blood cells (Lee et al. 2012), diatoms (Koh et al. 2015), oocytes (Lee et al. 2015) and mammalian cell lines (Kim et al. 2015), whereas suitable fish AFPs concentration is lower say 0.1 mg/mL. Suitable concentration of AFPs also helps protecting cells/organs such as rat livers (Lee et al. 1992) and mammalian hearts (Amir et al. 2004) during hypothermal storage (Lee et al. 1992). In view of the high commercial value of AFP (US$ 10/mg), intensive research has been pursued to enhance productivity of suitable AFP (https://www.antifreezeprotein.com/product).

Microfungi as a source of AFPs

AFPs have been reported more frequently in yeasts than in fungi. So far only two basidiomycetous psychrophilic fungi *Coprinus psychromorbidus* and *Typhula ishikariensis* (Xiao et al. 2010a) and two ascomycetous fungi *Antarctomyces psychrotrophicus* (psychrophilic) (Xiao et al. 2010a) and *Penicillium camemberti* (psychrotolerant) (Xio et al. 2010b) have been reported to produce AFPs in the extracellular space, though *Typhula* sp. along with *Sclerotia* sp. have also been found to secrete AFPs on the mycelial surface (Hoshino et al. 2009) Molecular masses of purified fungal AFPs of *C. psychromorbidus* and *A. psychrotrophicus* were approximately 22 and 28 kDa, respectively. The shape of ice crystal formed by *A. psychrotrophicus* antifreeze protein (ApAFP) has been found to be bipyramidal with a unique rugged surface. It exhibits thermal hysteresis activity equal to 0.42°C for a 0.48 mM solution as in case of fish antifreeze protein. A more studied AFP from *Typhula ishikariensis* (TiAFP) forms 'stone age knives' shaped ice crystal and exhibits thermal hysteresis is equal to 2°C as for an insect AFP.

Thus fungal AFPs generate no usual hexagonal ice crystal or exhibit regular binding pattern to ice crystal surface (Hoshino et al. 2003). Moreover, optimum thermal hysteresis of ApAFP is found under alkaline conditions, while that of the TiAFP under acidic conditions, and thus the two characterized fungal AFPs are significantly dissimilar ones. The applications of AFPs have been studied for both medical and industrial fields especially food industries. AFPs have been explored in some non-specified species of microfungi viz., *Hyphochytrium* sp. (Hypochytridiomycetes), *Chytriomyces* sp., *Kappamyces* sp., *Powellomyces* sp., *Rhizophydium* (Chytridiomycetes), unknown species (Balstocladiomycetes), *Pythium* sp. and other unknown species (Oomycetes), Mortierella (Zygomycetes) *Acremonium, Antarctcomyces, Embellisia, Geomyces* sp., *Leptosphaeria* sp., *Phialocephela* sp. and *Penicillium* sp. (Ascomycetes) with limited success. Of these, one unknown species of Oomycetes, *Antarctomyces psychrotrophicous* and *Peniciliiumm camemberti* have been reported to contain AFP like activity in their culture extract (Xiao et al. 2010a).

Unsaturated fatty acid

Polyunsaturated fatty acids (PUFAs) including both omega-6 (gama-linoleic acid GLA and arachidonic acid ADA) and omega-3 (alpha-linolenic acid ALA, eicosapentaenoic acid EPA and docosahexaenoic acid DHA) are essential for maintaining multiple biological functions in human beings (Bhagavan 1978), and medically recommended for reducing the risk of cardiac arrest and coronary artery disease and treat asthma, hypertension, rheumatoid arthritis and Crohn's disease (Vadivelan and Venkateswaran 2014). Omega-3 fatty acids are also known to help prevent breast and lung cancer (Merendino et al. 2013) and control mood disorders such as depression and bipolar disorder, schizophrenia, and dementia (Freeman et al. 2006). The human body cannot biosynthesize PUFAs but get them from dietary sources, for example marine fishes (salmon, mackerel, etc.) and some vegetable oils like flax, hemp, rapeseed (canola), hemp, and walnut. Since fishes are seasonal and known to contain toxic metals in their body while vegetable oils are costly, microbial sources are thus believed to be more potential ones. Microfungi *Mortierella* (Vadievelan and Venkateswaran 2014), *Umbelopsis* (Grantina-Ievina et al. 2014), *Rhizopus* (Suleiman et al. 2018), etc. have been reported to be promising producers of PUFAs.

The fungal genus *Rhizopus* is known as an efficient oleaginous fungus, whose culture conditions may be manipulated (especially by incubating the culture at 15°C) to obtain PUFAs in higher amounts at the expense of saturated fatty acids (Suleiman et al. 2018). Species of *Mortierella* such as *M. alpina, M. alliacea, M. elongata, M. spinosa* and *M. polycephala* have also been reported to be great source of PUFAs. Of these species *M. alpina* which is psychrotrophic and ubiquitous species reported from mesothermal and cold habitats including the Arctic region is currently known as higher lipid-producers as well as major producers of arachidonic acid, Linolenic Acid (LA), Gamma-Linolenic Acid (GLA), and dihomogamma-linolenic acid (DGLA) and minor producers of EPA and DHA under stress condition (Bajpai et al. 1991; Vadivelan and Venkateswaran 2014; Ho et al. 2017). Microfungus *Rhizopus* sp. GB2 has on the other hand, been found to show higher percentage of PUFAs when cultured at 15°C for three days following complete incubation under normal conditions (Suleiman et al. 2018).

Biodiesel production

To augment vegetable and animal fats for the production of biodiesel, alternative microbial sources (algae and fungi) are being explored and developed (Khot et al. 2012). In order to be an eligible source of oil feedstock for biodiesel production, microbial cells must produce/accumulate > 20% or lipid of their cell mass and the lipid must have higher neutral lipid content (Meng et al. 2009; Kosa and Ragauskas 2011). Microfungi belonging to zygomycetes are generally known as oleaginous one. Some of the members of this class, viz., *Mortierella* (Vadievelan and Venkateswaran 2014), *Umbelopsis* (Grantia-Levina et al. 2014), *Rhizopus* (Suleiman et al. 2018), etc. have been reported to be promising producers of oil with PUFAs. Species of *Mortierella*, viz. *M. alpina* and *M. isabellina* ATHUM have been reported to produce 40% (w/w) and 50.4% single cell oil respectively with glucose as the carbon source, thus believed to be a highly potential source for biodiesel production (Wynn et al. 2001; Papanikolaou et al. 2004). Another very potent fungus is *Cunninghamella echinulata* producing 92% (w/w) neutral lipids in its total lipid mass (Fakas et al. 2007). Moreover, fungal cells are believed to be superior over plant and algal resources because the former show faster, easier and more controllable (non-dependent on climatic variation) growth in laboratories, shorter life cycles, easier scaling up possibility and more flexibility to utilize a wide range of agro-forest-municipal residues as a carbon source or complete food (Rumbold et al. 2009; Subramaniam et al. 2010).

Natural colors

Risks attached with synthetic colorants is increasingly being recognized and therefore there is now much emphasis on the use of natural colors in foods, textile, etc. Natural colors may be obtained from plants and microbes, but since pigments from plants are season-dependent and costlier, interest in microbial sources is growing. Some very important pigments produced in the microfungi are given in Table 8.2. Fungi have been reported to produce a variety of pigments, viz., carotenoids, flavins, indigo, melanins, monascins, phenazines, quinines and violacein (Dufosse et al. 2014). For example, a single fungus *Monascus* sp. has been used to produce food colorants like rubropunctamine, monascorubramine (purple), rubropunctatine, monascorubrine (orange), and ankaflavine, monascine (yellow). This fungus is a source of at least 50 pigments and around 50 patents have been used related to the use of these fungal pigments as food colorant (Dufosse et al. 2005). Monascus pigments provide additional benefits to foods because of their anticancer, anti-mutagenic, anti-obesity and anti-microbial properties (Feng et al. 2012). A number of fungal species in the genera *Curvularia*, *Drechslera*, *Eurotium* and *Fusarium* have also been reported to produce pigments (Babitha 2009). Most of these genera have their psychrophilic species reported from cold habitats, thus are potential sources of pigments. For example cold deep marine habitats exert additional stresses such high pressure, salt and low light thus promoting production of defense related substances in them (Dufosse et al. 2014). Genera such as *Aspergillus* (He et al. 2017), *Eurotium* (Smetantina et al. 2007), *Penicillium* (Dhale and Vijay Raj 2009), *Trichoderma* (Blaszczyk et al. 2014) isolated from marine habitats with pigment producing ability. Though in general the pigments derived from marine fungi are similar to those derived from terrestrial counterparts (Capon et al. 2007); some specific pigments have also been found to be produced by marine fungi only, e.g., yellow pigment (anthracene-glycoside asperflavinribofuranoside) by Microsporum sp. (Li et al. 2006).

Carotenoid pigments

Carotenoids constitute an important class of pigments in plants, fungi and bacteria. In plants they supplement light harvesting during photosynthesis and in microbes they play a special role of protecting them from various stresses such as UV radiation, water and salt stresses and oxidative stress. So far more than 600 natural carotenoids have been identified (Sandmann 1994; Armstrong

Table 8.2 Chemical nature and uses of microfungal pigments.

Fungus	Pigment	Uses	Reference
Ashbya gossip	Riboflavin	Food and medicine	Ledesma-Amaro et al. 2015
Aspergillus versicolor	Asperversin	Antifungal agent	Miao et al. 2012
Blakeslea trispora	Lycopene	Food and medicine	Venil et al. 2013
Do	ß-carotene	Do	Do
Cordyceps unilateralis	Naphtoquinone	Do	Unagul et al. 2015
Monascus sp.	Monascorubramin	Food and medicine	Venil and Lakshmanaperumalsamy 2009
Do	Ankaflavin	Do	Do
Do	Rubropunctatin	Do	Do
Mucor circinelloides	ß-carotene	Do	Zhang et al. 2016
Penicillium oxalicum	Anthraquinone	Dyeing of wool fabrics	Wang et al. 2014
Talaromyces verruculosus	Red pigment	Dye textile	Chadni et al. 2017

* Psychrotolerant species

and Hearst 1996; Diaz-Sanchez et al. 2011) and more than 200 species of fungi have been found to produce carotenes (Dufosse et al. 2005). Fungal species in the genera *Blakeslea*, *Mucor* and *Phycomyces* (Mucorales), *Sclerotium*, *Sclerotinia* and *Ustilago* (Basidiomycetes) and *Aspergillus*, *Aschersonia*, *Cercospora*, *Penicillium* (Ascomycetes) have been found to produce carotenes. A mesophilic fungus *Neurospora crassa* has been reported to show enhanced synthesis of carotene at low temperature of 8°C (Castrillo et al. 2018). Cold-adapted *Penicillium* sp. isolated from alpine regions has been reported to low temperature (15°C) carotenes production. Carotenoid pigment has also been produced from a cold-tolerant fungal strain *Thelebolus microsporus* (Singh et al. 2014). The industrial applications of carotenoids include their uses in cosmetic and food coloring, nutritional supplements and pharmaceutical purposes. Extremophilic fungi have been reported to produce carotenoids with a greater variety and yield (Arcangeli and Cannistraro 2000).

Cold-active enzymes

Cold-active, also called 'cold-adapted' enzymes are enzymes that exhibit higher enzymatic activity at a low temperature vis-a-vis their mesophilic counterparts because the former exhibits a lower rate of activity-reduction with the lowering of temperature (Cavicchioli et al. 2011; Samie et al. 2012; Yadav et al. 2016a; Yadav et al. 2016b). Important adaptive features of these enzymes show enhanced flexibility in structure (Struvay and Feller 2012) due to decreased secondary structures and types of interaction that stabilize functional conformations such as disulphide bond, electrostatic interactions, metal binding sites (Cavicchioli et al. 2011), thermolability (Miyazaki et al. 2000), catalytic ability in organic solvent system (Gerday et al. 2000; Cavicchioli et al. 2002; Cavicchioli and Siddiqui 2006; Siddiqui and Cavicchioli 2006; Margesin and Feller 2010), variation in the types and sequence of amino acids mostly in the active site as compared to mesophilic counterpart (Siddiqui and Cavicchioli 2006). Due of these features, cold-active enzymes find special applications in some areas like cold cleaning and washing, food processing and molecular biology where they can enhance the quality and efficiency of the processes. Processes at low temperature helps conserving energy and avoiding side-reactions (Siddiqui 2015) while heat-lability enables their inactivation at moderate temperature *in situ* requiring no additional effort and cost for their extraction from the reaction-mix. Their heat-lability is thus especially desirable for application in fine-chemical synthesis, food processing and molecular biology (Cavicchioli et al. 2011). The action of cold-active alkaline phosphatase used to prevent self ligation of the vector by removing phosphate at 5' ends of DNA is a notable example of their application in molecular biology (Rina et al. 2000). Some of the

Table 8.3 Cold enzymes of microfungal sources.

Enzymes	Microfungi	Habitat	Optimum pH/temp°C/%	Reference
Acid phosphatase	*Penicillium citrinum*	Antarctica	4.2/40/90–100	Gawas-Sakhalkar et al. 2012
Alkaline protease	*Aspergillus ustus*	Deep sea	9.2/45/40	Damre et al. 2006
Do	*Penicillium chrysogenum*	Do	9/35/-	Zhu et al. 2009
Acidic protease	*Trichoderma atroviride*	-	6.2/25/-	Kredics et al. 2008
Serine protease	*Aspergillus terreus*	Deep sea	9.0/45/26% at 15	Damare et al. 2006
Amylase	*Mucor* sp.	Deep sea	5.0/60/-	Mohapatra et al. 1998
α-amylase	*Geomyces pannorum*	Antarctica	5.0/40/20	He et al. 2017
Chitinase	*Verticillium lacanii*	Antarctica	-/40/50	Fenice et al. 1998
Do	*Lecanicillium muscarium*	Antarctic	6.0/40/28	Nguyen et al. 2015
-glucanase	*Penicillium roquefortii*	Do	-/-/- (105 units at 4)	Do
Endo-PG I	*Achaetomium* sp.	-	6.0/45/0–80	Tu et al. 2013
Keratinase	*Geomyces pannorum*	Antarctica	-/-/-	Gradisar et al. 1999
Laccase	*Cerrena unicolor*	Deep sea	3–6/60/30	Michniewicz et al. 2006
Do	*Aspergillus nidulans*	Desert	5.0/60/65	Sahay et al. 2019
Do	*Mortierella echinosphaera*	-	6.6–7/25/90	Kotogán et al. 2018
Do	*Penicillium expansum*	Antarctica	8.0/30/95	Mohammed et al. 2013
Do	*Penicillium canescens* & Alpine	-	11/40/35	Sahay et al. 2019
Do	*Pseudogymnoascus roseus*	Do	9.0/40/70	Do
Do	*Rhizomucor endophyticus*	Do	6.0/40/> 75	Yan et al. 2016
Pectinases	*Geomyces* sp. & 15 others	Antarctica	-/30/-	Poveda et al. 2018
Do	*Mucor flavus*	Do	3.5–4.5/45/-	Gadre et al. 2003
PGA	*Achaetomium* sp.	-	6.0/45/10% at 0	Tu et al. 2013
	Sclerotinia borealis	-	4.5/40–50/30% at 5	Takasawa et al. 1997
EndoPGA I & II	Do	-	4–5/48&50/-	Do
Pullulanase	*Paenibacillus polymyxa*	-	6.0/35/40% at 10	Wei et al. 2015
mannosidase	*Aspergillus niger*	Do	5.0/45/30% at 5–60	Zhao et al. 2011
Xylanase	*Aspergillus niger*	Deep sea	4.5–8.5/28–30/	Raghukumar et al. 2004
Do	*Cladosporium* sp.	Antarctica sea	6.0/-/-	Del-Cid et al. 2014

characterized cold-enzymes produced by microfungi are given in Table 8.3. Cold-active enzymes are believed to be especially useful in following areas.

Food industries

Food industries are the largest utilizers of cold-active enzymes. Catalysis at low temperature is obviously advantageous for the sake of preservation of taste, aromatic profile and nutritional values. Cold-active pectinases are the largest group of enzymes used in fruit juice and wine industries. The cold-active and acidic pectinases facilitate various processes such as juice extraction, liquefaction and clarification in juice industries and in addition to color release in oenology at low temperatures (Sahay et al. 2019). Cold-active pectinases from psychrophilic fungus *Truncatella angustata* has been found as a potential enzyme for the juice and wine industry (Singh et al. 2012). Cold-active β-galactosidase can digest lactose in milk to glucose and galactose during storage at low temperature thus making available lactose-free milk to lactose-intolerant people without the risk of higher temperatures associated with spoilage (Hamid et al. 2013). The same enzyme can convert whey-lactose to D-tagatose-a sweetener with low caloric and glycemic index suitable for diabetic people (Struvay and Feller 2012). Suitable *β*-galactosidase has been reported from cold-active yeasts (Hamid et al. 2013) but there are hardly any reports from cold-active microfungus.

Xylanases are a group of enzymes that catalyze depolymerizing xylan molecules into xylose units. They have potential applications in beverage, food processing, livestock feed, paper and pulp, detergent and textile and 2G bioetahnol industries (Malik et al. 2018). Cold-active xylanases are especially useful in improving dough and resting at cold temperatures to upgrade the quality of bread (Bisht 2011). Cold-tolerant xylanases have been reported from marine-derived *Cladosporium* sp. (Del-Cid et al. 2014) and *Penicillium* sp. FS010 (Hou et al. 2006) the latter was also successfully expressed in *E. coli* (Hou et al. 2006). Cold-active xylanase with industrial potential has also been screened from three psychrotrophic fungi with the identification of the most potential fungus being *Truncatella angustata* (Malik et al. 2018).

Cold cleaning or washing

Cleaning or washing at low temperatures is advantageous in a general way since a 10°C reduction in temperature from 40°C is reported to reduce 100 g CO_2 per wash (Nielsen 2005), but in many specific instances cold cleaning has additional benefits, for examples using in automated dishwashers (Aehle 2007) or cleaning equipments used in brewing (Zahller et al. 2010) and dairy (Eide et al. 2003) industries, and membranes used in water treatment (Poele and van der Graaf 2005). Surface cleaning of larger objects such as carpets or benches or buildings (Valentini et al. 2010), cleaning of dairy and food processing equipment, heat exchanger etc at operating temperature (cold temperatures) (Marshall et al. 2003; Arizona Department of Health Services 2011) and cleaning of objects that require cleaner containing organic solvents (Flick 2006) are other potential areas that cold-active enzymes find in applications. In cold cleaning cold-active hydrolytic enzymes play important role (Cavicchioli with Siddiqui 2006; Chouhan and Sahay 2018). Novozymes made amylases, cellulases, lipases, proteses, lipases have commonly been used in cold-active (20°C) detergent formulations (Aehle 2007). Detergents are the single most important product that use almost 40% of enzymes (lipases, proteases and *α*-amylases) globally produced in its formulations. Enzymes in detergent were introduced to curtail the uses of chemicals injurious to the texture, colors and strength of fabrics, human and environmental health (Ranjan et al. 2016; Chouhan and Sahay 2018). Currently, detergent industries use cold-active alkaline proteases (subtilisins) isolated from Antarctic *Bacillus* sp. (Struvay and Feller 2012). Cold-active organic solvent tolerant lipases from *Penicillium canescense* and *Pseudogymnoascus roseus* are potential enzymes for cleaning formulations (Chouhan and Sahay 2018).

Textile industries

Cotton cloth with denim finishing is of the highest demand all over the world. This value addition depends on cellulases, the latter catalyzes removal of short fibers from the surface making the texture of the cloth smooth and glossy (Bhat 2000). The process if carried out could save much energy besides protecting the fiber from high temperatures leading to damages.

2G Bioethanol industries

2G bioethanol technology is facing discrepancy in optimum temperatures of commercial cellulases activity (50–65°C) and yeast fermentation (25–30°C) hindering simultaneous saccharification and fermentation process. Psychrophilic cellulase complex possessing both endo-b-1,4-D-glucanase and b-1,4-glucosidase activities from an earthworm can potentially remove such hurdles (Ueda et al. 2014). Extracellular production of endo-1,4-b-glucanases at low temperatures (4 and 15°C) by *Cadophora* sp., *Cladosporium* sp., *Geomyces* sp. and *Penicillium* sp., has been reported (Duncan et al. 2006), with *Cadophora malorum* and *Penicillium roquefortii* producing 120 (at 4°C) and 105 (at 15°C) units, the highest yield so far reported for any species. From the marine habitat, several fungi yielding cold-adapted endoglucanases have been reported (Baker et al. 2010). A potential source of suitable cold-active xylanase from *Pseudogymnoascus roseus* (Malik et al. 2018) and carboxymethyl cellulase from *Truncatella angustata* (Magrey et al. 2019) has been screened. As many as 571 CAZymes sequences identified and 23 xylanolytic enzymes have been studied in psychrotrophic fungal isolate *Cladosporium neopsychrotolerans* SL-16 isolated from alpine soil (Ma et al. 2018). A cold-active and highly robust laccase has been characterized from *Aspergillus nidulans* having the potential to apply 2G bioethanol technology (Sahay et al. 2019).

Organic synthesis

As the inbuilt flexibility leads to tolerance towards organic solvents (Gerday et al. 2000), cold-active enzymes (such as, lipases) may be applied in organic or aqueous-organic solvents catalysis especially for the synthetic processes of synthesis of biodiesel (Knežević et al. 2004; Krishna and Karanth 2002; Hassan et al. 2009). Cold-active lipases additionally exhibit substantial stereospecificity (Aurilia et al. 2008) and thus have potential applicability in the organic synthesis of chiral compounds (Jeon et al. 2009). The class of lipase called serine hydrolases are especially important for their ability to carry out esterification, interesterification and transesterification reactions in organic solvents and supercritical fluids (non-aqueous media) (Stergiou et al. 2013).

Microfungi such as *Aspergillus* sp., *Geotrichum* sp., *Humicola* sp., *Mucor* sp., *Penicillium* sp., *Rhizopus* sp., *Rhizomucor* sp. and *Rhizopus* sp. in general are the most important sources of industrial grade and stable lipases (Benjamin and Pandey 1998). For example, Lipozyme RM-IM is a very important commercially available lipase obtained from *Rhizomucor miehe* (Stergiou et al. 2013). Lipases from *Aspergillus niger*, *Geotrichum candidum*, *Penicillium cyclopium* and *Rhizopus delemar* have been used for esterification of oleic acid applying a number of primary alcohols. Most of the microfungi have their cold-active strains reported from relevant habitats, though their lipases from this angle have not been studied. Potential cold-active organic solvent tolerant lipases have been isolated and studied in *Mortierella echinosphaera* CBS 575.75 (Kotogan et al. 2018) and *Penicillium canescens* (Sahay et al. unpublished data).

Medical and pharmaceutical applications

Microfungi have been recognized as important sources of diverse types of medicines (Beekman and Barrow 2014) and cosmetics (Hyde et al. 2010). Antibacterial penicillin from *Penicillium chrysogenum* and antifungal griseofulvin from *Penicillium griseofulvin* are the pioneer products of their classes. Cold-active microfungi thus obviously need attention for two reasons; first they have innate potential to produce these molecules and second the varied stressful conditions they had been continuously exposed to during evolution are usually conducive to evolve diverse types of active biomolecules. On the contrary cold-active bacteria have long been studied as a source of new biomolecules of pharmaceutical and cosmetic importance (Tomova et al. 2015; Yadav et al. 2018a).

Psychrotolerant species of *Penicicillium* such as *P. jamesonlandense, P. soppii* and *P. lanosum* have been reported to produce a number of bioactive substances including cycloaspeptide and griseofulvin, while *P. ribium* producing precursor of griseofulvin, norlichexanthone (Frisvad et al. 2006). Polyketides, penilactones A and B (1 and 2), possess a novel structure and bioactivity reported from an Antarctic deep-sea derived fungus *Penicillium crustosum* PRB-2 (Wu et al. 2012). Four new chloro-eremophilane sesquiterpenes compounds obtained from an Antarctic deep sea fungus, *Penicillium* sp. PR19N-1A, of which one has been shown to possess cytotoxic activity against HL-60 and A549 cancer cell lines (Wu et al. 2013). Of the five fungal hybrid polyketides obtained from deep sea fungus *Cladosporium sphaerosporium* 2005-01-E3, one cladosine C (3) having 6-enamino-7(8)-en-10-ol has been shown to exhibit an anti-influenza AH1N1 virus effect

(Wu et al. 2014). Of the 40 polar soil fungal strains, 45% have shown antimicrobial activity against at least one of the tested pathogenic bacteria (Yogabaanu et al. 2017).

Conclusion

Cold-active microfungi exhibit considerable diversity in habitats and taxonomic positions. Generally they seem to be pandemic in nature and all the cold areas need to be extensively explored. Although psychrotrophic microbes are highly acclaimed as a source of a variety of precious biomolecules of biotechnological importance, cold-active microfungi have been studied very little from this point of view. This is especially true when we look at this fact against the background of a large number of microfungi isolated from diverse types of cold habitats.

References

Abneuf, M.A., A. Krishnan, M.G. Aravena, K.L. Pang, P. Convey, N. Mohamad-Fauzi et al. 2016. Antimicrobial activity of microfungi from maritime Antarctic soil. Czech. Polar Rep. 62: 141–154.

Aehle, W. 2007. Enzymes in Industry: Production and Applications. Wiley-VCH Verlag GmbH & Co, Weinheim, Germany.

Aislabie, J., R. Fraser, S. Duncan and R.L. Farrell. 2001. Effects of oil spills on microbial heterotrophs in Antarctic soils. Polar Biol. 24: 308–313.

Altunatmaz, S.S., G. Issa and A. Aydin. 2012. Detection of airborne psychrotrophic bacteria and fungi in food storage refrigerators. Braz. J. Microbiol. 43: 1436–1443.

Amir, G., L. Horowitz, B. Rubinsky, B.S. Yousif, J. Lavee and A.K. Smolinsky. 2004. Subzero nonfreezing cryopreservation of rat hearts using antifreeze protein I and antifreeze protein III. Cryobiology 48: 273–282.

Anupama, P.D., K.D. Praveen, R.K. Singh, S. Kumar, A.K. Srivastava and D.K. Arora. 2011. A psychrophilic and halotolerant strain of *Thelebolus* microspores from Pangong Lake, Himalaya. Mycosphere 2: 601–609.

Arcangeli, C. and S. Cannistraro. 2000. *In situ* raman microspectroscopic identification and localization of carotenoids: Approach to monitoring of UV-B irradiation stress on antarctic fungus. Biopolymers 57: 179–186.

Arifeen, M.Z.U. and C.H. Liu. 2018. Novel enzymes isolated from marine-derived fungi and its potential applications. J. Biotechnol. Bioeng. 1: 1–12.

Arizona Department of Health Services. 2011. Food equipment cleaning and sanitizing: Water chemistry and quality. http://www.azdhs.gov/phs/oeh/fses/fecs_wcq3.htm.

Armstrong, G.A. and J.E. Hearst. 1996. Carotenoids 2: Genetics and molecular biology of carotenoid pigment biosynthesis. FASEB J. 10: 228–237.

Aurilia, V., A. Parracino and S.D. Auria. 2008. Microbial carbohydrate esterases in cold adapted environments. Gene 410: 234–240.

Avalos, J. and M.C. Limon. 2015. Biological roles of fungal carotenoids. Curr. Genet. 61: 309–324.

Babitha, S. 2009. Microbial pigments. pp. 147–162. *In*: Singh, P. and A. Pandey (eds.). Biotechnology for Agro-industrial Resource Utilization. Springer, Dordrecht.

Baderschneider, B., R.W.R. Crevel, L.K. Earl, A. Lalljii, D.J. Sanders and I.J. Sanders. 2002. Sequence analysis and resistance to pepsin hydrolysis as part of an assessment of the potential allergenicity of ice structuring protein type III HPLC 12. Food Chem. Toxicol. 40: 965–978.

Baguisi, A., A. Arav, T.F. Crosby, J.F. Roche and M.P. Boland. 1997. Hypothermic storage of sheep embryos with antifreeze proteins: Development *in vitro* and *in vivo*. Theriogenology 48: 1017–1024.

Bajpai, P.K., P. Bajpai and O.P. Ward. 1991. Production of arachidonic acid by *Mortierella alpina* ATCC 32222. J. Ind. Microbiol. 8: 179–185.

Baker, P., J. Kennedy, J. Morrissey, F. O'Gara, A. Dobson and J. Marchesi. 2010. Endoglucanase activities and growth of marine-derived fungi isolated from the sponge *Haliclona simulans*. J. Appl. Microbiol. 108: 1668–1675.

Balamurugan, S., J.S. Ann, I.P. Varghese, S.P. Murugan, S.C. Harish, S.R. Kumar et al. 2018. Heterologous expression of *Lolium perenne* antifreeze protein confers chilling tolerance in tomato. J. Integ. Agri. 17: 1128–113.

Batista-Garcia, R.A., T. Sutton, S.A. Jackson, O.E. Tovar-Herrera, E. Balcaăzar-Loăpez, M.D.R. Saănchez-Carbente et al. 2017. Characterization of lignocellulolytic activities from fungi isolated from the deep-sea sponge *Stelletta normani*. PLoS ONE 12(3): e0173750.

Beekman, A.M. and R.A. Barrow. 2014. Fungal metabolites as pharmaceuticals. Aust. J. Chem. 67: 827–843.

Benjamin, S. and A. Pandey. 1998. Review: *Candida rugosa* lipases: molecular biology and versatility in biotechnology. Yeast 14: 1069–87.

Bhagavan, N.V. 1978. Medical Biochemistry 2nd edition, Jones and Bartlett Publishers International, London, UK.

Bhat, M.K. 2000. Cellulases and related enzymes in biotechnology. Biotechnol. Adv. 18: 355–383.

Bisht, S. 2011. Cold active proteins in food and pharmaceutical industry. http://www.biotecharticles.com/Biotechnology-products-Article/Cold-Active-Proteins-in-Food-and-Pharmaceutical-Industry-719.html.

Bjorbakmo, M.F.M., T. Carlsen, A. Brysting, T. Vralstad, K. Hoiland, K. Ugland et al. 2010. High diversity of root associated fungi in both alpine and arctic *Dryas octopetala*. BMC Plant Biol. 10: 244.

Blaalid, R., M.L. Davey, H. Kauserud, R. Halvorsen, K. Hoiland and P.B. Eidesen. 2014. Arctic root-associated fungal community composition reflects environmental filtering. Mol. Ecol. 23: 649–659.

Błaszczyk, L., M. Siwulski, K. Sobieralski, J. Lisiecka and M. Jedryczka. 2014. *Trichoderma* spp. application and prospects for use in organic farming and industry. J. Plant Prot. Res. 54: 309–317.

Boonsupthip, W. and T.C. Lee. 2003. Application of antifreeze protein for food preservation: effect of type III antifreeze protein for preservation of gel-forming of frozen and chilled actomyosin. J. Food Sci. 68: 1804–1809.

Botnen, S., U. Vik, T. Carlsen, P.B. Eidesen, M.L. Davey and H. Kauserud. 2014. Low host specificity of root-associated fungi at an Arctic site. Mol. Ecol. 23: 975–985.

Brunner, I., M. Plotze, S. Rieder, A. Zumsteg, G. Furrer and B. Frey. 2011. Pioneering fungi from the Damma glacier forefield in the Swiss Alps can promote granite weathering. Geobiology 9: 266–279.

Capon, R.J., M. Stewart, R. Ratnayake, E. Lacey and J.H. Gill. 2007. Citromycetins and bilains A–C: new aromatic polyketides and diketopiperazines from Australian marine-derived and terrestrial *Penicillium* sp. J. Nat Prod. 70: 1746–1752.

Carpenter, J.F. and T.N. Hansen. 1992. Antifreeze protein modulates cell survival during cryopreservation: Mediation through influence on ice crystal growth. Proc. Natl. Acad. Sci. USA 89: 8953–8957.

Castrillo, M., E.M. Luque, J. Pardo-Medina, M.C. Limon, L.M. Corrochano and J. Avalos. 2018. Transcription basis of enhanced photoinduction of carotenoid biosynthesis at low temperature in the fungus *Neurospora crassa*. Res. Microbiol. 169: 278–289.

Cavicchioli, R., K.S. Siddiqui, D. Andrews and K.R. Sowers. 2002. Low-temperature extremophiles and their applications. Curr. Opin. Biotechnol. 13: 253–261.

Cavicchioli, R. and K.S. Siddiqui. 2006. Cold-adapted enzymes (Chapter 31). pp. 615–638. *In*: Pandey, A., C. Webb, C.R. Soccol and C. Larroche (eds.). Enzyme Technology. New York, NY, USA: Springer Science.

Cavicchioli, R., T. Charlton, H. Ertan, S. Mohd Omar, K.S. Siddiqui and T.J. Williams. 2011. Biotechnological uses of enzymes from psychrophiles. Microb. Biotechnol. 4: 449–460.

Chadni, Z., M.H. Rahaman, I. Jerin, K.M.F. Hoque and M.A. Reza. 2017. Extraction and optimisation of red pigment production as secondary metabolites from Talaromyces verruculosus and its potential use in textile industries. Mycology 8: 48–57.

Chaikam, V. and D.T. Karlson. 2010. Comparison of structure, function and regulation of plant cold shock domain proteins to bacterial and animal cold shock domain proteins. BMB Rep. 43: 1–8.

Chao, H., P.L. Davies and J.F. Carpenter. 1996. Effects of antifreeze proteins on red blood cell survival during cryopreservation. J. Exp. Biol. 199: 2071–2076.

Chaves, D.F., I.S. Campelo, M.M.A.S. Silva, M.H. Bhat, D.I.A. Teixeira, L.M. Melo et al. 2016. The use of antifreeze protein type III for vitrification of *in vitro* matured bovine oocytes. Cryobiology 73: 324–328.

Chouhan, D., M. Shahnawaz, M.I. Butt and S. Sahay. 2017. Isolation and characterization of protease from *Pseudogymnoascus* sp. strain BPF6. Trends Biosci. 10: 8852–8859.

Chouhan, D. and S. Sahay. 2018. Detergent compatible cold-active lipases from psychrotrophic fungi for cold washing. J. Genet. Engin. Biotechnol. 16: 319–325.

Clemmensen, K.E. and A. Michelsen. 2006. Integrated long-term responses of an arctic-alpine willow and associated ectomycorrhizal fungi to an altered environment. Can. J. Bot. 84: 831–843.

Coleine, C., J.E. Stajich, L. Zucconi, S. Onofri, N. Pombubpa, E. Egidi et al. 2018. Antarctic Cryptoendolithic fungal communities are highly adapted and dominated by lecanoromycetes and dothideomycetes. Front. Microbiol. 9: 1392.

Connell, L., B. Segee, R. Redman, R.J. Rodriguez and H. Staudigel. 2018. Biodiversity and abundance of cultured microfungi from the permanently ice-covered lake fryxell, Antarctica. Life 8: 37.

Crevel, R.W.R., K.J. Cooper, L.K. Poulsen, L. Hummelshoj, C. Bindslev-Jensen, A.W. Burks et al. 2007. Lack of immunogenicity of ice structuring protein type III HPLC12 preparation administered by the oral route to human volunteers. Food Chem. Toxicol. 40: 79–87.

D'Souza-Ticlo, D., D. Sharma and C. Raghukumar. 2009. A thermostable metal-tolerant laccase with bioremediation potential from a marine-derived fungus. Mar. Biotechnol. 11: 725–737.

Damare, S., C. Raghukumar, U.D. Muraleedharan and S. Raghukumar. 2006. Deep-sea fungi as a source of alkaline and cold-tolerant proteases. Enz. Microbial. Technol. 39: 172–181.

Davey, M.L. and R.S. Currah. 2009. *Atradidymella muscivora* gen. et sp. nov. (Pleosporales) and its anamorph *Phoma muscivora* sp. nov: A new pleomorphic pathogen of boreal bryophytes. Am. J. Bot. 96: 1281–1288.

Del-Cid, A., P. Ubilla, M.C. Ravanal, E. Medina, I. Vaca, G. Levican et al. 2014. Cold-active xylanase produced by fungi associated with Antarctic marine sponges. Appl. Biochem. Biotechnol. 172: 524–532.

Dhale, M.A. and A.S. Vijay Raj. 2009. Pigment and amylase production in Penicillium sp. NIOM-02 and its radical scavenging activity. Int. J. Food Sci. Technol. 44: 2424–2430.

Diaz-Sanchez, V., A.F. Estrada, D. Trautmann, M.C. Limon, S. Al-Babili and J. Avalos. 2011. Analysis of al-2 mutations in *Neurospora*. PLoS ONE 6: e21948.

Dufosse, L., P. Galaup, A. Yaron, S.M. Arad, P. Blanc, K.N.C. Murthy et al. 2005. Microorganisms and microalgae as source of pigments for use: a scientific oddity or an industrial reality. Trends Food Sci. Technol. 16: 389–406.

Dufosse, L. 2006. Microbial production of food grade pigments. Food Technol. Biotechnol. 44: 313–321.

Dufosse, L., M. Fouillaud, Y. Caro, S.A. Mapari and N. Sutthiwong. 2014. Filamentous fungi are large-scale producers of pigments and colorants for the food industry. Curr. Opin. Biotechnol. 26: 56–61.

Duncan, S.M., R.L. Farrell, J.M. Thwaites, B.W. Held, B.E. Arenz, J.A. Jurgens et al. 2006. Endoglucanase producing fungi isolated from Cape Evans historic expedition hut on Ross Island, Antarctica. Environ. Microbiol. 8: 1212–1219.

Durán, P., P.J. Barra, M.A. Jorquera, S. Viscardi, C. Fernandez, C. Paz et al. 2019. Occurrence of soil fungi in Antarctic pristine environments. Front. Bioeng. Biotechnol. 7: 28.

Eide, M.H., J.P. Homleid and B. Mattsson. 2003. Life cycle assessment (LCA) of cleaning-in-place processes in dairies. Lebensm. Wiss. Technol. 36: 303–314.

European Commission. 1998. Consideration of the epidemiological basis for appropriate measures for the protection of the public health in respect of food allergy. SCOOP/NUTR/REPORT/2 European Commission, Brussels.

Fakas, S., M. Čertik, S. Papanikolaou, G. Aggelis, M. Komaitis and M. Galiotou-Panayotou. 2007. γ-linolenic acid production by *Cunninghamella echinulata* growing on complex organic nitrogen sources. Bioresour. Technol. 99: 5986–5990.

Fan, Y., B. Liu, H. Wang, S. Wang and J. Wang. 2002. Cloning of antifreeze protein gene from carrot and its influence on coldtolerance in transgenic tobacco plants. Plant Cell Rep. 2: 296–301.

Feng, Y., Y. Shao and F. Chen. 2012. Monascus pigments. Appl. Microbiol. Biotechnol. 96: 1421–1440.

Fenice, M., J. Leuba and F. Federici. 1998. Chitinolytic enzyme activity of *Penicillium janthinellum* P9 in bench-top bioreactor. J. Ferment. Bioeng. 86: 620–623.

Fenice, M., L. Selbmann, R.D. Giambattista and F. Federici. 1998. Chitinolytic activity at low temperature of an Antarctic strain (A3) of *Verticillium lecanii* Activité chitinolytique, à basse température, d'une souche antarctique (A3) de *Verticillium lecanii*. Res. Microbiol. 149: 289–300.

Fenice, M. 2016. The psychrotolerant Antarctic fungus *Lecanicillium muscarium* CCFEE 5003: A powerful producer of cold-tolerant chitinolytic enzymes. Molecules 21: 447.

Fletcher, G.L., S.V. Goddard and Y. Wu. 1999. Antifreeze proteins and their genes: From basic research to business opportunity. Chemtech. 30: 17–28.

Flick, E.W. 2006. Advanced cleaning products formulations. http://www.knovel.com/web/portal/main.

Food and Agriculture Organization of the United Nations. 2004. The State of the World Fisheries and Aquaculture, Rome.

Freeman, M.P., J.R. Hibbeln, K.L. Wisner, J.M. Davis, D. Mischoulon, M. Peet et al. 2006. Omega-3 fatty acids: evidence basis for treatment and future research in psychiatry. J. Clin. Psychiatry 67: 1954–1967.

Freeman, K.R., A.P. Martin, D. Karki, R.C. Lynch, M.S. Mitter, A.F. Meyer et al. 2009. Evidence that chytrids dominate fungal communities in high-elevation soils. Proc. Natl. Acad. Sci. U.S.A. 106: 18315–18320.

Frisvad, J.C., T.O. Larsen, P.W. Dalsgaard, K.A. Seifert, G. Louis-Seize, E.K. Lyhne et al. 2006. Four psychrotolerant species with high chemical diversity consistently producing cycloaspeptide A, *Penicillium jamesonlandense* sp. nov., *Penicillium ribium* sp. nov., *Penicillium soppii* and *Penicillium lanosum*. Int. J. Syst. Evo. Micro. 56: 1427–1437.

Gadre, R.V., G. Van Driessche, J. Van Beeumen and M.K. Bhat. 2003. Purification, characterisation and mode of action of an endo-polygalacturonase from the psychrophilic fungus *Mucor flavus*. Enzyme Microb. Technol. 32: 321–30.

Gardes, M. and A. Dahlberg. 1996. Mycorrhizal diversity in arctic and alpine tundra: an open question. New Phytol. 133: 147–157.

Gawas-Sakhalkar, P. and S.M. Singh. 2011. Fungal community associated with Arctic moss, *Tetraplodon mimoides* and its rhizosphere: bioprospecting for production of industrially useful enzymes. Curr. Sci. 100: 1701–1705.

Gawas-Sakhalkar, P., S.M. Singh, S. Naik and R. Ravindra. 2012. High-temperature optima phosphatases from the cold-tolerant Arctic fungus *Penicillium citrinum*. Polar Research 31: 11105.

Gené, J., J.L. Blanco, J. Cano, M.E. García and J. Guarro. 2003. New filamentous fungus *Sagenomella chlamydospora* responsible for a disseminated infection in a dog. J. Clin. Microbiol. 41: 1722–1725.

Gerday, C., M. Aittaleb, M. Bentahier, J.P. Chessa, P. Claverie and T. Collins. 2000. Cold-adapted enzymes: From fundamentals to biotechnology. Trends Biotechnol. 18: 103–107.

Gonçalves, V.N., A.B. Vaz, C.A. Rosa and L.H. Rosa. 2012. Diversity and distribution of fungal communities in lakes of Antarctica. FEMS Microbiol. Ecol. 82: 459–471.

Gradisar, H., D. Mandin and J.P. Chaoumont. 1999. Screening fungi for synthesis of keratinolytic enzymes. Lett. Appl. Microbiol. 28: 127–130.

Grantina-Ievina, L., A. Berzina, V. Nikolajeva, P. Mekss and I. Muiznieks. 2014. Production of fatty acids by *Mortierella* and *Umbelopsis* species isolated from temperate climate soils. Environ. Exp. Biol. 12: 15–27.

Griffith, M. and M.W. Yaish. 2004. Antifreeze proteins in overwintering plants: A tale of two activities. Trends Plants Sci. 9: 399–405.

Gupta, R. and R. Deswal. 2014. Antifreeze proteins enable plants to survive in freezing conditions. J. Biosci. 39: 931–944.

Hafizah, S.H., S.A. Alias, H.Y. Siang, J. Smykla, K.L. Pang, S.Y. Guo et al. 2013. Studies on diversity of soil microfungi in the Hornsund area, Spitsbergen. Pol. Polar Res. 34: 39–54.

Hamdan, A. 2018. Psychrophiles: Ecological significance and potential industrial application. S. Afr. J. Sci. 114: 1–6.

Hamid, B., P. Singh, F.A. Mohiddin and S. Sahay. 2013. Partial characterization of cold-active galactosidase activity produced by *Cystofilobasidium capitatum* SPY11 and *Rhodotorulla mucilaginosa* PT1. J. of Endocytobiosis and Cell Research 23: 23–26.

Harnpicharnchai, P., W. Pinngoen, W. Teanngam, W. Sornlake, K. Sae-Tang, P. Manitchotpisit et al. 2016. Production of high activity *Aspergillus niger* BCC4525 β-mannanase in *Pichia pastoris* and its application for mannooligosaccharides production from biomass hydrolysis. Biosci. Biotechnol. Biochem. 80: 2298–2305.

Hassan, F., A.A. Shah and A. Hameed. 2009. Methods for detection and characterization of lipases: a comprehensive review. Biotechnol. Adv. 27: 782–98.

Hassan, N., M. Rafiq, M. Hayat, A.A. Shah and F. Hasan. 2016. Psychrophilic and psychrotrophic fungi: a comprehensive review. Rev. Environ. Sci. Biotechnol. 15: 147–172.

He, L., Y. Mao, L. Zhang, H. Wang, S.A. Alias, B. Gao et al. 2017. Functional expression of a novel α-amylase from Antarctic psychrotolerant fungus for baking industry and its magnetic immobilization. BMC Biotechnol. 17: 22.

Hebraud, M. and P. Potier. 1999. Cold shock response and low temperature adaptation in psychrotrophic bacteria. J. Mol. Microbiol. Biotechnol. 1: 211–219.

Hew, C.L. and K.V. Ewart. 2002. Fish antifreeze proteins. World Scientific Publishing, 213–221.

Hightower, R., C. Baden, E. Penzes, P. Lund and P. Dunsmuir. 1991. Expression of antifreeze proteins in transgenic plants. Plant Mol. Biol. 17: 1013–1021.

Hirano, Y., Y. Nishimiya, S. Matsumoto, M. Matsushita, S. Todo, A. Miura et al. 2008. Hypothermic preservation effect on mammalian cells of type III antifreeze proteins from notched-fin eelpout. Cryobiology 57: 46–51.

Hoshino, T., M. Kiriaki, S. Ohgiya, M. Fujiwara, H. Kondo, Y. Nishimiya et al. 2003. Antifreeze proteins from snow mold fungi. Can. J. Bot. 81: 1175–1181.

Hoshino, T., N. Xiao and O.B. Tkachenko. 2009. Cold adaptation in the phytopathogenic fungi causing snow molds. Mycoscience 50: 26–38.

Hou, Y.-H., T.-H. Wang, H. Long and H.-Y. Zhu. 2006. Novel cold-adaptive *Penicillium strain* FS010 secreting thermo-labile xylanase isolated from Yellow Sea. Acta Biochimica et Biophysica Sinica 38(2): 142–149.

Huang, V.T., W.A. Barrier, L.H. Leake and S.G. Wittinger. 1992. Frozen dessert compositions and products. U. S. Patent # 5,175,013.

Hyde, K.D., A.H. Bahkali and M.A. Moslem. 2010. Fungi-an unusual source for cosmetics. Fungal Divers 43: 1–9.

Ideta, A., Y. Aoyagi, K. Tsuchiya, Y. Nakamura, K. Hayama, A. Shirasawa et al. 2015. Prolonging hypothermic storage (4°C) of bovine embryos with fish antifreeze protein. J. Reprod. Dev. 61: 1–6.

Jadhav, V., M. Jamle, P. Pawar, M. Devare and R. Bhadekar. 2010. Fatty acid profiles of PUFA producing Antarctic bacteria: Correlation with RAPD analysis. Ann. Microbiol. 60: 693–699.

Jeon, J., J.T. Kim, S. Kang, J.H. Lee and S.J. Kim. 2009. Characterization and its potential application of two esterases derived from the Arctic sediment metagenome. Mar. Biotechnol. 11: 307–316.

Jo, J.W., B.C. Jee, C.S. Suh and S.H. Kim. 2012. The beneficial effects of antifreeze proteins in the vitrification of immature mouse oocytes. PLoS ONE 7: e37043.

Johnston, D.J. and B. Williamson. 1992. Purification and characterization of four polygalacturonases from *Botrytis cinerea*. Mycol. Res. 96: 343–9.

Kamijima, T., M. Sakashita, A. Miura, Y. Nishimiya and S. Tsuda. 2013. Antifreeze protein prolongs the life-time of insulinoma cells during hypothermic preservation. PLoS ONE 8: e73643.

Khot, M.B., S. Kamat, S. Zingarde, A. Pant, B.A. Chpade and A. Ravikumar. 2012. Single cell oil of oleaginous fungi from the tropical mangrove wetlands as a potential feedstock for biodiesel. Microb. Cell Fact. 11: 71.

Kim, H.J., H.E. Shim, J.H. Lee, Y.C. Kang and Y.B. Hur. 2015. Ice-binding protein derived from *Glaciozyma* can improve the viability of cryopreserved mammalian cells. J. Microbiol. Biotechnol. 25: 1989–1996.

Kim, H.J., H.L. Jun, B.H. Young, C.W. Lee, S.H. Park and B.W. Koo. 2017. Marine antifreeze proteins: Structure, function, and application to cryopreservation as a potential cryoprotectant. Mar. Drugs 15: 27.

Kim, J.S., J.H. Yoon, G.H. Park, S.H. Bae, H.J. Kim, M.S. Kim et al. 2011. Influence of antifreeze proteins on boar sperm DNA damaging during cryopreservation. Dev. Biol. 356: 195–195.

Knežević, Z.D., S.I.S. Šiler-Marinković and L.V. Mojović. 2004. Immobilized lipases as practical catalysts—review. Acta Periodica Technologica 35: 151–64.

Ko, R., L.T. Smith and G.M. Smith. 1994. Glycine betaine confers enhanced osmotolerance and cryotolerance on *Listeria monocytogenes*. J. Bacteriol. 176: 426–431.

Koh, H.Y., J.H. Lee, S.J. Han, H. Park and S.G. Lee. 2015. Effect of the antifreeze protein from the Arctic yeast Leucosporidium sp. AY30 on cryopreservation of the marine diatom *Phaeodactylum tricornutum*. Appl. Biochem. Biotechnol. 175: 677–686.

Kosa, M. and A.J. Ragauskas. 2011. Lipids from heterotrophic microbes: advances in metabolism research. Trends Biotechnol. 29: 53–61.

Kotogán, A., C. Zambrano and A. Kecskeméti. 2018. An organic solvent-tolerant lipase with both hydrolytic and synthetic activities from the oleaginous fungus *Mortierella echinosphaera*. Int. J. Mol. Sci. 19: 11–29.

Kredics, L., K. Terecskei, Z. Antal, A. Szekeres, L. Hatvani, L. Manczinger et al. 2008. Purification and preliminary characterization of a cold-adapted extracellular proteinase from *Trichoderma atroviride*. Acta Biol. Hung. 59: 259–268.

Krishna, H.S. and N.G. Karanth. 2002. Lipases and lipase-catalyzed esterification reactions in nonaqueous media. Catal. Rev. 44: 499–591.

Kurek, E., T. Korniłowicz-Kowalska, A. Słomka and J. Melke. 2007. Characteristics of soil filamentous fungi communities isolated from various micro relief forms in the high Arctic tundra (Bellsund region, Spitsbergen). Pol. Polar. Res. 28: 57–73.

Lagneaux, D., M. Huhtinen, E. Koskinen and E. Palmer. 1997. Effect of anti-freeze protein (AFP) on the cooling and freezing of equine embryos as measured by DAPI-staining. Equine Vet. J. Suppl. 25: 85–87.

Ledesma-Amaro, R., C. Serrano-Amatriain, A. Jiménez and J.L. Revuelta. 2015. Metabolic engineering of riboflavin production in *Ashbya gossypii* through pathway optimization. Microb. Cell Fact. 514: 163.

Lee, C.Y., B. Rubinsky and G.L. Fletcher. 1992. Hypothermic preservation of whole mammalian organs with antifreeze proteins. Cryo-Lett. 13: 59–66.

Lee, S.G., H.Y. Koh, J.H. Lee, S.H. Kang and H.J. Kim. 2012. Cryopreservative effects of the recombinant ice-binding protein from the arctic yeast *Leucosporidium* sp. on red blood cells. Appl. Biochem. Biotechnol. 167: 824–834.

Lee, H.H., H.J. Lee, H.J. Kim, J.H. Lee, Y. Ko, S.M. Kim et al. 2015. Effects of antifreeze proteins on the vitrification of mouse oocytes: Comparison of three different antifreeze proteins. Hum. Reprod. 30: 2110–2119.

Li, Y., X. Li, U. Lee, J.S. Kang, H.D. Choi and B.W. Son. 2006. A new radical scavenging anthracene glycoside, asperflavin ribofuranoside, and polyketides from a marine isolate of the fungus *Microsporum*. Chem. Pharm. Bull. 54: 882–883.

Ma, R., H. Huang, Y. Bai, H. Luo, Y. Fan and B. Yao. 2018. Insight into the cold adaptation and hemicellulose utilization of *Cladosporium neopsychrotolerans* from genome analysis and biochemical characterization. Sci. Rep. 8: 6075.

Magrey, A., S. Sahay and R. Gothalwal. 2019. All climates carboxymethyl cellulase from psychrophilic fungus *Truncatella angustata*. Res. Rev. J. Microbiol. Virol. 9: 38–43.

Malik, S.M., F.A. Ahanger, N. Wani, S. Sahay and K. Jain. 2018. Finding potential source of cold-active xylanase. Int. J. Sci. Res. Biol. Sci. 5: 6–9.

Mancuso Nichols, C.A., J. Guezennec and J.P. Bowman. 2005. Bacterial exopolysaccharides from extreme marine environments with special consideration of the southern ocean, sea ice, and deep-sea hydrothermal vents: a review. Mar. Biotechnol. 7: 253–271.

Manning, T.H., M. Spurgeon, A.M. Wolfreys and A.P. Baldrick. 2004. Safety evaluation of ice-structuring protein (ISP) type III HPLC 12 preparation. Lack of genotoxicity and subchronic toxicity. Food Chem. Toxicol. 42: 321–333.

Margesin, R. and F. Schinner. 2001. Biodegradation and bioremediation of hydrocarbons in extreme environments. Appl. Microbiol. Biotechnol. 56: 650–663.

Margesin, R., P.A. Fonteyne, F. Schinner and J.P. Sampaio. 2007. Novel psychrophilic basidiomycetous yeasts from Alpine environments: *Rhodotorula psychrophila* sp. nov. *Rhodotorula psychrophenolica* sp. nov. and *Rhodotorula glacialis* sp. nov. Int. J. Syst. Evol. Microbiol. 57: 2179–2184.

Margesin, R. and G. Feller. 2010. Biotechnological applications of psychrophiles. Environ. Technol. 31: 835–844.

Margesin, R. and V. Miteva. 2011. Diversity and ecology of psychrophilic microorganisms. Res. Microbiol. 162: 346–361.

Marshall, R.T., H.D. Goff and R.W. Hartel. 2003. Ice Cream. Kluwer Academic/Plenum Publishers, New York, USA.

Marshall, W.A. 1998. Aerial transport of keratinaceous substrate and distribution of the fungus *Geomyces pannorum* in Antarctic soils. Microb. Ecol. 36: 212–219.

Martínez-Páramo, S., V. Barbosa, S. Pérez-Cerezales, V. Robles and M.P. Herraez. 2009. Cryoprotective effects of antifreeze proteins delivered into zebrafish embryos. Cryobiology 58: 128–133.

Marx, J.C., T. Collins, S. D'Amico, G. Feller and C. Gerday. 2007. Cold-adapted enzymes from marine Antarctic microorganisms. Mar. Biotechnol. 9: 293–304.

Mayordomo, I., F. Randez-Gil and J.A. Prieto. 2000. Isolation, purification, and characterization of a cold-active lipase from *Aspergillus nidulans*. J. Agric. Food Chem. 48: 105–109.

Meng, X., J. Yang, X. Xu, L. Zhang, Q. Nie and M. Xian. 2009. Biodiesel production from oleaginous microorganisms. Renew. Energ. 34: 1–5.

Merendino, N.L., L. Costantini, R. Manzi, D. Molinari, D.'Eliseo and F. Velotti. 2013. Dietary ω-3 polyunsaturated fatty acid DHA: a potential adjuvant in the treatment of cancer. Biomed Res. Int. 2013: 310186.

Miao, F.P., X.D. Li, X.H. Liu, R.H. Cichewicz and N.Y. Ji. 2012. Secondary metabolites from an algicolous *Aspergillus versicolor* strain. Mar. Drugs 10: 131–139.

Michniewicz, A., R. Ullrich, S. Ledakowicz and M. Hofrichter. 2006. The white-rot fungus *Cerrena unicolor* strain 137 produces two laccase isoforms with different physico-chemical and catalytic properties Appl. Microbiol. Biotechnol. 69: 682–688.

Miyazaki, K., P.L. Wintrode, R.A. Grayling, D.N. Rubingh and F.H. Arnold. 2000. Directed evolution study of temperature adaptation in a psychrophilic enzyme. J. Mol. Biol. 297: 1015–1026.

Mohammad, S., V.S.J. Te'o and H. Nevalainen. 2013. A gene encoding a new cold-active lipase from an Antarctic isolate of *Penicillium expansum*. Curr. Genet. 59: 129–137.

Mohapatra, B.R., U.C. Banerjee and M. Bapuji. 1998. Characterization of a fungal amylase from *Mucor* sp. associated with the marine sponge *Spirastrella* sp. J. Biotechn. 60: 113–117.

Morita, R.Y. 1975. Psychrophilic bacteria. Bacteriol. Rev. 39: 144–167.

Mugnano, J.A., T. Wang, J.R. Layne, A.L. Jr. DeVries and R.E. Lee. 1995. Antifreeze glycoproteins promote intracellular freezing of rat cardiomyocytes at high subzero temperatures. Am. J. Physiol. 269: R474–R479.

Newsham, K.K., R. Upson and D.J. Read. 2009. Mycorrhizas and dark septate root endophytes in polar regions. Fungal Ecol. 2: 10–20.

Nguyen, H.Q., D.T. Quyen, S.L.T. Nguyen and V.H. Vu. 2015. An extracellular antifungal chitinase from *Lecanicillium lecanii*: purification, properties, and application in biocontrol against plant pathogenic fungi. Turk. J. Biol. 39: 6–14.

Nielsen, P.H. 2005. Life cycle assessment supports cold-wash enzymes. Int. J. Appl. Sci. 10: 10.

Nishijima, K., M. Tanaka, Y. Sakai, C. Koshimoto, M. Morimoto, T. Watanabe et al. 2014. Effects of type III antifreeze protein on sperm and embryo cryopreservation in rabbit. Cryobiology 69: 22–25.

Onofri, S., L. Selbmann, L. Zucconi and S. Pagano. 2004. Antarctic microfungi as model exobiology. Planet Space Sci. 52: 229–237.

Panadero, J., F. Randez-Gil and J.A. Prieto. 2005. Heterologous expression of type I Antifreeze peptide GS-5 in Baker's yeast increases freeze tolerance and provides enhanced gas production in frozen dough. J. Agric. Food Chem. 53: 9966–9970.

Papanikolaou, S., M. Komaitis and G. Aggelis. 2004. Single cell oil (SCO) production by *Mortierella isabellina* grown on high-sugar content media. Bioresour. Technol. 95: 287–291.

Parker, J.C., R.K. McPherson, K.M. Andrews, C.B. Levy, J.S. Dubins, J.E. Chin et al. 2005. Heterologous expression of type I Antifreeze peptide GS-5 in Baker's yeast increases freeze tolerance and provides enhanced gas production in frozen dough 2005. J. Agric. Food Chem. 53: 9966–9970.

Pascual, S., P. Melgarejo and N. Magan. 2002. Water availability affects the growth, accumulation of compatible solutes and the viability of the biocontrol agent *Epicoccum nigrum*. Mycopathologia 156: 93–100.

Payne, S.R., J.E. Oliver and G.C. Upreti. 1994. Effect of antifreeze proteins on the motility of ram spermatozoa. Cryobiology 31: 180–184.

Petrovic, U., N. Gunde-Cimerman and P. Zaler. 2000. Xerotolerant mycobiota from high altitude Anapurna soils. Nepal. FEMS Microbiol. Lett. 182: 339–343.

Phadtare, S. 2004. Recent developments in bacterial cold-shock response. Curr. Iss. Mol. Biol. 6: 125–136.

Poele, S.T. and J. van der Graaf. 2005. Enzymatic cleaning in ultrafiltration of wastewater treatment plant effluent. Desalination 179: 73–81.

Poveda, G., C. Gil-Duran, I. Vaca, G. Levicán and R. Chávez. 2018. Cold-active pectinolytic activity produced by filamentous fungi associated with Antarctic marine sponges. Biol. Res. 51: 28.

Prathalingam, N.S., W.V. Holt, S.G. Revell, S. Mirczuk, R.A. Fleck and P.F. Watson. 2006. Impact of antifreeze proteins and antifreeze glycoproteins on bovine sperm during freeze-thaw. Theriogenology 66: 1894–1900.

Qadeer, S., M.A. Khan, Q. Shahzad, A. Azam, M.S. Ansari, B.A. Rakha et al. 2016. Efficiency of beetle (Dendroides canadensis) recombinant antifreeze protein for buffalo semen freezability and fertility. Theriogenology 86: 1662–1669.

Raghukumar, C., S. Raghukumar, G. Sheelu, S. Gupta, B. Nagendernath and B. Rao. 2004. Buried in time: culturable fungi in a deep-sea sediment core from the Chagos Trench, Indian Ocean. Deep Sea Res. 51: 1759–1768.

Raghukumar, C., U. Muraleedharan, V.R. Gaud and R. Mishra. 2004. Xylanases of marine fungi of potential use for biobleaching of paper pulp. J. Ind. Microbiol. Biotechnol. 31: 433–441.

Raghukumar, C. 2008. Marine fungal biotechnology: an ecological perspective. Fungal Divers 31: 19–35.

Ranjan, K., M.A. Lone and S. Sahay. 2016. Detergent compatible cold-active alkaline amylase from *Clavispora lusitaniae* CB13. J. Microbiol. Biotechnol. Food Sci. (JMBFS) 5: 306–310.

Regend, A. and H.D. Goff. 2006. Ice recrystallization inhibition in ice cream as affected by ice structuring proteins from winter wheat grass. J. Dairy Sci. 89: 49–57.

Rina, M., C. Pozidis, K. Mavromatis, M. Tzanodaskalaki, M. Kokkinidis and V. Bouriotis. 2000. Alkaline phosphatase from the Antarctic strain TAB5. Properties and psychrophilic adaptations. Eur. J. Biochem. 267: 1230–1238.

Robinson, C.H. 2001. Cold adaptation in Arctic and Antarctic fungi. New Phytol. 151: 341–353.

Rojas, J.L., J. Martin, J.R. Tormo, F. Vicente, M. Brunati, I. Ciciliato et al. 2009. Bacterial diversity from benthic mats of Antarctic lakes as a source of new bioactive metabolites. Mar. Genomics 2: 33–41.

Ruberto, L., S. Vazquez, A. Lobalbo and W. Mac Cormack. 2005. Psychrotolerant hydrocarbon-degrading *Rhodococcus* strains isolated from polluted Antarctic soils. Antarct. Sci. 17: 47–56.

Rubinsky, B., A. Arav and G.L. Fletcher. 1991. Hypothermic protection—a fundamental property of "antifreeze" proteins. Biochem. Biophys. Res. Commun. 180: 566–571.

Rumbold, K., H. Buijsen, J.J. Van, K.M. Overkamp, Groenestijn, J.W. Van et al. 2009. Werf MJ: Microbial production host selection for converting second-generation feedstocks into bioproducts. Microb. Cell Fact. 8: 64.

Sahay, S., M.A. Lone, P. Jain, P. Singh, D. Chouhan and F. Shehzad. 2013. Cold-active moulds from Jammu and Kashmir, India as potential source of cold-active enzymes. Am. J Curr. Microbiol. 1: 1–13.

Sahay, S., D. Chauhan and V. Chourse. 2019. Pb activated laccase from thermotolerant *Aspergillus nidulans* TTF6 showing Pb activation for smaller substrates and dyes remediation in all climates. Proc. Nat. Acad. Sci. India, Sect. B Biol. Sci. DOI: 10.1007/s40011-019-01092-y.

Samie, N., K. Noghabi, Z. Gharegozloo, H. Zahiri, G. Ahmadian and H. Sharafi. 2012. Psychrophilic α-amylase from *Aeromonas veronii* NS07 isolated from farm soils. Process Biochem. 47: 1381–1387.

Sandmann, G. 1994. Carotenoid biosynthesis in microorganisms and plants. Eur. J. Biochem. 223: 7–24.

Sati, S.C., R. Pathak and M. Belwal. 2014. Occurrence and distribution of Kumaun Himalayan aquatic hyphomycetes: *Lemonniera*. Mycosphere 5: 545–553.

Selbmann, L., S. Onofri, M. Fenice, F. Federici and M. Petruccioli. 2002. Production and structural characterization of the exopolysaccharide of the Antarctic fungus *Phoma herbarum* CCFEE 5080. Res. Microbiol. 153: 585–592.

Selbmann, L., G.S. de Hoog, A. Mazzaglia, E.I. Friedmann and S. Onofri. 2005. Fungi at the edge of life: cryptoendolithic black fungi from Antarctic desert. Stud. Mycol. 51: 1–32.

Shaw, J.M., C. Ward and A.O. Trounson. 1995. Survival of mouse blastocysts slow cooled in propanediol or ethylene glycol is influenced by the thawing procedure, sucrose and antifreeze proteins. Theriogenology 43: 1289–1300.

Shivaji, S. and J.S.S. Prakash. 2010. How do bacteria sense and respond to low temperature. Arch. Microbiol. 192: 85–95.

Siddiqui, K.S. and R. Cavicchioli. 2006. Cold-adapted enzymes. Annu. Rev. Biochem. 75: 403–433.

Siddiqui, K.S. 2015. Some like it hot, some like it cold: temperature dependent biotechnological applications and improvements in extremophilic enzymes. Biotechnol. Adv. 33: 1912–1922.

Singh, P., R. Raghukumar, P. Verma and Y. Shouche. 2010. Phylogenetic diversity of culturable fungi from the deep sea sediments of the Central Indian Basin and their growth characteristics. Fungal Divers. 40: 89–102.

Singh, P., B. Hamid, M.A. Lone, K. Ranjan, A. Khan, V.K. Chourse et al. 2012. Evaluation of pectinonase activity from the psychrophilic fungal strain *Truncatella angustata* BPF5 for wine industry. J. Endocyt. Cell Res. 22: 57–61.

Singh, S.M., P.M. Singh, S.K. Singh and P.K. Sharma. 2014. Pigments, fatty acids and extracellular enzymes Pigment, fatty acid and extracellular enzyme analysis of a fungal strain *Thelebolus microsporus* from Larsemann Hills, Antarctica. Polar Res. 50: 31–36.

Smetantina, O.F., A.I. Kalinovskii, Y.V. Khudyakova, N.N. Slinkina, M.V. Pivkin and T.A. Kuznetsova. 2007. Metabolites from the marine fungus *Eurotium repens*. Chem. Nat. Compd. 43: 395–398.

Sokolovski, S.G., A.A. Meharg and F.J.M. Maathuis. 2002. *Calluna vulgaris* root cells show increased capacity for amino acid uptake when colonized with the mycorrhizal fungus *Hymenoscyphus ericae*. New Phytol. 155: 525–530.

Sterflinger, K., D. Tesei and K. Zakharova. 2012. Fungi in hot and cold deserts with particular reference to microcolonial fungi. Fungal Ecol. 5: 453–462.

Stergiou, P.-Y., A. Foukis, M. Filippou, M. Koukouritaki, M. Parapouli, L.G. Theodorou et al. 2013. Advances in lipase-catalyzed esterification reaction. Biotechnol. Adv. 31: 1846–1859.

Struvay, C. and G. Feller. 2012. Optimization to low temperature activity in psychrophilic enzymes. Int. J. Mol. Sci. 13: 11643–11665.

Subramaniam, R., S. Dufreche, M. Zappi and R. Bajpai. 2010. Microbial lipids from renewable resources: production and characterization. J. Ind. Microbiol. Biotechnol. 37: 1271–1287.

Suleiman, W.B., H.H. El-Sheikh, G. Abu-Elreesh and A.H. Hashem. 2018. Isolation and screening of promising oleaginous *Rhizopus* sp. and designing of Taguchi method for increasing lipid production. J. Innov. Pharm. Biol. Sci. (JIPBS) 5: 08–15.

Takasawa, T., K. Sagisaka, K. Yagi, K. Uchiyama, A. Aoki, K. Takaoka et al. 1997. Polygalacturonase isolated from the culture of the psychrophilic fungus *Sclerotinia borealis*. Can. J. Microbiol. 43: 417–24.

Tibbett, M., F.E. Sanders and J.W.G. Cairney. 1998. The effect of temperature and inorganic phosphorus supply on growth and acid phosphatase production in arctic and temperate strains of ectomycorrhizal *Hebeloma* sp. in axenic culture. Mycol. Res. 102: 129–135.

Tibbett, M. and J.W. Cairney. 2007. The cooler side of mycorrhizas: their occurrence and functioning at low temperatures. Can. J. Bot. 85: 51–62.

Timling, I. and D.L. Taylor. 2012. Peeking through a frosty window: molecular insights into the ecology of Arctic soil fungi. Fungal Ecol. 5: 419–429.

Tomova, I., M. Stoilova-Disheva, I. Lazarkevich and E. Vasileva-Tonkova. 2015. Antimicrobial activity and resistance to heavy metals and antibiotics of heterotrophic bacteria isolated from sediment and soil samples collected from two Antarctic islands. Front. Life Sci. 8: 348–357.

Tu, T., K. Meng, Y. Bai, P. Shi, H. Luo, Y. Wng et al. 2013. High-yield production of a low-temperature-active polygalacturonase for papaya juice clarification. Food Chem. 141: 2974–2981.

Turchetti, B., P. Buzzini, M. Goretti, E. Branda, G. Diolaiuti, C. D'Agata et al. 2008. Psychrophilic yeasts in glacial environments of Alpine glaciers. FEMS Microbiol. Ecol. 63: 73–83.

Tursman, D. and J.G. Duman. 1995. Cryoprotective effects of thermal hysteresis protein on survivorship of frozen gut cells from the freeze-tolerant centipede *Lithobius forficatus*. J. Exp. Zool. 272: 249–257.

Ueda, M., A. Ito, M. Nakazawa, K. Miyatake, M. Sakaguchi and K. Inouye. 2014. Cloning and expression of the cold-adapted endo-1,4-beta-glucanase gene from *Eisenia fetida*. Carbohydr. Polym. 101: 511–516.

Unagul, P., P. Wongsa, P. Kittakoop, S. Intamas, P. Srikitikulchai and M. Tanticharoen. 2005. Production of red pigments by the insect pathogenic fungus *Cordyceps unilateralis* BCC 1869. J. Ind. Microbiol. Biotechnol. 32: 135–140.

Upson, R., D.J. Read and K.K. Newsham. 2007. Widespread association between the ericoid mycorrhizal fungus *Rhizoscyphus ericae* and a leafy liverwort in the maritime and sub-Antarctic. New Phytol. 176: 460–471.

Vadivelan, G. and G. Venkateswaran. 2014. Production and enhancement of omega-3 fatty acid from *Mortierella alpina* CFR-GV15: Its food and therapeutic application. BioMed Res. Int. 9.

Valentini, F., A. Diamantia and G. Palleschi. 2010. New bio-cleaning strategies on porous building materials affected by biodeterioration event. Appl. Surf. Sci. 256: 6550–6563.

Velmuruganm, N., D. Kalpana, J.H. Han, H.J. Cha and Y.S. Lee. 2011. A novel low temperature chitinase from the marine fungus *Plectosphaerella* sp. strain MF-1. Bot. Mar. 54: 75–81.

Venil, C.K. and P. Lakshmanaperumalsamy. 2009. An insightful overview on microbial pigment, Prodigiosin. Elect. J. Biol. 5: 49–61.

Venil, C.K., Z.A. Zakaria and W.A. Ahmad. 2013. Bacterial pigments and their applications. Process Biochem. 48: 1065–1079.

Venketesh, S. and C. Dayananda. 2008. Properties, potential and prospects of antifreeze proteins. Crit. Rev. Biotechnol. 28: 57–82.

Verma, P., A.N. Yadav, K.S. Khannam, N. Panjiar, S. Kumar, A.K. Saxena et al. 2015. Assessment of genetic diversity and plant growth promoting attributes of psychrotolerant bacteria allied with wheat (*Triticum aestivum*) from the northern hills zone of India. Ann. Microbiol. 65: 1885–1899.

Vestergaard, C.S. and G. Parolari, 1999. Lipid and cholesterol oxidation products in dry-cured ham. Meat Sci. 52: 397–401.

Vívian, N.G., A.B.M. Vaz, C.A. Rosa and L.H. Rosa. 2012. Diversity and distribution of fungal communities in lakes of Antarctica. FEMS Microbiol. Ecol. 82: 459–471.

Voets, I.K. 2017. From ice-binding proteins to bio-inspired antifreeze materials. Soft Matter 13: 4808–4823.

Wang, J.H. 2000. A comprehensive evaluation of the effects and mechanisms of antifreeze proteins during low-temperature preservation. Cryobiology 41: 1–9.

Wang, P.L., D.Y. Li, L.R. Xie, X. Wu, H.M. Hua and Z.L. Li. 2014. Two new compounds from a marine-derived fungus *Penicillium oxalicum*. Nat. Prod. Res. 28: 290–293.

Wang, T., Q. Zhu, X. Yang, J.R. Layne Jr and A.L. Devries. 1994. Antifreeze glycoproteins from antarctic notothenioid fishes fail to protect the rat cardiac explant during hypothermic and freezing preservation. Cryobiology 31: 185–192.

Warren, G.J., G.M. Mueller and R.L. Mcknown. 1992. Ice crystal growth polypeptides and method of making. U. S. Patent # 5,118,792.

Wei, W., J. Ma, S.Q. Chen, X.H. Cai and D.Z. Wei. 2015. A novel cold-adapted type I pullulanase of *Paenibacillus polymyxa* Nws-pp2: *in vivo* functional expression and biochemical characterization of glucans hydrolyzates analysis. BMC Biotechnol. 15: 96.

Wen, Y., S. Zhao, L. Chao, H. Yu, C. Song, Y. Shen et al. 2014. The protective role of antifreeze protein 3 on the structure and function of mature mouse oocytes in vitrification. Cryobiology 69: 394–401.

Whyte, L.G., L. Bourbonnière, C. Bellerose and C.W. Greer. 1999. Bioremediation assessment of hydrocarbon-contaminated soils from the high Arctic. Bioremed. J. 3: 69–79.

Widden, P. and D. Parkinson. 1979. Populations of fungi in a high arctic ecosystem. Can. J. Bot. 57: 2408–2417.

Wu, D.L., H.J. Li, D.R. Smith, J. Jaratsittisin, X.F.K.T.X.K. Er, W.Z. Ma et al. 2018. Polyketides and alkaloids from the marine-derived fungus *Dichotomomyces cejpii* F31-1 and the antiviral activity of scequinadoline A against dengue virus. Mar. Drugs 16: 229.

Wu, G., H. Ma, T. Zhu, J. Li, Q. Gu and D. Li. 2012. Penilactones A and B, two novel polyketides from Antarctic deep-sea derived fungus *Penicillium crustosum* PRB-2. Tetrahedron 68: 9745–9749.

Wu, G., A. Lin, Q. Gu, T. Zhu and D. Li. 2013. Four new choloro-eremophilane sesquiterpenes from an Antarctic deep-sea derived fungus, *Penicillium* sp. PR19N-1. Mar. Drug 11: 1399–1408.

Wu, G., X. Sun, G. Yu, W. Wang, T. Zhu, Q. Gu et al. 2014. Cladosins A-E, hybrid polyketides from a deep sea derived fungus *Cladosporium sphaerospermum*. J. Nat. Prod. 77: 270–275.

Wynn, J.P., A.A. Hamid, Y. Li and C. Ratledge. 2001. Biochemical events leading to the diversion of carbon into storage lipids in the oleaginous fungi *Mucor circinelloides* and *Mortierella alpina*. Microbiology 147: 2857–2864.

Xiao, N., K. Suzuki, Y. Nishimiya, H. Kondo, A. Miura, S. Tsuda et al. 2010. Comparison of functional properties of two fungal antifreeze proteins from Antarctomyces psychrotrophicus and *Typhula ishikariensis*. FEBS J. 277: 394–403.

Xiao, N., S. Inaba, M. Tojo, Y. Degawa, S. Fuju, Y. Hanada et al. 2010. Antifreeze activities of various fungi and Straminopila isolated from Antarctica. North Am. Fungi. 5: 215–220.

Yadav, A.N., S.G. Sachan, P. Verma and A.K. Saxena. 2015a. Prospecting cold deserts of north western Himalayas for microbial diversity and plant growth promoting attributes. J. Biosci. Bioeng. 119: 683–693.

Yadav, A.N., S.G. Sachan, P. Verma, S.P. Tyagi, R. Kaushik and A.K. Saxena. 2015b. Culturable diversity and functional annotation of psychrotrophic bacteria from cold desert of Leh Ladakh (India). World J. Microbiol. Biotechnol. 31: 95–108.

Yadav, A.N., S.G. Sachan, P. Verma, R. Kaushik and A.K. Saxena. 2016a. Cold active hydrolytic enzymes production by psychrotrophic Bacilli isolated from three sub-glacial lakes of NW Indian Himalayas. J. Basic Microbiol. 56: 294–307.

Yadav, A.N., S.G. Sachan, P. Verma and A.K. Saxena. 2016b. Bioprospecting of plant growth promoting psychrotrophic Bacilli from cold desert of north western Indian Himalayas. Indian J. Exp. Biol. 54: 142–150.

Yadav, A.N., P. Verma, V. Kumar, P. Sangwan, S. Mishra, N. Panjiar et al. 2018a. Biodiversity of the genus *Penicillium* in different habitats. pp. 3–18. *In*: Gupta, V.K. and S. Rodriguez-Couto (eds.). New and Future Developments in Microbial Biotechnology and Bioengineering, *Penicillium* System Properties and Applications. Elsevier, Amsterdam, doi:10.1016/B978-0-444-63501-3.00001-6.

Yadav, A.N., P. Verma, S.G. Sachan, R. Kaushik and A.K. Saxena. 2018b. Psychrotrophic microbiomes: molecular diversity and beneficial role in plant growth promotion and soil health. pp. 197–240. *In*: Panpatte, D.G., Y.K. Jhala, H.N. Shelat and R.V. Vyas (eds.). Microorganisms for Green Revolution-Volume 2: Microbes for Sustainable Agro-ecosystem. Springer, Singapore, doi:10.1007/978-981-10-7146-1_11.

Yan, Q., X. Duan, Y. Liu, Z. Jiang and S. Yang. 2016. Expression and characterization of a novel 1,3-regioselective cold-adapted lipase from *Rhizomucor endophyticus* suitable for biodiesel synthesis. Biotechnol. Biofuels 9: 86.

Yogabaanu, U., J.F.F. Weber, P. Convey, M. Rizman-Idid and A. Siti. 2017. Antimicrobial properties and the influence of temperature on secondary metabolite production in cold environment soil fungi. Pol. Sci. 14: 60–67.

Zahller, M., D. Daggett and D. Noble. 2010. Life cycle assessment of detergent for brewery clean in place applications. The American center for life cycle assessment conference X. [URL http://www.lcacenter.org/LCAX/presentations/181.pdf].

Zhang, C., S.Z. Fei, R. Arora and D. Hannapel. 2010. Ice recrystallization inhibition proteins of perennial ryegrass enhance freezing tolerance. Planta 232: 155–164.

Zhang, T., X.L. Wei, Y.Q. Zhang, H.Y. Liu and L.Y. Yu. 2015. Diversity and distribution of lichen associated fungi in the Ny-Ålesund Region (Svalbard, High Arctic) as revealed by 454 pyrosequencing. Sci. Rep. 5: 14850.

Zhang, Y., E. Navarro, J.T. Canovas-Marquez, L. Almagro, H. Chen, Y.Q. Chen et al. 2016. A new regulatory mechanism controlling carotenogenesis in the fungus Mucor circinelloides as a target to generate β-carotene over-producing strains by genetic engineering. Microb. Cell Fact. 7: 99.

Zhao, W., J. Zheng and H. Zhou. 2011. A thermotolerant and cold-active mannan endo-1,4-β-mannosidase from *Aspergillus niger* CBS 51388: constitutive overexpression and high-density fermentation in *Pichia pastoris*. Bioresour. Technol. 102: 7538–7547.

Zhou, Z., J. Qing, M. Jinlai and L. Fangming. 2008. Antarctic psychrophile bacteria screening for oil degradation and their degrading characteristics. Mar. Sci. Bull. 10: 50–57.

Zhu, H.Y., Y. Tian, Y.H. Hou and T.H. Wang. 2009. Purification and characterization of the cold-active alkaline protease from marine cold-adaptive *Penicillium chrysogenum* FS010. Mol. Biol. Rep. 36: 2169–2174.

Chapter 9

Cold Adapted Microorganisms
Survival Mechanisms and Applications

Deep Chandra Suyal,[1] *Ravindra Soni,*[2] *Ajar Nath Yadav*[3] and *Reeta Goel*[4,]*

Introduction

Microorganisms are ubiquitous, possess enormous metabolic versatility and are inevitable to almost all biogeochemical cycling processes. They can inhabit many of the extreme environments, where otherwise no other life exists. Most part of the Earth is under cold, *viz.* deep sea, subterranean caverns, alpine regions, permafrost, and the polar regions. These environments are predominantly inhabited by cold-adapted microorganisms including bacteria, viruses, archaea, yeast and algae. Microorganisms never act alone, they function as populations or as communities, and interact with other organisms and their environment, thereby, contributing to the functioning of ecosystems. Understanding microbial interactions is not an easy task as most microscopically observable microorganisms cannot be grown. However, modern molecular detection techniques are very precise in searching for the hidden microbial wealth which otherwise cannot be cultured from the extreme environments.

Microorganisms are prone to temperature fluctuations. Their growth shows characteristic temperature dependence with distinct cardinal temperatures—maximum, optimum and minimum growth temperatures (Madigan et al. 2018), respectively. Furthermore, on the basis of their growth temperature ranges, they can be divided into the following two groups.

Psychrophiles

Psychrophiles are microorganisms that showed temperature optima in the range of 15°C or lower. They are considered as true extremophiles as they exert not only cold stress, but also other environmental constraints, viz. high pressure at ocean depths, strong ultraviolet radiation at polar caps, etc.

[1] Department of Microbiology, Akal College of Basic Sciences, Eternal University, Baru Sahib, Sirmour, Himachal Pradesh, India.

[2] Department of Agricultural Microbiology, College of Agriculture, Indira Gandhi Krishi Vishwa Vidyalaya, Raipur, Chhatisgarh, India.

[3] Department of Biotechnology, Dr. KSG Akal College of Agriculture, Eternal University, Baru Sahib, Sirmour, Himachal Pradesh, India.

[4] Department of Microbiology, College of Basic Sciences & Humanities, GBPUAT Pantnagar, Uttarakhand, India.

* Corresponding author: rg55@rediffmail.com

Psychrotrophs

Psychrotrophs can grow up to 0°C even though they have optima between 20 to 30°C. They are widely distributed in nature and can be detected in temperate soils and water samples, dairy products, vegetables and fruits stored at ~ 4°C.

The lowest growth temperature for life was observed around –20°C, for bacterial life living under permafrost soil and sea ice (Gilichinsky et al. 2008; Margesin and Collins 2019). The cold habitats possess small amounts of unfrozen water that contains high salt concentrations, organic matter and/or minerals. Moreover, they may harbor aerobic as well as anaerobic microorganisms which are greatly affected by several abiotic constraints, viz. radiations, nutritive stresses, oxidative stresses and hydrostatic and osmotic pressures. Despite all of these challenges, remarkable microbial biodiversity was observed in such environments. The microorganisms include bacteria affiliated to Actinobacteria, Bacteroidetes, Chloroflexi, Planctomycetes, Proteobacteria (Alpha-, Beta-, Delta-, Epsilon-, and Gammaproteobacteria), and Verrucomicrobia (Yadav et al. 2015a; Yadav et al. 2015b). Archaeal members of Crenarchaeota, Euryarchaeota, and Methanopyri were also detected in cold habitats (Yadav et al. 2016a). Among the psychrophiles, archaea (*Halorubrum, Methanococcoides,* and *Methanogenium*), bacteria (*Arthrobacter, Bacillus, Micrococcus, Moraxella, Polaribacter, Polaromonas, Pseudoalteromonas, Pseudomonas, Psychrobacter, Psychroflexus, Shewanella,* and *Vibrio*), algae (*Chlamydomonas nivalis*), cyanobacteria (*Nostoc, Oscillatoria,* and *Phormidium*), fungi (*Penicillium,* and *Cladosporium*) and yeast (*Candida,* and *Cryptococcus*) are prominent in these habitats (Kumar et al. 2011; Gesheva and Negoiţa 2011; Yadav et al. 2020a). Riley et al. (2008) reported a novel species of *Psychromonas* which was able to grow at –12°C. Microbial diversity from different cold environments is given in Table 9.1 (Jain et al. 2010; Suyal et al. 2014; 2015a, b, 2018; Saxena et al. 2016; Verma et al. 2016; Verma et al. 2015). It has been observed that

Table 9.1 A glance to the microbial diversity from the cold environments.

Cold environments	Dominant groups	Examples
Polar environments		
Pacific cold deep seawater	Crenarchaeota, Proteobacteria	*Polaromonas, Psychrobacter,*
Arctic sea water	Bacteroidetes, Proteobacteria	*Flavobacterium, Polaribacter*
Arctic marine surface sediments	Bacteroidetes, Proteobacteria, Sulfate-reducers	*Colwellia psychrerythraea, Desulfotalea psychrophila*
Antarctic cold deep seawater	Proteobacteria (especially Gammaproteobacteria)	*Colwellia, Glaciecola, Psychrobacter*
Deep Lake sediments, Antarctica	Euryarchaeota, Proteobacteria	*Halorubrum lacusprofundi, Methanococcoides burtonii*
Permafrost, Siberia, Russia	Firmicutes, Microalgae, Proteobacteria	*Chlamydomonas nivalis, Psychrobacter arcticus, Psychrobacter cryohalentis*
Sea ice, Antarctica	Bacteroidetes, Proteobacteria	*Antarcticus, Haloplanktis, Octadecabacter, Polaribacter, Pseudoalteromonas*
Non-polar cold environments		
Himalayan Glaciers	Proteobacteria, Firmicutes, Bacteroidetes, Actinobacteria, Cyanobacteria	*Bacillus, Microcoleus, Phormidium, Pseudomonas*
Himalayan Agroecosystems	Proteobacteria, Firmicutes, Bacteroidetes, Actinobacteria	*Arthrobacter, Bacillus, Dyadobacter, Pseudomonas, Rhizobium, Rhodococcus*
Bench glacier, Alaska	Proteobacteria, Firmicutes	*Acinetobacter, Bacillus, Carnobacterium Pleistocenium, Pseudomonas*
Siberian tundra	Proteobacteria, Cyanobacteria	*Bacillus, Chamaesiphon, Pseudomonas, Synechococcus*

temperatures lower than –20°C prevent the growth of microorganisms and do not cause microbial death every time. Microorganisms can even metabolize at temperatures which are much lower than their optima. They can do so because they are able to use various survival strategies, *viz.* conversion of actively growing cells into the viable but non-culturable state under environmental stresses and non-growing but culturable state under nutritional stress (Asakura et al. 2007).

Role of microbes in cold environments

Microbes are essential components of every ecosystem. Cold adapted microorganisms play an important role in the biogeochemical cycling and the entire polar food web including terrestrial and aquatic habitats. Some of the important microbial groups which play a key role in cold environments are described below:

- Diatoms (*Actinocyclus, Asteromphalus,* and *Corethron criophilum*) are the major primary producers in the polar regions and make the principal component of the respective food web (Buffen et al. 2007).
- Methanogens, *viz. Methanogenium frigidum, Methanolobus psychrophilus, Methanosarcina lacustris* are crucial for methane emission processes in the wetlands (Kolton et al. 2019; Kumar et al. 2019b).
- Acetogens, *viz. Acetobacterium bakii, A. carbinolicum,* and *A. paludosum* can grow as lithotrophs and compete with other similar groups for the substrates under polar oligotrophic conditions (Nozhevnikova et al. 2001).
- Sulfur oxidizing microbes, *viz. Thiomicrospira arctica* and sulfate-reducing microbes, *viz. Desulfobacter, Desulfobacterium, Desulfobulbus, Desulfotomaculum, Desulfosarcina,* and *Desulfovibrio* (Tian et al. 2017).

Further, rhizospheric microorganisms from cold habitats have proved useful for native crops for their survival and productivity (Suyal et al. 2015a, b; Rana et al. 2020; Verma et al. 2015; Yadav et al. 2016b). Therefore, cold-adapted bio-inoculants can be of great interest in the near future for agricultural sustainability at higher altitudes (Joshi et al. 2019; Kumar et al. 2014; Tomer et al. 2017; Rajwar et al. 2018). Cold adaptive microbiomes play an important role in plant growth promotion and soil health for agricultural sustainability through different plant growth promoting mechanisms (Yadav et al. 2018a; Yadav et al. 2017a; Yadav et al. 2017d).

Cold habitats

The varieties of cold habitats that have been observed on Earth are listed below:

Cryoconite holes

These are water-filled differently shaped holes seen on glacier surfaces. Due to the accumulation of soil particles and/or minerals from the surroundings, they become organically rich (Telling et al. 2014). Sun warming melts the surface ice and creates the cylindrical holes which are inoculated with the microorganisms attached on the surface. Photosynthetic organisms are the primary producers which provide nutrition to the microorganisms. Cryconites can be seen as two distinct zones—(i) the pelagic zone (upper)—containing meltwater and (ii) the benthic zone (lower)—consisting of dark-colored, tiny inorganic and organic particulate matters along with complex microbial population (Anesio et al. 2017). These structures are seen only in summers and become permanently frozen during the winters.

Glacial ice

In the high altitude polar and non-polar regions, snowfall accumulates along with various non-biological (soil and mineral particles) and biological material (plant fragments, insects, pollen grains, fungal spores, bacteria, and viruses). Permafrost or cryotic soil is soil that exerts the subzero temperature for years but still shows a low level of microbial associated metabolic activities (Gilichinsky et al. 2008; Margesin and Collins 2019).

Sub glacial lakes

A sub glacial lake is a lake under a glacier. There are so many sub glacial lakes, such as Lake Vostok in Antarctica which is the largest known so far. The water below the glacier ice remains in a liquid state due to the tremendous pressure caused by the accumulated ice layers. It is found to supersaturate with oxygen and geothermally heated (Zaragotas et al. 2016). There are many reports on microbial diversity from sub glacial lakes (Singh et al. 2016; Yadav et al. 2016a; Yadav et al. 2015c; Yadav et al. 2018c).

Brinicle

In cold habitats, when seawater freezes, salt is forced out from the ice crystals and therefore, the surrounding water becomes more saline. Due to this, the density is increased while the freezing temperature goes down. Further, the water present around the surroundings does not convert into ice and thus, small brine tunnels are formed all through the ice as this supercooled, super saline water sinks away from the frozen pure water (Miteva 2008).

Supercooled cloud droplets

Supercooling is the process in which the temperature of a gas or liquid lowers down below their freezing point without reaching a solid-state. The microorganisms residing under these cloud droplets are known to be metabolically active at subzero temperatures.

Ice cores

These are blocks of accumulated ice during a long period of time. Each time, the deposition of a newer layer of ice settles down on the microorganisms present on the older one. Therefore, microbial assemblages found in the ice cores represent chronologically different microbial populations. However, their concentration may differ due to the climatic conditions, topography and ecosystem proximity during the respective time period (Tian et al. 2017).

Extraterrestrial

Extraterrestrial represents life 'other from Earth'. Recent studies and space missions have increased the curiosity about the possibilities of the water bodies/ice in the planets other than Earth. It has led to the concept of extraterrestrial life forms especially due to the vast survival strategies of the extremophilic microorganisms. In fact, the extremophiles are considered as 'the ultimate survivors' and have been found at 121°C to –20°C. In this scenario, astrobiology is emerging as a recent field of science (Siddiqui et al. 2013) (Table 9.2).

Physico-chemical constraints at the cold environment

Microorganisms exert several physico-chemical constraints besides various stresses from the cold. Among them, H bond strengthening, fluidity reduction in cell membranes, metabolite stability, enhanced viscosity, decreased catalysis and increased rate of gas solubility are important. Not only

Table 9.2 Characteristics of some planets and moons from the Earth's solar system with the potential to harbor psychrophilic life.

Planet/moon	Atmospheric, surface, and subsurface Composition	Surface temperature
Earth	CO_2, N_2, O_2 Water exists in all three states (gas, liquid, solid)	−89 to 58°C
Mars	CO_2, N_2 Polar water and CO_2 ice caps	−140 to 20°C
Europa (Jupiter)	O_2 Liquid water ocean may exist under surface ice sheet	−223 to −148°C
Ganymede (Jupiter)	O_2 Water ice	−203 to −121°C
Callisto (Jupiter)	CO_2 (99%), O_2 (1%) Liquid water ocean may exist beneath its surface	−193 to −108°C
Titan (Saturn)	CH_4, N_2, H_2 CH_4 and C_2H_6 exist in all three states as gas, liquid, and solid	−179°C
Enceladus (Saturn)	CH_4, CO_2, H_2O, N_2 Water ice	−240 to −128°C

this, cold reduces the reaction rates of most of the biochemical reactions occurring inside the cells as evident from the following rate equation:

$$K_{cat} = \kappa \frac{K_B T}{h} \exp(\Delta G^{\#}/RT)$$

where, k denotes the reaction rate, κ transmission coefficient (~ 1), $\Delta G^{\#}$ represent activation energy, T denotes absolute temperature, R is the gas constant while, h and K_B are Planck and Boltzmann constant, respectively (Rodrigues and Tiedje 2008). Enzymes of the cold-adapted microbes are found to have higher specific activity due to a reduction in several non-covalent interactions and thus increased flexibility. Moreover, they have lower activation energy and enthalpy change in comparison to their mesophilic counterparts and thus, are highly active.

Temperature below 0°C causes the freezing of water, resulting in a shortage of a solvent system for enzymes. Furthermore, the formation of ice crystals increases the chances of damage to the cell membranes. However, water can be maintained in a liquid state below 0°C by different processes, viz. supercooling (supercooled cloud droplets), pressure (sub glacial lakes), freezing point depressions (brinicles), etc. (Zaragotas et al. 2016). Microorganisms frequently take advantage of these phenomena and inhabit such extreme places successfully using various adaptive mechanisms.

Cold adaptation

Physiological adaptation

The struggle against low temperatures has compelled microbes to adopt various strategies beginning right at the level of individual molecules to that of the whole cell. The cell's physiology is always determined by its genes and their expression/regulation in response to any stimuli. Lowering in the environmental temperature is a kind of abiotic stimuli that greatly influences the cellular physiology and thus survival of the cell. Cold adapted microorganisms have numerous mechanisms to counter the physiological adversities like alter enzyme kinetics, modulate the membrane fluidity, polymerize microtubules, regularize the permeability of ion channels, and undergo seasonal dormancy (Tribelli and López 2018). The major adaptive mechanisms are discussed below.

Cell membranes and fluidity

The maintenance of the cellular membrane fluidity is one of the major tasks for the cold-adapted microorganisms as a decrease in temperature causes a reduction in membrane fluidity resulting in loss of its functionality. Cold loving microbes can modify the fluidity and flexibility of their membranous lipid bilayer by increased production of unsaturated fatty acids with double bonds (cis), etc. (de Pascale et al. 2012).

Compatible solutes, cryoprotectants and antifreeze proteins

Cold adapted microorganisms may use additional strategies to cope with low temperatures. Compatible solutes are the cellular moieties that act as a protectant against a range of stresses including freezing and osmotic stress. Betaine, glycine, glycerol, mannitol and trehalose are among the most common compatible solutes. Moreover, exopolysaccharides (EPSs) and trehalose are known to prevent protein aggregation and denaturation. EPSs increase the water retention capacity and nutrient sequestering ability of the cells. Besides these, there are some special class of the proteins, i.e., Ice-Binding Proteins (IBPs) and antifreeze proteins which sustains the cells under cold conditions by inhibiting ice-crystal formation (Voets 2017).

Cold shock response

It can be defined as a cellular response towards the cold. It increases the specialized proteins inside the cells called Cold shock proteins (Csps). Csps contain a DNA binding domain called Cold-Shock Domain (CSD) which prevents any possible damage to the cellular genetic material which may occur due to the cold. CspA from *Escherichia coli* was the first cold shock protein that was studied in detail. It was found to link with eight other proteins, i.e., CspB–CspI. CspA, B, E, G and I are responsible for cold adaptation while CspC and E help in cell division with UspA and RpoS regulation (Phadtare and Inouye 2001; Kandasamy et al. 2013). Besides these, some of them may act as chaperones (Hofweber et al. 2005). They are distributed in several bacterial genera including *Arthrobacter, Bacillus, Enterobacter, Escherichia, Listeria, Pseudomonas, Streptococcus,* and *Thermotoga.*

Cold-active proteins and enzymes

Microorganisms that grow at lower temperatures are able to produce cold-active enzymes and proteins and therefore, they can perform all the vital functions under severe stress. Such proteins are characterized by higher catalytic rate and increased flexibility which is probably due to their specialized amino acid composition (Feller et al. 2013; Yadav et al. 2017b; Yadav et al. 2017c; Yadav et al. 2019c). Some of the important adaptive features of cold-adapted proteins in comparison to their mesophilic counterparts are summarized below:

- presence of more glycine residues—for greater conformational mobility
- reduction of proline residues in loop regions—for conformational rigidity
- reduction in arginine residues—for conformational rigidity
- reduction in salt bridge and hydrogen bonds—for increased flexibility
- reduced size of nonpolar residues—to create weaker hydrophobic interactions
- reduced frequency of surface—for increased conformational flexibility
- reduced internal hydrophobic interactions—reduced hydrophobic effect

Metabolic adaptations

Cold-adapted microorganisms possess a slower growth rate but higher metabolic rates than their mesophilic counterparts. It has been observed that cold-adapted microbes increase ATP concentration intracellularly to cope with reduced thermal energy (Bajerski et al. 2018). Moreover, AMP synthetic pathways have been observed to increase during purine biosynthesis in comparison to the mesophiles where AMP degradative pathways are dominating. Additionally, cold-adapted enzymes exhibit amino acid-specific substitutions to deal with the problem of metabolic flux globally (Williams et al. 2016). Further, most of the cold-inducible proteins, viz. DeaD exonucleases, PNPase, RNA helicase and RNase are involved in RNA metabolism which is crucial for cold adaptation (Awano et al. 2010). At low temperatures, the production of toxic Reactive Oxygen Species (ROS) increases significantly. To cope with this, *Pseudomonas haloplanktis* lacks the entire molybdopterin metabolism pathway which is responsible for the production of toxic ROS (Médigue et al. 2005).

Molecular adaptation

The major barriers to protein synthesis under cold conditions are reduced protein folding, decreased transcription and translation rate; and a stabilization of the secondary structures in the nucleic acids (Dalluge et al. 1997). In psychrophiles, specialized elongation factors, RNA polymerases, helicases and isomerases take care of this problem (Masullo et al. 2000; Bjerga et al. 2016; Brunet et al. 2018).

The global analysis of cold-adapted microbial genomes using genomics, proteomics (Jain et al. 2010; Suyal et al. 2014; 2015a, b; 2018) and metagenomics (Jeon et al. 2009) approaches are in progress. Some of the psychrophiles which have been completely sequenced are *Colwellia psychrerythraea* 34H (Methé et al. 2005), *Desulfotalea psychrophila* (Rabus et al. 2004), *Idiomarina loihiensis L2TR* (Hou et al. 2004), *Pseudoalteromonas haloplanktis* TAC125 (Médigue et al. 2005), and *Psychromonas ingrahamii* 37 (Riley et al. 2008).

Accessing microbial diversity from a cold environment

In spite of having a large range of synthetic growth media as well as metagenomic approaches (Soni and Goel 2011; Soni et al. 2016; Suyal et al. 2019b), researchers were unable to scan the entire microbial diversity and community structure (Table 9.3). Therefore, a mixed approach is needed to explore the microbial treasure for the sustainability and betterment of humankind. The following methods can be adopted to study microbial diversity.

Table 9.3 Common methodologies used by the researchers to explore psychrophilic microbial diversity.

Method	Comments
Whole-cell fatty acid analysis	Antarctic sea ice
fosmid libraries	Psychrophiles isolated from sponges—collected from offshore of Santa Barbara, CA
PCR-RFLP and rRNA hybridization	Psychrophilic Archaeon found living in association with a marine sponge, *Cenarchaeum symbiosum*
Fluorescent *In situ* hybridization	Bacterial Activity at –2 to –20°C in Arctic Wintertime Sea Ice
ARDRA	Anaerobic Psychrophilic Enrichment Cultures Obtained from a Greenland Glacier Ice Core
2D gel electrophoresis	Western Indian Himalaya

Approaches for culturable microorganisms

This approach generally involves two preliminary steps, i.e., isolation of microorganisms on growth medium and then getting their pure culture. A growth medium may be a solid or liquid preparation containing chemical ingredients that are essential for the growth and development of microorganisms. Growth media may be of selective, differential or an enriched type depending on the requirements.

Phenotypic approach

It completely relies on the morphological characteristics of the bacteria, viz. cell shape, size, gram-staining, colony characteristics, pigmentation, sporulation, biofilm production, etc.

Biochemical approach

This approach is based on the ability of the microorganisms to utilize various compounds as a carbon Source for their growth and development. It involves two basic methods—carbon source utilization tests and Fatty Acid Methyl Esters (FAME) analysis. These techniques are very useful in developing the metabolic profile of the microorganisms.

Molecular approach

This approach uses several tools and techniques for microbial diversity and community analysis, viz. analysis of G+C content, plasmid fingerprinting, PFGE, RFLP, RAPD, AFLP, MLVNTSSLP, SSCP, DGGE, RFLP, ribotyping, serotyping, mass spectrometry, protein-based methods, etc. (Singh et al. 2010; Soni and Goel 2010; 2011; Soni et al. 2010).

Approaches for unculturable microbial communities

It is a well-known fact that only 0.1–1% of bacteria can be cultured by traditional laboratory cultivation methods. Thus, culture-independent approaches are very popular and important for microbial identification and subsequently, in the exploration of their products (Suyal et al. 2014; 2015a, b; 2017; 2019b). This approach does not involve pure cultures and depends upon the isolation of genetic materials and/or proteins directly from environments. Some of the frequently used techniques in accessing the unculturable microbial communities are—fluorescent *in situ* hybridizations, isotope array, micro-autoradiography, phylogenetic markers and stable isotope probing.

Metagenomics

The word 'Metagenomics', i.e., direct extraction of genetic material from environmental samples was coined by Handelsman and coworkers. Some of the techniques and methods used by various researchers to explore the cold-adapted microorganisms are given in Table 9.1. Metagenomics can also be termed as community genomics, environmental genomics or microbial economics. It involves the following steps:

 i) isolation of total DNA from an environmental sample
 ii) cloning of PCR amplified products (PCR cloning) and/or of random DNA fragments (shotgun cloning) into a suitable vector
 iii) transforming the clones into a host bacterium
 iv) screening for positive clones
 v) sequence-driven analysis using high-throughput sequencing techniques or function-driven analysis of expressed phenotypes

Metagenomics is an extremely powerful omics technique which primarily focuses on gene clusters or genes encoding enzymes for commercially important microbial products.

Biotechnological potential and prospects

Microorganisms have a vital role to play in today's materialistic world. The high metabolic activity of psychrophiles at low and moderate temperatures offers several economic benefits which can be divided into the following broad categories.

Enzymes

Microorganisms thriving in cold habitats are a huge reservoir of cold-active enzymes that may be applied successfully by different industries (Table 9.4). Some of the examples are, industrial 'peeling' of leather by proteases, which can be performed at the temperature of tap water using cold-active enzymes instead of warming it to 37°C suitable for mesophilic enzymes (de Pascale et al. 2012). Further, the enzymes of the detergent industry, viz. lipase, glycosidases and subtilisin are much less active at the temperature of tap water, while, warming the water effects the color of fabrics, therefore, psychrophilic enzymes could be a better alternative (Qoura et al. 2014; Yadav et al. 2019a; Yadav et al. 2020c; Yadav et al. 2018b). Similarly, these enzymes can help in the transformation and/or refinement of heat-sensitive food products. Mesophilic glycosidases can be replaced by their psychrophilic version in the baking industry so that their residual activity in final products can be avoided. Lactose removal from milk by using a cold-active β-galactosidase has recently been patented. Psychrophilic pectinases can help to purify fruit juices at low temperatures. Cold-active esterases and lipases exhibit a high level of stereospecificity that may be used for developing chiral drugs (Jeon et al. 2009; Mayordomo et al. 2000; Kour et al. 2019). In molecular biology experiments, the application of heat-labile Antarctic alkaline phosphatase is being proposed as it does not interfere with end labeling after heat treatment. Recently, a recombinant γ-glutamyl-cysteine ligase from the psychrophilic *Pseudoalteromonas haloplanktis* (rPhGshA II) was produced and characterized (Albino et al. 2014). However, higher cost of production and processing of psychrophilic enzymes acts as a barrier for their large scale production and utilization. Cold-adapted microorganisms and their products are summarized in Tables 9.5 and 9.6.

Table 9.4 Cold-active enzymes and their biotechnological applications.

Enzyme	Organisms	Applications
Lipase	*Aspergilus nidulans* WG 312	Food, detergents, cosmetics
Serine peptidase	PA-43 subarctic bacterium	Food, detergents, molecular biology
Metalloprotease	*Sphingomonas paucimobilis*	Food, detergents, molecular biology
Alcohol dehydrogenase	*Moraxella* sp. TAE 123	Asymmetric chemical synthesis
3-Isopropylmate dehydrogenase	*Vibrio* sp. I5	Asymmetric chemical synthesis
Alkaline phosphatase	*Vibrio* sp. G15-21	Molecular biology
Valine dehydrogenase	*Cytophaga* sp. KUC-1	Biotransformation
β-Galactosidase	*Arthrobacter* sp. 20B, *Alkalilactibacillus ikkense*, *Carnobacterium piscicola* BA	Dairy industries
RNA polymerase	*Pseudomonas syringae*	Molecular biology
DNA polymerase	*Cenarchaeum symbiosum*	Molecular biology
DNA ligase	*Pseudoalteromonas haloplanktis*	Molecular biology
Esterase	Arctic sediment	Food, textile industries
α-amylase	*Pseudoalteromonas arctica* GS230	Food, detergent industries
Chitinase	*Glaciozyma antarctica* PI12	Food, health products
Uracil-DNA glycosylase	*Gadus morhua*	Molecular biology
Restriction endonuclease UnbI	Antarctic bacterium	Molecular biology
Triose phosphate isomerase	*Vibrio marinus*	Biotransformation
Glucanase	*Fibrobacter succinogenes* S85	Animal feed, textiles, detergents

Table 9.5 Products/general application of cold-adapted microorganisms.

Microorganism or product	Application
Polyunsaturated fatty acids	Dietary supplements for humans, livestock and fish
Ice nucleation proteins	Food industry, synthetic snow
Antifreeze proteins and solutes	Cryoprotectants, cold-active catalysts
Cold-adapted bacteria and fungi	Food industry (cheese and yoghurt manufacture, flavor modification, lactose removal AND meat tenderizing from milk), bioremediation of ocean oil spills, contaminated ground water and toxic waste, bio-inoculants for high altitude agro-ecosystems
'Ice-minus' bacteria	Frost protection for plants
Methanogenic Archaea	Methane production, low-temperature waste treatment
Microorganism or product	Application
Polyunsaturated fatty acids	Dietary supplements for humans, livestock and fish
Ice nucleation proteins	Food industry, synthetic snow
Antifreeze proteins and solutes	Cryoprotectants, cold-active catalysts
Cold-adapted bacteria and fungi	Food industry (cheese and yoghurt manufacture, flavor modification, lactose removal from milk, and meat tenderizing), bioremediation of ocean oil spills, contaminated ground water and toxic waste, bio-inoculants for high altitude agro-ecosystems
'Ice-minus' bacteria	Frost protection for plants
Methanogenic Archaea	Methane production, low temperature waste treatment

Table 9.6 Commercially available products of the cold adapted microorganisms.

Enzyme	Source	Company
Glicoprotein	*Pseudoalteromonas antarctica*	Lipotec S.A. [SP]
Beta-galactosidase	*Pseudoalteromonas haloplanktis*	University of Liege [BE]
Alkaline phosphatase	*Bacterium* HK-47	Patent n. US4720458
Antifreeze lipoprotein	*Moraxella* sp.	Kansai University [JP]
Lipase-catalyzed ether hydrolysis	*Pseudomonas* sp.	Nippon Paper Industries [JP], Novozymes [DK]
Anti-freeze proteins	*Marinomonas, Pseudomonas* sp.	Unilever [UK]
Enzymes xylanolytic activity	*Pseudoalteromonas haloplanktis*	Puratos Naamloze Vennootschap [BE]
Polyunsaturated fatty acid (PUFA) synthase systems	*Shewanella japonica and Shewanella olleyana*	Martek Biosciences Corporation [US]
Dehydrogenases	*Arthrobacter* sp., *Micrococcus* sp.	University of London [UK]
Detergent compositions enzymes	*Psychrophilic bacteria* sp.	Procter & Gamble [US]
Chlamysin B antibacterial protein	*Chlamys islandica*	Biotec ASA [NO]
Thermostable isomerase	*Thermoanaerobacter mathranii*	Bioneer A/S [DK]

Bioremediation

Psychrophilic microorganisms can be explored for the bioremediation of polluted soils and wastewaters under cold environmental conditions. Moreover, a recombinant DNA technology can be used to engineer a bacterium in such a way that the bioremediation-related genes could be expressed under natural conditions. The biodegradation of a wide range of hydrocarbons, heavy metals and polyethylenes using cold-adapted microorganisms has been studied in several extreme habitats (Kumar et al. 2019a; Yadav 2019; Yadav et al. 2019b).

Development of dietary supplements

Psychrophilic microorganisms are also able to produce polyunsaturated fatty acids (PUFA). They produce these for their cell membrane fluidity which is an adaptive feature for surviving under extreme cold conditions (D'Amico et al. 2006). These bacteria may represent an alternative source for human use. A significant advantage is that bacteria generally contain only one long-chain PUFA (usually eicosapentaenoic or docosahexaenoic acid) instead of multiple chains present in fish or algal oils.

Antifreeze proteins and cryoprotectants

Psychrophiles have evolved several strategies to cope with extremely cold conditions. One of them is the production of antifreeze proteins (AFPs), which facilitate maintaining their body fluids in the liquid state at very low temperatures. The AFPs helps in improving the cold storage and cryopreservation properties of cells and/or tissue, thereby reducing microbial contamination of frozen foods (Voets 2017). Some commercialized enzymes and molecules from polar environments are summarized in Table 9.6.

Cold-adapted plant growth-promoting bacteria

It is of interest that despite the effect of cold stress on nodulation and thus nitrogen fixation (Dixon and Khan 2004), native legumes from the high Arctic regions can nodulate and perform at low temperatures (Kour et al. 2020; Saxena and Goel 1999; Bordeleau and Prevost 1994). Therefore, the interest in using cold-adapted plant growth-promoting bacteria in the agriculture and horticulture sector is increasing day by day (Singh and Yadav 2020; Yadav et al. 2020a). *Burkholderia phytofirmans* PsJN was found to increase grapevine root growth and physiological activity at temperatures as low as 4°C. Coinoculation of *Serratia proteamaculans* with *Bradyrhizobium japonicum* is reported to stimulate soybean growth at 15°C (Zhang et al. 1996). Further, mutants of cold-tolerant *P. fluorescens* strains were developed to promote plant growth under cold (Das et al. 2003; Katiyar and Goel 2003; 2004). In temperate regions, legumes rhizosphere are inhabited by diverse nitrogen-fixing microbial assemblages which can be explored for the isolation of plant growth-promoting psychrophiles and associated adaptive mechanisms (Rastegari et al. 2020; Yadav et al. 2020b).

Conclusion and future prospects

Cold environments are inhabited by a diverse range of microbes including archaea, bacteria and eukarya. They can adapt their cellular and metabolic processes to cope with cold stress. The range of psychrophilic products and their extent of utilization highlight the further need for exploiting novel cold-adapted microorganisms and associated processes for advancement in biotechnology. Both culture-dependent and culture-independent molecular approaches including metagenomics can be explored to unravel the cold-adapted microbial diversity and their products.

References

Albino, A., A. De Angelis, S. Marco, V. Severino, A. Chambery, A. Di Maro et al. 2014. The cold-adapted γ-glutamyl-cysteine ligase from the psychrophile *Pseudoalteromonas haloplanktis*. Biochimie. 104: 50–60.

Anesio, A.M., S. Lutz, N.A.M. Chrismas and L.G. Benning. 2017. The microbiome of glaciers and ice sheets. NPJ Biofilms Microbi. 3: 10.

Asakura, H., A. Ishiwa, E. Arakawa, S.I. Makino, Y. Okada, S. Yamamoto et al. 2007. Gene expression profile of *Vibrio cholerae* in the cold stress-induced viable but non-culturable state. Environ. Microbiol. 9: 869–879.

Awano, N., V. Rajagopal, M. Arbing, S. Patel, J. Hunt, M. Inouye et al. 2010. *Escherichia coli* RNase R has dual activities, helicase and RNase. J. Bac. 192(5): 1344–1352.

Bajerski, F., J. Stock, B. Hanf, T. Darienko, E. Heine-Dobbernack, M. Lorenz et al. 2018. ATP content and cell viability as indicators for cryostress across the diversity of life. Front. Physiol. 9: 921.

Bjerga, G.E., R. Lale and A.K. Williamson. 2016. Engineering low-temperature expression systems for heterologous production of cold-adapted enzymes. Bioengineered 7(1): 33–38.

Bordeleau, L. and D. Prevost. 1994. Nodulation and nitrogen fixation in extreme environments. Plant Soil. 161: 115–125.

Brunet, A., L. Salomé, P. Rousseau, N. Destainville, M. Manghi and C. Tardin. 2018. How does temperature impact the conformation of single DNA molecules below melting temperature. Nucl. Acid Res. 46(4): 2074–2081.

Buffen, A., A. Leventer, A. Rubin and T. Hutchins. 2007. Diatom assemblages in surface sediments from the northwestern weddell Sea, Antarctic Peninsula. Mar. Micropaleontol. 62: 7–30.

Dalluge, J.J., T. Hamamoto, K. Horikoshi, R.Y. Morita, K.O. Stetter and J.A. McCloskey. 1997. Posttranscriptional modification of tRNA in psychrophilic bacteria. J. Bacteriol. 179: 1918–1923.

D'Amico, S., T. Collins, J.C. Marx, G. Feller and C. Gerday. 2006. Psychrophilic microorganisms: challenges for life. EMBO Rep. 7(4): 385–389.

Das, K., V. Katiyar and R. Goel. 2003. 'P' solubilization potential of plant growth promoting *Pseudomonas* mutants at low temperature. Microbiol. Res. 158: 359–362.

de Pascale, D., C. De Santi, J. Fu and B. Landfald. 2012. The microbial diversity of Polar environments is a fertile ground for bioprospecting. Mar. Genomics 8: 15–22.

Dixon, R. and D. Khan. 2004. Genetic regulation of biological nitrogen fixation. Nature 2: 621–631.

Feller, G. 2013. Psychrophilic enzymes: from folding to function and biotechnology. Scientifica 512840: 1–28.

Gesheva, V. and T.G. Negoita. 2011. Psychrotrophic microorganism communities in soils of Haswell Island, Antarctica, and their biosynthetic potential. Polar Biol. 35: 291–297.

Gilichinsky, D., T. Vishnivetskaya, M. Petrova, E. Spirina, V. Mamykin and E. Rivkina. 2008. Bacteria in permafrost. pp. 83–102. *In*: Margesin, R., F. Schinner, J.C. Marx and C. Gerday (eds.). Psychrophiles: From Biodiversity to Biotechnology. Springer, Berlin, Heidelberg.

Hofweber, R., G. Horn, T. Langmann, J. Balbach, W. Kremer, G. Schmitz et al. 2005. The influence of cold shock proteins on transcription and translation studied in cell-free model systems. FEBS J. 272: 4691–4702.

Hou, S., J.H. Saw, K.S. Lee, T.A. Freitas, C. Belisle, Y. Kawarabayasi et al. 2004. Genome sequence of the deep-sea gamma-proteobacterium Idiomarina loihiensis reveals amino acid fermentation as a source of carbon and energy. Proc. Nat. Acad. Sci. U.S.A. 101(52): 18036–18041.

Jain, S., A. Rani, S. Marla and R. Goel. 2010. Differential proteomic analysis of psychrotolerant *Pseudomonas putida* 710A and alkaliphilic *P. monteilli* 97AN for cadmium stress. Int. J. Biol. Med. Res. 1(4): 234–241.

Jeon, J.H., J.T. Kim, S.G. Kang, J.H. Lee and S.J. Kim. 2009. Characterization and its potential application of two esterases derived from the arctic sediment metagenome. Mar. Biotechnol. 11: 307–316.

Joshi, D., R. Chandra, D.C. Suyal, S. Kumar and R. Goel. 2019. Impact of bioinoculants *Pseudomonas jesenii* MP1 and *Rhodococcus qingshengii* S10107 on *Cicer arietinum* yield and soil nitrogen status. Pedosphere. 29(3): 388–399.

Kandasamy, P., S. Gahoi, S.S. Marla and R. Goel. 2013. Expression of *CspE* by a psychrotrophic bacterium *Enterobacter ludwigii* PAS1, isolated from Indian Himalayan soil and *in silico* protein modelling, prediction of conserved residues and active sites. Curr. Microbiol. 66: 507–514.

Katiyar, V. and R. Goel. 2003. Solubilization of inorganic phosphate and plant growth promotion by cold tolerant mutants of *Pseudomonas fluorescens*. Microbiol. Res. 158: 163–168.

Katiyar, V. and R. Goel. 2004. Improved plant growth from seed bacterization of a siderophore overproducing cold resistant mutant of fluorescent pseudomonad. J. Microbiol. Biotechnol. 14(4): 653–657.

Kolton, M., A. Marks, R.M. Wilson, J.P. Chanton and J.E. Kostka. 2019. Impact of warming on greenhouse gas production and microbial diversity in anoxic peat from a *Sphagnum*-dominated bog (Grand Rapids, Minnesota, United States). Front. Microbiol. 10(870): 1–9.

Kour, D., K.L. Rana, T. Kaur, B. Singh, V.S. Chauhan, A. Kumar et al. 2019. Extremophiles for hydrolytic enzymes productions: biodiversity and potential biotechnological applications. pp. 321–372. *In*: Molina, G., V.K. Gupta, B. Singh and N. Gathergood (eds.). Bioprocessing for Biomolecules Production. doi:10.1002/9781119434436.ch16.

Kour, D., K.L. Rana, A.N. Yadav, N. Yadav, M. Kumar, V. Kumar et al. 2020. Microbial biofertilizers: Bioresources and eco-friendly technologies for agricultural and environmental sustainability. Biocatal. Agric. Biotechnol. 23: 101487. doi:https://doi.org/10.1016/j.bcab.2019.101487.

Kumar, A., A.K. Chaturvedi, K. Yadav, K.P. Arunkumar, S.K. Malyan, P. Raja et al. 2019a. Fungal phytoremediation of heavy metal-contaminated resources: current scenario and future prospects. pp. 437–461. *In*: Yadav, A.N., S. Singh, S. Mishra and A. Gupta (eds.). Recent Advancement in White Biotechnology Through Fungi: Volume 3: Perspective for Sustainable Environments. Springer International Publishing, Cham. doi:10.1007/978-3-030-25506-0_18.

Kumar, L., G. Awasthi and B. Singh. 2011. Extremophiles: A novel source of industrially important enzymes. Biotechnology 10: 121–135.

Kumar, M., D. Kour, A.N. Yadav, R. Saxena, P.K. Rai, A. Jyoti et al. 2019b. Biodiversity of methylotrophic microbial communities and their potential role in mitigation of abiotic stresses in plants. Biologia 74: 287–308. doi:10.2478/s11756-019-00190-6.

Kumar, S., D.C. Suyal, N. Dhauni, M. Bhoriyal and R. Goel. 2014. Relative plant growth promoting potential of Himalayan psychrotolerant *Pseudomonas jesenii* strain MP1 against native *Cicer arietinum* L., *Vigna mungo* (L.) Hepper; *Vigna radiata* (L.) Wilczek., *Cajanus cajan* (L.) Millsp. and *Eleusine coracana* (L.) Gaertn. Afri. J. Microbiol. 8(50): 3931–3943.

Madigan, M.T., K.S. Bender, D.H. Buckley, W.M. Sattley and D.A. Stahl. 2018. Microbial Growth and Its Control, Chapter 5, Brock Biology of Microorganisms, 15th Global Edition.

Margesin, R. and T. Collins. 2019. Microbial ecology of the cryosphere (glacial and permafrost habitats): current knowledge. Appl. Microbiol. Biotechnol. 103: 2537.

Masullo, M., P. Arcari, B. De Paola, A. Parmeggiani and V. Bocchini. 2000. Psychrophilic elongation factor Tu from the antarctic Moraxella sp. Tac II 25: biochemical characterization and cloning of the encoding gene. Biochemistry 39(50): 15531–15539.

Mayordomo, I., F. Randez-Gil and J.A. Prieto. 2000. Isolation, purification and characterization of a cold-active lipase from *Aspergillus nidulans.* J. Agric. Food Chem. 48: 105–109.

Médigue, C., E. Krin, G. Pascal, V. Barbe, A. Bernsel, P.N. Bertin et al. 2005. Coping with cold: the genome of the versatile marine Antarctica bacterium *Pseudoalteromonas haloplanktis* TAC125. Genome Res. 15(10): 1325–1335.

Methé, B.A., K.E. Nelson, J.W. Deming, B. Momen, E. Melamud, X. Zhang et al. 2005. The psychrophilic lifestyle as revealed by the genome sequence of *Colwellia psychrerythraea* 34H through genomic and proteomic analyses. Proc. Natl. Acad. Sci. U.S.A. 102(31): 10913–8.

Miteva, V.I. 2008. Bacteria in snow and glacier ice. pp. 31–50. *In*: Margesin, R., F. Schinner, J.C. Marx and C. Gerday (eds.). Psychrophiles: From Biodiversity to Biotechnology. Springer-Verlag, Heidelberg, Germany.

Nozhevnikova, A.N., M.V. Simankova, S.N. Parshina and O.R. Kotsyurbenko. 2001. Temperature characteristics of methanogenic archaea and acetogenic bacteria isolated from cold environments. Water Sci. Technol. 44: 41–48.

Phadtare, S. and M. Inouye. 2001. Role CspC and CspE in regulation of expression of RpoS and UspA, the stress response proteins in *Escherichia coli.* J. Bacteriol. 183: 1205–1214.

Qoura, F., T. Brueck and G. Antranikian. 2014. The psychrophile *Shewanella arctica* sp. nov: A new source of industrially important enzyme systems. JSM Biotechnol. Bioeng. 2(1): 1–10.

Rabus, R., A. Ruepp, T. Frickey, T. Rattei, B. Fartmann, M. Stark et al. 2004. The genome of *Desulfotalea psychrophila,* a sulfate-reducing bacterium from permanently cold Arctic sediments. Environ. Microbiol. 6: 887–902.

Rajwar, J., R. Chandra, D.C. Suyal, S. Tomer, S. Kumar and R. Goel. 2018. Comparative phosphate solubilizing efficiency of psychrotolerant *Pseudomonas jesenii* MP1 and *Acinetobacter* sp. ST02 against chickpea for sustainable hill agriculture. Biologia 73(8): 793–802.

Rana, K.L., D. Kour, T. Kaur, I. Sheikh, A.N. Yadav, V. Kumar et al. 2020. Endophytic microbes from diverse wheat genotypes and their potential biotechnological applications in plant growth promotion and nutrient uptake. Proc. Natl. Acad. Sci. India, Sect. B Biol. Sci. doi:10.1007/s40011-020-01168-0.

Rastegari, A.A., A.N. Yadav, A.A. Awasthi and N. Yadav. 2020. Trends of Microbial Biotechnology for Sustainable Agriculture and Biomedicine Systems: Diversity and Functional Perspectives. Elsevier, Cambridge, USA.

Riley, M., J.T. Staley, A. Danchin, T.Z. Wang, T.S. Brettin, L.J. Hauser et al. 2008. Genomics of an extreme psychrophile, *Psychromonas ingrahamii.* BMC Genom. 9: 210–217.

Rodrigues, D.F. and J.M. Tiedje. 2008. Coping with our cold planet. Appl. Environ. Microbiol. 74(6): 1677–1686.

Saxena, A. and R. Goel. 1999. Growth promotory activities of Standard and Antarctic Pseudomonads. Physiol. Mol. Biol. Plants 57: 181–184.

Saxena, A.K., A.N. Yadav, M. Rajawat, R. Kaushik, R. Kumar, M. Kumar et al. 2016. Microbial diversity of extreme regions: An unseen heritage and wealth. Indian J. Plant Genet. Resour. 29: 246–248.

Siddiqui, K.S., T.J. Williams, D. Wilkins, S. Yau, M.A. Allen, M.V. Brown et al. 2013. Psychrophiles. Annu. Rev. Earth Planet. Sci. 41: 87–115.

Singh, C., R. Soni, S. Jain, S. Roy and R. Goel. 2010. Diversification of nitrogen fixing bacterial community using *nif*H gene as a biomarker in different geographical soils of Western Indian Himalayas. J. Environ. Biol. 31: 553–556.

Singh, J. and A.N. Yadav. 2020. Natural Bioactive Products in Sustainable Agriculture. Springer, Singapore.

Singh, R.N., S. Gaba, A.N. Yadav, P. Gaur, S. Gulati, R. Kaushik et al. 2016. First, high quality draft genome sequence of a plant growth promoting and cold active enzymes producing psychrotrophic *Arthrobacter agilis* strain L77. Stand Genomic. Sci. 11: 54. doi:10.1186/s40793-016-0176-4.

Soni, R. and R. Goel. 2010. Triphasic approach for assessment of bacterial population in different soil systems. Ekologija 56(3-4): 94–98.

Soni, R., B. Saluja and R. Goel. 2010. Bacterial community analysis using temporal temperature gradient gel electrophoresis (TTGE) of 16s rDNA PCR products of soil metagenome. Ekologija. 56(3-4): 99–104.

Soni, R. and R. Goel. 2011. *nif*H homologs from soil Metagenome. Ekologija. 57(3): 87–95.

Soni, R., D.C. Suyal, S. Sai and R. Goel. 2016. Exploration of *nif*H gene through soil metagenomes of the western Indian Himalayas. 3Biotech. 6(1): 25–31.

Suyal, D.C., A. Yadav, Y. Shouche and R. Goel. 2014. Differential proteomics in response to low temperature diazotrophy of Himalayan psychrophilic nitrogen fixing *Pseudomonas migulae* S10724 strain. Curr. Microbiol. 68: 543–550.

Suyal, D.C., A. Yadav, Y. Shouche and R. Goel. 2015a. Diversified diazotrophs associated with the rhizosphere of Western Indian Himalayan native red kidney beans (*Phaseolus vulgaris* L.). 3Biotech 5: 433–441.

Suyal, D.C., A. Yadav, Y. Shouche and R. Goel. 2015b. Bacterial diversity and community structure of Western Indian Himalayan red kidney bean (*Phaseolus vulgaris* L.) rhizosphere as revealed by 16S rRNA gene sequences. Biologia 70(3): 305–313.

Suyal, D.C., S. Kumar, A. Yadav, Y. Shouche and R. Goel. 2017. Cold stress and nitrogen deficiency affected protein expression of psychrotrophic *Dyadobacter psychrophilus* B2 and *Pseudomonas jessenii* MP1. Front. Microbiol. 8(430): 1–6.

Suyal, D.C., S. Kumar, D. Joshi, R. Soni and R. Goel. 2018. Quantitative proteomics of psychotrophic diazotroph in response to nitrogen deficiency and cold stress. J. Proteomics 187: 235–242.

Suyal, D.C., S. Kumar, D. Joshi, A. Yadav, Y. Shouche and R. Goel. 2019a. Comparative overview of red kidney bean (*Phaseolus valgaris*) rhizospheric bacterial diversity in perspective of altitudinal variations. Biologia. 74: 1405–1413.

Suyal, D.C., D. Joshi, S. Kumar, R. Soni and R. Goel. 2019b. Differential protein profiling of soil diazotroph *Rhodococcus qingshengii* S10107 towards low-temperature and nitrogen deficiency. Sci. Rep. 20378: 1–9.

Telling, J., A.M. Anesio, M. Tranter, A.G. Fountain, T. Nylen, J. Hawkings et al. 2014. Spring thaw ionic pulses boost nutrient availability and microbial growth in entombed Antarctic Dry Valley cryoconite holes. Front. Microbiol. 11(5): 694–690.

Tian, H., P. Gao, Z. Chen, Y. Li, Y. Li, Y. Wang et al. 2017. Compositions and abundances of sulfate-reducing and sulfur-oxidizing microorganisms in water-flooded petroleum reservoirs with different temperatures in China. Front. Microbiol. 8: 143.

Tomer, S., D.C. Suyal, J. Rajwar, A. Yadav, Y. Shouche and R. Goel. 2017. Isolation and characterization of phosphate solubilizing bacteria from Western Indian Himalayan soils. 3Biotech. 7(95): 1–5.

Tribelli, P.M. and N.I. López. 2018. Reporting key features in cold-adapted bacteria. Life Basel. 8(1): 8–11.

Verma, P., A.N. Yadav, K.S. Khannam, N. Panjiar, S. Kumar, A.K. Saxena et al. 2015. Assessment of genetic diversity and plant growth promoting attributes of psychrotolerant bacteria allied with wheat (*Triticum aestivum*) from the northern hills zone of India. Ann. Microbiol. 65: 1885–1899.

Verma, P., A.N. Yadav, K.S. Khannam, S. Kumar, A.K. Saxena and A. Suman. 2016. Molecular diversity and multifarious plant growth promoting attributes of Bacilli associated with wheat (*Triticum aestivum* L.) rhizosphere from six diverse agro-ecological zones of India. J. Basic Microbiol. 56: 44–58.

Voets, I.K. 2017. From ice-binding proteins to bio-inspired antifreeze materials. Soft Matter. 13(28): 4808–4823.

Williams, C.M., M.D. McCue, N.E. Sunny, A. Szejner-Sigal, T.J. Morgan, D.B. Allison et al. 2016. Cold adaptation increases rates of nutrient flow and metabolic plasticity during cold exposure in *Drosophila melanogaster*. Proc. R. Soc. B 283: 1–9.

Yadav, A.N., S.G. Sachan, P. Verma and A.K. Saxena. 2015a. Prospecting cold deserts of north western Himalayas for microbial diversity and plant growth promoting attributes. J. Biosci. Bioeng. 119: 683–693.

Yadav, A.N., S.G. Sachan, P. Verma, S.P. Tyagi, R. Kaushik and A.K. Saxena. 2015b. Culturable diversity and functional annotation of psychrotrophic bacteria from cold desert of Leh Ladakh (India). World J. Microbiol. Biotechnol. 31: 95–108.

Yadav, A.N., P. Verma, M. Kumar, K.K. Pal, R. Dey, A. Gupta et al. 2015c. Diversity and phylogenetic profiling of niche-specific Bacilli from extreme environments of India. Ann. Microbiol. 65: 611–629.

Yadav, A.N., S.G. Sachan, P. Verma, R. Kaushik and A.K. Saxena. 2016a. Cold active hydrolytic enzymes production by psychrotrophic Bacilli isolated from three sub-glacial lakes of NW Indian Himalayas. J. Basic Microbiol. 56: 294–307.

Yadav, A.N., S.G. Sachan, P. Verma and A.K. Saxena. 2016b. Bioprospecting of plant growth promoting psychrotrophic Bacilli from cold desert of north western Indian Himalayas. Indian J. Exp. Biol. 54: 142–150.

Yadav, A.N., P. Verma, D. Kour, K.L. Rana, V. Kumar, B. Singh et al. 2017a. Plant microbiomes and its beneficial multifunctional plant growth promoting attributes. Int. J. Environ. Sci. Nat. Resour. 3: 1–8. doi:10.19080/IJESNR.2017.03.555601.

Yadav, A.N., P. Verma, V. Kumar, S.G. Sachan and A.K. Saxena. 2017b. Extreme cold environments: A suitable niche for selection of novel psychrotrophic microbes for biotechnological applications. Adv. Biotechnol. Microbiol. 2: 1–4.

Yadav, A.N., P. Verma, S.G. Sachan and A.K. Saxena. 2017c. Biodiversity and biotechnological applications of psychrotrophic microbes isolated from Indian Himalayan regions. EC Microbiol. ECO 01: 48–54.

Yadav, A.N., P. Verma, B. Singh, V.S. Chauhan, A. Suman and A.K. Saxena. 2017d. Plant growth promoting bacteria: biodiversity and multifunctional attributes for sustainable agriculture. Adv. Biotechnol. Microbiol. 5: 1–16.

Yadav, A.N., V. Kumar, R. Prasad, A.K. Saxena and H.S. Dhaliwal. 2018a. Microbiome in crops: diversity, distribution and potential role in crops improvements. pp. 305–332. *In*: Prasad, R., S.S. Gill and N. Tuteja (eds.). Crop Improvement through Microbial Biotechnology. Elsevier, USA.

Yadav, A.N., P. Verma, S.G. Sachan, R. Kaushik and A.K. Saxena. 2018b. Psychrotrophic microbiomes: molecular diversity and beneficial role in plant growth promotion and soil health. pp. 197–240. *In*: Panpatte, D.G., Y.K. Jhala, H.N. Shelat and R.V. Vyas (eds.). Microorganisms for Green Revolution-Volume 2: Microbes for Sustainable Agro-ecosystem. Springer, Singapore. doi:10.1007/978-981-10-7146-1_11.

Yadav, A.N. 2019. Fungal white biotechnology: conclusion and future prospects. pp. 491–498. *In*: Yadav, A.N., S. Singh, S. Mishra and A. Gupta (eds.). Recent Advancement in White Biotechnology Through Fungi: Volume 3: Perspective for Sustainable Environments. Springer International Publishing, Cham. doi:10.1007/978-3-030-25506-0_20.

Yadav, A.N., D. Kour, S. Sharma, S.G. Sachan, B. Singh, V.S. Chauhan et al. 2019a. Psychrotrophic microbes: biodiversity, mechanisms of adaptation, and biotechnological implications in alleviation of cold stress in plants. pp. 219–253. *In*: Sayyed, R.Z., N.K. Arora and M.S. Reddy (eds.). Plant Growth Promoting Rhizobacteria for Sustainable Stress

Management: Volume 1: Rhizobacteria in Abiotic Stress Management. Springer Singapore, Singapore. doi:10.1007/978-981-13-6536-2_12.

Yadav, A.N., S. Singh, S. Mishra and A. Gupta. 2019b. Recent Advancement in White Biotechnology Through Fungi. Volume 3: Perspective for Sustainable Environments. Springer International Publishing, Cham.

Yadav, A.N., N. Yadav, S.G. Sachan and A.K. Saxena. 2019c. Biodiversity of psychrotrophic microbes and their biotechnological applications. J. Appl. Biol. Biotechnol. 7: 99–108.

Yadav, A.N., S. Mishra, D. Kour, N. Yadav and A. Kumar. 2020a. Agriculturally Important Fungi for Sustainable Agriculture, Volume 1: Perspective for Diversity and Crop Productivity. Springer International Publishing, Cham.

Yadav, A.N., A.A. Rastegari, N. Yadav and D. Kour. 2020b. Advances in Plant Microbiome and Sustainable Agriculture: Diversity and Biotechnological Applications. Springer, Singapore.

Yadav, A.N., J. Singh, A.A. Rastegari and N. Yadav. 2020c. Plant Microbiomes for Sustainable Agriculture. Springer International Publishing, Cham.

Yadav, N., D. Kour and A.N. Yadav. 2018c. Microbiomes of freshwater lake ecosystems. J. Microbiol. Exp. 6: 245–248.

Zaragotas, D., N.T. Liolios and E. Anastassopoulos. 2016. Supercooling, Ice Nucleation and Crystal Growth: A Systematic Study in Plant Samples. Cryobiology 72: 239–243.

Zhang, F., N. Dashti, R. Hynes and D.L. Smith. 1996. Plant growth promoting rhizobacteria and soybean [*Glycine max* (L.) Merr.] nodulation and nitrogen fixation at suboptimal root zone temperatures. Ann. Bot. 77: 453–460.

Chapter 10

Psychrophilic Microbiomes

Unravelling the Molecular Adaptation Strategies using *In Silico* Approaches

Abhigyan Nath

Introduction

Psychrophilic organisms come under those extremophiles which can successfully thrive in surroundings of extremely low temperatures (Siddiqui and Cavicchioli 2006). Seventy five to eighty percent of the Earth is under cold temperature conditions and psychrophilic organisms are widely distributed from oceans to permafrost (De Maayer et al. 2014; Deming 2007). Psychrophiles also play important roles in biogeochemical cycles (Achberger et al. 2017; Deming and Collins 2017) and in the polar food web (Junge et al. 2019). Some of the psychrophilic organisms are also polyextremophilic, surviving in fluctuations of more than one extremophilic environment (Yadav et al. 2015c). Notable among them are the psychrophiles which can flourish under extremes of pressure (psychrophilic-piezophilic extremophiles) (Fang et al. 2010). The psychrophilic microbiomes have potential biotechnological applications in agriculture, the environment and medicine. The psychrotrophic microbiome from hilly regions having plant growth promoting attributes that could be utilized for plant growth promotion and soil fertility under the low temperature conditions (Rana et al. 2020; Verma et al. 2015; Yadav et al. 2019; Yadav et al. 2016b; Yadav et al. 2018).

Metabolic activities can remain functional in psychrophiles at temperatures as low as −17°C (Collins and Buick 1989; Schroeter and Scheidegger 1995) and low temperature surroundings affect many cellular and molecular processes such as fluidity of membranes, enzyme kinetics, etc. (Piette et al. 2011; Rodrigues and Tiedje 2008). For the maintenance of membrane fluidity a varied range of adaptations are present in psychrophiles, for example, an increase in saturated fatty acid content (Casanueva et al. 2010; D'Amico et al. 2006; Deming 2002; Guan et al. 2013). A number of cryoprotectants are also present for maintaining the osmotic balance and preventing the formation of ice crystals, notable among them are betaine, mannitol and carnitine (Cowan 2009; Klähn and Hagemann 2011). Apart from cryoprotectants which helps in the survival of psychrophiles, protein adaptation at the low temperature surroundings is also one of the important adaptations from the vast pool of cellular adjustments responsible for survival of psychrophilic organisms (Yadav et al. 2015a; Yadav et al. 2015b). With the advent of cost effective and fast sequencing technologies there

Department of Biochemistry, Pt. Jawahar Lal Nehru Memorial Medical College, Raipur-492001, India.
Email: abhigyannath01@gmail.com

is an explosion of sequences of extremophilic organisms being deposited in public databases. These deposited sequences provide a wealth of information related to their adaptation and provides an opportunity for the application of *in silico* methods for mining these databases which can provide insights into their successful survival strategies.

Cold-adapted proteins are also found to be useful in a number of industrial and laboratory processes (Barroca et al. 2017; van den Burg 2003; Yadav 2015) for example in bioremediation (Brouchkov et al. 2017; Santiago et al. 2016; Timmis and Pieper 1999) and in cold washing as detergent additives (Feller et al. 1996; Yadav et al. 2016a). Analysis of psychrophilic protein sequences can facilitate in increasing the knowledgebase of sequence-structure-function and adaptation relationships responsible for providing the ability to survive under the low temperature surroundings. Psychrophilic proteins have evolved to combat the problems of cold denaturation and local flexibility which are prevalent at low temperature (Siddiqui and Cavicchioli 2006). Maintaining flexibility at low temperature surroundings is an important aspect of cold-adapted enzymes (Georlette et al. 2003). Knowledge of molecular basis of cold adaptation can help in designing of biocatalysts which can remain functional at low temperatures. Understanding the molecular adaptation at the amino acid level can facilitate the rational designing of industrially important psychrophilic proteins in the laboratory. In comparison to thermophilic and hyperthermophilic proteins (protein sequences from organisms which can thrive under high temperature surroundings), there is still much less understanding of the molecular features responsible for protein cold adaptation (Nath et al. 2012).

The present chapter will explain broadly two types of *in silico* analysis that are prevalent in the research community: Comparative statistical analysis using sequences and structures and machine learning based predictive models for the extraction of psychrophilic signatures. Sequence and structure based comparative statistical analysis between psychrophilic and homologous mesophilic protein sequences forms the basis for extraction of adaptation parameters at the molecular level. The availability of psychrophilic protein structures is still quite less as compared to the available sequences, consequently most of the previous research works have utilized either full proteomes of psychrophiles or protein sequences filtered based on some criteria for comparative analysis. The application of machine learning algorithms in psychrophilic research opens the door for their prediction using sequence or structural properties. The predictive models generated through machine learning algorithms give the ability to annotate a protein sequence with psychrophilicity information. The rule induction algorithms can further provide an explanation to the classification, giving insights into their adaptation parameters.

In silico approaches

The major requirement for application of data mining and statistical comparative approaches is the availability of psychrophilic protein sequences and structures. Authentic structural data for psychrophilic protein sequences can be obtained from Protein Data Bank (PDB) (Berman et al. 2000). The corresponding sequences can be accessed from either the PDB itself or from the NCBI protein database (https://www.ncbi.nlm.nih.gov/protein/), while the complete proteome and genome sequences can be downloaded from the NCBI ftp repository (ftp://ftp.ncbi.nlm.nih.gov/genomes/). Specialized databases such as, Prokaryotic Growth Temperature database (PGTdb) (Huang et al. 2004) provides the growth temperature data of extremophilic organisms. PGTdb classifies each organism in its database based on Optimal Growth Temperature (OGT), psychrophilic organisms being classified as having OGT less than 20°C. Another notable database which has found its value in protein stability and engineering studies is the ProtDataTherm (Pezeshgi Modarres et al. 2018). It consists of protein family-wise categorization of extremophilic protein sequences including psychrophiles. Full genome and metagenomic sequences of psychrophilic organisms can also be retrieved from databases such as Genomics Online Database (GOLD) (Mukherjee et al. 2018), MG-RAST (Wilke et al. 2016), ENA (Leinonen et al. 2011) and IMG/M (Chen et al. 2017). The basic flow for an *in silico* approach for psychrophilic protein analysis is depicted in Fig. 10.1.

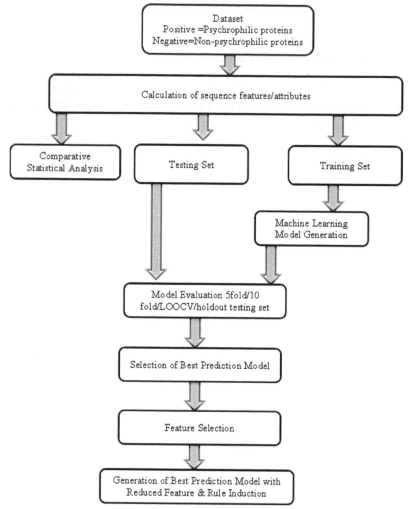

Figure 10.1 Schematic representation of an *in silico* approach for comparative and predictive analysis.

Statistical approaches using sequence features

Comparative protein sequence analysis forms the foundation to unravel the specific sequence features pertaining to psychrophilic adaptation. In this approach a set of psychrophilic protein sequences are compared with homologous mesophilic protein sequences using some statistical hypothesis testing method. The most popular among the statistical tests is the independent sample two-tailed student's t-test. Previously the whole proteome-wise comparative analysis was performed by Metpally and Reddy (2009) (Psychrophilic proteomes against homologous mesophilic proteomes) and reported statistically significant differences among amino acid composition, amino acid property group composition and differential preferences of amino acid residues in secondary structure elements. A similar type of comparative analysis was also performed in Jahandideh et al. (2007a).

All these studies pointed at the subtle compositional changes at the amino acid level to be significant and contributing to psychrophilic protein adaptation. Comparative analysis also revealed that flexibility is an important parameter for the proper catalytic activity of psychrophilic enzymes, which they achieve by weakening of intramolecular forces, e.g., psychrophilic DNA ligases show avoidance of charged residues in the active sites when compared to its homologous thermophilic equivalent for the maintenance of optimal local flexibility (Georlette et al. 2003). In comparison to psychrophilic proteins, thermophilic proteins are more rigid which is responsible for maintaining the tertiary structure at higher temperatures, while higher local flexibility has been observed in psychrophilic proteins (Georlette et al. 2003; Paredes et al. 2011).

Statistical approaches using structural features

Psychrophilic protein involves a diverse array of structural adaptation. As the structures of psychrophilic proteins are still quite limited, only a few comparative studies have been performed using protein tertiary structure. Comparative analysis of psychrophilic protein structures with their non-psychrophilic counterpart facilitated in filtering out probable cold adaptation parameters related to different structural elements. For example, there exists a differential preference of amino acids on the surface of psychrophilic and mesophilic proteins. Earlier studies also reported preference/ avoidance of specific amino acid residues and amino acid property groups in different regions of the protein structure, the regions being divided into buried (core), intermediate and exposed regions based on relative solvent accessibility (Jahandideh et al. 2008; Jahandideh et al. 2007a; Jahandideh et al. 2007b). A comparative analysis using protein structures was also conducted by Shokrollahzade et al (2015). Paredes et al. (2011) confirming the importance of flexibility in psychrophilic proteins. Most of the comparative methods take the advantage of the independent sample two-tailed student's t-test for deriving the preference/avoidance of sequence and structural features, which can be defined by the equation given below:

$$t = \frac{SFM_{\text{Psychrophilic-sequence/structural features}} - SFM_{\text{Non-psychrophilic-sequence/structural features}}}{\sqrt{\dfrac{V_{\text{Psychrophilic}}}{\text{tot}_{\text{Psychrophilic}}} + \dfrac{V_{\text{Non-psychrophilic}}}{\text{tot}_{\text{Non-psychrophilic}}}}} \tag{1}$$

where

$SFM_{\text{Psychrophilic-sequence/structural features}}$ are the mean values of psychrophilic sequence/structural features

$S_{\text{Non-psychrophilic-sequence/structural features}}$ are the mean values of homologous non-psychrophilic sequence/ structural features

$V_{\text{Psychrophilic}}$ are the variances of sequence/structural features for psychrophilic protein sequences/ structures

$V_{\text{Non-psychrophilic}}$ are the variances of sequence/structural features for homologous non-psychrophilic protein sequences/structures

$\text{tot}_{\text{Psychrophilic}}$ is the total number of psychrophilic protein sequences/structures

$\text{tot}_{\text{Non-psychrophilic}}$ is the total number of non-psychrophilic proteins sequences/structures

Comparative structural analysis also revealed important differences in hydrophobic packing in adenylate kinases of *Bacillus globisporous* and its non-psychrophilic homologue *Bacillus subtilis*. The hydrophobic packing in the CORE region is quite loose which resulted in low thermal stability and increased flexibility (Moon et al. 2019). A similar study also confirmed the importance of hydrophobic packing in a comparative structural analysis involving Adenylate Kinases (AK) of *Nototheniacoriiceps* (psychrophile) and *Danio rerio* (mesophile) (Moon et al. 2017). It was also observed that in spite of having striking similarities of psychrophilic and mesophilic AKs, have different stabilities and noticeable difference in sequence composition at N and C terminal regions.

Inferences from comparative proteomics, genomics and molecular dynamics

Advances in computational proteomics and genomics also have a large impact in facilitating the understanding of psychrophilic adaptation (De Maayer et al. 2014). There exists some differences in the genomic structure of psychrophiles and non-psychrophiles and which are related to the various cold adaptive strategies (Ayala-del-Río et al. 2010; Rabus et al. 2004). Comparative genomics can also shed light on the presence or absence of some genomic structures, e.g., as is evident from the comparison of psychrophilic *Alteromonas* with its non-psychrophilic counterpart (Math et al. 2012). The advent of new techniques in genomics and proteomics has also facilitated in revealing the differential expression of genes, i.e., which genes are up-regulated or down-regulated according to the temperature surroundings, e.g., genes related to DNA replication, membrane adaptation are found to be differentially regulated (Rodrigues et al. 2008). There is an up-regulation of genes pertaining to the cell membrane formation, membrane transport and down-regulation of genes related to chemotaxis (Bakermans et al. 2007; Cacace et al. 2010; Durack et al. 2013).

Cold temperature generally results in lower reaction rates and to mitigate these effects, psychrophilic enzymes have developed higher flexibility as compared to their mesophilic counterpart, which can be localized either to the catalytic active sites or throughout the protein (D'Amico et al. 2006; Paredes et al. 2011; Rodrigues and Tiedje 2008). Arginine and proline residues are avoided as these are involved in conferring stability to the protein structure thereby reducing the flexibility (Ayala-del-Río et al. 2010; Zhao et al. 2010). Further favourable substitutions are those which resulted in the decrease in hydrogen bonding and salt bridges (Michaux et al. 2008; Siddiqui et al. 2006). A reduced number of hydrophobic contacts is also attributed to protein cold adaptation (Feller 2013; Gerday 2013; Santiago et al. 2016; Struvay and Feller 2012). Overall there is a decrease in the strength of various intramolecular interactions (Struvay and Feller 2012). A larger cleft at the catalytic site is also observed among psychrophilic enzymes (Russell et al. 1998).

Molecular Dynamics (MD) simulation studies can be very fruitful in facilitating the protein flexibility at different temperature ranges (Hansson et al. 2002). A MD simulation study was performed using homologues of psychrophilic, mesophilic and hyperthermophilic Tryptophan Synthase (TS). The study involved simulation at four different temperatures. Higher fluctuations were observed in psychrophilic TS in comparison to non-mesophilic and hyperthermophilic TSs. These fluctuations are more localized to the loop regions, which were found to be important in catalysis (Khan et al. 2016). Another important observation from the same study is the decrease in the number of hydrogen bonds with the increase in temperature in psychrophilic TSs and a reduction in the number of salt bridges and rigidity.

Machine learning based approaches

Machine learning algorithms are used for pattern extraction, which are then further employed for classification/recognition tasks. These techniques help to mine biologically relevant knowledge. Broadly machine learning techniques can be applied in two modes: Supervised learning mode, in which the class labels are known and are used for training the learning algorithms for prediction purposes (Kotsiantis 2007; Witten et al. 2011). The other is the unsupervised learning mode which does not require class labels and aims to find the natural groupings present in the dataset (e.g., clustering algorithms) (Ghahramani 2004; Nath and Subbiah 2016c; Witten et al. 2011). The process of clustering reads from metagenomic data and allocating them to operational taxonomic units is called binning. Unsupervised machine learning algorithms are also used for binning of microbial genomes (Nissen et al. 2018). In case of psychrophilic proteins, machine learning algorithms can be trained to predict psychrophilic proteins from a pool of unknown proteins or for discrimination of psychrophilic proteins from mesophilic/thermophilic proteins. The discrimination of psychrophilic proteins from its mesophilic/thermophilic counterpart can be defined as a two class classification problem (also known as binary classification problem), in which the class of interest is the psychrophilic protein being the positive class and all other non-psychrophilic proteins being the negative class.

Machine learning platforms

A plethora of machine learning algorithms are now available in various machine learning platforms for developing accurate prediction models that can be applied for prediction of psychrophilic proteins. Notable among the machine learning platform is WEKA (Hall et al. 2009), which is a java based platform and provides for implementation of a large number of both supervised and unsupervised algorithms. R and Python programming language also provides a rich set of machine learning platforms like Rattle (Maindonald 2012) and scikit-learn (Pedregosa et al. 2011) respectively. Both WEKA and Rattle provides graphical user interface apart from scripting which is very a practical property that can be used by biologists (non-programmers) to test different supervised/unsupervised algorithms with a click of a button.

Black box approaches and white box approaches

Most of the machine learning algorithms are like black boxes as they donot provide us with the explanation of how the classification is being performed, i.e., we do not get to know the mechanism of classification. These types of machine learning algorithms constitute the black box approach. Rule induction algorithms like Jrip (Cohen 1995), C4.5 (Salzberg 1994), Partial Decision Trees (PART) (Frank and Witten 1998), Classification and Regression Trees (CART) (Breiman et al. 1984), etc. constitute the white box approach, as they provide an explanation for their classification mechanism. These rule-induction algorithms provide a set of human interpretable rules which can aid in understanding how the classification is being made. Consequently, these rules constitute the interaction of sequence/structural features which can facilitate in the understanding of low temperature protein adaptation parameters.

Previously, rule induction has been applied for extraction of psychrophilic sequence signatures in (Nath and Subbiah 2014; Nath and Subbiah 2016a). Both of these methods consisted of two stages:

I stage (Predictive Modelling): In this step the best performing classifier was selected to filter out only true positives and true negatives to remove the noise present in the dataset selecting only true positives (psychrophilic proteins predicted as psychrophilic proteins) and true negatives (non-psychrophilic proteins predicted as non-psychrophilic proteins).

II stage (Rule Induction): The filtered sequences were subjected to rule induction using PART algorithm.

Most of the rule-induction algorithms generate rules having the representation (Verma et al. 2016) 'IF Antecedent 1, IF Antecedent 2, IF Antecedent 3, ..., IF Antecedent K" then Consequent' (here Antecedent is the Pre-Condition and Consequent is the Predicted Class). Where Antecedent 1 to K are the sequence/ structural features or both with their respective numerical values/ranges (if the features were discretized) and the consequent is the class (psychrophilic/non-psychrophilic). A sample output from J48 decision tree (based on C4.5 algorithm) is depicted in Fig. 10.2.

Each level splits the data based on the different feature values. Here the numbers 16, 12, 1, 6 20, 3 are the sequence or structural features. The rule from the sample tree can be interpreted as follows:

Rule: 1. If the feature **16** > 5.775764 and if feature **12** > 7.894734 and if feature **16** <= 8.885299 and if feature **20** <= 5.357143 then Class = **Mesophile**.

Rule: 2. If the feature **16** > 5.775764 and if feature **12** > 7.894734 and if feature **16** > 8.885299 then Class = **Psychrophile**.

Ensemble learning is an important paradigm of machine learning. The superiority of ensemble classifiers over individual classifiers have been demonstrated in previous research works (Kandaswamy et al. 2011; Nath et al. 2012; Nath and Subbiah 2014; Nath and Subbiah 2016c; Verma et al. 2016). Bagging (Breiman 1996), boosting (Freund and Schapire 1997; Schapire 1990)

Figure 10.2 Output tree from J48 decision trees (based on C4.5 algorithm) for the classification of psychrophilic and mesophilic protein sequences.

and random forest (Breiman 2001) are the noticeable ensemble learning techniques, which have found application in many biological classification problems. An ensemble learner consists of a group of individual base learners, trained on different regions of the input feature space (Polikar 2006). The final decision is reached by combining the outcomes of all the base learners either by majority voting or by some other suitable method (Kuncheva 2004). The psychrophilic protein prediction accuracy can be further improved using suitable ensemble learners.

Representation of protein sequences

One of the important parameters in the development of accurate prediction models is the representation of protein sequences that can then be fed into the machine learning algorithms. The type of representation that is used for training of the machine learning algorithms has a profound effect on its accuracy to classify sequences into the right classes. As the learning algorithms cannot be trained with sequence of alphabets (amino acid residue symbols), they have to be first changed into some fixed length numerical representation. Any measurable quantity of a protein sequence is known as its feature or attribute and can be used for its representation. For instance, amino acid composition is the simplest depiction of any protein sequence, giving the frequency of each of the 20 different amino acid residues. Many different types of numerical representation can be calculated for protein sequences ranging from dipeptide composition, tripeptide composition, biophysical properties, evolutionary information and pseudo amino acid composition (Chou 2004). Many user friendly servers are now available which can calculate a large number of features given a protein sequence, for ex. PROFEAT (Li et al. 2006), iFeature (Chen et al. 2018), propy (python package) (Cao et al. 2013), etc.

Feature reduction

If the number of features calculated for a given set of protein sequences is of a large dimension, then machine learning algorithms suffer from 'Curse of dimensionality' (Trunk 1979), which primarily results in deterioration in the performance of the machine learning algorithms. Not all the calculated features/attributes are equally important in classification, some of the features are non-informative or highly correlated with other features. To solve these issues, feature reduction or feature selection algorithms are used. The primary aim of feature selection algorithms is to select an optimal subset of features from the full set of calculated features which can result in enhanced prediction accuracy. These feature selection algorithms can be broadly divided into two categories: Filter methods (ReliefF (Kira and Rendell 1992), minimal redundancy, maximum relevance (mRMR) (Radovic et al. 2017), etc.), which calculates a feature importance score and ranks the features as per their discriminative ability and Wrapper methods (e.g., Sequential forward selection, Sequential backward selection) in which features are added or removed based on the enhancement in the classifier performance (Witten et al. 2011).

Performance evaluation methods and metrics

The assessment of machine learning models is an important step in the predictive analytic pipeline which helps us to select the best performing algorithm on a given dataset. K-fold cross validation or holdout testing set methods are mostly used for the purpose of performance evaluation. In K-fold cross validation, out of K folds, one fold is kept aside as a testing set and the K–1 folds are used as the training set. The method should be repeated until all the folds are used once as a test set. K fold variation of 5 and 10 fold cross validation are most popular in the research community. Leave One Out Cross Validation (LOOCV) is similar to K fold cross validation which involves taking all the samples but one as the training set and the left out single sample is used as the testing set. The process is repeated until all the samples are used once as a testing set. If the dataset is large enough, holdout testing set based evaluation can then applied. In the hold out method, a proportion of data is kept aside as a testing set which is mutually exclusive to the training dataset (usually 20–30% of data is used for the holdout testing dataset). After training of the machine learning algorithms on the training dataset, the hold out testing set is used for the evaluation. It is imperative to mention that if the dataset is small then either K folds cross validation or LOOCV is the preferred method.

Confusion matrix

The prediction output of machine learning algorithms can be described using a confusion matrix (also known as error matrix) (Ting 2017). Table 10.1 presents a representation of a confusion matrix for psychrophilic and non-psychrophilic protein sequence classifications. It consists of prediction results in the form of counts for the two classes. A psychrophilic protein classified as psychrophilic is a true positive, a psychrophilic protein classified as non-psychrophilic is the false negative, a non-psychrophilic protein classified as psychrophilic is the false positive, while a non-psychrophilic protein classified as non-psychrophilic is considered as the true negative. Various performance evaluation metrics can then be calculated from the confusion matrixes which are given below:

Sensitivity: the % of correctly predicted psychrophilic proteins

$$Sensitivity = \frac{TP}{(TP + FN)} \times 100 \tag{2}$$

Specificity: the % of correctly predicted non-psychrophilic proteins

$$Specificity = \frac{TN}{(TN + FP)} \times 100 \tag{3}$$

Accuracy: the % of correctly predicted psychrophilic and non-psychrophilic proteins

$$Accuracy = \frac{TP + TN}{TP + FP + TN + FN} \times 100 \tag{4}$$

Table 10.1 Confusion matrix for the prediction of psychrophilic and non-psychrophilic scenario.

	Predicted Class-Psychrophilic	**Predicted Class-Non-psychrophilic**
Actual Class Psychrophilic	TP	FN
Actual Class Non-Psychrophilic	FP	TN

Curios case of antifreeze proteins

Some organisms have developed a specialized group of proteins known as Antifreeze proteins (also known as ice nucleating proteins) (AFPs) which can inhibit the growth of ice crystals in freezing temperature surroundings and results in a phenomenon called thermal hysteresis (Nath et al. 2013). According to one hypothesis, these proteins evolved by the process of convergent evolution from different protein families (Fletcher et al. 2001; Logsdon and Doolittle 1997). They are found in a wide range of organisms from animals, plants, bacteria and fungi. AFPs were first reported by DeVries et al. (DeVries et al. 1970; DeVries and Wohlschlag 1969). In fishes alone there are five different subtypes of AFPs (type I to IV and Antifreeze Glycoproteins (AFGP)) which have no distinct similarities among them in terms of sequence and structure. Using the t-test, previously Nath et al. 2013 compared the sequence features of type II, type III and type IV AFP subtypes with their non-AFP homologues (C-lectins, proteins belonging to SAF superfamily and apolipoproteins respectively). A common theme of preference and avoidance is observed in type II, type III and type IV AFP subtypes, aromatic amino acid residues were found to be avoided, while small residues were preferred. The absence of significant structural and sequence similarity among the various subtypes makes the process of their characterization difficult. It is imperative to mention that sequence similarity-based search methods like BLAST (Altschul et al. 1990) and PSI-BLAST (Altschul et al. 1997) fail to predict new AFPs in the absence of detectable sequence and structural similarities. Machine learning methods are a suitable alternative in cases where text-based sequence similarity methods fail.

Sequence information in the form of amino acid composition, dipeptide composition, physicochemical properties, etc. and amino acid conservation (evolutionary information) in the form of Position Specific Scoring Matrices (PSSM) profiles were used for AFP prediction by earlier researchers (Doxey et al. 2006; Huan et al. 2013; Kandaswamy et al. 2011; Yu and Lu 2011; Zhao et al. 2012). In most of the biological classification problems, the datasets are imbalanced. Imbalanced datasets tend to have different ratios of instances belonging to different classes, e.g., AFP vs. Non-AFP classification presents a classical scenario of imbalanced datasets where the number of Non-AFPs (proteins other than AFPs) is quite large in comparison to AFPs. If such an imbalanced dataset is used for training of machine learning algorithms, the trained model becomes a majority class classifier (Nath and Subbiah 2016b; Wei and Dunbrack 2013). In such cases the learned classifier has diminished discriminating/predicting power for the minority class (i.e., AFP proteins).

There are two main approaches for dealing with class imbalance problems; (i) algorithmic level approaches, which involves changes at the algorithm implementation level to handle the imbalanced data, e.g., cost sensitive algorithms. (ii) data level, approaches which involves various resampling techniques. Among the resampling techniques, random under-sampling of the majority class or random over-sampling of the minority class are the simplest of the methods. Both random under-sampling and random over-sampling are prone to information loss. It may happen that during random under-sampling some informative instances are removed or during random oversampling some instances are overrepresented resulting in redundancy (Nath and Karthikeyan 2017). The construction of representative dataset free from redundancy is an important necessity for the proper training and evaluation of machine learning algorithms. The representative training dataset consists of samples from the full input space. Informative oversampling methods such as Synthetic Minority Oversampling Technique (SMOTE) (Chawla et al. 2002) and its variants (Han et al. 2005; Kovács 2019) can also be used to minimize the information loss and can mitigate the problem of imbalance. Other effective ways to handle dataset imbalance and representativeness is by using unsupervised K-means (Jain 2010; Jain et al. 1999) clustering algorithm. K-means based sampling allows for the allocation of representative samples in the training dataset, apart from handling the problem of difference in the number of samples belonging to the two classes.

Conclusion and future prospects

Cold habitats dominate the planet Earth and the source of psychrophilic proteins can range from algae, bacteria, archaea to higher animals and plants. With the ever-increasing number of sequences and structures of psychrophilic proteins in the public databases, there is an immediate need to implement data mining and statistical techniques for analyses and extraction of biologically relevant knowledge pertaining to cold temperature adaptation. Comparative statistical analysis between psychrophilic proteome/protein sequences/structures against homologous mesophilic proteomes/protein sequences/structures is the primary method for filtering out cold adaptation sequence and structural parameters. Machine learning techniques facilitate in annotation of biological sequences and are applied for developing accurate prediction systems. Further the rules generated from the rule induction algorithms can also facilitate in understanding the feature interaction and dominance of certain features in a particular prediction class. In comparison to thermophilic and hyperthermophilic proteins, there is still less understanding of protein adaptation to extremes of cold temperature surroundings. Both of the above *in silico* methods (comparative statistical analysis and machine learning based predictive and rule induction approaches) can facilitate in filtering the cold adaptation parameters and will broaden the understanding of their structural and functional stabilities. The knowledge of psychrophilic protein adaptation will help in the rational design of industrially important proteins which can remain functional and structurally stable at extremes of low temperature environment. With the advent of deep learning algorithms (LeCun et al. 2015; Nath and Karthikeyan 2018), it is hoped that the accuracy for psychrophilic protein prediction will further expand.

References

Achberger, A.M., A.B. Michaud, T.J. Vick-Majors, B.C. Christner, M.L. Skidmore, J.C. Priscu et al. 2017. Microbiology of subglacial environments. pp. 83–110. *In*: Margesin, R. (ed.). Psychrophiles: From Biodiversity to Biotechnology. Springer International Publishing, Cham. doi:10.1007/978-3-319-57057-0_5.

Altschul, S.F., W. Gish, W. Miller, E.W. Myers and D.J. Lipman. 1990. Basic local alignment search tool. J. Mol. Biol. 215: 403–410. doi:https://doi.org/10.1016/S0022-2836(05)80360-2.

Altschul, S.F., T.L. Madden, A.A. Schäffer, J. Zhang, Z. Zhang, W. Miller et al. 1997. Gapped BLAST and PSI-BLAST: a new generation of protein database search programs. Nucleic Acids Res. 25: 3389–3402. doi:10.1093/nar/25.17.3389.

Ayala-del-Río, H.L., P.S. Chain, J.J. Grzymski, M.A. Ponder, N. Ivanova, P.W. Bergholz et al. 2010. The genome sequence of *Psychrobacter arcticus* 273-4, a psychroactive siberian permafrost bacterium, reveals mechanisms for adaptation to low-temperature growth. Appl. Environ. Microbiol. 76: 2304–2312. doi:10.1128/aem.02101-09.

Bakermans, C., S.L. Tollaksen, C.S. Giometti, C. Wilkerson, J.M. Tiedje and M.F. Thomashow. 2007. Proteomic analysis of *Psychrobacter cryohalolentis* K5 during growth at subzero temperatures. Extremophiles 11: 343–354. doi:10.1007/s00792-006-0042-1.

Barroca, M., G. Santos, C. Gerday and T. Collins. 2017. Biotechnological aspects of cold-active enzymes. pp. 461–475. *In*: Margesin, R. (ed.). Psychrophiles: From Biodiversity to Biotechnology. Springer International Publishing, Cham. doi:10.1007/978-3-319-57057-0_19.

Berman, H.M., J. Westbrook, Z. Feng, G. Gilliland, T.N. Bhat, H. Weissig et al. 2000. The protein data bank. Nucleic Acids Res. 28: 235–242. doi:10.1093/nar/28.1.235.

Breiman, L., J. Friedman, C.J. Stone and R.A. Olshen. 1984. Classification and Regression Trees. Taylor & Francis.

Breiman, L. 1996. Bagging predictors. Machine Learning 24: 123–140. doi:10.1023/a:1018054314350.

Breiman, L. 2001. Random forests. Machine Learning 45: 5–32. doi:10.1023/a:1010933404324.

Brouchkov, A., V. Melnikov, L. Kalenova, O. Fursova, G. Pogorelko, V. Potapov et al. 2017. Permafrost bacteria in biotechnology: biomedical applications. pp. 541–554. *In*: Margesin, R. (ed.). Psychrophiles: From Biodiversity to Biotechnology. Springer International Publishing, Cham. doi:10.1007/978-3-319-57057-0_23.

Cacace, G., M.F. Mazzeo, A. Sorrentino, V. Spada, A. Malorni and R.A. Siciliano. 2010. Proteomics for the elucidation of cold adaptation mechanisms in *Listeria monocytogenes*. J. Proteomics 73: 2021–2030. doi:https://doi.org/10.1016/j.jprot.2010.06.011.

Cao, D.-S., Q.-S. Xu and Y.-Z. Liang. 2013. Propy: a tool to generate various modes of Chou's PseAAC. Bioinformatics 29: 960–962. doi:10.1093/bioinformatics/btt072.

Casanueva, A., M. Tuffin, C. Cary and D.A. Cowan. 2010. Molecular adaptations to psychrophily: the impact of omic technologies. Trends Microbiol. 18 (8): 374–381.

Chawla, N.V., K.W. Bowyer, L.O. Hall and W.P. Kegelmeyer. 2002. SMOTE: synthetic minority over-sampling technique. J. Artif. Int. Res. 16: 321–357.

Chen, I.M.A., V.M. Markowitz, K. Chu, K. Palaniappan, E. Szeto, M. Pillay et al. 2017. IMG/M: integrated genome and metagenome comparative data analysis system. Nucleic Acids Res. 45: D507–D516. doi:10.1093/nar/gkw929.

Chen, Z., P. Zhao, F. Li, A. Leier, T.T. Marquez-Lago, Y. Wang et al. 2018. iFeature: a Python package and web server for features extraction and selection from protein and peptide sequences. Bioinformatics 34: 2499–2502. doi:10.1093/bioinformatics/bty140.

Chou, K.-C. 2004. Using amphiphilic pseudo amino acid composition to predict enzyme subfamily classes. Bioinformatics 21: 10–19. doi:10.1093/bioinformatics/bth466.

Cohen, William W. 1995. Fast effective rule induction. *In*: A. Prieditis, and S. Russell (eds.). Machine Learning Proceedings 1995. San Francisco (CA): Morgan Kaufmann.

Collins, M.A. and R.K. Buick. 1989. Effect of temperature on the spoilage of stored peas by Rhodotorula glutinis. Food Microbiol. 6: 135–141. doi:https://doi.org/10.1016/S0740-0020(89)80021-8.

Cowan, D.A. 2009. Cryptic microbial communities in Antarctic deserts. Proc. Natl. Acad. Sci. U. S. A. 106: 19749–19750. doi:10.1073/pnas.0911628106.

D'Amico, S., T. Collins, J.-C. Marx, G. Feller, C. Gerday and C. Gerday. 2006. Psychrophilic microorganisms: challenges for life. EMBO Rep. 7: 385–389. doi:10.1038/sj.embor.7400662.

De Maayer, P., D. Anderson, C. Cary and D.A. Cowan. 2014. Some like it cold: understanding the survival strategies of psychrophiles. EMBO Rep. 15: 508–517. doi:10.1002/embr.201338170.

Deming, J.W. 2002. Psychrophiles and polar regions. Curr. Opin. Microbiol. 5: 301–309. doi:https://doi.org/10.1016/S1369-5274(02)00329-6.

Deming, J.W. 2007. Life in ice formations at very cold temperatures. *In*: Physiology and Biochemistry of Extremophiles. American Soc. Microbiol. doi:doi:https://doi.org/10.1128/9781555815813.ch10.

Deming, J.W. and R.E. Collins. 2017. Sea ice as a habitat for Bacteria, Archaea and viruses. pp. 326–351. *In*: Sea Ice. doi:10.1002/9781118778371.ch13.

DeVries, A.L. and D.E. Wohlschlag. 1969. Freezing resistance in some Antarctic fishes. Science 163: 1073–1075. doi:10.1126/science.163.3871.1073.

DeVries, A.L., S.K. Komatsu and R.E. Feeney. 1970. Chemical and physical properties of freezing point-depressing glycoproteins from Antarctic fishes. J. Biol. Chem. 245: 2901–2908.

Doxey, A.C., M.W. Yaish, M. Griffith and B.J. McConkey. 2006. Ordered surface carbons distinguish antifreeze proteins and their ice-binding regions. Nat. Biotechnol. 24: 852–855. doi:10.1038/nbt1224.

Durack, J., T. Ross and J.P. Bowman. 2013. Characterisation of the transcriptomes of genetically diverse *Listeria monocytogenes* exposed to hyperosmotic and low temperature conditions reveal global stress-adaptation mechanisms. PLOS ONE 8: e73603. doi:10.1371/journal.pone.0073603.

Fang, J., L. Zhang and D.A. Bazylinski. 2010. Deep-sea piezosphere and piezophiles: geomicrobiology and biogeochemistry. Trends Microbiol. 18: 413–422. doi:10.1016/j.tim.2010.06.006.

Feller, G., E. Narinx, J.L. Arpigny, M. Aittaleb, E. Baise, S. Genicot et al. 1996. Enzymes from psychrophilic organisms. FEMS Microbiol. Rev. 18: 189–202. doi:https://doi.org/10.1016/0168-6445(96)00011-3.

Feller, G. 2013. Psychrophilic enzymes: from folding to function and biotechnology. Scientifica 2013: 28. doi:10.1155/2013/512840.

Fletcher, G.L., C.L. Hew and P.L. Davies. 2001. Antifreeze proteins of teleost fishes. Annu. Rev. Physiol. 63: 359–390. doi:10.1146/annurev.physiol.63.1.359.

Frank, E. and I.H. Witten. 1998. Generating Accurate Rule Sets Without Global Optimization. Paper presented at the Proceedings of the Fifteenth International Conference on Machine Learning. Morgan Kaufmann Publishers Inc.

Freund, Y. and R.E. Schapire. 1997. A decision-theoretic generalization of on-line learning and an application to boosting. J. Comp. Syst. Sci. 55: 119–139. doi:https://doi.org/10.1006/jcss.1997.1504.

Georlette, D., B. Damien, V. Blaise, E. Depiereux, V.N. Uversky, C. Gerday et al. 2003. Structural and functional adaptations to extreme temperatures in psychrophilic, mesophilic, and thermophilic DNA ligases. J. Biol. Chem. 278: 37015–37023. doi:10.1074/jbc.M305142200.

Gerday, C. 2013. Psychrophily and catalysis. Biology 2: 719–741.

Ghahramani, Z. 2004. Unsupervised learning. pp. 72–112. *In*: Bousquet, O., U. von Luxburg and G. Rätsch (eds.). Advanced Lectures on Machine Learning: ML Summer Schools 2003, Canberra, Australia, February 2–14, 2003, Tübingen, Germany, August 4–16, 2003, Revised Lectures. Springer Berlin Heidelberg, Berlin, Heidelberg. doi:10.1007/978-3-540-28650-9_5.

Guan, Z., B. Tian, A. Perfumo and H. Goldfine. 2013. The polar lipids of Clostridium psychrophilum, an anaerobic psychrophile. Biochim. Biophys. Acta (BBA) – Molecular and Cell Biology of Lipids 1831: 1108–1112. doi:https://doi.org/10.1016/j.bbalip.2013.02.004.

Hall, M., E. Frank, G. Holmes, B. Pfahringer, P. Reutemann and I.H. Witten. 2009. The WEKA data mining software: an update. SIGKDD Explor. Newsl. 11: 10–18. doi:10.1145/1656274.1656278.

Han, H., W.-Y. Wang, B.-H. Mao. 2005. Borderline-SMOTE: A New Over-Sampling Method in Imbalanced Data Sets Learning. In, Berlin, Heidelberg. Advances in Intelligent Computing. Springer Berlin Heidelberg, pp. 878–887.

Hansson, T., C. Oostenbrink and W. van Gunsteren. 2002. Molecular dynamics simulations. Curr. Opin. Struct. Biol. 12: 190–196. doi:https://doi.org/10.1016/S0959-440X(02)00308-1.

Huan, W., L. Jun-Jie and L. Qian-Zhong. 2013. Motif analysis and identification of antifreeze protein sequences. *In*: Proceedings of the 2nd International Conference on Computer Science and Electronics Engineering, 2013/03 2013. Atlantis Press. doi:https://doi.org/10.2991/iccsee.2013.236.

Huang, S.-L., L.-C. Wu, H.-K. Liang, K.-T. Pan, J.-T. Horng and M.-T. Ko. 2004. PGTdb: a database providing growth temperatures of prokaryotes. Bioinformatics 20: 276–278. doi:10.1093/bioinformatics/btg403.

Jahandideh, M., S.M.H. Barkooie, S. Jahandideh, P. Abdolmaleki, M.M. Movahedi, S. Hoseini et al. 2008. Elucidating the protein cold-adaptation: Investigation of the parameters enhancing protein psychrophilicity. J. Theor. Biol. 255: 113–118. doi:https://doi.org/10.1016/j.jtbi.2008.07.034.

Jahandideh, S., P. Abdolmaleki, M. Jahandideh and E. Barzegari Asadabadi. 2007a. Sequence and structural parameters enhancing adaptation of proteins to low temperatures. J. Theor. Biol. 246: 159–166. doi:https://doi.org/10.1016/j.jtbi.2006.12.008.

Jahandideh, S., E. Barzegari Asadabadi, P. Abdolmaleki, M. Jahandideh and S. Hoseini. 2007b. Protein psychrophilicity: Role of residual structural properties in adaptation of proteins to low temperatures. J. Theor. Biol. 248: 721–726. doi:https://doi.org/10.1016/j.jtbi.2007.06.019.

Jain, A.K., M.N. Murty and P.J. Flynn. 1999. Data clustering: a review. ACM Comput. Surv. 31: 264–323. doi:10.1145/331499.331504.

Jain, A.K. 2010. Data clustering: 50 years beyond K-means. Pattern Recog. Lett. 31: 651–666. doi:https://doi.org/10.1016/j.patrec.2009.09.011.

Junge, K., K. Cameron and B. Nunn. 2019. Chapter 12—Diversity of psychrophilic bacteria in sea and glacier ice environments—insights through genomics, metagenomics, and proteomics approaches. pp. 197–216. *In*: Das, S. and H.R. Dash (eds.). Microbial Diversity in the Genomic Era. Academic Press, doi:https://doi.org/10.1016/B978-0-12-814849-5.00012-5.

Kandaswamy, K.K., K.-C. Chou, T. Martinetz, S. Möller, P.N. Suganthan, S. Sridharan et al. 2011. AFP-Pred: A random forest approach for predicting antifreeze proteins from sequence-derived properties. J. Theor. Biol. 270: 56–62. doi:https://doi.org/10.1016/j.jtbi.2010.10.037.

Khan, S., U. Farooq and M. Kurnikova. 2016. Exploring protein stability by comparative molecular dynamics simulations of homologous hyperthermophilic, mesophilic, and psychrophilic proteins. Journal of Chemical Information and Modeling 56: 2129–2139. doi:10.1021/acs.jcim.6b00305.

Kira, K. and L.A. Rendell. 1992. A practical approach to feature selection. pp. 249–256. *In*: Sleeman, D. and P. Edwards (eds.). Machine Learning Proceedings 1992. Morgan Kaufmann, San Francisco (CA). doi:https://doi.org/10.1016/B978-1-55860-247-2.50037-1.

Klähn, S. and M. Hagemann. 2011. Compatible solute biosynthesis in cyanobacteria. Environ. Microbiol. 13: 551–562. doi:10.1111/j.1462-2920.2010.02366.x.

Kotsiantis, S.B. 2007. Supervised Machine Learning: A Review of Classification Techniques. Paper presented at the Proceedings of the 2007 conference on Emerging Artificial Intelligence Applications in Computer Engineering: Real Word AI Systems with Applications in eHealth, HCI, Information Retrieval and Pervasive Technologies. IOS Press.

Kovács, G. 2019. Smote-variants: A python implementation of 85 minority oversampling techniques. Neurocomputing 366: 352–354. doi:https://doi.org/10.1016/j.neucom.2019.06.100.

Kuncheva, L.I. 2004. Combining Pattern Classifiers: Methods and Algorithms. Wiley-Interscience.

LeCun, Y., Y. Bengio and G. Hinton. 2015. Deep learning. Nature 521: 436. doi:10.1038/nature14539.

Leinonen, R., R. Akhtar, E. Birney, L. Bower, A. Cerdeno-Tárraga, Y. Cheng et al. 2011. The European nucleotide archive. Nucleic Acids Res. 39: D28–D31. doi:10.1093/nar/gkq967.

Li, Z.R., H.H. Lin, L.Y. Han, L. Jiang, X. Chen and Y.Z. Chen. 2006. PROFEAT: a web server for computing structural and physicochemical features of proteins and peptides from amino acid sequence. Nucleic Acids Res. 34: W32–W37. doi:10.1093/nar/gkl305.

Logsdon, J.M. and W.F. Doolittle. 1997. Origin of antifreeze protein genes: A cool tale in molecular evolution. Proceedings of the National Academy of Sciences 94: 3485–3487. doi:10.1073/pnas.94.8.3485.

Maindonald, J.H. 2012. Data mining with rattle and R: The art of excavating data for knowledge discovery by Graham Williams. International Statistical Review 80: 199–200. doi:10.1111/j.1751-5823.2012.00179_23.x.

Math, R.K., H.M. Jin, J.M. Kim, Y. Hahn, W. Park, E.L. Madsen et al. 2012. Comparative genomics reveals adaptation by Alteromonas sp. SN2 to marine tidal-flat conditions: cold tolerance and aromatic hydrocarbon metabolism. PLOS ONE 7: e35784. doi:10.1371/journal.pone.0035784.

Metpally, R.P.R. and B.V.B. Reddy. 2009. Comparative proteome analysis of psychrophilic versus mesophilic bacterial species: Insights into the molecular basis of cold adaptation of proteins. BMC Genomics 10: 11. doi:10.1186/1471-2164-10-11.

Michaux, C., J. Massant, F. Kerff, J.-M. Frère, J.-D. Docquier, I. Vandenberghe et al. 2008. Crystal structure of a cold-adapted class C β-lactamase. The FEBS Journal 275: 1687–1697. doi:10.1111/j.1742-4658.2008.06324.x.

Moon, S., J. Kim and E. Bae. 2017. Structural analyses of adenylate kinases from Antarctic and tropical fishes for understanding cold adaptation of enzymes. Scientific Reports 7: 16027. doi:10.1038/s41598-017-16266-9.

Moon, S., J. Kim, J. Koo and E. Bae. 2019. Structural and mutational analyses of psychrophilic and mesophilic adenylate kinases highlight the role of hydrophobic interactions in protein thermal stability. Structural Dynamics 6: 024702. doi:10.1063/1.5089707.

Mukherjee, S., D. Stamatis, J. Bertsch, G. Ovchinnikova, H.Y. Katta, A. Mojica et al. 2018. Genomes OnLine database (GOLD) v.7: updates and new features. Nucleic Acids Res. 47: D649–D659. doi:10.1093/nar/gky977.

Nath, A., R. Chaube and S. Karthikeyan. 2012. Discrimination of psychrophilic and mesophilic proteins using random forest algorithm. pp. 179–182. *In*: 2012 International Conference on Biomedical Engineering and Biotechnology, 28–30 May 2012. doi:10.1109/iCBEB.2012.151.

Nath, A., R. Chaube and K. Subbiah. 2013. An insight into the molecular basis for convergent evolution in fish antifreeze Proteins. Comput. Biol. Med. 43: 817–821. doi:https://doi.org/10.1016/j.compbiomed.2013.04.013.

Nath, A. and K. Subbiah. 2014. Inferring biological basis about psychrophilicity by interpreting the rules generated from the correctly classified input instances by a classifier. Comput. Biol. Chem. 53: 198–203. doi:https://doi.org/10.1016/j.compbiolchem.2014.10.002.

Nath, A. and K. Subbiah. 2016a. Insights into the molecular basis of piezophilic adaptation: Extraction of piezophilic signatures. J. Theor. Biol. 390: 117–126. doi:https://doi.org/10.1016/j.jtbi.2015.11.021.

Nath, A. and K. Subbiah. 2016b. Probing an optimal class distribution for enhancing prediction and feature characterization of plant virus-encoded RNA-silencing suppressors. 3 Biotech 6: 93–93. doi:10.1007/s13205-016-0410-1.

Nath, A. and K. Subbiah. 2016c. Unsupervised learning assisted robust prediction of bioluminescent proteins. Comput. Biol. Med. 68: 27–36. doi:https://doi.org/10.1016/j.compbiomed.2015.10.013.

Nath, A. and S. Karthikeyan. 2017. Enhanced prediction and characterization of CDK inhibitors using optimal class distribution. Interdisciplinary Sciences: Computational Life Sciences 9: 292–303. doi:10.1007/s12539-016-0151-1.

Nath, A. and S. Karthikeyan. 2018. Enhanced prediction of recombination hotspots using input features extracted by class specific autoencoders. J. Theor. Biol. 444: 73–82. doi:https://doi.org/10.1016/j.jtbi.2018.02.016.

Nissen, J.N., C.K. Sønderby, J.J.A. Armenteros, C.H. Grønbech, H.B. Nielsen, T.N. Petersen et al. 2018. Binning microbial genomes using deep learning. bioRxiv: 490078. doi:10.1101/490078.

Paredes, D.I., K. Watters, D.J. Pitman, C. Bystroff and J.S. Dordick. 2011. Comparative void-volume analysis of psychrophilic and mesophilic enzymes: Structural bioinformatics of psychrophilic enzymes reveals sources of core flexibility. BMC Struct. Biol. 11: 42. doi:10.1186/1472-6807-11-42.

Pedregosa, F., G. Varoquaux, A. Gramfort, V. Michel and B. Thirion. 2011. Scikit-learn: machine learning in python. J. Mach. Learn. Res. 12: 2825–2830.

Pezeshgi Modarres, H., M.R. Mofrad and A. Sanati-Nezhad. 2018. ProtDataTherm: A database for thermostability analysis and engineering of proteins. PLOS ONE 13: e0191222. doi:10.1371/journal.pone.0191222.

Piette, F., S. Amico, G. Mazzucchelli, A. Danchin, P. Leprince and G. Feller. 2011. Life in the Cold: a Proteomic Study of Cold-Repressed Proteins in the Antarctic Bacterium *Pseudoalteromonas haloplanktis* TAC125. Appl. Environ. Microbiol. 77(11): 3881.

Polikar, R. 2006. Ensemble based systems in decision making. IEEE Circuits and Systems Magazine 6: 21–45. doi:10.1109/MCAS.2006.1688199.

Rabus, R., A. Ruepp, T. Frickey, T. Rattei, B. Fartmann, M. Stark et al. 2004. The genome of *Desulfotalea psychrophila*, a sulfate-reducing bacterium from permanently cold Arctic sediments. Environ. Microbiol. 6: 887–902. doi:10.1111/j.1462-2920.2004.00665.x.

Radovic, M., M. Ghalwash, N. Filipovic and Z. Obradovic. 2017. Minimum redundancy maximum relevance feature selection approach for temporal gene expression data. BMC Bioinformatics 18: 9–9. doi:10.1186/s12859-016-1423-9.

Rana, K.L., D. Kour, T. Kaur, I. Sheikh, A.N. Yadav, V. Kumar et al. 2020. Endophytic microbes from diverse wheat genotypes and their potential biotechnological applications in plant growth promotion and nutrient uptake. Proc. Natl. Acad. Sci. India, Sect. B Biol. Sci. doi:10.1007/s40011-020-01168-0.

Rodrigues, D.F., N. Ivanova, Z. He, M. Huebner, J. Zhou and J.M. Tiedje. 2008. Architecture of thermal adaptation in an Exiguobacterium sibiricum strain isolated from 3 million year old permafrost: A genome and transcriptome approach. BMC Genomics 9: 547. doi:10.1186/1471-2164-9-547.

Rodrigues, D.F. and J.M. Tiedje. 2008. Coping with our cold planet. Appl. Environ. Microbiol. 74: 1677. doi:10.1128/AEM.02000-07.

Russell, R.J.M., U. Gerike, M.J. Danson, D.W. Hough and G.L. Taylor. 1998. Structural adaptations of the cold-active citrate synthase from an Antarctic bacterium. Structure 6: 351–361. doi:10.1016/S0969-2126(98)00037-9.

Salzberg, S.L. 1994. C4.5: Programs for Machine Learning by J. Ross Quinlan. Morgan Kaufmann Publishers, Inc., 1993. Machine Learning 16: 235–240. doi:10.1007/bf00993309.

Santiago, M., C.A. Ramírez-Sarmiento, R.A. Zamora and L.P. Parra. 2016. Discovery, molecular mechanisms, and industrial applications of cold-active enzymes. Front. Microbiol. 7: doi:10.3389/fmicb.2016.01408.

Schapire, R.E. 1990. The strength of weak learnability. Machine Learning 5: 197–227. doi:10.1007/bf00116037.

Schroeter, B. and C. Scheidegger. 1995. Water relations in lichens at subzero temperatures: structural changes and carbon dioxide exchange in the lichen Umbilicaria aprina from Continental Antarctica. New Phytol. 131: 273–285.

Shokrollahzade, S., F. Sharifi, A. Vaseghi, M. Faridounnia and S. Jahandideh. 2015. Protein cold adaptation: Role of physico-chemical parameters in adaptation of proteins to low temperatures. J. Theor. Biol. 383: 130–137. doi:https://doi.org/10.1016/j.jtbi.2015.07.013.

Siddiqui, K.S. and R. Cavicchioli. 2006. Cold-adapted enzymes. Annu. Rev. Biochem. 75: 403–433. doi:10.1146/annurev.biochem.75.103004.142723.

Siddiqui, K.S., A. Poljak, M. Guilhaus, D. De Francisci, P.M.G. Curmi, G. Feller et al. 2006. Role of lysine versus arginine in enzyme cold-adaptation: Modifying lysine to homo-arginine stabilizes the cold-adapted α-amylase from *Pseudoalteramonas haloplanktis*. Proteins: Structure, Function, and Bioinformatics 64: 486–501. doi:10.1002/prot.20989.

Struvay, C. and G. Feller. 2012. Optimization to low temperature activity in psychrophilic enzymes. International Journal of Molecular Sciences 13: 11643–11665.

Timmis, K.N. and D.H. Pieper. 1999. Bacteria designed for bioremediation. Trends Biotechnol. 17: 201–204. doi:https://doi.org/10.1016/S0167-7799(98)01295-5.

Ting, K.M. 2017. Confusion matrix. pp. 260–260. *In*: Sammut, C. and G.I. Webb (eds.). Encyclopedia of Machine Learning and Data Mining. Springer US, Boston, MA. doi:10.1007/978-1-4899-7687-1_50.

Trunk, G.V. 1979. A problem of dimensionality: a simple example. IEEE Trans. Pattern Anal. Mach. Intell. PAMI-1: 306–307. doi:10.1109/TPAMI.1979.4766926.

van den Burg, B. 2003. Extremophiles as a source for novel enzymes. Curr. Opin. Microbiol. 6: 213–218. doi.https://doi. org/10.1016/S1369-5274(03)00060-2.

Verma, P., A.N. Yadav, K.S. Khannam, N. Panjiar, S. Kumar, A.K. Saxena et al. 2015. Assessment of genetic diversity and plant growth promoting attributes of psychrotolerant bacteria allied with wheat (*Triticum aestivum*) from the northern hills zone of India. Ann. Microbiol. 65: 1885–1899.

Verma, P., A.N. Yadav, K.S. Khannam, S. Kumar, A.K. Saxena and A. Suman. 2016. Molecular diversity and multifarious plant growth promoting attributes of Bacilli associated with wheat (*Triticum aestivum* L.) rhizosphere from six diverse agro-ecological zones of India. J. Basic Microbiol. 56: 44–58.

Wei, Q. and R.L. Dunbrack, Jr. 2013. The role of balanced training and testing data sets for binary classifiers in bioinformatics. PloS One 8: e67863–e67863. doi:10.1371/journal.pone.0067863.

Wilke, A., J. Bischof, W. Gerlach, E. Glass, T. Harrison, K.P. Keegan et al. 2016. The MG-RAST metagenomics database and portal in 2015. Nucleic Acids Res. 44: D590–D594. doi:10.1093/nar/gkv1322.

Witten, I.H., E. Frank and M.A. Hall. 2011. Data Mining: Practical Machine Learning Tools and Techniques. Morgan Kaufmann Publishers Inc.

Yadav, A.N. 2015. Bacterial diversity of cold deserts and mining of genes for low temperature tolerance. Ph.D. Thesis, IARI, New Delhi/BIT, Ranchi pp. 234, DOI: 10.13140/RG.2.1.2948.1283/2.

Yadav, A.N., S.G. Sachan, P. Verma and A.K. Saxena. 2015a. Prospecting cold deserts of north western Himalayas for microbial diversity and plant growth promoting attributes. J. Biosci. Bioeng. 119: 683–693.

Yadav, A.N., S.G. Sachan, P. Verma, S.P. Tyagi, R. Kaushik and A.K. Saxena. 2015b. Culturable diversity and functional annotation of psychrotrophic bacteria from cold desert of Leh Ladakh (India). World J. Microbiol. Biotechnol. 31: 95–108.

Yadav, A.N., P. Verma, M. Kumar, K.K. Pal, R. Dey, A. Gupta et al. 2015c. Diversity and phylogenetic profiling of niche-specific Bacilli from extreme environments of India. Ann. Microbiol. 65: 611–629.

Yadav, A.N., S.G. Sachan, P. Verma, R. Kaushik and A.K. Saxena. 2016a. Cold active hydrolytic enzymes production by psychrotrophic Bacilli isolated from three sub-glacial lakes of NW Indian Himalayas. J. Basic Microbiol. 56: 294–307.

Yadav, A.N., S.G. Sachan, P. Verma and A.K. Saxena. 2016b. Bioprospecting of plant growth promoting psychrotrophic Bacilli from cold desert of north western Indian Himalayas. Indian J. Exp. Biol. 54: 142–150.

Yadav, A.N., D. Kour, S. Sharma, S.G. Sachan, B. Singh, V.S. Chauhan et al. 2019. Psychrotrophic microbes: biodiversity, mechanisms of adaptation, and biotechnological implications in alleviation of cold stress in plants. pp. 219–253. *In*: Sayyed, R.Z., N.K. Arora and M.S. Reddy (eds.). Plant Growth Promoting Rhizobacteria for Sustainable Stress Management: Volume 1: Rhizobacteria in Abiotic Stress Management. Springer Singapore, Singapore. doi:10.1007/978-981-13-6536-2_12.

Yadav, A.N., P. Verma, S.G. Sachan, R. Kaushik and A.K. Saxena. 2018. Psychrotrophic microbiomes: molecular diversity and beneficial role in plant growth promotion and soil health. pp. 197–240. *In*: Panpatte, D.G., Y.K. Jhala, H.N. Shelat and R.V. Vyas (eds.). Microorganisms for Green Revolution-Volume 2: Microbes for Sustainable Agro-ecosystem. Springer, Singapore. doi:10.1007/978-981-10-7146-1_11.

Yu, C.-S. and C.-H. Lu. 2011. Identification of antifreeze proteins and their functional residues by support vector machine and genetic algorithms based on n-peptide compositions. PLOS ONE 6: e20445. doi:10.1371/journal.pone.0020445.

Zhao, J.-S., Y. Deng, D. Manno and J. Hawari. 2010. Shewanella spp. genomic evolution for a cold marine lifestyle and *in-situ* explosive biodegradation. PLOS ONE 5: e9109. doi:10.1371/journal.pone.0009109.

Zhao, X., Z. Ma and M. Yin. 2012. Using support vector machine and evolutionary profiles to predict antifreeze protein sequences. Int. J. Mol. Sci. 13: 2196–2207. doi:10.3390/ijms13022196.

Chapter 11

Psychrophilic Microbiomes

Biodiversity, Molecular Adaptations and Applications

K. Kamala,[1] P. Sivaperumal,[2,]* S. Manjunath Kamath,[3] P. Kumar[4] and Richard Thilagaraj[1]

IIIIIIIIIIIIIIIIIIIIIIIIIIIIIIIIIIIIIIIIIIIIIIIIIIIIIIIIIIIIIIIIIIIIIIIIIIIIIIIIIIIIIIIIIIIIIIIIIIIIIIIIIIIIIIIIIIIIIIIIIIIIIIIIIIIIIIIIIIIIIIIIIIIIIIIIIIIIIIIIIIIIIIIIIIII

Introduction

Microbes living in the lowest temperature ranges are called psychrophiles. As most of our universal is largely cold (lower than 5°C), this indicates that extremophiles are common within an extensive range of habitats. To survive in extreme cold environments, microbes have developed their cellular components by a number of adaptations such as, protein synthesis, enzyme production, energy generating system, limited influx and out flow of the nutrient. These adaptive modification of proteins and lipids of cell membranes can be considered as prime adaptation. Moreover, microbial thriving in any environment which is not viable for existence is known as extremophiles. These environments are classified by physical and chemical factors like temperature (Thermophiles and psychrophiles), pH (Acidophiles and Alkaliphiles), Pressure (Pizophiles), radiation (Radioresistant) Dessication (Xerophiles), Salinity (Halophiles), density (Barophiles) and oxygen (Anaerobes). Now recent studies have revealed that these organisms are polyextremophiles like the microbes in hydrothermal vent/volcano which has an extreme temperature in high pressure and density with an alkaline condition (Rampelotto 2013; Rausk et al. 2017). Generally microbes are easily adapted to any environment, among them a few microbes can adapt to various extreme environments (Yadav et al. 2019a; Yadav et al. 2018). The present chapter concurs to the brief indication of the extremophilic explanation and diversity and description of habitats, the importance on the adaptive alteration in lipids, enzymes and proteins that could be commercially useful for future prospects.

---

[1] Department of Biotechnology, School of Bioengineering, SRM Institute of Science and Technology, Kattankulathur-602203, India.

[2] Department of Pharmacology, Saveetha Dental College and Hospitals, Saveetha Institute of Medical & Technical Sciences, Chennai, Tamil Nadu, India.

[3] Department of Translational Medicine and Research, SRM Medical College, SRMIST, Kattankulathur-603203, Tamilnadu, India.

[4] Center for Environmental Nuclear Research, Directorate of Research, SRMIST, Kattankulathur-603203, Tamilnadu, India.

* Corresponding author: marinesiva86@gmail.com

Biodiversity of psychrophilic microbes

Culture-dependent psychrophilic microbial diversity

The study of microbes from the Antarctica is important because it is an extreme environment, which has not yet been extensively investigated but is believed to have a large diversity of organisms that have not yet been discovered (Tindall 2004). Kochkina et al. (2001) investigated the microbial flora of the Arctic and Antarctic permafrost sediments and most of the bacterial group consisted of *Actinobacter, Micrococcus, Rhodococcus, Microbacterium, Streptomyces* and *Cryobacterium* at 15 to 18°C. They reported that the unique structural, biochemical and molecular biological organization of these microbes were responsible for their higher viability in the permafrost sediments. A totally 47 actinobacterial isolates were obtained from the Antarctic soils, among which, 19 showed activity of antagonistic against both gram positive and negative bacteria. Among them, 6 isolates showed extreme antibacterial activities. The physiological and biochemical results of 39 characteristics confirm that the isolates belonged to *Streptomyces, Actinomadura* and *Kitasatospora* (Moncheva et al. 2002).

Novel and rare genera were isolated from the Antarctic samples and most of them belonged to *Streptomyces, Nocardia* and *Micromonospora,* which possessed pharmaceutically active compounds (Nichols et al. 2002). Similarly, Willerslev et al. (2004) isolated 17 genera from various permafrost samples with the prevalence of gram positive representative of *Firmicutes* and *Actinobacteria.* Densita and Mariana (2005) reported 40 acatinobacterial strains from the Antarctica which was tested for antagonistic activity against gram positive, gram negative, yeast and phytopathogenic fungi. They found that four strains (8013, 3787, 3718 and 8007) showed good antibacterial activity and so, they were recommended for plant protection.

Miteva (2008) reported different types of marine bacterium of *Rhodococcus, Arthrobacter, Sphingomonas, Exiguobacterium, Frigoribacterium, Janthinobacterium Methylobacterium, Chryseobacterium* and *Acinetobacter* from glacier ice. Christner et al. (2008) analyzed the microbial diversity of universal non polar and polar region using the 16S rRNA gene sequence and constantly distinguished representatives of similar genera of proteobacteria, cytophaga-Flavobacteria and actinobacteria. With this finding, they suggested that these microbes might possess similar survival mechanisms. Microbial diverse and their metabolism on ice shelves in the Arctic were detected by Bottos et al. (2008) at –10°C. A major fraction of bacteria, constituted by *Bacteroides, Proteobacteria, Actinobateria* and *Euryarchaeota* represented the archeal communities.

In the Antarctic cold desert, actinobacteria comprised the major phylogenetic group and the maximum isolates were *Streptomycetes* that had low frequency in the metagenomics analysis (Babalola et al. 2009). Gesheva et al. (2010) reported one psychrophilic *Streptomyces* sp. possessing antibiotic activity against gram positive bacteria, yeasts and fungi. This strain produced three antibiotics (azalomycin, nigericin and nonpolymeric macrolide). In the same year, Zenova et al. (2010) isolated psychrotolerant *Streptomyces* from peat bog soils of tundra at a temperature below 10°C. This isolate demonstrated antagonism against bacteria (*Bacillus, Pseudomonas* and *Rhodococcus*) yeast (*Saccharomyces cerevisae*) and psychrotolerant fungi (*Mucor hiemalis*) and *Cladosporium herbarum* at 5°C. This strain also possessed pectinolytic and amylolytic activities.

Halo tolerant *Rhodococcus fascians* was identified from the Antarctic soil by Gesheva et al. (2010) and synthesized glycolipid biosurfactant and that the surfactant was used for bioremediation of hydrocarbon contaminated sediments and soils. Further, the isolates were characterized by the presence of hydrocarbons using kerosene and glucose as the only carbon sources (in the medium) for biosurfactant synthesis. Molecular identification was done based on the 16S rDNA gene sequencing and the strain was found to be closely matched with *Rhodococcus fascians* with 100% sequence similarity. Similarly, psychrophilic and mesophilic microorganisms, viz. *Streptomyces, Nocardia* and *Geodermatophilus,* identified from the Antarctic soils and glycolipids was biosynthesized with potential emulsifying performance and was used for bioremediation of oil spills in ice temperatures (Gesheva and Negotia 2012).

Recently, Echauri et al. (2011) obtained 260 culturable psychrophilic prokaryotes from the sediment samples (glacier), melted ice (glacier) and seaside mud from the Antarctica region. Taxonomic classification, based on the molecular approach, revealed that the most dominant group of β-proteobacteria had 35.2%, followed by γ-proteobacteria which had 18.5%, α-proteobacteria had 16.6% and Gram-positive microbes had the highest content of GC (13%) and CFB (13%), and only gram-positive microbes with low GC content (3.7%) were also observed. The high GC content having Gram positive group belonged to *Nakamurellaceae* and *Micobacteriaceae*. Li et al. (2011) isolated a novel actinobacterial strain GW25-5T from west Antarctica. On polyphasic taxonomic evidence, this strain was assigned to *Streptomyces fildesensis* sp. nov.

Prospecting the cold habitats of the Indian Himalayas has led to the isolation of a great diversity of microorganisms which was found to be useful for different applications in agriculture and industry. Culturable bacterial diversity of psychrophilic and psychrotrophic microbes from high altitude regions of the Indian Himalayas were studied (Verma et al. 2015; Yadav et al. 2015a; Yadav et al. 2015b; Yadav et al. 2015c; Yadav et al. 2019b). There are many reports on complete genome sequences to know the different genes responsible for diverse attributes including *Colwellia chukchiensi* (Zhang et al. 2018), *Exiguobacterium oxidotolerans* (Cai et al. 2017), *Arthrobacter agilis* (Singh et al. 2016), *Paenibacillus* sp. (Dhar et al. 2016), *Clavibacter* sp. (Du et al. 2015), *Planomicrobium glaciei* (Salwan et al. 2014), *Octadecabacter antarcticus* (Vollmers et al. 2013), *Exiguobacterium antarcticum* (Carneiro et al. 2012), *Rheinheimera* sp. (Gupta et al. 2011), *Methanococcoides burtonii* (Allen et al. 2009), *Exiguobacterium sibiricum* (Rodrigues et al. 2008), *Cenarchaeum symbiosum* (Hallam et al. 2006) and *Colwellia psychrerythraea* (Methé et al. 2005).

Culture-independent psychrophilic microbial diversity

With the beginning of molecular phylogeny, assessment of the unculturable microorganisms in the environment led to biodiversity surveys, based on the 16S rRNA gene sequences of bacteria and archea across the polar and other cold environments (Karner et al. 2001; Mock and Thomas 2005). The study of the actinobacterial strains from the Antarctica is important because it is an extreme environment that has not been extensively studied, but is believed to have a large diversity of organisms that have yet to be discovered (Tindall 2004). Abundant bacterial species have been identified by some authors around Antarctic soil and those microbes are fall under similar taxonomic classes. But, in latest years, with the initiation of metagenomic technology (Delmont et al. 2011; Eisen 2007; Hugenholz et al. 1998; Handelsman et al. 1998) were developed for identifying the microbes. It is also possible to make a further inclusive assessment of the psychrophilic microbial biodiversity available in these particular environments. Moreover, to determine some of the potential geochemical functions, these microbial communities might be used (Pearce et al. 2012).

Use of molecular approaches for describing microbial diversity has greatly enhanced the knowledge of the population structure in natural microbial communities. In recent years, application of rRNA sequence analysis to study the systematics of the actinobacteria has improved the understanding of the taxonomy of this phylum. Venter et al. (2004) identified nine major bacterial phyla (*Actinobacteria, Proteobacteria, Cyanobacteria, Bacteroidetes, Firmicutes, Spirochaetes, Chloroflexi, Fusobacteria* and *Deinococcus-Thermus*) and two phyla of archeal (*Euryarchaeota* and *Crenarchaeota*) from the surface water samples of the Sargasso Sea. The majority of the sequences from 1800 bacterial isolates comprising 148 new phylotypes have been clustered with alpha and gamma division of the *Proteobacteria* phylum. Wegley et al. (2007) identified the microbial isolates from coral reef environment samples through metagenomic analysis and the sequences were aligned, among them, bacteria comprised 7%, archaea have 1%, eukaryotic viruses covered with 2% and phages were comprised with 3%. The bacterial isolates were identified and sequences were comparable to *Actinobacteria, Proteobacteria, Cyanobacteria* and *Firmicutes*.

Babalola et al. (2009) evaluated the actinobacterial diversity around Antarctic desert soils using traditional cultivable techniques as well as the metagenomic approach. Phylogenetic analysis was done and the clone showed similarity with actinobacteria. The streptomycete-specific PCR primers were revealed and most of the phylotypes was closely related to uncultured *Nocardioides* and *Pseudonocardia* species and some rare actinobacterial genera containing *Modestobacter, Geodermatophilus* and *Sporichthya*. Simon et al. (2009) assessed phylogenetic alignment of the prokaryotic community by evaluation of pyrosequenced data set from 16S rRNA gene from glacial ice of the Northern Scheneeferner, Germany. The predominant groups were *Proteobacteria* (mainly *Betaproteobacteria*), *Bacteroidetes* and *Actinobacteria* in addition to 13 different bacterial groups.

Pearce et al. (2010) isolated air-borne microbes from the continental shelf of the Antarctica and very little microbial diversity was noticed with many sequence replicates and many uncultivable sequences. In their study, they found that assembling a synthetic bacterial community and sequencing with the metagenome approach showed partial results, which suggested that DNA preparation methodology as well as samples prepared with different protocols will not be appropriate for relative metagenomics (Morgan et al. 2010).

Prokaryotic communities were also identified by Bruneel et al. (2011) from creek samples based on the metaproteomic approach of the sediment samples, using 16S rRNA gene encoding library. Many of them were beta proteobacteria such as *Thiomonas* and *Gallionella*. Gama proteobacteria such as *ferrooxidans* and *Acidithiobacillus* and alpha proteobacteria such as *Actinobacteria, Acidiphilium* and *Firmicutes* were also identified. Similarly, Cerritos et al. (2011) identified 89 actinobacterial strains and 6 strains from thermo resistant *Proteobacteria*, observed in 22 different clusters based on the phylogenetic analysis.

Emilio et al. (2012) determined the microbial community alignment using microscopic counts, pigment analysis and further 16S rRNA gene finger printing in Spain. On an average, actinobacteria were frequently recorded from the oxygenated water masses than the anoxic water masses. Rastogi et al. (2012) obtained a total of 84 bacterial 16S rRNA gene sequences when subjected to phylogenetic study and were dispersed into 60 OTUs spanning 6 diverse phyla such as Proteobacteria (11 clones), Actinobacteria (9 clones), Bacteroidetes (8 clones), Acidobacteria (6 clones), Chloroflexi (5 clones) and Firmicutes (2 clones).

Functional genotypic metagenomic screening of cold-adapted enzymes, viz. lipase, esterase, amylase and cellulose from Antarctic soil samples was done by Berlemont et al. (2011). Jimenez-Lopez et al. (2012) constructed 20,000 clones for searching new lipolytic enzymes from high Andean forest soil with *Escherichia coli*, using plasmid p-Bluescript II SK+, with the library containing 80 kb metagenomic DNA of *Proteobacteria, Actinobacteria* and *Acidobacteria*. They also identified a monomeric esterase from an undescribed *Actinobacterium* that showed a preference for short chain fatty acids, active at low and high temperatures and with potential for protein engineering and biotechnological use.

Indigenous marine actinobacteria has been identified and confirmed by culture-independent molecular approach (Sun et al. 2010; Das et al. 2007; Monciardini et al. 2002). These culture-independent approaches engaged straight extraction of nucleic acids from the samples (Mincer et al. 2005). It frequently encompasses the DNA extracted from environmental samples amplification done by PCR, further subsequent estimation of the microbial diversity of the amplified molecules (Bull et al. 2005). Otherwise, amplified DNA products could be cloned and further sequenced to identify and enumerate rare or new microbes present in particular sample (Riedlinger et al. 2004; Stach et al. 2003). Moreover, Monciardini et al. (2002) established a selective set of primers for PCR amplification of 16S rDNA gene from the Actinomycetales families such as Streptomycetaceae, Streptosporangiaceae, Micromonosporaceae and Thermomonosporaceae. Each primer set, assessed on DNA from reference isolates, showed good sensitivity and the highest specificity. Application of particular primers for environmental samples analysis showed a great presence of these actinobacterial group and revealed that these sequences could be recognized to novel groups of actinobacteria (Monciardini et al. 2002).

Now, it is usually accepted that < 1% of microbes might be isolated and identified using conventional techniques, while the majority of microorganisms were missing and their biochemical pathways were unreachable (Rath et al. 2011; Kennedy et al. 2010). Development of metagenomic techniques and culture-independent methods provides us with the tools to analyze the highest extent of the analysis of uncultured microbial assembly and allows access to the pathways (biochemical) within uncultured microbes (Venter et al. 2004; Riesenfeld et al. 2004). Interestingly, Yang et al. (2013) identified 19 actinobacterial genera from two soft corals using both culture-independent and culture-dependent approaches and suggested that highly diverse actinobacteria were associated with different types of corals. In particular, 5 actinobacterial genera (*Gordonia, Serinicoccus, Cellulomonas, Candidatus microthrix* and *Dermacoccus*) were recorded for the first time from corals reef area, encompassing the known coral-associated actinobacterial diversity. Roger et al. (2013) studied metagenomic and metatranscriptomic microbial diversity from the accretion ice of the Antarctica and found predominant bacterial sequences, closer to *Firmicutes, Proteobacteria* and *Actinobacteria* and the predominant eukaryotic sequences were similar to the ascomycetous and basidiomycetous fungi.

Biotechnological application from psychrophilic microbes

Psychrophilic enzymes

The function of enzymes obtained from psychrophiles which covered a dominant role in living organisms present in cold environment have been mainly focused here. Psychrophilic enzymes having crystal structures were of significance to examine the properties of cold-active catalysts (Yadav et al. 2016a). Till date, only a few reports regarding NMR structure analysis for determination of psychrophilic enzymes crystal structures were published by Collins et al. (2010) for a thiol-disulfide oxidoreductase from bacterium isolated from the Antarctic. The experiment for a thermophilic enzyme isolation and analysis were easily understood; to continue to be stable and lively extreme temperature conditions. The challenge for a psychrophilic enzyme has been mostly understated. However in very low temperature conditions, to a large degree, lead decreases of the all enzyme catalyzed reaction rates and, further, protein function and molecular signals also will reduce. In order to redress for this procrastinating reaction rates at cold temperatures condition, psychrophiles synthesize enzymes have an up to tenfold upper specific activity in this cold temperature range. This is the condition of the major physiological adaptation to cold temperature at the enzyme level and the temperature for apparent highest activity for cold-active enzymes moves towards in low temperatures, repeating the low stability of these proteins and their recounting and inactivation at reasonable temperatures (Collins et al. 2008).

Psychrophilic enzyme with a faster activity is cost effective and saves energy and time of the reaction in industries. Cold-active enzyme uracil DNA N-glycosylases was used in molecular biology for the digestion of RNA and DNA. Similarly, lipase, protease and amylase enzymes are used in the detergent industry to breakdown the lipid, protein and starch, respectively. The chitinase enzymes used as microbes have a biocontrol activity for different pathogenic microbes for agricultural sustainability (Yadav et al. 2016b). Cellulase enzyme has been used to wash cotton fabrics and pectate lyases were used for stain removal. Food and beverages industries use pectinase enzyme for the fermentation of wine, beer, fruit juice processing and bread making. Similarly amylase and cellulose also used in textile industries for desizing of woven fabrics and biofinishing combined with cellulosic fabrics and mannanses enzyme used to degrade the gum (Felipe et al. 2015).

Molecular adaptation of psychrophilic microbes

It is necessary to maintain a regular cell function for microbial survival in extreme environmental conditions. Such a condition leads to produce more extra cellular polymeric substances which protect the cell from an unfavorable environment, provide a nutrient and some of the extrapolysaccharides could act as cryoprotectant. Along with this, the cell membrane also produces unsaturated fatty acids and carotenoids to maintain the flexibility and water content. Hence, protein and membrane lipids play an important role for the survival of extremophiles in a polyextremophilic condition. These protein and membrane lipid contents are dependent on temperature and environmental factors, consequently the physical or chemical factor of the macromolecule structure may vary.

Protein structure

Protein structure of psychrophilic microorganisms are one of the main reasons to survive in zero and lesser temperatures. These protein and enzymes structures are not temperature dependent but a few cellular enzymes may change their activity which leads to phenotypic changes. Therefore, phenotypic and genotypic changes in the protein structures are largely effective for growth temperature, for example, psychrophiles, mesophiles and thermophile protein structure are genotypically different from each other (Schlatter et al. 1987; Rentier-Delrue et al. 1993; Arpigny et al. 1994). Researchers have focused on them due to the biotechnological and industrial significance of psychrophilic enzymes. In 1996, Aghajari et al. crystallized the psychrophilic enzyme and investigated it by x-ray diffraction methods. Following this, lactase dehydrogenase and protease producing psychrophilic *Bacillus* spp., were cloned and sequenced (Vckovski et al. 1990; Davail et al. 1994; Davail et al. 1992), further α-amylase producing *Alteromonas* sp. Triosephosphate isomerase, lipase and citrate synthase from *Moraxcella* spp. (Rentier et al. 1993; Feller et al. 1991; Gerike et al. 1997) β-Galactosidase from *Arthrobacter* sp. (Trimbur et al. 1994; Gutshall et al. 1995) Isocitrate dehydrogenase from *vibrio* sp. (Ishii et al. 1993) B-lactamase and lipase from *Psychrobacter* sp. (Feller et al. 1996; Arpigny et al. 1993) and esterase from *Pseudomonas* spp. (Mckay et al. 1992) molecular adaptation and their biotechnological applications were studied. During the study, the sequence of lactase dehydrogenase enzyme from psychrophilic *Bacillus* sp., had a low level of hydrophobic and iron residues and more polar and charged residues than the enzyme from mesophilic and thermophilic *Bacillus* sp., which provided flexibility and stability of at low temperature (Jaenicke 1991; Feller et al. 1996). Exothermal electrostatic interaction and hydrogen bond formation will strengthen the protein but endothermic formation of hydrophobic bonds will cause denaturation of enzymes at low temperature (Arpigny et al. 1994; Feller et al. 1996; Davail et al. 1994; Feller et al. 1997). Cold-active subtilisin genes from psychrotolerant *Bacillus* spp., isolated from Antarctica seawater had 92% identity among 309 amino acids (each) considered as isozymes which were encoded by adjacent genes (Feller et al. 1996; Davil et al. 1992; Narinx et al. 1992; Siezen et al. 1991).

Retain activity of RNA polymerase, ribosaomal extract, peptidyl-prolylcistrans isomerase and the elongation factor was observed in several psychrophilic microorganisms. In addition, over expression and higher activity of the enzyme catalyses cistransprolyl isomerization might be important for maintaining protein folding rate at lower temperatures (Berger et al. 1996; Lim et al. 2000; D'Amico et al. 2006). The thermosensor of the cell membrane in psychrophiles is the base for changes in the fluidity. For example, Cold shock proteins (Csps) in mesophiles act as Cold Assimilation Proteins (Caps) in psychrophiles at a low temperature (Ray et al. 1994).

The binding ability of antifreeze protein (AFP) and ice crystals will create a thermal hysteresis and favor the growth of the organism (Jia and Davis 2002). Protein aggregation and degradation was prevented by the colligative effect of trehalose (Phadtare 2004). In addition exopolysaccharides also have a crucial role as cryoprotectant, these Eps will bind the cell surfaces that can alter the physico-chemical environment of the cell; favor the concentration of nutrients, retention of water and sequestration (Nichols et al. 2005). These extracellular secretions prevent the cold denaturation and act as cryoprotectant (Krembs et al. 2002; Mancuso Nichols et al. 2005).

However, various stabilizing factors are observed in the crystallographic study of psychrophilic proteins which includes: glycine residues, subunit interactions, hydrogen bonds, number of iron pairs, hydrophobic interactions, solvent interaction, cofactor binding, proline and arginine content (Collins et al. 2003; Wintrode et al. 2001; Violot et al. 2005). Cold shock protein and other proteins are the main reason for the adaptation of the controlled nutrient influx and outflow of waste products, macromolecule assemblies, nuclic acid dynamics and appropriate folding, stability of substrate and products, transcription and translation. Apart from all these, a large number of α-helix relative to the β-sheets is one of the key factors to retain flexibility at low temperatures (Madigan et al. 2014). In addition, a higher proportion (52%) of unsaturated fatty acids support the maintenance of the semi-fluid state of membrane (Deming 2009).

Membrane lipids

Microbes in extreme environments largely depend on the cell for rapid adjustment of the membrane lipid composition (Siliakus et al. 2017). Fatty acids composition of membrane lipids depends on temperature fluctuation (McGibbon and Russell 1983) and isomeric fatty acid distribution among acyl lipids influenced membrane fluidity (Sutton et al. 1990). The low temperature causes the following changes: increasing methyl branching, isobranching, fatty acid unsaturation and decrease in average chain length (Russell 1984; 1990). Di-unsaturated phospholipids lower the temperature of the membrane by the formation of liquid crystalline from gel (Russell 1992). B-ketoacyl-ACP synthatase II is responsible for thermal regulation which is known as molecular thermometer (Russell 1984).

The chemical stability of membrane lipids of bacteria are different from archaea based on the archaetidic acid and phosphatidic acid in bacteria and eukaryotes (Siliakus et al. 2017). The membrane lipid structure of bacteria has two fatty acid hydrocarbon chains, ester linkages and glycerol triphosphate, while archaea has glycerol mono phosphate, ether linkage and isoperinoid hydrocarbon chains (Koga 2012). A few of them have bipolar lipids known as glycerol dialkyl glycerol tetraether and membrane spanning ether lipids are also produced sporadically (Weijers et al. 2006). Most of the thermophilic microbe have chemically stabled ether lipids which are fatty alcohol containing monobranched diether lipids (Langworthy et al. 1983). Tetra ether lipids are the main reason for membrane flexibility in thermophilic and thermoacidophilic microbes and the same lipid composition is also found in mesophiles. The most important function of the cell membrane is to panel the inner cytoplasmic compartment away from their surroundings and to control the low molecular weight compound from both inside and outside the cell. This is very vital role of cell membranes (Blocher et al. 1984).

Conclusion and future prospects

While reviewing literature, it is evident that there is a wide gap in the understanding of microbial taxonomy and metagenomics and the disparity is high in the case of polar regions. Moreover, there are no such studies pertaining to culture-independent diversity of microbes in the polar oceanic waters. Considering these facts, the present chapter was undertaken to bridge the gap in knowledge by studying the biodiversity of the psychrophilic microbes and their potential adaptation behavior and biotechnological application have been listed.

Acknowledgement

The corresponding author would be like to acknowledge and is thankful to SERB-DST, Govt. of India for providing funds through TARE scheme (File No.: TAR/2019/000143) and the first author is grateful to National Post-Doctoral Fellowship programme (File No.: PDF/2015/000680). The authors also wish to convey their gratitude to the Directorate of Research, SRM IST, Tamil Nadu for providing facilities.

References

Aghajari, N., R. Haser, G. Feller and C. Gerday. 1996. Crystallization and preliminary X-ray diffraction studies of α-amylase from the antarctic psychrophile *Alteromonas haloplanctis* A23. Protein Science 5(10): 2128–2129.

Alcaman, M.E., J. Alcorta, B. Bergman, M. Vasquez, M. Polz and B. Diez. 2017. Physiological and gene expression responses to nitrogen regimes and temperatures in *Mastigocladus* sp. strain CHP1, a predominant thermotolerant cyanobacterium of hot springs. Syst. Appl. Microbiol. 40: 102–113.

Allen, M.A., F.M. Lauro, T.J. Williams, D. Burg, K.S. Siddiqui, D. De Francisci et al. 2009. The genome sequence of the psychrophilic archaeon, Methanococcoides burtonii: the role of genome evolution in cold adaptation. The ISME Journal 3: 1012.

Arpigny, J.L., G. Feller and C. Gerday. 1993. Cloning, sequence and structural features of a lipase from the Antarctic facultative psychrophile *Psychrobacter immobilis* B10. Biochim. Biophys. Acta. 1171: 331–333.

Arpigny, J.L., G. Feller, S. Davail, E. Narinx, Z. Zekhnini and C. Gerday. 1994. Molecular adaptations of enzymes from thermophilic and psychrophilic organisms. Adv. Comp. Environ. Physiol. 20: 269–295.

Babalola, O.O., B.M. Kirby, M. Le Roes-Hill, A.E. Cook, S.C. Cary, S.G. Burton et al. 2008. Phylogenetic analysis of actinobacterial populations associated with Antarctic Dry Valley mineral soils. Environ. Microbiol. 11: 566–576.

Babalola, O.O., B.M. Kirby, M.Le. Roes-Hill, A.E. Cook, S.C. Cary, S.G. Burton et al. 2009. Phylogenetic analysis of actinobacterial populations associated with Antarctic Dry Valley mineral soils. Environ. Microbiol. 11(3): 566–576.

Berger, F., N. Morellet, F. menu and P. Potier. 1996. Cold shock and cold assimilation proteins in the psychrophilic bacterium *Arthrobacter globiformis* S155. J. Bacteriol. 178: 2999–3007.

Berlemont, R., D. Pipers, M. Delsaute, F. Angiono, G. Feller, M. Galleni et al. 2011. Exploring the Antarctic soil metagenome as a source of novel cold-adapted enzymes and genetic mobile elements. Rev. Argent. Microbiol. 43(2): 94–103.

Blocher, D., R. Gutermann, B. Henkel and K. Ring. 1984. Physicochemical characterization of tetra ether lipids from *Thermoplasma acidophilum* differential scanning calorimetry studies on glycolipids and glycophospholipids. Biochimica Biophysica Acta. 778(1): 74–80.

Bottos, E.M., W.F. Vincent, C.W. Greer and L.G. Whyte. 2008. Prokaryotic diversity of arctic ice shelf microbial mats. Environ. Microbiol. 10: 950–966.

Brock, T.D., K.M. Brock, R.T. Belly and R.L. Weiss. 1972. Sulfolobus: a new genus of sulfur-oxidizing bacteria living at low pH and high temperature. Arch. Microbiol. 84: 54–68.

Bruneel, O., A. Volant, S. Gallien, B. Chaumande, C. Casiot and C. Carapito. 2011. Characterization of the active bacterial community involved in natural attenuation processes in arsenic-rich creek sediments. Microb. Ecol. 61: 793–810.

Bull, A.T., J.E.M. Stach, A.C. Ward and M. Goodfellow. 2005. Marine actinobacteria: Perspectives, challenges, future directions. Antonie Van Leeuwenhoek, 87: 65–79.

Cai, Q., X. Ye, B. Chen and B. Zhang. 2017. Complete genome sequence of Exiguobacterium sp. strain N4-1P, a psychrophilic bioemulsifier producer isolated from a cold marine environment in North Atlantic Canada. Genome Announcements 5: e01248–01217. doi:10.1128/genomeA.01248-17.

Carneiro, A.R., R.T.J. Ramos, H. Dall'Agnol, A.C. Pinto, S. de Castro Soares, A.R. Santos et al. 2012. Genome sequence of Exiguobacterium antarcticum B7, isolated from a biofilm in Ginger Lake, King George Island, Antarctica. Journal of Bacteriology 194: 6689–6690. doi:10.1128/JB.01791-12.

Cerritos, R., L. Equiarte, A. Luis, S. Morena, J. Siefert, M. Travisano et al. 2011. Diversity of culturable thermo-resistant aquatic bacteria along an environmental gradient in Cuatro Ciénegas, Coahuila, México. Antonie van Leeuwenhoek 99(2): 303–318.

Christner, B., M. Skidmore, J. Priscu, M. Tranter and C. Foreman. 2008. Bacteria in subglacial environments. pp. 51–71. *In*: Margesin, R., F. Schinner, J-C. Marx and C. Gerday (eds.). Psychrophiles: From Biodiversity to Biotechnology. Springer-Verlag, Berlin.

Collins, T., M.A. Meuwis, C. Gerday and G. Feller. 2003. Activity, stability and flexibility in glycosidases adapted to extreme thermal environments. J. Mol. Biol. 328: 419–428.

Collins, T., F. Roulling and F. Piette, 2008. Fundamentals of cold-adapted enzymes. pp. 211–227. *In*: Margesin, R., F. Schinner, J.C. Marx and C. Gerday (ed.). Psychrophiles, from Biodiversity to Biotechnology. Springer, Berlin, Germany.

Collins, T., M. Matzapetakis and H. Santos. 2010. Backbone and side chain 1H, 15N & 13C assignments for a thiol-disulphide oxidoreductase from the Antarctic bacterium *Pseudoalteromonas haloplanktis* TAC125. Biomol. NMR Assign. 4(2): 151–154.

D'Amico, S., T. Collins, J.C. Marx, G. Feller and C. Gerday. 2006. Psychrophilic microorganisms challenges for life. EMBO Reports 7(4): 385–389.

Das, M., T.V. Royer and L.G. Leff. 2007. Diversity of fungi, bacteria, and actinomycetes on leaves decomposing in a stream. Appl. Environ. Microbiol. 73(3): 756–767.

Davail, S., G. Feller, E. Narinx and C. Gerday. 1992. Sequence of the subtilisin-encoding gene from an antarctic psychrotroph *Bacillus* TA41. Gene 119: 143–144.

Davail, S., G. Feller, E. Narinx and C. Gerday. 1994. Cold adaptation of proteins. J. Biol. Chem. 269: 17448–17453.

Delmont, T.O., P.R. Robe, I. Clark, P. Simonet and T.M. Vogel. 2011. Metagenomic comparison of direct and in direct soil DNA extraction approaches. J. Microbiol. Methods 86: 397–400.

Deming, J.W. 2009. Extremophiles: Cold environments. pp.147–157. *In*: Schaechter, M. (ed.). The Desk Encyclopedia of Microbiology. Academic Press. Oxford. UK.

Densita, N. and N. Mariana. 2005. Screening the antimicrobial activity of actinomycetes strains isolated from Antarctica. J. Cult. Coll. 4: 29–35.

Dhar, H., M.K. Swarnkar, A. Rana, K. Kaushal, A.K. Singh, R.C. Kasana et al. 2016. Complete genome sequence of a low-temperature active and alkaline-stable endoglucanase-producing Paenibacillus sp. strain IHB B 3084 from the Indian Trans-Himalayas. Journal of Biotechnology 230: 1–2.

Du, Y., B. Yuan, Y. Zeng, J. Meng, H. Li, R. Wang et al. 2015. Draft genome sequence of the cellulolytic bacterium Clavibacter sp. CF11, a strain producing cold-active cellulase. Genome Announcements 3: e01304–01314.

Echauri, J.A., J.F. Montalvo, M.A. Martinez and J.M. Azcarate. 2011. Personality disorders in Batterer Men: Differential profile between Aggressors in prison and aggressors with a suspended sentence. Anuario de Psicología Jurídica 21: 97–105.

Eisen, J.A. 2007. Environmental shot-gun sequencing: its potential and challenges for studying the hidden world of microbes. PLoS Biol. 5: 82.

Emilio, O.C., L. Marc, P. Antoni, B. Albert, B.M. Carles and C. Antonio. 2012. Contribution of deep dark fixation processes to overall CO_2 incorporation and large vertical changes of microbial populations in stratified Karstic lakes. Aquat. Sci. 74: 61–75.

Escalante, G., V.L. Campos, C. Valenzuela, J. Yañez, C. Zaror and M.A. Mondaca. 2009. Arsenic resistant bacteria isolated from arsenic contaminated river in the Atacama Desert (Chile). Bull. Environ. Contam. Toxicol. 83: 657–661.

Felipe, S., R. Peralta and J.M. Blamey. 2015. Cold and hot extremozymes: industrial relevance and current trends. Front. Bioeng. Biotechnol. 3:148. doi: 10.3389/fbioe.2015.00148.

Feller, G., M. Thiry and C. Gerday. 1991. Nucleotide sequence of the lipase gene lip2 from the antarctic psychrotroph Moraxella TA144 and site-specific mutagenesis of the conserved serine and histidine residues. DNA Cell Biol. 10: 381–388.

Feller, G., E. Narinx, J.L. Arpigny, M. Aittaleb, E. Baise, S. Genicot et al. 1996. Enzymes from psychrophilic organisms. FEMS Microbiol. Lett. 18: 189–202.

Feller, G., Z. Zekhnini, J. Lamotte-Brasseur and C. Gerday. 1997. Enzymes from cold adapted microorganisms. The class C beta-lactamase from the Antarctic psychrophile *Psychrobacter immobilis*. Eur. J. Biochem. 244: 186–191.

Gerike, U., M.J. Danson, N.J. Russell and D.W. Hough. 1997. Sequencing and expression of the genes encoding a cold active citrate synthase from Antarctic bacterium strain DS-3R. European J. Biochem. 248: 49–57.

Gesheva, V. 2010. Production of antibiotics and enzymes by soil microbes from the windmill Islands region, Wilkes Land, East Antarctica. Polar Boil. 33(10): 1351–1357.

Gesheva, V., E. Stackebrandt and E. Vasileva-Tonkova. 2010. Biosurfactant production by halotolerant *Rhodococcus fascians* from Casey Station, Wilkes Land, Antarctica. Curr. Microbial. 61: 112–117.

Gesheva, V. and T. Negoita. 2012. Psychrotrophic microorganism communities in soils of Haswell Island, Antarctica, and their biosynthetic potential. Polar Biol. 35(2): 291–297.

Gupta, H.K., R.D. Gupta, A. Singh, N.S. Chauhan and R. Sharma. 2011. Genome sequence of Rheinheimera sp. strain A13L, isolated from Pangong Lake, India. Journal of Bacteriology 193: 5873–5874.

Gutshall, K., D. Trimbur, J. Kasmir and J. Brenchley. 1995. Analysis of a novel gene and β-galactosidase isozyme from a psychrotrophic Arthrobacter isolate. J. Bacteriol. 177: 1981–1988.

Hallam, S.J., K.T. Konstantinidis, N. Putnam, C. Schleper, Y.-i. Watanabe, J. Sugahara et al. 2006. Genomic analysis of the uncultivated marine crenarchaeote Cenarchaeum symbiosum. Proceedings of the National Academy of Sciences of the United States of America 103: 18296–18301. doi:10.1073/pnas.0608549103.

Handelsman, J., M.R. Rondon, S.F. Brady, J. Clardy and R.M. Goodman. 1998. Molecular biological access to the chemistry of unknown soil microbes: a new frontier for natural products. Chem. Biol. 5: 245–249.

Hugenholz, P., B.M. Goebel and N.R. Pace. 1998. Impact of culture-independent studies on the emerging phylogenetic view of bacterial diversity. J. Bacteriol. 180: 4765–4774.

Ishii, A., T. Ochiai, S. Imagawa, N. Fukunaga, S. Sasaki, O. Minowa et al. 1987. Isozymes of isocitrate dehydrogenase from an obligately psychrophilic bacterium, *Vibrio* sp. strain ABE-1: purification, and modulation of activities by growth conditions. J. Biochem. 102(6): 1489–1498.

Jaenicke, R.1991. Protein stability and molecular adaptation to extreme conditions. Eur. J. Biochem. 202(3): 715–728.

Jia, Z. and P.L. Davis. 2002. Antifreeze proteins: an unusual receptor-ligand interaction. Trend. Biochem. Sci. 27: 101–106.

Jimenez-Lopez, J.C., S. Morales, A.J. Castro, D. Volkmann, M.I. Rodriguez-Garcia and J.D. Alche. 2012. Characterization of profilin polymorphism in pollen with a focus on multifunctionality. PLoS One 7(2): 1–13.

Karner, M.B., E.F. De Long and D.M. Karl. 2001. Archaeal dominance in the mesopelagic zone of the Pacific Ocean. Nat. 409: 507–510.

Kennedy, J., B. Flemer, S.A. Jackson, D.P.H. Lejon, J.P. Morrissey and F. O'Gara. 2010. Marine metagenomics: new tools for the study and exploitation of marine microbial metabolism. Mar. Drugs 8(3): 608–628.

Kochkina, G.A., N.E. Ivanushkina, S.G. Karasev, E.Y. Gavrish, L.V. Gurina, L.I. Evtushenko et al. 2001. Survival of micromycetes and actinobacteria under conditions of long-term natural cryopreservation. Microbiol. 70: 356–364.

Koga, Y. and M. Nakano. 2008. A dendrogram of archaea based on lipid component parts composition and its relationship to rRNA phylogeny. Syst. App. Microbiol. 31(3): 169–182.

Koga, Y. 2012. Thermal adaptation of the archaeal and bacterial lipid membranes. Archaea. 1: 1–6.

Krembs, C., H. Eicken, K. Junge and J.W. Deming. 2002. High concentrations of exopolymeric substances in arctic winter sea ice: implications for the polar ocean carbon cycle and cryoprotection of diatoms. Deep-Sea Res. 49: 2163–2181.

Langworthy, T.A., G. Holzer, J.G. Zeikus and T.G. Tornabene. 1983. Iso- and anteiso-branched glycerol diethers of the thermophilic anaerobe *Thermodesulfoto-bacterium commune*. System. App. Microbiol. 4(1): 1–17.

Li, J., X.P. Tian, T.J. Zhu, L.L. Yang and W.J. Li. 2011. *Streptomyces fildensis* sp. nov., a novel *Streptomycete* isolated from Antarctic soil. Antonie Van Leeuwenhoek 100(4): 537–543.

Lim, J., T. Thomas and R. Cavicchioli. 2000. Low temperature regulated DEAD-box RNA helicase from the Antarctic archeaon, *Methanococcoides Burtonii*. J. Mol. Biol. 297: 533–567.

Madigan, M.T., J.M. Martinko, P.V. Dunlap and D.P. Clarck, 2014. Biology of Microorganisms, 14th ed.; Benjamin Cummings: San Francisco, NC, USA.

Mancuso Nichols, C.A., J. Guezennec and J.P. Bowman. 2005. Bacterial exopolysaccarides from extreme marine environments with special consideration of the southern ocean, sea ice and deep sea hydrothermal vents: a review. Mar. Biotechnol. 7: 253–271.

McGibbon, L. and N.J. Russell. 1983. Fatty acid positional distribution in phospholipids of a psychrophilic bacterium during changes in growth temperature. Current Microbial. 9: 241–244.

McKay, D.B., M. P.Jennings, E.A. Godfrey, I.C. MacRae, P.J. Rogers and I.R. Beacham. 1992. Molecular analysis of an eaterase–encoding gene from a lipolytic psychrotrophic pseudomonad. J. Gen. Microbiol. 138(4): 701–708.

Methé, B.A., K.E. Nelson, J.W. Deming, B. Momen, E. Melamud, X. Zhang et al. 2005. The psychrophilic lifestyle as revealed by the genome sequence of Colwellia psychrerythraea 34H through genomic and proteomic analyses. Proceedings of the National Academy of Sciences of the United States of America 102: 10913–10918. doi:10.1073/pnas.0504766102.

Mincer, T.J., W. Fenical and P.R. Jensen. 2005. Culture-dependent and culture-independent diversity within the obligate marine actinomycete genus *Salinispora*. Appl. Environ. Microbiol. 71(11): 7019–7028.

Miteva, V.I. 2008. Bacteria in snow and glacier ice. pp. 31–50. *In*: Margesin, R., F. Schinner, J.-C. Marx and C. Gerday (eds.). Psychrophiles: From Biodiversity to Biotechnology. Springer-Verlag, Heidelberg, Germany.

Mock, T. and D.N. Thomas. 2005. Recent advances in sea-ice microbiology. Environ. Microbiol. 7(5): 605–619.

Moncheva, P., S. Tishkov, N. Dimitrova, V. Chipeva, S.A. Nicolova and N. Bogatzevska. 2002. Characteristics of soil actinomycetes from Antarctica. J. Cult. Coll. 3: 3–14.

Monciardini, P., M. Sosio, L. Cavaletti, C. Chiocchini and S. Donadio. 2002. New PCR primers for the selective amplification of 16S rDNA from different groups of actinomycetes1. FEMS Microbiol. Ecol. 42(3): 419–429.

Morgan, J.L., A.E. Darling and J.A. Eisen. 2010. Metagenomic sequencing of an *in vitro*-simulated microbial community. PLoS One 5(4): 1–10.

Narinx, E., S. Davail, G. Feller and C. Gerday. 1992. Nucleotide and derived amino acid sequence of the antarctic psychrotroph *Bacillus* TA39. Biochim. Biophys. Acta 1131(1): 111–113.

Nichols, C.M., S.G. Lardiere, P. Bowman, P.D. Nichols, J.A.E. Gibson and J. Guezennec. 2005. Chemical characterization of exopolysaccharides from Antarctic marine bacteria. Microb. Ecol. 49: 578–589.

Nichols, D.S., K. Sanderson, A. Buia, J.V. Kamp, P. Holloway, J.P. Bowman et al. 2002. Bioprospecting and biotechnology in Antarctica. pp. 85–103. *In*: Jabour-Green, J. and M. Haward (eds.). The Antarctic: Past, Present and Future. Antarctic CRC Research Report 28. Hobart, Tasmania, Australia.

Pearce, D.A., K.A. Hughes, S.A. Harangozo, T.A. Lachlan-Cope and A.E. Jones. 2010. Biodiversity of air-borne microorganisms at Halley station, Antarctica. Extremophiles 14: 145–159.

Pearce, D.A., K.K. Newsham, M.A.S. Throne, L.C. Bado, M. Krsek, P. Laskaris et al. 2012. Metagenomic analysis of a southern maritime Antarctic soil. Front. Microbiol. 3: 1–13.

Phadtare, S. 2004. Recent developments in bacterial cold shock response. Curr. Issues Mol. Biol. 6: 125–136.

Quatrini, R., L.V. Escudero, A. MayoBeltrain, P.A. Galleguillos, F. Issotta, M. Acosta et al. 2017. Draft genome sequence of *Acidithiobacillus thiooxidans* CLST isolated from the acidic hypersaline Gorbea salt flat in northern Chile. Stand. Genomic Sci. 12(84): 2–8.

Rampelotto, P.H. 2013. Extemophiles and extreme environments. Life 3: 482–485.

Rastogi, G., A. Sbodio, J.J. Tech, T.V. Suslow, G.L. Coaker and J.H. Leveau. 2012. Leaf microbiota in an agroecosystem: spatiotemporal variation in bacterial community composition on field-grown lettuce. ISME J. 6: 1812–1822.

Rath, C.M., B. Janto, J. Earl, A. Ahmed, F.H. Hu and L. Hiller. 2011. Metagenomic characterization of the marine invertebrate microbial consortium that produces the chemotherapeutic natural product ET-743. Chem. Biol. 6(11): 1244–56.

Rausk, M.C., G.M. Ferrer, J.R. Moreno, M.E. Farias and V.H. Albarracin. 2017. The diversity of microbial extremphiles. pp. 87–126. *In*: Rodrigues, T.B. and A.M.E.T. silva (eds.). Molecular Diversity of Environmental Prokaryotes. CRC Press, United States.

Ray, M.K., G.S. Kumar and S. Shivaji. 1994. Phosphorylation of membrane proteins in response to temperature in an Antarctic *Pseudomonas syringae*. Microbiol. 140: 3217–3223.

Rentier-Delrue, F., S.C. Mande, S. Moyens, P. Terpstra, V. Mainfroid, K. Goraj et al. 1993. Cloning and over-expression of the triosephosphate isomerase genes from psychrophilic and thermophilic bacteria. J. Mol. Biol. 229: 85–93.

Riedlinger, J., A. Reicke, H. Zahner, B. Krismer, A.T. Bull and L.A. Maldonado. 2004. Abyssomicins, inhibitors of the para-aminobenzoic acid pathway produced by the marine *Verrucosispora* strain AB-18-032. J. Antibiot. 57(4): 271.

Riesenfeld, C.S., P.D. Schloss and J. Handelsman. 2004. Metagenomics: genomic analysis of microbial communities. Annual Rev. Genet. 38: 525–552.

Rodrigues, D.F., N. Ivanova, Z. He, M. Huebner, J. Zhou and J.M. Tiedje. 2008. Architecture of thermal adaptation in an Exiguobacterium sibiricum strain isolated from 3 million year old permafrost: A genome and transcriptome approach. BMC Genomics 9: 547–547. doi:10.1186/1471-2164-9-547.

Roger, O.S., Y.M. Shtarkman, Z.A. Kocer, R. Edgar, R. Veerapaneni and T. D'Elia. 2013. Ecology of subglacial lake Vostok (Antarctica), based on metagenomic analysis of accretion ice. Biol. 2: 629–650.

Russell, N.J. 1984. Mechanisms of thermal adaptation in bacteria: Blueprints for survival. Trends Biochem. Sci. 9: 108–112.

Russell, N.J. 1990. Cold adaptation of microorganisms. Phil. Trans. Roy. Sot. London Series B 329: 595–611.

Russell, N.J. 1992. Physiology and molecular biology of psychrophilic micro-organisms. pp. 203–224. *In*: Herbert, R.A. and R.J. Sharp (eds.). Molecular Biology and Biotechnology of Extremophiles. Glasgow and London: Blackie, Glasgow and London.

Salwan, R., M.K. Swarnkar, A.K. Singh and R.C. Kasana. 2014. First draft genome sequence of a member of the genus Planomicrobium, isolated from the Chandra River, India. Genome Announcements 2: e01259–01213.

Sarmiento, F., R. Peralta and J.M. Blamey. 2015. Cold and hot extremozymes: Industrial relevance and current trends. Frontiers Bioengineering Biotechnol. 3(148): 1–15.

Schlatter, D., O. Kriech, F. Suter and H. Zuber. 1987. The primary structure of the psychrorphilic lactate dehydrogenase from *Bacillus psychrosaccharolyticus*. Biol. Chem. Hoppe-Seyler 368: 1435–1446.

Siezen, R.J., W.M. De Vos, J.A.M. Leunissen and B.W. Dijkstra. 1991. Homology modeling and protein engineering strategy of subtiases, the family of subtilisin-like srine proteinases. Protein Eng. 4(7): 719–737.

Siliakus, M.F., John van der Oost and W.M.K. Servé. 2017. Adaptations of archaeal and bacterial membranes to variations in temperature, pH and pressure. Extremophiles 21: 651–670.

Simon, C., A. Wiezer, A.W. Strittmatter and R. Daniel. 2009. Phylogenetic diversity and metabolic potential revealed in a glacier ice metagenome. Appl. Environ. Microbiol. 75: 7519 –7526.

Singh, R.N., S. Gaba, A.N. Yadav, P. Gaur, S. Gulati, R. Kaushik et al. 2016. First, High quality draft genome sequence of a plant growth promoting and cold active enzymes producing psychrotrophic *Arthrobacter agilis* strain L77. Stand Genomic Sci. 11: 54. doi:10.1186/s40793-016-0176-4.

Stach, J.E.M., L.A. Maldonado, A.C. Ward, M. Goodfellow and A.T. Bull. 2003. New primers for the class Actinobacteria: application to marine and terrestrial environments. Environ. Microbiol. 5(10): 828–841.

Sun, W., S. Dai, S. Jiang, G. Wang, G. Liu and H. Wu. 2010. Culture-dependent and culture independent diversity of Actinobacteria associated with the marine sponge *Hymeniacidon perleve* from the South China Sea. Antonie Van Leeuwenhoek 98(1): 65–75.

Sutton, G.C., P.J. Quinn and N.J. Russell. 1990. The effect of salinity on the composition of fatty acid double-bond isomers and snl/m-2 positional distribution in membrane phospholipids of a moderately halophilic eubacterium. Curr. Microbial. 20: 143–146.

Tindall, B.J. 2004. Prokaryotic diversity in the Antarctic: the tip of the iceberg. Microb. Ecol. 47(3): 271–283.

Trimbur, D.E., K.R. Gutshall, P. Prema and J.E. Brenchley. 1994. Characterization of a psychrotrophic Arthrobacter gene and its cold-active β-galactosidase. Appl. Environ. Microbiol. 60: 4544–4552.

Vckovski, V., D. Schlatter and H. Zuber. 1990. Structure and function of L-lactate dehydrogenases from thermophilic, mesophilic and psychrophilic bacteria, IX. Biol. Chem. 371: 103–110.

Venter, J.C., K. Remington, J.F. Heidelberg, A.L. Halpern, D. Rusch and J.A. Eisen. 2004. Environmental genome shotgun sequencing of the Sargasso Sea. Science 304(5667): 66–74.

Verma, P., A.N. Yadav, K.S. Khannam, N. Panjiar, S. Kumar, A.K. Saxena et al. 2015. Assessment of genetic diversity and plant growth promoting attributes of psychrotolerant bacteria allied with wheat (*Triticum aestivum*) from the northern hills zone of India. Ann. Microbiol. 65: 1885–1899.

Violot, S., N. Aghajari, M. Czjzek, G. Feller, G.K. Sonan, P. Gouet et al. 2005. Structure of a full length psychrophilic cellulose from pseudomonas haloplanktis revealed by x-ray diffraction and small angle x-ray scattering. J. Mol. Biol. 348: 1211–1224.

Vollmers, J., S. Voget, S. Dietrich, K. Gollnow, M. Smits, K. Meyer et al. 2013. Poles apart: Arctic and Antarctic Octadecabacter strains share high genome plasticity and a new type of xanthorhodopsin. PLoS ONE 8: e63422. doi:10.1371/journal.pone.0063422.

Wegley, L., R. Edwards, B. Rodriguez-Brito, H. Liu and F. Rohwer. 2007. Metagenomic analysis of the microbial community associated with the coral *Porites astreoides*. Environ. Microbiol. 9: 2707–2719.

Weijers, J., S. Schouten, E.C. Hopemans, J.A. Geenevasen, O.R. David, J.M. Coleman et al. 2006. Membrane lipids of mesophilic anaerobic bacteria thriving in peats have typical archaeal traits. Environ. Microbiol. 8: 648–657.

Willerslev, E., A.J. Hansen and H.N. Poinar. 2004. Isolation of nucleic acids and cultures from fossil ice and permafrost. Trend. Ecol. Evol. 19: 141–147.

Wintrode, P.L., K. Miyazaki and F.H. Arnold. 2001. Patterns of adaptation in a laboratory evolved thermophilic enzyme. Biochem. Biophys. Acta. 1549: 1–8.

Yadav, A.N., S.G. Sachan, P. Verma and A.K. Saxena. 2015a. Prospecting cold deserts of north western Himalayas for microbial diversity and plant growth promoting attributes. J. Biosci. Bioeng. 119: 683–693.

Yadav, A.N., S.G. Sachan, P. Verma, S.P. Tyagi, R. Kaushik and A.K. Saxena. 2015b. Culturable diversity and functional annotation of psychrotrophic bacteria from cold desert of Leh Ladakh (India). World J. Microbiol. Biotechnol. 31: 95–108.

Yadav, A.N., P. Verma, M. Kumar, K.K. Pal, R. Dey, A. Gupta et al. 2015c. Diversity and phylogenetic profiling of niche-specific Bacilli from extreme environments of India. Ann. Microbiol. 65: 611–629.

Yadav, A.N., S.G. Sachan, P. Verma, R. Kaushik and A.K. Saxena. 2016a. Cold active hydrolytic enzymes production by psychrotrophic Bacilli isolated from three sub-glacial lakes of NW Indian Himalayas. J. Basic Microbiol. 56: 294–307.

Yadav, A.N., S.G. Sachan, P. Verma and A.K. Saxena. 2016b. Bioprospecting of plant growth promoting psychrotrophic Bacilli from cold desert of north western Indian Himalayas. Indian J. Exp. Biol. 54: 142–150.

Yadav, A.N., P. Verma, S.G. Sachan, R. Kaushik and A.K. Saxena. 2018. Psychrotrophic microbiomes: molecular diversity and beneficial role in plant growth promotion and soil health. pp. 197–240. *In*: Panpatte, D.G., Y.K. Jhala, H.N. Shelat and R.V. Vyas (eds.). Microorganisms for Green Revolution-Volume 2: Microbes for Sustainable Agro-ecosystem. Springer, Singapore. doi:10.1007/978-981-10-7146-1_11.

Yadav, A.N., D. Kour, S. Sharma, S.G. Sachan, B. Singh, V.S. Chauhan et al. 2019a. Psychrotrophic microbes: biodiversity, mechanisms of adaptation, and biotechnological implications in alleviation of cold stress in plants. pp. 219–253. *In*: Sayyed, R.Z., N.K. Arora and M.S. Reddy (eds.). Plant Growth Promoting Rhizobacteria for Sustainable Stress Management: Volume 1: Rhizobacteria in Abiotic Stress Management. Springer Singapore, Singapore. doi:10.1007/978-981-13-6536-2_12.

Yadav, A.N., N. Yadav, D. Kour, A. Kumar, K. Yadav, A. Kumar et al. 2019b. Bacterial community composition in lakes. pp. 1–71. *In*: Bandh, S.A., S. Shafi and N. Shameem (eds.). Freshwater Microbiology. Academic Press. doi:https://doi.org/10.1016/B978-0-12-817495-1.00001-3.

Yang, S., W. Sun, C. Tang, L. Jin, F. Zhang and Z. Li. 2013. Phylogenetic diversity of actinobacteria associated with soft coral *Alcyonium gracllimum* and stony coral *Tubastraea coccinea* in the East China Sea. Microb. Ecol. 66(1): 189–199.

Zenova, G.M., M.S. Dubrova and D.G. Zvyagintsev. 2010. Structural-functional specificity of the complexes of Psychrotolerant soil actinomycetes. Eurasian Soil Science 43(4): 447–452.

Zhang, C., W. Guo, Y. Wang and X. Chen. 2018. Draft genome sequences of two psychrotolerant strains, Colwellia polaris MCCC 1C00015T and Colwellia chukchiensis CGMCC 1.9127 T. Genome Announcements 6: e01575–01517.

Chapter 12

Extremophilic Microbes
Blooms the Biological Synthesis of Nanoparticles

Aathira Sreevalsan, Jamseel Moopantakath and *Ranjith Kumavath**

Introduction

The exploitation of atoms, paved the way for the inception of nanotechnology in the year 1959 (Feynman 1959). The book 'Engine of Creation: Coming Era of Nanotechnology' by Drexler's connected the conceptual framework that drew attention (Eric 1986). Nanotechnology provides the ability to engineer properties of materials by monitoring their size and has led research to a multitude of potential uses, encompassing knowledge from various fields such as, material Science, health Science, biology, chemistry and physics (Hull et al. 2018; Mozetic et al. 2018; Hu et al. 2019). Nanoparticles are solid and colloidal particles with accepted sizes ranging between 10–100 nm and physicochemical properties relatively different from their bulk counterparts (Batista et al. 2015). Traditionally synthesis of nanoparticles can be achieved by using a physical and chemical method such as ultraviolet irradiations, aerosol technologies, laser ablation and photochemical reduction techniques (Remya et al. 2017). Physical and chemical processes fabricate large quantities of nanoparticles with a specific size and shape during a comparatively short period. Although it has disadvantages such as highly complicated, expensive, and an outdated method which are inadequate and dangerous to the environment (Iravani et al. 2014). There are several microbial species that are capable of synthesizing different nanoparticles using their enzymes both intracellularly and extracellularly such as gold, silver and copper that have high biological activities (Kumar et al. 2011).

The common side effect of microbial fermentation is contamination. Subsequently a 'Green Route' was chosen as the best alternative since microorganisms, plants and algae can survive at ambient conditions such as temperature, pH and pressure (Durvasula and Subba Rao 2018). Biogenic green synthesis of a nanoparticle can be prepared in an eco-friendly, cost-effective manner with higher catalytic reactivity and greater surface area (Abdelghany et al. 2018). Halophilic microorganisms are the most dominant species present in saline waters and with multiple biochemical reactions providing great benefits to pharmaceutical and biological applications (Thombre et al. 2016; Rodrigo-Banos et al. 2015). These are extremely tolerant enzymatic systems providing high pharmaceutical application. This chapter illustrates the importance of nanoparticles for human beings in general, the physical parameters of halophiles, biological application of halophilic compounds, using various microorganisms and their application synthesis of nanoparticles.

Department of Genomic Science, Central University of Kerala, Tejaswini Hills, Periya, Kasaragod-671320.
* Corresponding author: RNKumavath@gmail.com, RNKumavath@cukerala.edu.in

Applications on welfare of mankind

Nanotechnology plays a significant role in diverse fields of life science such as pharmaceuticals, agriculture and industrial purposes (Neethirajan et al. 2011). Nanomedicine is an upcoming multidisciplinary medical field that allows nanoparticles to study, diagnose, treat and prevent a number of communicable and non-communicable disease (Wagner et al. 2006). Coherently dispersed nanoparticles are commonly used in nanomedicine. Nanoparticles have distinct physicochemical properties such as thermal, optical, electrical that support biosensors and molecule imaging in the diagnostic field (Kolluru et al. 2013). Biosensors accept diverse varieties of DNA sensors concerned with diagnosis of diseases, mutations and detection of specific DNA sequences (Bursten et al. 2016). Glucose sensors play an important role in observing diabetes fitness due to excellence in selectivity, reliability and effectiveness (Luo et al. 2006). Fluorescent markers have many disadvantages such as, fading fluorescence per time, restricted use of specific dyes due to bleeding effect and colour matching are prime causes to consider in nanoparticles (Wolfbeis 2015). Nowadays, a major problem in the therapeutic field is a lack of drug delivery system to a specific receptor and biomolecule in a safe and precise manner (Steichen et al. 2013).

Targeted nano carriers must cross blood tissues barriers, to enter specific cells to contact cytoplasmic targets besides the help of specific endocytosis and transcytotic transport mechanism (Fadeel et al. 2010). In the state of drug delivery system, the size of nanoparticles results in an accelerated settlement of the drug on a forward surface and aids in more durable drug release as compared with bulk molecules (Roduner 2006). The targeted-drug delivery decreases drug-related toxicity and increases the patient's compliance against the drug by less frequent doses (Rizvi et al. 2018). Molecules such as those conjugated with proteins or nucleic-acid carriers should be efficient and able to protect them from unwanted degradation (Jayakumar et al. 2010). Cancer is a complex biological phenomenon and termed as a disease of many diseases. Chemotherapy principally focuses on destroying every dividing cell in a body, which is one of hallmark of cancer cells and side effects of chemotherapy are harsh and affect the entire body due to non-specificity of the chemotherapeutic agent (Hanahan and Weinberg 2011).

Nanoparticles have shown to increase the delivery precision of various drugs and nutraceuticals with several biological applications. Drug delivery with the aid of specific nanoparticles provide enhanced availability of the drug in localized tissue without adverse effects to the normal function of the cells (Shen et al. 2016; Huang et al. 2015). These are lipophilic molecules, fat-soluble vitamins A, D, E and K, polysaturated lipids and phytochemicals usually used for dissolution mechanisms of nutraceuticals via nanoparticle formulations (Acosta 2009; McClements 2015). Halotolerant organisms are classified based on the domain known as halotolerant bacteria and haloarchaeal species. Halobacteria are bacteria, which can survive salt conditions and haloarchaea, bloom at extreme salt conditions represented among the Halobacteriaceae family, Euryarchaeota phylum, Archaea domain (Gunde-Cimerman et al. 2018). Halotolerant organisms have an excellent ability to adjust from low to higher salt concentration under external environment, while the halophilic archaeal organisms are frequently observed at high salinities for proper growth and metabolism (Oren et al. 2001). They are most extensively observed in salt lakes, brines, saline soils, cold saline habitat and saline food products (Singh et al. 2019). Based on the salt concentration required for optimum growth, these can be differentiated into three class (Thombre et al. 2016). Microorganisms have become one of most promising organisms for synthesizing pigments such as chlorophyll and carotenoids. Interestingly carotenoids synthesized through the isoprenoid pathway are found to be a biologically active compound against cancer and microbial infection (Nisar et al. 2015). Carotenoids are C40–C50 hydrophobic compounds with a long conjugated double bond, bilaterally symmetrical around the central double bond including the colour ranging from red, orange, yellow or even colourless (Mackinnon et al. 2011). With the appearance of oxygen and chemical compositions, carotenoids can be classified into carotenes or carotenoid hydrocarbons containing carbon and hydrogen atom alone and oxygenated carotenoids such as xanthophyll containing oxygenated mixtures with methoxy, carboxylic or additional functional groups (Rivera et al. 1998) (Fig. 12.1).

In the past few decades, scientists have shown a keen interest in carotenoid products due to their distinct biological functions that include acting as oxygen transporters, light energy absorbers, antitumour, and scavenging free radical (Vílchez et al. 2011). Free radical scavengers are compounds, which can scavenge free radicals, that are formed during the metabolic process. Bacterioruberin has 13 pairs of conjugated double bonds providing higher scavenging activities compared to other carotenoids (Fig. 12.2). Carotenoids play a significant role in embryonic development and morphogenesis and are primarily used as precursors of vitamin A and retinoid compounds (Zile 1998).

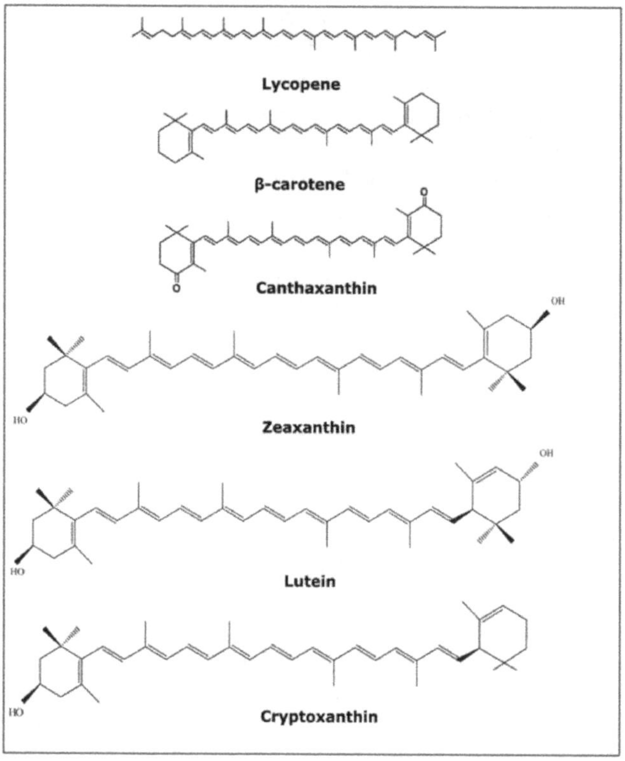

Figure 12.1 Different highly active carotenoid pigments synthesized by halophilic organisms and their structure characteristics (Gradelet et al. 1998; Torregrosa-Crespo et al. 2018; DasSarma et al. 2014; Bhosale et al. 2005).

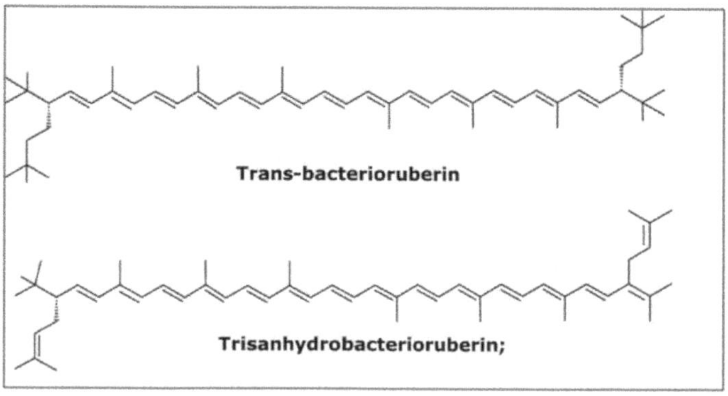

Figure 12.2 The structure of unique carotenoid synthesized by haloarchaeal species (Mandelli et al. 2012).

Animals and humans are unable to synthesize carotenoids due to the lack of enzymes and precursors and are usually administrated in the oral form. Various carotenes such as lycopene, β-carotene and oxygenated carotenoids including zeaxanthin, asthaxanthin and canthaxanthin are extensively used in the field of biotechnology as well as biomedicine (Fassett et al. 2012; Jehlička et al. 2013; Tanaka et al. 2012).

Halophiles and salt tolerance

Halophiles are capable of tolerating highly concentrated salt due to two major mechanisms; primarily, maintenance of cytoplasm osmolality regulation (Saum and Müller 2008). According to his explanation, accumulation of molar concentration of potassium and chloride in the cytoplasm leads to comprehensive adaptations of intracellular enzymatic machinery. It is allows unusual conformation and activity of proteins at near saturating salt concentrations. The second approach is to eliminate salt from the cytoplasm for synthesis or accumulate organic compatible solutes, which do not interfere in enzymatic activity (Oren 2008). The salt requirement and tolerance of species change from each other by their growth conditions such as temperature and media composition (Ventosa et al. 1998; Yadav et al. 2019; Yadav et al. 2015). *Halomonaselongata* is extreme halotolerant and needs at least 0.5 M salt concentration for growth, and the optimum growth condition of *Marinococcushalophilis*, is at 0–5.5 M concentration at 35°C (Bowers et al. 2011). *Salinivibriocosticola, Halomonaselongata*, and *Halomonasisrealensis* are significant organisms widely used for studying physiological and biochemical features for tolerance of extreme salt conditions (Ventosa et al. 1998).

Internal ion concentration

Halophilic organisms can withstand high external osmotic pressure with various mechanisms that include osmoprotectants present in the cytoplasm and selective influx of potassium ion in to the cytoplasm. For optimization of growth, inorganic ions and low molecular weight organic compounds such as Na, K, Cl, S, and carbonates are present in the external environment. The most significant elements in the cell such as Na, Cl, and K and osmotic pressure is sustained by substitution of Na and K^+ ions along with ATP utilization. K^+ ions are the principal osmoprotectants accumulated in the cell due to low osmotic stress level to restore turgor pressure. If NaCl concentration reaches up to 1.6 M concentration, the sum of the Na^+ and K^+ ions are higher than the external environmental condition. If the concentration of the internal ions is increased with external NaCl concentration, ratio of K^+/Na^+ ions are decreased (Hamaide et al. 1983). Cellular solute concentration is analyzed using radioactively marked molecular markers and the number of cell pellets invaded in the cell, so quantifying osmotic concentration of cell can succeed with periplasmic space which excludes total water space in the cytoplasm and cell. Specific intracellular and extracellular water spaces are essential for maintaining equilibrium and sodium plus chloride is generally present in a large quantity in the media. The cell wall comprising of polar lipid layer performs an important role in the ion exchange process. As salt concentration rises, a number of uncharged phosphatidylethanolamine could be reduced, which would increase negatively charged phosphatidylglycerol or cardiolipin. The transfer to high negatively charged ions in membrane intensified salt concentration with the aid of anionic lipids at the membrane provides high Na^+ concentration (Ventosa et al. 1998).

Compatible solutes

Compatible solutes are organic molecules with neutral charge which provides osmotic regulation in high osmotic stress conditions. Most of the halophilic organisms have compatible solutes such as specific amino acids (glycine, betaine, and ectoine). Compatible organic solutes provide two model mechanisms, i.e., preferential exclusion and preferential interaction model (Roberts 2005).

Preferential exclusion model

Halophiles having intracellular organic compounds such as, amino acids, glycine betaine, ectoine, hydroectoine and provide osmotic balance without interfering with the metabolic functions of the cell. Organic solutes play a significant role in protein stabilization by excluding it from the hydration shell as it helps to protect proteins from destabilizing, drying or freezing due to environmental changes (Roberts et al. 2005).

Preferential interaction model

Osmotic solute transport from higher to lower concentrations with increase in the surface water tension both the osmolality mechanism plays significant role in maintaining ion equilibrium inside and outside the cell, which are expressed more actively to provide osmotic balance and decrease the concentration enzymes are at an inactive stage due to regulations of mechanism (Oren et al. 2002).

Synthesis of nanoparticles

Biological synthesis of nanoparticles has advantages such as cost effectiveness, eco-friendly, no harsh chemicals are used, reproducibility and low expenditure as compared to chemical synthesis. Nanoparticles synthesis can be achieved by using two approaches which include the top down approach and bottom up approach. The top down method involves the reduction of various bulk nanoparticles to a smaller size using different mechanisms such as grinding, cutting, etching of materials whereas bottom up process works on building up of materials from its atomic level, molecule by molecule or in clusters (Huang et al. 2005).

Biosynthesis of nanoparticles by microorganisms

Microorganisms have the capacity to detoxify toxic chemicals and heavy metals that enter into the body due to their chemical detoxification technique using reductase enzyme and energy dependent efflux from the cell by membrane proteins, thus reducing metal ions into metal nanoparticles (Das et al. 2017). The reductase enzymes and energy-dependent membrane protein efflux play an important role in the detoxification of heavy metals and convert them into useful compound nanoparticles such as, silver and copper nanoparticles. Microbial synthesis nanoparticles can be achieved by using intracellular enzymes/biomass and extracellular enzymes/supernatant or derived compounds (Hamad 2019).

Intracellular method ions are transported into the cell through the bacterial cell wall composed of peptidoglycan, which helps in binding of metal ions with metal binding sites and leads to chemical modification of the cell wall that in turn results in penetration of the ions to the cell with help of various enzymes (Li et al. 2011). The extracellular synthesis of nanoparticles involves trapping metal ions on the surface of the cell and reducing them with assistance of NADH dependent reductase enzymes (Fig. 12.3).

Scientists mainly prefer the extracellular method of biological synthesis so as to avoid the down streaming process during the recovery of nanoparticles such as, sonication for lysis of the cell wall, washing and centrifugation for purification of synthesized nanoparticles. The metal resistant genes and reductase enzymes, are cofactors that help in synthesis of nanoparticles along with natural capping agents also provide an additional stability to nanoparticles formation. The extreme halophilic organisms such *Halobacterium salinarium*, *Halococcus salifodinae* are also used for the synthesis of biologically active silver nanoparticles (Table 12.1).

Synthesis of nanoparticles using halophiles

The nanoparticle synthesizes have mainly focused on metallic nanoparticles including silver, gold and copper, etc. Halophilic microorganisms are largely taken into consideration because of their new biomolecules for industrial interest for improving human life by utilizing enzymes,

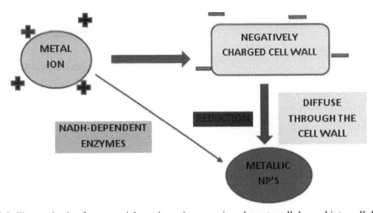

Figure 12.3 The synthesis of nanoparticles using microorganisms by extracellular and intracellular method.

Table 12.1 The nanoparticles synthesized from microorganisms.

Organisms	Nanoparticle	Location	Size	Reference
Halobacterium salinarium	Ag	Extracellular	20 nm	Mochalov et al. 2016
Halococcus salifodinae BK$_3$	Ag	Extracellular	12 nm	Srivastava et al. 2013
Aspergillus fumigatus	Ag	Extracellular	5–25 nm	Bhainsa et al. 2006
Bacillus cereus	Ag	Intracellular	4–5 nm	Babu et al. 2009
Bacillus licheniformis	Ag	Extracellular	50 nm	Kalimuthu et al. 2008
Escherichia coli	Ag	Extracellular	50 nm	Gurunathan et al. 2009
Lactobacillus	CdS	Intracellular	4.9 nm	Prasad and Jha 2010

exopolysaccharides and phytohormones. Extreme environments help in the secretion of these biomolecules. *Haloferax, Haloarcula, Halococcus, Natronococcus* and *Halobacterium* are species that can secrete exopolysaccharides and can be used in various fields like cosmetics, petroleum industries, biotechnological, food, etc. (Des Marais and David 2005). The *Halomonas* species synthesize exopolysaccharides comprising of high sulphate content and substantial amounts of uronic acid. Sulphated exopolysaccharides are mainly used as anti-coagulant, anti-thrombotic, anti-inflammatory, antioxidant, anti-viral, anti-proliferative, etc. (Llamas et al. 2012).

The novel halophilic organisms *Natronotaleasambharensis* (Nat AK-103[T]) were located in Sambhar Lake, Rajasthan and were used for purification, extraction and characterization of exopolysaccharides and synthesis of gold nanoparticles. Exopolysaccharide Nat-103 have high contents of uronic acid which makes them strong antioxidants and can produce stable gold nanoparticles which in turn can be used for therapeutically applications (Singh et al. 2019). Another important criterion is biosynthesis of quantum dots from halophilic organisms. Fifteen bacterial isolates were collected from different parts of the world including Atacama salt flat, Uyuni salt flat and the Dead Sea. The *Halobacillus* DS2 strain was able to extracellularly synthesize the CdS nanoparticles at high salt concentration (3–22%). The behaviour of the CdS nanoparticle as quantum dots was observed and showed changes in colour at different times from blue to red after 32 hours (Bruna et al. 2019).

Conjugated nanoparticles using various metabolites such as microbial metabolites play an important role in the drug delivery system such a antibody conjugated nanoparticle and polymer conjugated nanoparticles. Research has focused on the haloarchaeal or extremophilic organism cell materials such as gas vesicle used for reduction of *Salmonella SopB antigen* with the administration of attenuated bacteria (DasSarma et al. 2014). However, haloarchaeal biological compounds such as carotenoids are not involved for nanoparticle synthesis. Although research has focused on the boom of haloarchaea and its importance. So it can be hypothesized that haloarchaeal carotenoid conjugated nanoparticle might be able to provide better opportunities in different branches of life especially in the pharmaceutical field (Fig. 12.4). Extraction and purification procedures were simple as compared with other microbes because the Cell wall lysis with less salt conditions (Rodrigo-Baños et al. 2015).

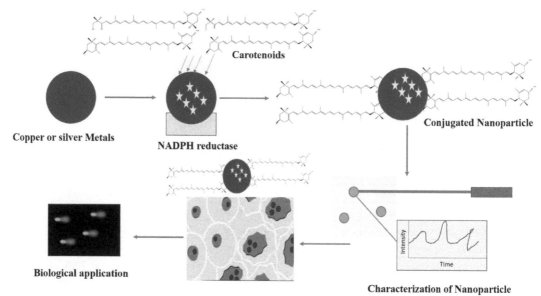

Figure 12.4 Schematic representation of conjugated haloarchaeal carotenoid nanoparticle.

Factors affecting the biological synthesis

Several factors influence synthesized nanoparticles such as nature, shape and the size:

i) **pH:** The size of nanoparticles can be controlled by altering pH of the solution. The size and shape of the synthesized nanoparticles may differ with pH (Armendariz et al. 2004).

ii) **Temperature:** It is one of most essential factors that affect synthesis of nanoparticles, as the temperature of reactive medium determines the nature of the nanoparticles. In biological synthesis temperature can be below 50°C or even at room temperature, but in the case of the physical process of Ag nanoparticle synthesis, ethylene glycol is treated along with silver nitrate ($AgNO_3$) at a temperature ranging from 300–350°C (Rai et al. 2006).

iii) **Pressure:** The shape and size of nanoparticles are determined by pressure conditions in the reaction medium (Dinh et al. 2015). The ambient pressure provided is directly proportional to the reduction rate of metal ions.

iv) **Environment:** The surrounding environment plays an important role for nanoparticles in both the physical and chemical structures. The oxidation or corrosion of metals in the environment results in the formation of core shell nanoparticles from single nanoparticles and the environment plays a vital role in thickening and large sized nanoparticles (Nowack et al. 2007).

v) **Time:** The quality and nature of the nanoparticles are greatly influenced by the length of time in the reaction medium. Variation in time may result in shrinking or the aggregation of the nanoparticles due to a long incubation period. The exposure to light, storage condition can also affect the synthesized nanoparticles (Kuchibhatla et al. 2012).

Biological applications of halophilic nanoparticles and synthesis

Halotolerant organisms can survive stress conditions providing a better application in the reduction of various effluent produced by industries (Margesin et al. 2001). Silver nanoparticles are potent and broad-spectrum inhibiting the agent so is used in the biosynthesis of nanoparticles from halophiles (Srivastava et al. 2013). From ancient civilization to now, silver and silver salts have being used for the treatment of eye infections using oligo dynamic properties of silver (Russell and Hugo 1994). The use of silver in the field of agriculture and medicine as antimicrobial, antibacterial and antioxidant (Siddiqi et al. 2018) are commonly seen. The nanoparticles are quite small which helps them to diffuse easily into the cell wall of the microorganisms. AgNPs inhibits the growth and multiplication of various bacteria including *Staphylococcus aureus, Pseudomonas aeruginosa, Enterococcus faecalis,* etc. by binding Ag/Ag^{2+} with biomolecules (Singh et al. 2015). While the antibacterial activity of silver nanoparticles is not clearly understood, it is convincingly explained through free radical formation. The free radical penetration into a living organisms attacks the lipid membrane that leads to dissociation, damage and finally death of the organism. Ag^+ ions permeates into the cell by rupturing the cell wall ultimately leading to protein denaturation and death, since Ag ions are positively charged and can easily bind to a cell wall containing N, S and P (Niraimathi et al. 2013) (Fig. 12.5).

The size of the nanoparticles is also important, as smaller nanoparticles are more effective than larger ones. Biofilms are surface-bound microbial communities that help in persistent bacterial life. Biofilm formation results in various infections around world; for example, oral biofilms are the major cause for tooth decay or dental caries by dissolving the enamel and turning them slightly acidic and are extensively seen due to their higher prevalence in humans (Bowen 2013). The bacteria is able to produce exopolysaccharides which protect them from antibacterial drug action. There are various nanoparticles that have been seen as highly active against biofilms and related organisms (Kalishwaralal et al. 2010). Their highly active conjugated haloarchaeal and carotenoid nanoparticles might have more antibacterial and anticancer activities.

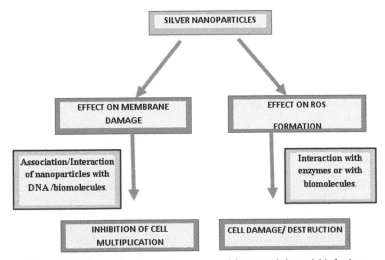

Figure 12.5 The mechanism of sliver nanoparticles towards bacterial infection.

Conclusion and future prospects

Nanotechnology is a burgeoning field in science that deals with various aspects of life that enables us in building a suitable and healthy life. Nanoparticles are solid or colloidal suspensions with a size ranging from 10–100 nm. Scientists are reproductively choosing the green route, by using ambient environmental conditions, as they are eco-friendly and cost effective. Halophilic organisms can survive at extreme salt conditions with two mechanisms, internal ion concentration and compatible solutes. Halophilic organisms have an excessive enzymatic system, which can contribute for synthesis nanoparticles. Nanoparticles have an ambient affect to act as antimicrobial, antibacterial, antioxidant and antibiofilm activities. The size, morphology of the nanoparticles and time taken for synthesis during green synthesis must be studied in more detail. The extremophilic organisms need be taken into more consideration for the betterment of human life.

Acknowledgements

The authors thanks to SERB-EMEQ/051/2014 and EEQ/2018/001085/2019 for financial assistance. J. Moopantakath wishes to thank ICMR-SRF (45/21/2019-BIO/BMS) fellowship Govt. of India and the research facilities supported by Central University of Kerala.

References

Abdelghany, T.M., A.M. Al-Rajhi, M.A. Al-Abboud, M.M. Alawlaqi, A.G. Magdah, E.A. Helmy et al. 2018. Recent advances in green synthesis of silver nanoparticles and their applications: About future directions. A review. Bionanoscience. 8: 5–16.

Acosta, E. 2009. Bioavailability of nanoparticles in nutrient and nutraceutical delivery. Curr. Opin. Colloid Interface Sci. 14: 3–15.

Armendariz, V., I. Herrera, M. Jose-yacaman, H. Troiani, P. Santiago and J.L. Gardea-Torresdey. 2004. Size controlled gold nanoparticle formation by *Avenasativa* biomass: use of plants in nanobiotechnology. J. Nanopart. Res. 6: 377–382.

Babu, M.G. and P. Gunasekaran. 2009. Production and structural characterization of crystalline silver nanoparticles from *Bacillus cereus* isolate. Colloid Surface B. 74: 191–195.

Bansal, Vipul, Debabrata Rautaray, Atul Bharde, Keda Ahire, Ambarish Sanyal, Absar Ahmad et al. 2005. Fungus-mediated biosynthesis of silica and titania particles. J. Mater. Chem. 15: 2583–2589.

Batista, Carlos A. Silvera, Ronald G. Larson and Nicholas A. Kotov. 2015. Nonadditivity of nanoparticle interactions. Science 350: 1242477.

Bhainsa, C. Kuber and S.F. D'souza. 2006. Extracellular biosynthesis of silver nanoparticles using the fungus *Aspergillus fumigatus*. Colloid Surface B. 47: 160–164.

Bhaskar, Sonu, Furong Tian, Tobias Stoeger, Wolfgang Kreyling, Jesús M. de la Fuente, Valeria Grazú et al. 2010. Multifunctional nanocarriers for diagnostics, drug delivery and targeted treatment across blood-brain barrier: perspectives on tracking and neuroimaging. Part Fibre Toxicol. 7: 3.

Bhosale, Prakash and Paul S. Bernstein. 2005. Microbial xanthophylls. Appl. Microbiol. Biot. 68: 445–455.

Bouchotroch, Samir, Emilia Quesada, Ana del Moral, Inmaculada Llamas and Victoria Bejar. 2001. *Halomonasmaura* sp. nov., a novel moderately halophilic, exopolysaccharide-producing bacterium. Int. J. Syst. Evol. Microbiol. 51: 1625–1632.

Bowen, H. William. 2013. The Stephan curve revisited: Odontology 101: 2–8.

Bowers, J. Karen and Juergen Wiegel. 2011. Temperature and pH optima of extremely halophilic archaea: a mini-review. Extremophiles 15: 119–128.

Bruna, N., Bernardo Collao, A. Tello, P. Caravantes, N. Díaz-Silva, J.P. Monrás, N. Órdenes-Aenishanslinset et al. 2019. Synthesis of salt-stable fluorescent nanoparticles (quantum dots) by polyextremophile halophilic bacteria. Sci. Rep. 9(1): 1953.

Bursten, Julia R., Mihail C. Roco, Wei Yang, Yuliang Zhao, Chunying Chen, Kai Savolainen, Christoph Gerber et al. 2016. Nano on reflection A number of experts from different areas of nanotechnology describe how the field has evolved in the last ten years. Nat. Nanotechnol. 11: 828–834.

Das, Ratul Kumar, Vinayak Laxman Pachapur, Linson Lonappan, Mitra Naghdi, Rama Pulicharla, Sampa Maiti et al. 2017. Biological synthesis of metallic nanoparticles: plants, animals and microbial aspects. Nanotechnol. Environ. Eng. 2: 18.

Dasarahally-Huligowda, L.K., M.R. Goyal and H.A.R. Suleria. 2009. Halophiles, industrial applications. *In*: Encyclopedia of Industrial Biotechnology: Bioprocess, Bioseparation, and Cell Technology 1–43.

Dasarahally-Huligowda, Lohith Kumar, Megh R. Goyal and Hafiz Ansar Rasul Suleria (eds.). 2019. Nanotechnology Applications in Dairy Science: Packaging, Processing, and Preservation. CRC Press.

DasSarma, Priya, Vidya Devi Negi, Arjun Balakrishnan, Ram Karan, Susan Barnes, Folasade Ekulona et al. 2014. Haloarchaeal gas vesicle nanoparticles displaying *Salmonella* SopB antigen reduce bacterial burden when administered with live attenuated bacteria. Vaccine 32: 4543–4549.

DES MARAIS and J. David. 2005. The potential for habitable environments in the basaltic plains and columbia hills of gusev crater, mars. Salt Lake City Annual Meeting.

Dharahaas, C. and R.V. Geetha. 2018. Antibacterial action of bionanoparticles on *Streptococcus* mutants and *Enterococcus faecalis*—An *in vitro* study. Drug Invent. Today 10: 12.

Dinh, Ngo Xuan, Nguyen Van Quy, Tran QuangHuy and Anh-Tuan Le. 2015. Decoration of silver nanoparticles on multiwalled carbon nanotubes: antibacterial mechanism and ultrastructural analysis. J. Nanomater. J. Nanomater. 16: 63.

Drexled, K. Eric. 1986. Engines of Creation: The Coming Era of Nanotechnology.

Durvasula, Ravi and D.V. Subba Rao. 2018. Extremophiles: Nature's amazing adapters. In Extremophiles 1–18. CRC Press.

Edelstein, R.L., C.R. Tamanaha, P.E. Sheehan, M.M. Miller, D.R. Baselt, LJetal Whitman et al. 2000. The BARC biosensor applied to the detection of biological warfare agents. Biosens. Bioelectron. 14: 805–813.

Eric, D.K. 1986. Engines of creation: the coming era of nanotechnology. Anchor Book.

Fadeel, Bengt and Alfonso E. Garcia-Bennett. 2010. Better safe than sorry: understanding the toxicological properties of inorganic nanoparticles manufactured for biomedical applications. Adv. Drug Deliv. Rev. 62: 362–374.

Fassett, R.G. and J.S. Coombes. 2012. Astaxanthin in cardiovascular health and disease. Molecules. 17: 2030–2048 .

Feynman, R.P. 1959. December. Plenty of Room at the Bottom. In APS Annual Meeting.

Gamez, G., J.L. Gardea-Torresdey, K.J. Tiemann, J. Parsons, K. Dokken and M. Jose Yacaman. 2003. Recovery of gold (III) from multi-elemental solutions by alfalfa biomass. Adv. Environ. Res. 7: 563–571.

Gericke, M. and A. Pinches. 2006. Biological synthesis of metal nanoparticles. Hydrometallurgy 83: 132–140.

Gochnauer, B. Margaret, Gary G. Leppard, Prayad Komaratat, Morris Kates, Thomas Novitsky and Donn J. Kushner. 1975. Isolation and characterization of *Actinopolysporahalophila*, gen. et sp. nov. an extremely halophilic actinomycete, an extremely halophilic actinomycete. Can. J. Microbiol. 21: 1500–1511.

Gradelet, S., A.M. Le Bon, R. Bergès, M. Suschetet and P. Astorg. 1998. Dietary carotenoids inhibit aflatoxin B1-induced liver preneoplastic foci and DNA damage in the rat: role of the modulation of aflatoxin B1 metabolism. Carcinogenesis. 19: 403–411 .

Gunde-Cimerman, N., A. Plemenitaš and A. Oren. 2018. Strategies of adaptation of microorganisms of the three domains of life to high salt concentrations. FEMS Microbiol Rev. 42: 353–375.

Gurunathan, Sangiliyandi, Kalimuthu Kalishwaralal, Ramanathan Vaidyanathan, Deepak Venkataraman, Sureshbabu Ram Kumar Pandian, Jeyaraj Muniyandi et al. 2009. Biosynthesis, purification and characterization of silver nanoparticles using *Escherichia coli*. Colloid Surface B. 74: 328–335.

Gurunathan, Sangiliyandi, JaeWoong Han, Jung Hyun Park and Jin-Hoi Kim. 2014. A green chemistry approach for synthesizing biocompatible gold nanoparticles. Nanoscale Res. Lett. 9: 248.

Hamaide, F.R.A.N.C.E.T.T.E., DONN J. Kushner and G. DENNIS Sprott. 1983. Proton motive force and Na$^+$/H$^+$ antiport in a moderate halophile. J. Bacteriol. Res. 156: 537–544.

Hamad, M.T. 2019. Biosynthesis of silver nanoparticles by fungi and their antibacterial activity. International Journal of Environmental Science and Technology. 16: 1015–1024.

Hanahan, D. and R.A. Weinberg. 2011. Hallmarks of cancer: the next generation. Cell. 144: 646–674.

Hu, Yong and Christof M. Niemeyer. 2019. From DNA nanotechnology to material systems engineering. Adv. Mater. 31: 1806294.

Huang, Jiaxing, Franklin Kim, Andrea R. Tao, Stephen Connor and Peidong Yang. 2005. Spontaneous formation of nanoparticle stripe patterns through dewetting. Nat. Mater. 4: 896.

Huang, B., W.D. Abraham, Y. Zheng, S.C.B. López, S.S. Luo and D.J. Irvine. 2015. Active targeting of chemotherapy to disseminated tumors using nanoparticle-carrying T cells. Sci. Transl. Med. 7: 291ra94.

Hull, Matthew and Diana Bowman (eds.). 2018. Nanotechnology Environmental Health and Safety: Risks, Regulation, and Management. William Andrew.

Iravani, Siavach, Hassan Korbekandi, Seyed Vahid Mirmohammadi and Behzad Zolfaghari. 2014. Synthesis of silver nanoparticles: chemical, physical and biological methods. Res. Pharm. Sci. 9: 385.

Jayakumar, R., K.P. Chennazhi, R.A.A. Muzzarelli, H. Tamura, S.V. Nair and N. Selvamurugan. 2010. Chitosan conjugated DNA nanoparticles in gene therapy. Carbohydr. Polym. 79: 1–8.

Jehlička, J., H.G.M. Edwards and A. Oren. 2013. Bacterioruberin and salinixanthin carotenoids of extremely halophilic Archaea and Bacteria: a Raman spectroscopic study. Spectrochim. Acta A. 106: 99–103.

Jones, A. Samantha, Philip G. Bowler, Michael Walker and David Parsons. 2004. Controlling wound bioburden with a novel silver-containing Hydrofiber® dressing. Wound Repair and Regeneration 12: 288–294.

Kalimuthu, Kalishwaralal, Ramkumarpandian Suresh Babu, Deepak Venkataraman, Mohd Bilal and Sangiliyandi Gurunathan. 2008. Biosynthesis of silver nanocrystals by *Bacillus licheniformis*. Colloid Surface B. 65: 150–153.

Kalishwaralal, Kalimuthu, Selvaraj Barath Mani Kanth, Sureshbabu Ram Kumar Pandian, Venkataraman Deepak and Sangiliyandi Gurunathan. 2010. Silver nanoparticles impede the biofilm formation by *Pseudomonas aeruginosa* and *Staphylococcus epidermidis*. Colloid Surface B: Biointerfaces 79: 340–344.

Kolluru, L.P., S.A. Rizvi, M. D'Souza and M.J. D'Souza. 2013. Formulation development of albumin based theragnostic nanoparticles as a potential delivery system for tumor targeting. J. Drug Target. 21: 77–86.

Kuchibhatla, Satyanarayana V.N.T., Ajay S. Karakoti, Donald R. Baer, Saritha Samudrala, Mark H. Engelhard, James E. Amonette et al. 2012. Influence of aging and environment on nanoparticle chemistry: implication to confinement effects in nanoceria. J. Phys. Chem. C. 116: 14108–14114.

Kumar, C. Ganesh and Suman Kumar Mamidyala. 2011. Extracellular synthesis of silver nanoparticles using culture supernatant of *Pseudomonas aeruginosa*. Colloid Surface B 84: 462–466.

Li, Xiangqian, Huizhong Xu, Zhe-Sheng Chen and Guofang Chen. 2011. Biosynthesis of nanoparticles by microorganisms and their applications. J. Nanomater.

Llamas, Inmaculada, Hakima Amjres, Juan Antonio Mata, Emilia Quesada and Victoria Béjar. 2012. The potential biotechnological applications of the exopolysaccharide produced by the halophilic bacterium *Halomonasalmeriensis*. Molecules 17: 7103–7120.

Luo, Xiliang, Aoife Morrin, Anthony J. Killard and Malcolm R. Smyth. 2006. Application of nanoparticles in electrochemical sensors and biosensors. Electroanalysis: An International Journal Devoted to Fundamental and Practical Aspects of Electroanalysis 18: 319–326.

Mackinnon, E.S., A. Venket Rao and L.G. Rao. 2011. Dietary restriction of lycopene for a period of one month resulted in significantly increased biomarkers of oxidative stress and bone resorption in postmenopausal women. J Nutr. Health Aging. 15: 133–138.

Mandelli, Fernanda, Viviane S. Miranda, Eliseu Rodrigues and Adriana Z. Mercadante. 2012. Identification of carotenoids with high antioxidant capacity produced by extremophile microorganisms. World J. Microb. Biot. 28: 1781–1790.

Margesin, Rosa and Franz Schinner. 2001. Potential of halotolerant and halophilic microorganisms for biotechnology. Extremophiles 5: 73–83.

McClements, D.J. 2015. Nanoscale nutrient delivery systems for food applications: improving bioactive dispersibility, stability, and bioavailability. J. Food Sci. 80: N1602–N1611.

Mezzomo, Natália and Sandra R.S. Ferreira. 2016. Carotenoids functionality, sources, and processing by supercritical technology: a review. J. Chem-Ny.

Mochalov, Konstantin, Daria Solovyeva, Anton Chistyakov, Boris Zimka, EugeniLukashev, Igor Nabiev et al. 2016. Silver nanoparticles strongly affect the properties of bacteriorhodopsin, a photosensitive protein of *Halobacterium salinarium* purple membranes. Mater. Today, Proc. 3: 502–506.

Mohanta, Yugal Kishore, Kunal Biswas, Jaya Bandyopadhyay, Abiral Tamang, Debashis De, Dambarudhar Mohanta, Sujogya Kumar Panda et al. 2018. Abutilon indicum (L.) sweet leaf extracts assisted bio-inspired synthesis of electronically charged silver nano-particles with potential antimicrobial, antioxidant and cytotoxic properties. Mater. Focus 7: 94–100.

Monopoli, P. Marco, ChristofferÅberg, Anna Salvati and Kenneth A. Dawson. 2012. Biomolecular coronas provide the biological identity of nanosized materials. Nat. Nanotechnol. 7: 779.

Mozetič, Miran, AlenkaVesel, Gregor Primc, C. Eisenmenger-Sittner, J. Bauer, A. Eder, G.H.S. Schmidet et al. 2018. Recent developments in surface science and engineering, thin films, nanoscience, biomaterials, plasma science, and vacuum technology. Thin Solid Films 660: 120–160.

Neethirajan, Suresh and Digvir S. Jayas. 2011. Nanotechnology for the food and bioprocessing industries. Food Bioproc. Tech. 4: 39–47.

Niraimathi, K.L., V. Sudha, R. Lavanya and P. Brindha. 2013. Biosynthesis of silver nanoparticles using *Alternantherasessilis* (Linn.) extract and their antimicrobial, antioxidant activities. Colloid Surface B 102: 288–291.

Nisar, Nazia, Li Li, Shan Lu, Nay Chi Khin and Barry J. Pogson. 2015. Carotenoid metabolism in plants. Mol. Plant. 8: 68–82.

Nowack, Bernd and Thomas D. Bucheli. 2007. Occurrence, behavior and effects of nanoparticles in the environment. Environ. Pollut. 150: 5–22.

Oren, A. 2001. The bioenergetic basis for the decrease in metabolic diversity at increasing salt concentrations: implications for the functioning of salt lake ecosystems. In Saline Lakes. 61–72. Springer, Dordrecht.

Oren, Aharon, Mikal Heldal, Svein Norland and Erwin A. Galinski. 2002. Intracellular ion and organic solute concentrations of the extremely halophilic bacterium *Salinibacterruber*. Extremophiles 6: 491–498.

Oren, Aharon. 2008. Microbial life at high salt concentrations: phylogenetic and metabolic diversity. Saline Systems 4: 2.

Pandey, B.D. 2012. Synthesis of zinc-based nanomaterials: a biological perspective. IET Nanobiotechnol. 6: 144–148.

Pantarotto, Davide, Charalambos D. Partidos, Johan Hoebeke, Fred Brown, E.D. Kramer, Jean-Paul Briand, Sylviane Muller et al. 2003. Immunization with peptide-functionalized carbon nanotubes enhances virus-specific neutralizing antibody responses. Chem. Biol. 10: 961–966.

Patil, Sushama, Julio Fernandes, RudreshwarTangasali and Irene Furtado. 2014. Exploitation of *Haloferaxalexandrinus* for biogenic synthesis of silver nanoparticles antagonistic to human and lower mammalian pathogens. J. Clust. Sci. 25: 423–433.

Patra, Jayanta Kumar and Kwang-Hyun Baek. 2014. Green nanobiotechnology: factors affecting synthesis and characterization techniques. J. Nanomater. 219.

Prasad, K. and Anal K. Jha. 2010. Biosynthesis of CdS nanoparticles: an improved green and rapid procedure. J. Colloid Interface Sci. 342: 68–72.

Rai, Akhilesh, Amit Singh, Absar Ahmad and Murali Sastry. 2006. Role of halide ions and temperature on the morphology of biologically synthesized gold nanotriangles. Langmuir. 22: 736–741.

Remya, V.R., V.K. Abitha, P.S. Rajput, A.V. Rane and A. Dutta. 2017. Silver nanoparticles green synthesis: a mini review. Chemistry International 3: 165–171.

Rivera, S.M. and R. Canela-Garayoa. 2012. Analytical tools for the analysis of carotenoids in diverse materials. J. Chromatogr. A. 1224: 1–10.

Rizvi, A.A. Syed and Ayman M. Saleh. 2018. Applications of nanoparticle systems in drug delivery technology. Saudi Pharm. J. 26: 64–70.

Roberts, F. Mary. 2005. Organic compatible solutes of halotolerant and halophilic microorganisms. Saline Systems 1: 5.

Rodrigo-Baños, Montserrat, Inés Garbayo, Carlos Vílchez, María José Bonete and Rosa María Martínez-Espinosa. 2015. Carotenoids from haloarchaea and their potential in biotechnology. Mar. Drugs 13: 5508–5532.

Roduner, Emil. 2006. Size matters: why nanomaterials are different. Chem. Soc. Rev. 35: 583–592.

Russell, A.D. and W.B. Hugo. 1994. 7 antimicrobial activity and action of silver. Prog. Med. Chem. 31: 351–370. Elsevier.

Salomoni, Roseli, P. Léo, A.F. Montemor, B.G. Rinaldi and M.F.A. Rodrigues. 2017. Antibacterial effect of silver nanoparticles in *Pseudomonas aeruginosa*. Nanotechnology, Science and Applications 10: 115.

Saum, S.H. and V. Müller. 2008. Regulation of osmoadaptation in the moderate halophile Halobacillus halophilus: chloride, glutamate and switching osmolyte strategies. Saline systems. 4 : 1–15.

Sarathy, Vaishnavi, Paul G. Tratnyek, James T. Nurmi, Donald R. Baer, James E. Amonette, Chan Lan Chun et al. 2008. Aging of iron nanoparticles in aqueous solution: effects on structure and reactivity. J Phys. Chem. C 112: 2286–2293.

Shen, B., Y. Ma, S. Yu and C. Ji. 2016. Smart multifunctional magnetic nanoparticle-based drug delivery system for cancer thermo-chemotherapy and intracellular imaging. Acs Appl Mater Inter. 8: 24502–24508.

Siddiqi, Khwaja Salahuddin, Azamal Husen and Rifaqat A.K. Rao. 2018. A review on biosynthesis of silver nanoparticles and their biocidal properties. J. Nanobiotechnology 16: 14.

Siddiqi, Khwaja Salahuddin, Azamal Husen and Rifaqat A.K. Rao. 2018. Mycosynthesis of bactericidal silver and polymorphic gold nanoparticles: physicochemical variation effects and mechanism. Nanomedicine 13: 191–207.

Singh, Prachi, Kunal Jain, Chirayu Desai, Onkar Tiwari and Datta Madamwar. 2019. Microbial community dynamics of extremophiles/extreme environment. In Microbial Diversity in the Genomic Era 323–332. Academic Press.

Singh, Richa, Utkarsha U. Shedbalkar, Sweety A. Wadhwani and Balu A. Chopade. 2015. Bacteriagenic silver nanoparticles: synthesis, mechanism, and applications. Appl. Microbiol. Biot. 99: 4579–4593.

Singh, Shweta, Kulwinder Singh Sran, Anil Kumar Pinnaka and Anirban Roy Choudhury. 2019. Purification, characterization and functional properties of exopolysaccharide from a novel halophilic *Natronotaleasambharensis* sp. nov. Int. J. Biol. Macromol. 136: 547–558.

Sintubin, Liesje, Willy Verstraete and Nico Boon. 2012. Biologically produced nanosilver: current state and future perspectives. Biotechnol. Bioeng. 109: 2422–2436.

Srivastava, Pallavee, Judith Bragança, Sutapa Roy Ramanan and Meenal Kowshik. 2013. Synthesis of silver nanoparticles using haloarchaeal isolate Halococcussalifodinae BK 3. Extremophiles 17: 821–831.

Steichen, Stephanie D., Mary Caldorera-Moore and Nicholas A. Peppas. 2013. A review of current nanoparticle and targeting moieties for the delivery of cancer therapeutics. Eur. J. Pharm. Sci. 48: 416–427.

Tanaka, T., M. Shnimizu and H. Moriwaki. 2012. Cancer chemoprevention by carotenoids. Molecules 17: 3202–3242.

Thombre, S. Rebecca, Vinaya D. Shinde, Radhika S. Oke, Sunil Kumar Dhar and Yogesh S. Shouche. 2016. Biology and survival of extremely halophilic archaeon *Haloarcula marismortui* RR12 isolated from Mumbai salterns, India in response to salinity stress. Sci. Rep. 6: 25642.

Torregrosa-Crespo, Javier, Zaida Montero, Juan Luis Fuentes, Manuel ReigGarcía-Galbis, Inés Garbayo, Carlos Vílchez et al. 2018. Exploring the valuable carotenoids for the large-scale production by marine microorganisms. Mar. Drugs 16: 203.

Ventosa, A., J.J. Nieto and A. Oren. 1998. Biology of moderately halophilic aerobic bacteria. Microbiol Mol Biol R. 62: 504–544.

Vílchez, Carlos, Eduardo Forján, María Cuaresma, Francisco Bédmar, Inés Garbayo and José M. Vega. 2011. Marine carotenoids: biological functions and commercial applications. Mar. Drugs 9: 319–333.

Wagner, Volker, Anwyn Dullaart, Anne-Katrin Bock and Axel Zweck. 2006. The emerging nanomedicine landscape. Nat. Biotechnol. 24: 1211.

Wolfbeis, O.S. 2015. An overview of nanoparticles commonly used in fluorescent bioimaging. Chem. Soc. Rev. 44: 4743–4768.

Yadav, A.N., D. Sharma, S. Gulati, S. Singh, R. Dey, K.K. Pal et al. 2015. Haloarchaea endowed with phosphorus solubilization attribute implicated in phosphorus cycle. Sci. Rep. 5: 12293.

Yadav, A.N., S. Gulati, D. Sharma, R.N. Singh, M.V.S. Rajawat, R. Kumar et al. 2019. Seasonal variations in culturable archaea and their plant growth promoting attributes to predict their role in establishment of vegetation in Rann of Kutch. Biologia 74: 1031–1043. doi:10.2478/s11756-019-00259-2.

Zile, M.H. 1998. Vitamin A and embryonic development: an overview. J. Nutr. 128: 455S–458S.

Chapter 13

Biocontrol Potential and Applications of Extremophiles for Sustainable Agriculture

Gajanan Mehetre,[1] *Vincent Vineeth Leo,*[1] *Garima Singh,*[2] *Prashant Dhawre,*[1] *Igor Maksimov,*[4] *Mukesh Yadav,*[5] *Kalidas Upadhyaya*[6] and *Bhim Pratap Singh*[3,*]

Introduction

Agriculture forms the base of human survival and remains the mainstay of the livelihood for the world. A major workforce in many countries including India is mostly dependent on agriculture and its related activities (Bruinsma 2017). Apart from providing food, agriculture is the main source of raw materials for various industries including textile and fibres, sugar and beverages and many others (Boehlje and Bröring 2011). Although in modern agricultural practices, different techniques are being used to increase crop productivity, plant pest and pathogens are the major biotic agents responsible for significant loss of crop productivity and damage to the plant. To manage pathogens and plant pests, several strategies are being used in agriculture. The most extensive and effective is the use of chemical pesticides. However, the widespread use of chemical pesticides and fertilizers in agriculture is a major health concern due to the potentially harmful effects of various chemical compounds on humans, other living organisms and the environment (Nicolopoulou-Stamati et al. 2016). Plant pathogens such as bacteria, fungi and viruses cause plant diseases resulting in a significant loss to crop productivity (Flood 2010; Oerke et al. 2012).

The use of chemical pesticides is also a matter of controversy, as their excessive use may lead to the development of resistance among pathogenic microorganisms and some of the chemicals also possess carcinogenic effects (Edgerton 2009; Damalas and Eleftherohorinos 2011). Some of the chemical compounds of pesticide formulations are also absorbed by most crops and their consumption may lead to systematic disorders in animals as well as in humans (Aktar et al. 2009; Alavanja et al. 2013). As a consequence, there is a need to implicate more environmentally friendly methods in agriculture without compromising on the economic viability of the crop (Tracy 2015). Biological control of phytopathogens has been considered the most viable option to confront plant

[1] Department of Biotechnology, Aizawl, Mizoram University, Mizoram 796004, India.

[2] Department of Botany, Pachhunga University College, College Veng, Mizoram 796001, India.

[3] Department of Agriculture and Environmental Sciences, National Institute of Food Technology Entrepreneurship & Management (NIFTEM), Industrial Estate, Kundli 131028, Sonepat, Haryana, India.

[4] Institute of Biochemistry and Genetics, Ufa Scientific Center, Russian Academy of Sciences, pr. Oktyabrya 71, 450054 Ufa, Russia.

[5] Department of Forestry, Mizoram University, Mizoram 796004, India.

[6] Department of Biotechnology, Pachhunga University College, College Veng, Mizoram 796001, India.

* Corresponding author: bhimpratap@gmail.com

diseases. Biocontrol is based on living organisms, especially microbes and their products, which can eliminate or suppress the growth of infectious pathogens of the plant.

A variety of microbial species has the potential to be used as biocontrol agents. Different research groups have shown the application of a diverse range of bacterial and fungal species in plant-growth promotion and pest management (Sivasakthi et al. 2014; Gowdar et al. 2018; Köhl et al. 2019). However, for effective adoption of microbes as a biocontrol agent, there is a need for a proper understanding of the complex interactions occurring among host plants, microbes and the environmental conditions (Heydari and Pessarakli 2010). Environmental conditions in which plants are growing are also important for their associate microbiome, which helps the plant to sustain various abiotic stresses (Grover et al. 2011). Microorganisms associated with such extreme environmental conditions are called extremophiles, which are important sources for various biotechnological applications. Recently, there is a growing interest in the use of such extremophilic microorganisms for sustainable agriculture, as most of these organisms are a great source for the development of effective biocontrol agents for the management of phytopathogens and plant health (Yadav 2017; Verma et al. 2017).

Extremophiles able to tolerate a wide range of adverse environmental conditions such as high to low temperatures, different pH range, acidic, alkaline and high salinity conditions. Due to the their remarkable properties of different bacterial members such as *Proteobacteria*, *Actinobacteria*, *Firmicutes* and also diverse members of archaeal, they are likely to have broad implications in biotechnology, including both environmental and industrial applications (Singh et al. 2018; Mehetre et al. 2019a; Yadav et al. 2019). They are classified according to the extreme conditions under which they can survive and grow (Rampelotto 2013). Thermophiles and hyper-thermophiles are microorganisms able to grow at high or very high temperatures respectively and therefore have the potential for promoting plant growth and could act as a biocontrol for crops growing under arid and semi-arid environments. Psychrophiles are known to grow at very low temperatures and could be beneficial for plant growth under cold climatic conditions (Schultz and Rosado 2019; Yadav et al. 2015a; Yadav et al. 2015b). Acidophiles and alkaliphiles are adapted to acidic or basic pH values and could sustain extreme pH conditions. Most importantly, halophiles can grow under high salinity conditions and therefore are more helpful for plant growth under these conditions (Yadav et al. 2015c). Importantly, most of these microorganisms could tolerate different extreme conditions and therefore are represented as poly-extremophiles. They can adapt to diverse environmental conditions such as high temperature hot springs which are acidic or alkaline in nature or very low nutrient and high pressure as well as saline conditions of deep oceans. Extremophiles include members of all three domains of life, i.e., bacteria, archaea and eukarya (Rothschild and Mancinelli 2001). Extremophiles have a wide range of applications in various biotechnological industries in the production of important biomolecules of human importance, industrial enzymes and biofuel industries (Coker 2016), as shown in Fig. 13.1. Nevertheless, their potential use for bio-control activity has been rarely studied. Only a few reports are available regarding the use of extremophiles as potential biocontrol agents for crop disease management and in improving crop productivity. In this chapter, the biocontrol potential and applications of extremophiles for managing phytopathogens as well as for plant-growth promotion are reviewed and discussed for improving the crop productivity through sustainable agriculture.

Extremophiles as a biocontrol agent for sustainable agriculture

Extremophiles can adapt against environmental change which is probably through the expression/ regulation of the specific genes. The phenomenon is likely to have broad implications in biotechnology, including both environmental and industrial applications. Therefore their use for biological disease control is an attractive alternative strategy for the control of plant diseases (Verma et al. 2017). Various parameters are being considered in deciding whether a biological system is feasible or not for the control of a plant pathogen with the ability to grow on host plants (Naga 2014). Plants

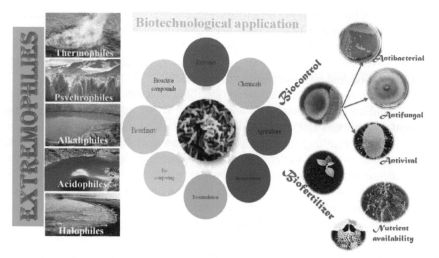

Figure 13.1 Extremophiles as a source for potential biotechnological application in various fields including agriculture as a biocontrol agent and plant growth promoting bacteria.

growing under extreme environmental conditions have their associated microbiome colonized as rhizosphere, phyllosphere and also in the internal tissue as an entophytic microorganisms (Yadav 2017). This plant microbiome is represented by a different class of microorganisms including fungi, archaea, bacteria and viruses. Most of these extremophilic microbes help the plant to grow under environmental stresses by making soluble nutrients available, fighting against different plant pathogens and also able to produce various secondary metabolites and phytohormones. These extremophilic microbes have been isolated from the plant growing under the diverse abiotic stress conditions, which help in plant-growth promotion and adaptations under harsh environments of temperatures, salinity, pH and drought stresses. Different microbial groups were reported as a biocontrol agent against a variety of plant pathogens from diverse extreme environments for the protection of plant health are shown in Table 13.1. Thermophilic microbes are important due to their high-temperature tolerance and production of thermostable enzymes and other microbial products. It has been reported that temperature is one of the important stress factors for the production of various antagonistic compounds to inhibit the growth of other microorganisms. Therefore thermophilic microorganisms could also act as a biocontrol agent for the management of phytopathogens. Nevertheless, their potential against phytopathogenic organisms has been rarely studied. A few of the strains belonging to different microbial species were isolated from diverse high-temperature environments such hot springs, compost and from the rhizospheric soil of arid regions to access their biocontrol potential against different types of plant pathogens. Saikia et al. (2011) reported the isolation of a bacterial strain belonging to *Brevibacillus laterosporus* from the mud of a natural hot water spring of Nambar Wild Life Sanctuary, Assam, India. This bacterial species has strong antimicrobial activity against phytopathogenic fungi such as *Fusarium oxysporum f.* sp. *ciceri*, *F. semitectum*, *Magnaporthe grisea* and *Rhizoctonia oryzae* with the suppressed blast disease of rice by 30–67% and protected the weight loss by 35–56.5%.

In a study for the development of plant growth-promoting microbial consortium for drought tolerance, Verma et al. (2018) isolated three *Bacillus* species, *Bacillus subtilis*, *Bacillus* sp. and *B. licheniformis* from the hot spring of Chumathang area of Leh, India. All three species were found capable of growing in the consortium. The treatment of consortium with Vigna radiate plant showed enhancement in the growth and was also found to be effective in enhancing the production of mung bean under field conditions. Thermophilic species of *Bacillus amyloliquefaciens* isolated from the compost extracts were reported to have a broad spectrum antifungal activity against various plant pathogenic fungi (Zouari et al. 2016). In this study, the author showed that *Bacillus* species possess

Table 13.1 Extremophilic microorganisms with potential biocontrol activity against various phytopathogens.

Extreme environment	Biological organisms	Targeted disease/pathogens	Diseased crop plants	Mechanism of control	Reference
Hot spring	*Brevibacillus laterosporus*	*F. oxysporum, F. Semitectum, Magnaporthe grisea, Rhizoctonia oryzae*	Blast disease of rice	Antimicrobial compounds	Saikia et al. 2011
Hot spring	*B. subtilis, B. licheniformis*	Not mentioned	Plant growth attributes of *Vigna radiata*	Plant growth promotion	Verma et al. 2018
Compost	*B. amyloliquefaciens*	*Pythium aphanidermatum*	Damping-off, tomato	Antifungal by cyclic lipopeptides	Zouari et al. 2016
Field soil (High temperature)	*B. subtilis, B. licheniformis*	*Phytophtora capsici*	Red peppers and suppress phytophtora blight	Antimicrobial activity	Lim and Kim 2010
Tunisian oasis ecosystem (drought and high salinity)	*Bacillus* sp.	*Botrytis cinerea, Galactomyces geotrichum, F. Oxysporum, Rhizoctonia solani, Verticillium longisporum*	Date palm	Antifungal by cyclic lipopeptides	El Arbi et al. 2016
Moss tissues (extremophilic endophyte)	*Pseudomonas* sp.	*Rizoctonia spp., Fusarium spp., Penicillium spp., Phy-tophthora spp., Phoma exiqua*	Potato tuber fungal diseases	Fungicidal and bactericidal activity	Shcherbakov et al. 2017
Andean tropical glaciers	*Pseudomonas* spp.	*Pythium ultimum*	*Triticum aestivum*	Production of phytohormones and antagonism	Rondón et al. 2019
Soil of a decayed mushroom	*Pedobacter* sp.	*Rhizoctonia solani, Botrytis cinerea,*	Not mentioned	Chitinolytic activity	Song et al. 2017
Orchard soil	*B. subtilis, Enterobacter aerogenes*	*Alternaria alternata, Armillariella mellea, Botrytis allii, Botrytis cinerea, Colletotrichum lindemuthianum, Monilinia fructicola, Penicillium expansum, Phytophthora cactorum, Pythium ultimum, Rhizoctonia solani, Verticillium dahliae, Venturia inequalis*	Cherry fruit brown rot and alternaria rot	Antagonistic activity	Utkhede and Sholberg 1986
High salt environment	*T. harzianum*	*F. oxysporum*	Tomato wilt	Antagonistic	Mohamed and Haggag 2006
Soil rice straw waste (high salinity)	*T. asperellum*	*F. solani, F. oxysporum, Sclerotium rolfesii*	Cowpea plant	Antifungal compounds	Hamed et al. 2015

Table 13.1 Contd. ...

...Table 13.1 Contd.

Extreme environment	Biological organisms	Targeted disease/pathogens	Diseased crop plants	Mechanism of control	Reference
Hypersaline ecosystem	*Virgibacillus marismortui, B. subtilis. B. pumilus, B. licheniformis, Terribacillus halophilus, Halomonas elongata, Planococcus rifietoensis, Staphylococcus equorum*	*Botrytis cinerea*	Grey mould of Strawberries	Hydrolytic enzymes, anti-Botrytis metabolites and volatiles	Essghaier et al. 2009
Saline soils of Howze-Soltan playa	*Bacillus sp., Paenibacillus sp., Staphylococcus sp.*	*Aspergillus parasiticus*	Not mentioned	Antifungal and antiaflatoxigenic activity	Jafari et al. 2018
Extreme environments (low temperature, high salinity, drought)	*Aureobasidium pullulans, Debayomyces hansenii, Meyerozyma guilliermondii, Metschnikowia fructicola, Rhodotorula mucilaginosa*	Not mentioned	Different plants	Lytic enzymes	Zajc et al. 2019
Salinity and alkaline environment	*T. asperellum*	*Alternaria alternate, Fusarium oxysporum*	*Populus davidiana × P. alba* var. pyramidalis	Antifungal activity	Guo et al. 2018

*itu*C and *itu*D, *srf*P and fend genes involved in the production of iturin, surfactin and fengycin, which are effective against the phytopathogens for the biocontrol of damping-off diseases of tomato.

Similarly, microorganisms growing at low temperatures were also isolated from various cold environments and reported antimicrobial activities against various plant pathogens. Psychrophilic bacteria with plant-growth-promoting attributes are valuable tools to develop cold-active biofertilizers and/or biopesticides. Recently, Rondón et al. (2019) showed that the potential of bacteria confined within glacial ice for plant-growth promotion and also in protecting plants from pathogens at low temperatures. In their study, among the four selected isolates of *Pseudomonas* sp., all the isolates showed solubilization of inorganic phosphates, production of phytohormones and antagonism against a phytopathogenic oomycete (*Pythium ultimum*). These cold-active biocontrol agents were also found to be effective in elongation of root and shoot length of the *T. Aestivum*. Psychrophilic microbes were also found to produce lytic enzymes such as chitinase effective against different fungal phytopathogens. A study by Song et al. (2017) reported the isolation of a novel psychrotolerant bacterium *Pedobacter* sp. isolated from decayed mushroom soil with the ability to use colloidal chitin as a substrate. Purification of enzymes led to the identification of six chitinase isozymes (*chi*I, *chi*II, *chi*III, *chi*IV, *chi*V, and *chi*VI) from C25-fractions expressed on SDS-PAGE gels at 25°C. The bacterium showed a high inhibition rate of 60.9 and 57.5% against *Rhizoctonia solani* and *Botrytis cinerea*, respectively indicating that *Pedobacter* sp. may be used as an effective biocontrol agent for the management of agricultural phytopathogens at lower temperatures.

Salinity is also one of the important abiotic factors, where the plant has to adapt to high salt concentration and low water activity. Microorganisms inhabiting such an environment are called halophiles, which are an important source to understand the mechanism involved in salinity stress adaptation. Halophiles mostly found in hypersaline environments such as solar salt, soda lakes and desert soils. They are important sources for halophilic hydrolytic enzymes such as amylases, cellulases, lipases and proteases which could be used as potential biocontrol agents in agriculture irrigated by saline water (Chamekh et al. 2019). Earlier gamma irradiation-induced salt-tolerant mutant strains *Trichoderma harzianum* were found effective in suppressing the growth of phytopathogenic fungi *Fusarium oxysporum*, the causal agent of tomato wilt disease (Mohamed and Haggag 2006). The mutants strain of *Trichodarma* was reported in a study to produce enzymes such as chitinases, cellulases, b-galactosidases and also some active compounds such as trichodermin, gliotoxin and gliovirin. The strain was shown to reduce wilt disease incidence with improved yield and mineral contents of tomato plants under both saline and non-saline soil conditions (Mohamed and Haggag 2006). Similarly, Guo et al. (2018) investigated the development of the saline and alkaline tolerant mutants of *Trichoderma asperellum* to mediate the effects of salinity or alkalinity stresses on *Populus davidiana* × *P. alba* var. *pyramidalis* (PdPap poplar) seedlings. The strains generated in this study were found to be effective in decreasing the incidence of pathogenic fungi infection under saline or alkaline stress (Guo et al. 2018). Halophiles isolated from a hyper-saline ecosystem were also reported to act as a biological control against pathogenic fungi during the post-harvest of storage of strawberries (Essghaier et al. 2009). Various halophilic strains of *Virgibacillus marismortui*, *B. subtilis*, *B. pumilus*, *B. licheniformis*, *Terribacillus halophilus*, *Halomonas elongata*, *Planococcus rifietoensis*, *Staphylococcus equorum* and *Staphylococcus* sp. have shown antifungal activity against *Botrytis cinerea*; a causative agent of the grey mould disease of the economically important plant of strawberries. Such halophilic bacteria could be exploited in commercial production and application of the effective strains under storage and greenhouse conditions for some economically important fruit and crop plants (Essghaier et al. 2009).

Furthermore, several other studies have also reported the potential application of various types of extremophilic microorganisms in improving plant health by plant growth-promoting attributes (Verma et al. 2015). Moreover, extremophiles also help the plant to tolerate different abiotic stress conditions such as drought, high to low temperature, high salinity, and low water and alkaline soil environments (Grover et al. 2011; Vurukonda et al. 2016). The ever increasing studies on extremophiles for their possible exploration in the management of plant diseases and in improving the plant efficiency to withstand harsh environmental conditions may lead to the development of suitable and environmentally friendly technologies to confront phytopathogens and pest for sustainable agricultural practices.

Plant diseases caused by different microbial groups

Bacterial diseases

Plant pathogenic bacteria cause many serious diseases of different crop plants causing damage and economic loss in plant productivity (Vidaver and Lambrecht 2004). Different bacterial species such as *Pseudomonas syringae, Ralstonia solanacearum, Agrobacterium tumefaciens, Xanthomonas oryzae* pv. *Oryzae, Xanthomonas campestris pathovars, Erwinia amylovora, Xylella fastidiosa, Pectobacterium carotovorum* were well-known plant pathogens causing different diseases in the plant (Mansfield et al. 2012).

Fungal diseases

Many fungal species are known to cause different diseases in plants. A plant disease caused by a fungal pathogen is often recognized from the particular plant organ infected and the type of symptom produced. Various diseases such as damping-off diseases, root and foot rots, vascular wilts, downy mildews, powdery mildews, leaf spots and blights, rusts, smut and some others found on plants at different stages of cropping (Brown and Ogle 1997).

Viral diseases

Many plant viruses cause significant economic losses in a wide range of crop plants. Viruses such as tobacco mosaic virus, tomato spotted wilt virus, tomato yellow leaf curl virus, cucumber mosaic virus, potato virus Y, cauliflower mosaic virus, African cassava mosaic virus, plum pox virus, brome mosaic virus and potato virus X are well known causative agents of different plants such as tobacco, pepper, potato, tomato, eggplant, cucumber and petunia (Scholthof et al. 2011).

Mechanisms of biocontrol and stress tolerance

Biocontrol mechanisms

A plant always has interactions with its associated microbiome, which includes a wide variety of microorganisms. Most of the organisms have a significant impact on plant health. Beneficial microorganisms interact with plants and also with other pathogenic bacteria in a variety of ways. Different mechanisms are used by microorganisms to suppress the growth of pathogens (Chowdhury et al. 2015). A correct understanding of the various mechanisms used by biocontrol agents could help to formulate more effective tools and techniques with more useful strategies for the formulation of the successful biocontrol agent (Heydari and Pessarakli 2010). Different microbial groups use diverse types of mechanisms for inhibition of the growth of different plant pathogenic microorganisms such as bacteria, fungi and viruses. Microbial biocontrol agents use several modes of action such as competition, antibiosis using the production of lytic enzymes and several secondary metabolites. Also, some of the biocontrol agents infect pathogens itself (parasitism), whereas others could be able to induce host response against pathogens. Microbial species isolated from extreme environments were also able to utilize several such mechanisms to suppress the growth of phytopathogen (Fig. 13.2). In general, most of these modes of actions are mutually exclusive and often differently used based on the type of pathogens. A few of such mechanisms used by biocontrol agents are described below.

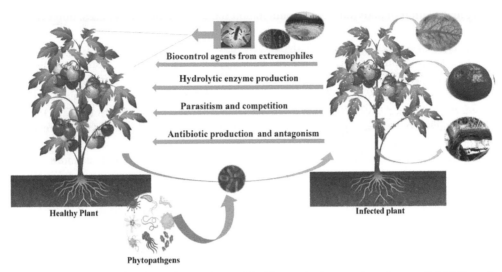

Figure 13.2 Various modes of action and strategies used by biocontrol microorganisms for growth inhibition of plant pathogens.

Using hydrolytic enzymes

Different microbial species including bacteria and fungi secrete some hydrolytic enzymes for the growth inhibition of plant pathogens (Beneduzi et al. 2013; Wang et al. 2019). Diverse microbial groups are known to produce different extracellular enzymes acting as biocontrol agents and could suppress the growth of phytopathogens by hydrolyzing the various polymeric compounds like chitin, proteins, cellulose and nucleic acids (Jadhav and Sayyed 2016; Sokurenko et al. 2016). Microbial species inhabiting extreme environments such as hot springs, submarine, alkaline and saline environments are a great source for production of such hydrolytic enzymes (Raddadi et al. 2015; Mehetre et al. 2018). Several enzymes such as chitinase, glucanase, protease, lipase were reported to be effective against different fungal pathogens.

Chitinase act on a fungal cell wall by degrading chitin polymer present in the cell wall of fungal phytopathogens. Other types of such enzymes acting on the cell wall are glucanase. They can be classified either as exo or endo-β-1,3-glucanases (β-1,3-glucan:glucanohydrolase) and able to lyse the pathogen fungal species (Marco and Felix 2016; Bashan and de-Bashan 2005). Various species of *Bacillus* like *S. marcescens*, *B. subtilis*, *B. cereus*, *B. subtilius*, *B. thuringiensis* and many more have the potential to produce hydrolytic enzymes for the biocontrol of phytopathogens like *R. solani*, *F. oxysporum*, *S. rolfsii*, *P. ultimum* (Jadhav and Sayyed 2016). Fungi are also known to produce several lytic enzymes to inhibit the growth of other fungal species. Cell wall-degrading enzymes like exo-β-1,3-glucanase, chitinase and protease are also reported to produce in yeast cells (Bar-Shimon et al. 2004). A mycoparasitic fungus *Trichoderma* also produces lytic enzymes, which are highly active as compared to the same class of enzymes from other microorganisms and plants (Viterbo et al. 2002). Besides, many rhizosphere microorganisms also produce some phosphatides for nutrient availability to the plant and act as plant growth-promoting bacteria (Azeem et al. 2015).

Microbial ribonucleases (binase) for viral control

RNA viruses cause the most diseases in humans, animals, as well as in plants. The development of successive agents against RNA viruses is more difficult, due to their high mutation rates, which enable them to quickly evade selective pressures and adapt new host (Müller et al. 2017). Therefore extensive efforts are needed to target the specific virus to prevent the development of viral resistance. In recent times, there are no confirmed measures available to combat viral diseases of animals and plants. A formulation of antiviral agents could inhibit the development of viruses; however, for the complete elimination of viruses, degradation of the viral RNA is the most viable approach that is being investigated with great effort due to their selective toxicity for the virally infected cells (Shah et al. 2017). The enzymes RNases (binase and burnase) are mostly isolated from microbial species belonging to *Bacillus* group. There are different types of RNases being produced by various *Bacillus* species such as *B. amyloliquefaciens* (termed barnase, P00648), *B. pumilus*, (binase, P00649), *B. altitudinis* (balnase, A0A0J1IDI7), *B. circulans* (P35078), and *B. coagulans* (P37203), however, the most studied have been produced by *B. pumilus* and *B. amyloliquefaciens* respectively (Sokurenko et al. 2016).

Nevertheless, most of the RNases of bacterial origin are considered as more promising therapeutics for anticancer treatment and less attention has been paid for plant viral diseases. Very scanty reports are available for the inhibitory effect of bacterial RNases on plant viruses. A new type of antiviral RNase active against tobacco mosaic virus has been isolated from the *Bacillus cereus* ZH14 strain (Zhou et al. 2008). RNases effective as antiviral agents are more commonly expressed in the genome of the plant to inhibit viral-based infections (Zhou et al. 2010; Cao et al. 2013). A study performed by Sharipova et al. (2015) showed the effect the *B. pumilus* ribonuclease on different phytopathogenic viruses to eliminate them from plant explants under *in vitro* conditions. Surprisingly, to date very few reports are available for isolation of microbial RNases from the extremophilic sources for their possible use as antiviral in control of plant viral diseases. In the future, there is a need to pay more attention to the isolation of extremophilic RNases from extreme microbial species of archea and bacteria.

Production of secondary metabolites

Biocontrol actions are also exerted through the production of various secondary metabolites. Secondary metabolites are naturally produced substances, originating indirectly from the primary metabolism. Several species of bacteria, fungi and yeast can survive under a competitive atmosphere in the environment. Therefore, antagonism among different species of bacteria, fungi or different species of fungi is observed in the microbial ecosystem (Reino et al. 2008). Among the several strategies used by microbial biocontrol agents, antibiosis is one of the important modes of action for inhibiting the growth of phytopathogens. In this way, microbes have the ability to synthesis different types of secondary metabolites having either a broad-spectrum or specific antimicrobial activity against different types of pathogens (Reino et al. 2008; Mishra and Arora 2018). Among the bacteria, *Bacillus* sp., *Streptomyces* sp., *Pseudomonas* sp., *Pantoea* sp., *Stenotrophomonas* sp., *Agrobacterium* sp., and *Serratia* sp. were reported for the production of secondary metabolites with a broad-spectrum activity against different types of pathogens (Köhl et al. 2019).

A few of the bacterial antibiotics such as agrocin 84 from *A. radiobacter* active against crown gall caused by virulent *A. tumefaciens* and phenazine antibiotic produced by *P. fluorescens* used to control wheat diseases caused by *Gaeumannomyces graminis* were the best-reported examples of effective antibiotics for biocontrol (Lo 1998). Lipopeptides as surfactin, iturin and fengycin are the special class of antibiotics mostly produced by *Bacillus* species. Different species of *Bacillus* were also isolated from various extreme environments such as hot springs, marine ecosystems was known to produced different types of such compounds (Jackson et al. 2016; Mehetre et al. 2019b). Such extremophilic lipopeptides could be used to inhibit the growth of the phytopathogens. Numerous microbial species were known to produce such secondary metabolites; however, their role as potential biocontrol agents requires more extensive research in this field.

Induction of host systemic resistance

Several studies have shown that biocontrol agents could enhance in suppressing diseased pathogens by inducing a systemic resistance in the host plant. Plants have been associated with a variety of microbial members including some pathogens, to combat with them, plants have their own defence system to recognize and defend various phytopathogen (Köhl et al. 2019). The plant defence system includes physical and chemical barriers through innate immune system such as induction of defence-related proteins and enzymes, which become active during infection. Biocontrol agents also exert similar types of inducible responses in the plant without harming it. Several enzymes involved in these processes such as peroxidise, polyphenol oxidase, phenylalanine ammonia-lyase, lipoxygenase and β-1, 3-glucanase, which are primarily involved in the host defence (Ngadze et al. 2012).

Among members of *Firmicultes, Proteobacteria* and *Actinobacteria*, species of *Bacillus* are well known for inducing the systemic resistance against pathogen infection (Shafi et al. 2017). *Bacillus* species such as, *B. subtilis, B. amyloliquefaciens, B. pasteurii, B. mycoides, B. cereus, B. pumilus, B. sphaericus, B. cereus* and *B. pumilus* were reported with a reduction of the disease using induced systematic resistance in several plants such as tomato, watermelon, sugar beet, tobacco, Arabidopsis sp., bell pepper, muskmelon, cucumber, and in tropical crops (Choudhary and Johri 2009; Santoyo et al. 2012). Most of these biocontrol agents are recognized in the same way as pathogens; however they induce a different defence response (Khare et al. 2018).

Plant growth promoting rhizobacteria

The PGPR are bacterial members associated with the plant rhizosphere and play a vital role during plant development in utilization of nutrient and water for plant growth and also in the biocontrol of plant pathogens (Liu et al. 2018). Many of the rhizospheric microbes such as nitrogen-fixing bacteria, mycorrhizal fungi, mycoparasitic fungi and protozoa have several beneficial effects on plant health (Rastegari et al. 2020; Yadav et al. 2020). PGPR are well known bacteria having characteristics to suppress the growth of various pathogens with a broad spectrum of bacterial, fungal and nematode diseases as well also to provide protection against viral diseases. Although a significant control of plant pathogens has been demonstrated by several PGPR by different studies, more studies about understanding their diversity, colonizing ability, mechanisms of action, formulation and application are further necessary to use them as a reliable biocontrol agent against plant pathogens (Mendes et al. 2013).

Conclusion and future prospective

Several extremophilic microbes have been isolated from different extreme environmental habitats. Such extremophilic microbes have attracted significant consideration of the scientific community due to their potential biotechnological applications in industry, pharmaceuticals, medicine, food for humans. The extremophiles growing under arid regions, under drought and high salinity conditions, have multifarious attributes to promote and enhance the productivity of crops under abiotic stress conditions. Therefore, extremophiles could also act as biofertilizers and biocontrol agents for crops growing in hilly and low-temperature conditions, high salinity conditions for enhancing crop production and soil health for sustainable agriculture. To identify such beneficial microbes from extreme environments, more investigations should be directed towards the exploration of these extreme environmental microbes in this field. A contentious effort in exploring different extreme environments for isolation of extremophiles with improved biocontrol activities may find some useful applications in the management of phytopathogens. Moreover, for understanding the complexity of such processes involved in biosupression, implementation of the use of high throughput techniques will be worthwhile in future studies. Such studies may lay the foundation for the development of more effective biocontrol methods to increase the efficacy of plant productivity for sustainable agriculture. Successful application of biological control strategies requires more knowledge-intensive management; therefore, more extensive research effort is needed in this field. In the future, biocontrol of plant diseases using extreme microorganisms will be more promising for the development of more effective biocontrol strategies to manage plant diseases for the sustainable agricultural system.

Acknowledgements

The work was carried out by a joint international grant from the Russian Science Foundation and the Department of Science and Technology (DST) of the Government of India No. 19-46-02004.

References

Aktar, W., D. Sengupta and A. Chowdhury. 2009. Impact of pesticides use in agriculture: their benefits and hazards. Interdiscipl. Toxicol. 2: 1–2.

Alavanja, M.C., M.K. Ross and M.R. Bonner. 2013. Increased cancer burden among pesticide applicators and others due to pesticide exposure. CA: A Cancer J. Clinicians 63: 120–142.

Arguelles-Arias, A., M. Ongena, B. Halimi, Y. Lara, A. Brans, B. Joris et al. 2009. *Bacillus amyloliquefaciens* GA1 as a source of potent antibiotics and other secondary metabolites for biocontrol of plant pathogens. Microbiol. Cell. Fact. 8: 63.

Azeem, M., A. Riaz, A.N. Chaudhary, R. Hayat, Q. Hussain, M.I. Tahir et al. 2015. Microbial phytase activity and their role in organic P mineralization. Arch. Agro. Soil. Sci. 61: 751–766.

Bar-Shimon, M., H. Yehuda, L. Cohen, B. Weiss, A. Kobeshnikov, A. Daus et al. 2004. Characterization of extracellular lytic enzymes produced by the yeast biocontrol agent *Candida oleophila*. Curr. Gen. 45: 140–148.

Bashan, Y. and L.E. de-Bashan. 2005. Bacteria/plant growth-promotion. pp. 103–115. *In*: Hillel, D. (ed.). Encyclopedia of Soils in the Environment. Vol. 1, Elsevier, Oxford, UK.

Beneduzi, A., A. Ambrosini and L.M. Passaglia. 2012. Plant growth-promoting rhizobacteria (PGPR): their potential as antagonists and biocontrol agents. Gen. Mol. Biol. 35: 1044–1051.

Beneduzi, A, F. Moreira, P.B. Costa, L.K. Vargas, B.B. Lisboa, R. Favreto et al. 2013. Diversity and plant growth promoting evaluation abilities of bacteria isolated from sugarcane cultivated in the South of Brazil. Appl. Soil Ecol. 63: 94–104.

Boehlje, M. and S. Bröring. 2011. The increasing multi-functionality of agricultural raw materials: Three dilemmas for innovation and adoption. Int. Food Agribus. Man. Rev. 14: 1–6.

Brown, J.F. and H.J. Ogle. 1997. Fungal diseases and their control. Plant Pathogens and Plant Diseases. APPS, Australia. 1: 443–467.

Bruinsma, J. 2017. World Agriculture: Towards 2015/2030: An FAO Study. Routledge.

Cao, X., Y. Lu, D. Di, Z. Zhang, H. Liu, L. Tian et al. 2013. Enhanced virus resistance in transgenic maize expressing a dsRNA-specific endoribonuclease gene from *E. coli*. PloS One 8: e60829.

Chamekh, R., F. Deniel, C. Donot, J.L. Jany, P. Nodet and L. Belabid. 2019. Isolation, identification and enzymatic activity of halotolerant and halophilic fungi from the Great Sebkha of Oran in Northwestern of Algeria. Mycobiology 14: 1–2.

Choudhary, D.K. and B.N. Johri. 2009. Interactions of Bacillus spp. and plants—with special reference to induced systemic resistance (ISR). Microbiol. Res. 164: 493–513.

Chowdhury, S.P., A. Hartmann, X. Gao and R. Borriss. 2015. Biocontrol mechanism by root-associated *Bacillus amyloliquefaciens* FZB42—a review. Front. Microbiol. 6: 780.

Coker, J.A. 2016. Extremophiles and biotechnology: current uses and prospects. F1000Res. 5:F1000 Faculty Rev-396. doi:10.12688/f1000research.7432.1.

Damalas, C.A. and I.G. Eleftherohorinos. 2011. Pesticide exposure, safety issues, and risk assessment indicators. Int. J. Environ. Res. Public Health. 8: 1402–1419.

Edgerton, M.D. 2009. Increasing crop productivity to meet global needs for feed, food, and fuel. Plant Physiol. 149: 7–13.

El Arbi, A., A. Rochex, G. Chataigné, M. Béchet, D. Lecouturier, S. Arnauld et al. 2016. The Tunisian oasis ecosystem is a source of antagonistic *Bacillus* spp. producing diverse antifungal lipopeptides. Res. Microbiol. 167: 46–57.

Essghaier, B., M.L. Fardeau, J.L. Cayol, M.R. Hajlaoui, A. Boudabous, H. Jijakli et al. 2009. Biological control of grey mould in strawberry fruits by halophilic bacteria. J. Appl. Microbiol. 106: 833–846.

Flood, J. 2010. The importance of plant health to food security. Food Secur. 2: 215–231.

Gowdar, S.B., H. Deepa and Y.S. Amaresh. 2018. A brief review on biocontrol potential and PGPR traits of *Streptomyces* sp. for the management of plant diseases. J. Pharmacogn. Phytochem. 7: 3–7.

Grover, M., S.Z. Ali, V. Sandhya, A. Rasul and B. Venkateswarlu. 2011. Role of microorganisms in adaptation of agriculture crops to abiotic stresses. World J. Microbiol. Biotechnol. 27: 1231–1240.

Guo, R., Z. Wang, Y. Huang, H. Fan and Z. Liu. 2018. Biocontrol potential of saline-or alkaline-tolerant *Trichoderma asperellum* mutants against three pathogenic fungi under saline or alkaline stress conditions. Braz. J. Microbiol. 49: 236–245.

Hamed, E.R., H.M. Awad, E.A. Ghazi, N.G. El-Gamal and H.S. Shehata. 2015. *Trichoderma asperellum* isolated from salinity soil using rice straw waste as biocontrol agent for cowpea plant pathogens. J. Appl. Pharm. Sci. 5: 91–98.

Heydari, A. and M. Pessarakli. 2010. A review on biological control of fungal plant pathogens using microbial antagonists. J. Biol. Sci. 10: 273–290.

Jackson, S.A., E. Borchert, F. O'Gara and A.D. Dobson. 2015. Metagenomics for the discovery of novel biosurfactants of environmental interest from marine ecosystems. Curr. Opi. Biotechnol. 33: 176–182.

Jaelwen, M., C. Endeforth, D. Schowanek, T. Delfosse, A. Riddle and N. Budgen. 2016. Comprehensive review of several surfactants in marine environments: Fate and ecotoxicity. Environ. Toxicol. Chem. 35: 1077–1086.

Jadhav, H.P. and R.Z. Sayyed. 2016. Hydrolytic enzymes of rhizospheric microbes in crop protection. MOJ Cell. Sci. Rep. 3: 135–136.

Jafari, S., S.S. Aghaei, H. Afifi-Sabet, M. Shams-Ghahfarokhi, Z. Jahanshiri, M. Gholami-Shabani et al. 2018. Exploration, antifungal and antiaflatoxigenic activity of halophilic bacteria communities from saline soils of Howze-Soltan playa in Iran. Extremophiles 22: 87–98.

Khare, E., J. Mishra and N.K. Arora. 2018. Multifaceted interactions between endophytes and plant: developments and prospects. Front. Microbiol. 9–2732.

Köhl, J., R. Kolnaar and W.J. Ravensberg. 2019. Mode of action of microbial biological control agents against plant diseases: relevance beyond efficacy. Front. Plant Sci. 10: 845.

Lim, J.H. and S.D. Kim. 2010. Biocontrol of phytophthora blight of red pepper caused by *Phytophthora capsici* using *Bacillus subtilis* AH18 and *B. licheniformis* K11 formulations. J. Korean. Soc. Appl. Bi. 53: 766–773.

Liu, K., J.A. McInroy, C.H. Hu and J.W. Kloepper. 2018. Mixtures of plant-growth-promoting rhizobacteria enhance biological control of multiple plant diseases and plant-growth promotion in the presence of pathogens. Plant Dis. 102: 67–72.

Lo, C.T. 1998. General mechanisms of action of microbial biocontrol agents. Plant. Pathol. Bull. 7: 155–166.

Mansfield, J., S. Genin, S. Magori, V. Citovsky, M. Sriariyanum, P. Ronald et al. 2012. Top 10 plant pathogenic bacteria in molecular plant pathology. Mol. Plant Pathol. 13: 614–629.

Marco, J.L. and C.R. Felix. 2007. Purification and characterization of a beta-Glucanase produced by *Trichoderma harzianum* showing biocontrol potential. Braz. Arch. Biol. Technol. 50: 21–29.

Mehetre, G.T., M. Shah, S.G. Dastager and M.S. Dharne. 2018. Untapped bacterial diversity and metabolic potential within Unkeshwar hot springs. India. Arch. Microbiol. 200: 753–770.

Mehetre, G.T., S.G. Dastager and M.S. Dharne. 2019a. Biodegradation of mixed polycyclic aromatic hydrocarbons by pure and mixed cultures of biosurfactant producing thermophilic and thermo-tolerant bacteria. Sci. Total Environ. 679: 52–60.

Mehetre, G.T., J.S. Vinodh, B.B. Burkul, D. Desai, B. Santhakumari, M.S. Dharne et al. 2019b. Bioactivities and molecular networking-based elucidation of metabolites of potent actinobacterial strains isolated from the Unkeshwar geothermal springs in India. RSC Adv. 9: 9850–9859.

Mendes, R., P. Garbeva and J.M. Raaijmakers. 2013. The rhizosphere microbiome: significance of plant beneficial, plant pathogenic, and human pathogenic microorganisms. FEMS Microbiol. Rev. 37: 634–663.

Mishra, J. and N.K. Arora. 2018. Secondary metabolites of fluorescent *pseudomonads* in biocontrol of phytopathogens for sustainable agriculture. Appl. Soil Ecol. 125: 35–45.

Mohamed, H.A. and W.M. Haggag. 2006. Biocontrol potential of salinity tolerant mutants of *Trichoderma harzianum* against *Fusarium oxysporum*. Braz. J. Microbiol. 37: 181–191.

Müller, C., V. Ulyanova, O. Ilinskaya, S. Pleschka and R.S. Mahmud. 2017. A novel antiviral strategy against MERS-CoV and HCoV-229E using binase to target viral genome replication. BioNanoScience 7: 294–299.

Nega, A. 2014. Review on concepts in biological control of plant pathogens. J. Biol. Agric. Healthcare. 4: 33–54.

Ngadze, E., D. Icishahayo, T.A. Coutinho and J.E. Van der Waals. 2012. Role of polyphenol oxidase, peroxidase, phenylalanine ammonia lyase, chlorogenic acid, and total soluble phenols in resistance of potatoes to soft rot. Plant Dis. 96: 186–192.

Nicolopoulou-Stamati, P., S. Maipas, C. Kotampasi, P. Stamatis and L. Hens. 2016. Chemical pesticides and human health: the urgent need for a new concept in agriculture. Front. Public Health. 4: 148.

Oerke, E.C., H.W. Dehne, F. Schönbeck and A. Weber. 2012. Crop Production and Crop Protection: Estimated Losses in Major Food and Cash Crops. Elsevier.

Raddadi, N., A. Cherif, D. Daffonchio, M. Neifar and F. Fava. 2015. Biotechnological applications of extremophiles, extremozymes and extremolytes. Appl. Microbiol. Biotechnol. 99: 7907–7913.

Rampelotto, P. 2013. Extremophiles and extreme environments. 482–485.

Rastegari, A.A., A.N. Yadav, A.A. Awasthi and N. Yadav. 2020. Trends of Microbial Biotechnology for Sustainable Agriculture and Biomedicine Systems: Diversity and Functional Perspectives. Elsevier, Cambridge, USA.

Reino, J.L., R.F. Guerrero, R. Hernández-Galán and I.G. Collado. 2008. Secondary metabolites from species of the biocontrol agent *Trichoderma*. Phytochem. Rev. 7: 89–123.

Rondón, J.J., M.M. Ball, L.T. Castro and L.A. Yarzábal. 2019. Eurypsychrophilic *Pseudomonas* spp. isolated from Venezuelan tropical glaciers as promoters of wheat growth and biocontrol agents of plant pathogens at low temperatures. Environ. Sust. 1–1.

Rothschild, L.J. and R.L. Mancinelli. 2001. Life in extreme environments. Nature 409: 1092.

Saikia, R., D.K. Gogoi, S. Mazumder, A. Yadav, R.K. Sarma, T.C. Bora et al. 2011. *Brevibacillus laterosporus* strain BPM3, a potential biocontrol agent isolated from a natural hot water spring of Assam, India. Microbiol. Res. 166: 216–225.

Santoyo, G., M.D. Orozco-Mosqueda and M. Govindappa. 2012. Mechanisms of biocontrol and plant growth-promoting activity in soil bacterial species of *Bacillus* and *Pseudomonas*: a review. Biocontrol Sci. Technol. 22: 855–872.

Scholthof, K.B., S. Adkins, H. Czosnek, P. Palukaitis, E. Jacquot, T. Hohn et al. 2011. Top 10 plant viruses in molecular plant pathology. Mol. Plant Pathol. 12: 938–954.

Schultz, J. and A.S. Rosado. 2019. Microbial role in the ecology of Antarctic plants. pp. 257–275. *In*: The Ecological Role of Micro-organisms in the Antarctic Environment Springer, Cham.

Shafi, J., H. Tian and M. Ji. 2017. *Bacillus* species as versatile weapons for plant pathogens: a review. Biotechnol. Biotechnol. Equip. 31: 446–459.

Shah Mahmud, R., C. Müller, Y. Romanova, A. Mostafa, V. Ulyanova, S. Pleschka et al. 2017. Ribonuclease from *Bacillus* acts as an antiviral agent against negative-and positive-sense single stranded human respiratory RNA viruses. BioMed Res. Intl. 2017: 5279065.

Sharipova, M., A. Rockstroh, N. Balaban, A. Mardanova, A. Toymentseva, A. Tikhonova et al. 2015. Antiviral effect of ribonuclease from *Bacillus pumilus* against phytopathogenic RNA-viruses. Agri. Sci. 6: 1357.

Shcherbakov, A.V., E.N. Shcherbakova, S.A. Mulina, P.Y. Rots, R.F. Dar'yu, E.I. Kiprushkina et al. 2017. Psychrophilic endophytic *Pseudomonas* as potential agents in biocontrol of phytopathogenic and putrefactive microorganisms during potato storage. Sel'skokhozyaistvennaya Biol. 52: 116–128.

Singh, B.P., V.K. Gupta and A.K. Passari (eds.). 2018. Actinobacteria: Diversity and Biotechnological Applications: New and Future Developments in Microbial Biotechnology and Bioengineering. Elsevier.

Sivasakthi, S., G. Usharani and P. Saranraj. 2014. Biocontrol potentiality of plant growth promoting bacteria (PGPR)-*Pseudomonas fluorescens* and *Bacillus subtilis*: a review. African J. Agri. Res. 9: 1265–1277.

Sokurenko, Y., A. Nadyrova, V. Ulyanova and O. Ilinskaya. 2016. Extracellular ribonuclease from *Bacillus licheniformis* (Balifase), a New member of the N1/T1 RNase superfamily. BioMed. Res. Int. 2016: 4239375.

Song, Y.S., D.J. Seo and W.J. Jung. 2017. Identification, purification, and expression patterns of chitinase from psychrotolerant *Pedobacter* sp. PR-M6 and antifungal activity *in vitro*. Microbial. Pathogenes 107: 62–68.

Tracy, E.F. 2015. The promise of biological control for sustainable agriculture: a stakeholder-based analysis. J. Sci. Policy Governance. 5: 1.

Utkhede, R.S. and P.L. Sholberg. 1986. *In vitro* inhibition of plant pathogens by *Bacillus subtilis* and *Enterobacter aerogenes* and *in vivo* control of two postharvest cherry diseases. Can. J. Microbiol. 32: 963–967.

Verma, J.P., D.K. DJaiswal, R. Krishna, S. Prakash, J. Yadav and V. Singh. 2018. Characterization and screening of thermophilic Bacillus strains for developing plant growth promoting consortium from hot spring of Leh and Ladakh region of India. Front. Microbiol. 9: 1293.

Verma, P., A.N. Yadav, K.S. Khannam, N. Panjiar, S. Kumar, A.K. Saxena et al. 2015. Assessment of genetic diversity and plant growth promoting attributes of psychrotolerant bacteria allied with wheat (*Triticum aestivum*) from the northern hills zone of India. Ann. Microbiol. 65: 1885–1899.

Verma, P., A.N. Yadav, V. Kumar, D.P. Singh and A.K. Saxena. 2017. Beneficial plant-microbes interactions: biodiversity of microbes from diverse extreme environments and its impact for crop improvement. pp. 543–580. *In*: Plant-Microbe Interactions in Agro-Ecological Perspectives Springer, Singapore.

Vidaver, A.K. and P.A. Lambrecht. 2004. Bacteria as plant pathogens. Plant Health Instructor 10.

Viterbo, A., O. Ramot, L. Chernin and I. Chet. 2002. Significance of lytic enzymes from *Trichoderma* spp. in the biocontrol of fungal plant pathogens. Antonie Van Leeuwenhoek 81: 549–56.

Vurukonda, S.S., S. Vardharajula, M. Shrivastava and A. SkZ. 2016. Enhancement of drought stress tolerance in crops by plant growth promoting rhizobacteria. Microbiol. Res. 184: 13–24.

Wang, X., Q. Li, J. Sui, J. Zhang, Z. Liu, J. Du et al. 2019. Isolation and characterization of antagonistic bacteria *Paenibacillus jamilae* HS-26 and their effects on plant growth. BioMed Res. Int. 2019: 3638926.

Yadav, A.N., S.G. Sachan, P. Verma and A.K. Saxena. 2015a. Prospecting cold deserts of north western Himalayas for microbial diversity and plant growth promoting attributes. J. Biosci. Bioeng. 119: 683–693.

Yadav, A.N., S.G. Sachan, P. Verma, S.P. Tyagi, R. Kaushik and A.K. Saxena. 2015b. Culturable diversity and functional annotation of psychrotrophic bacteria from cold desert of Leh Ladakh (India). World J. Microbiol. Biotechnol. 31: 95–108.

Yadav, A.N., D. Sharma, S. Gulati, S. Singh, R. Dey, K.K. Pal et al. 2015c. Haloarchaea endowed with phosphorus solubilization attribute implicated in phosphorus cycle. Sci. Rep. 5: 12293.

Yadav, A.N. 2017. Beneficial role of extremophilic microbes for plant health and soil fertility. J. Agric. Sci. Bot. 1: 9–12.

Yadav, A.N., S. Gulati, D. Sharma, R.N. Singh, M.V.S. Rajawat, R. Kumar et al. 2019. Seasonal variations in culturable archaea and their plant growth promoting attributes to predict their role in establishment of vegetation in Rann of Kutch. Biologia 74: 1031–1043. doi:10.2478/s11756-019-00259-2.

Yadav, A.N., J. Singh, A.A. Rastegari and N. Yadav. 2020. Plant Microbiomes for Sustainable Agriculture. Springer International Publishing, Cham.

Zajc, J., C. Gostinčar, A. Černoša and N. Gunde-Cimerman. 2019. Stress-tolerant yeasts: opportunistic pathogenicity versus biocontrol potential. Genes 10: 42.

Zhou, W.W., L.X. Zhang, B. Zhang, F. Wang, Z.H. Liang and T.G. Niu. 2008. Isolation and characterization of ZH14 with antiviral activity against Tobacco mosaic virus. Can. J. Microbiol. 54: 441–449.

Zhou, W.W., Y.L. He, T.G. Niu and J.J. Zhong. 2010. Optimization of fermentation conditions for production of anti-TMV extracellular ribonuclease by *Bacillus cereus* using response surface methodology. Biopro. Biosyst. Eng. 33: 657–663.

Zouari, I., L. Jlaiel, S. Tounsi and M. Trigui. 2016. Biocontrol activity of the endophytic *Bacillus amyloliquefaciens* strain CEIZ-11 against *Pythium aphanidermatum* and purification of its bioactive compounds. Biol. control. 100: 54–62.

Chapter 14

Bioalcohol and Biohydrogen Production by Hyperthermophiles

Kesen Ma,* Sarah Danielle Kim and *Vivian Serena Chu*

Introduction

Dependency and rapid consumption of fossil fuels result in significantly negative impacts on the environment. To combat such deteriorating trends and maintain sustainability, biofuels are rapidly considered to be a major alternative source of fuel. Biofuels are defined as a liquid, gas, or solid fuel that are predominantly produced from biomass (Savaliya et al. 2015), and can be classified as either primary (unprocessed form) or secondary (produced by processing of biomass) biofuels. The production of fuels and chemicals using renewable resources has garnered particular interest, among which bio-alcohol and bio-hydrogen show great potential as alternative fuels (Ranjan and Moholkar 2012; Kengen et al. 2009; Rodionova et al. 2017; Rastegari et al. 2019). Microbial bio-alcohol production is achieved by fermentation mainly using carbohydrates as substrates, which are major components of biomass. Microbial bio-hydrogen formation can be a light-dependent or a light-independent (dark fermentation) process. Large-scale production, however, appears to be feasible only by dark fermentation when non-food biomass are largely available. Various types of microorganisms have the capacity to produce bio-alcohol and/or bio-hydrogen, however, for this chapter only hyperthermophiles (microorganisms that can grow optimally at 80°C and above, or are capable of growing at 90°C and above) will be discussed (Tse and Ma 2016). The metabolic relationship between bio-alcohol and bio-hydrogen production will also be discussed.

Alcohol fermentation

A biological or chemical process is needed to convert biomass into simple sugars that can be fermented by a particular microorganism. Depending on the microorganism, the type of alcohol fermentation can vary. A widely studied type of alcohol fermentation is ethanol fermentation. Ethanol is the major end-product of alcohol fermentation by yeast (*Saccharomyces cerevisae*) and some *Zymomonas* species (Muller 2008). In ethanol fermentation, sugars are converted to the central intermediate pyruvate via carbohydrate catabolism pathways, such as Embden-Meyerhof (EM), Entner-Doudoroff (ED) and Pentose Phosphate (PP) pathways, where pyruvate is then converted to various fermentation products (Hoelzle et al. 2014). *Clostridium* species, particularly *Clostridium acetobutylicum*, are well-studied and utilized for Acetone-Butanol-Ethanol (ABE) production from the fermentation of carbohydrates (Lee et al. 2008; Lee et al. 2012).

Department of Biology, University of Waterloo, 200 University Avenue West, Waterloo, Ontario N2L 3G1, Canada.
* Corresponding author: kma@uwaterloo.ca

Pathways of ethanol fermentation

For ethanol production from the intermediate pyruvate, two different pathways are involved: a two-step pathway and a three-step pathway (Fig. 14.1, Eram and Ma 2013).

The two-step pathway is present in yeast and *Zymomonas* species, while the three-step pathway is used by *E. coli* and *Clostridium* species. Pyruvate decarboxylase (PDC) catalyzes the non-oxidative decarboxylation of pyruvate to acetaldehyde that is reduced to ethanol by alcohol dehydrogenase (ADH). Either pyruvate ferredoxin oxidoreductase (POR) or Pyruvate formate lyase (PFL) catalyzes the oxidative decarboxylation of pyruvate to acetyl-CoA that is then reduced to acetaldehyde by a CoA-dependent aldehyde dehydrogenase (AlDH).

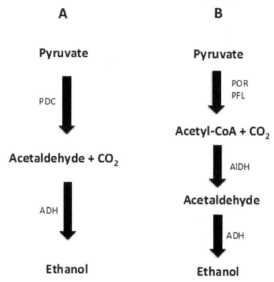

Figure 14.1 Pathways of ethanol production from pyruvate. (A) two-step pathway; (B) three-step pathway. PDC, pyruvate decarboxylase; ADH, alcohol dehydrogenase; POR, pyruvate ferredoxin oxidoreductase; PFL, pyruvate format lyase; AlDH, CoA-dependent aldehyde dehydrogenase.

Pathways of butanol fermentation

Compared to ethanol production, the pathway for ABE production is more complex as butanol and acetone are produced in addition to ethanol (Fig. 14.2, Garcia et al. 2011).

Acetyl-CoA produced after the oxidative decarboxylation of pyruvate by POR is converted to acetoacetyl-CoA by acetyl-CoA acetyltransferase (AAT, thiolase). Acetoacetyl-CoA is reduced to 3-hydroxybutyryl-CoA by hydroxybutyryl-CoA dehydrogenase (HBDH). Crontonyl-CoA produced after the dehydration of 3-hydroxybutyryl-CoA, which is catalyzed by crontonase, will be reduced to butyryl-CoA by butyryl-CoA dehydrogenase (BCDH). Butyryl-CoA is reduced to butyraldehyde by CoA-dependent butyraldehyde dehydrogenase (BADH). Butanol is produced by the reduction of butyraldehyde by butanol dehydrogenase (BDH). For the production of acetone, acetoacetyl-CoA is converted to acetoacetate by acetoacetyl-CoA: acetate/butyrate: CoA transferase (AACT). Acetone is then produced by the decarboxylation of acetoacetate, which is catalyzed by acetoacetate decarboxylase (AADC). In *C. acetobutylicum*, ABE production is regulated by the change in pH as a result of excessive production of acetate and butyrate (Jones and Woods 1986). There is a butanol production pathway naturally present in yeast which uses amino acid glycine as the substrate (Branduardi et al. 2013; Kour et al. 2019).

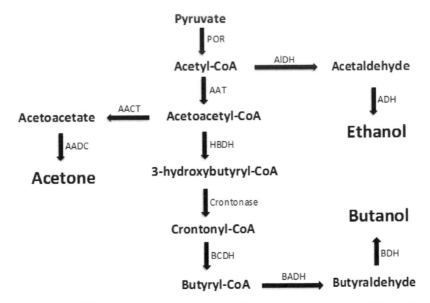

Figure 14.2 Pathways of ABE production from pyruvate in *Clostridium acetobutylicum*. POR, pyruvate ferredoxin oxidoreductase; AlDH, CoA-dependent aldehyde dehydrogenase; ADH, alcohol dehydrogenase; AAT, acetyl-CoA acetyltransferase (thiolase); AACT, acetoacetyl-CoA: acetate: CoA transferase; AADC, acetoacetate decarboxylase; HBDH, hydroxybutyryl-CoA dehydrogenase; BCDH, butyryl-CoA dehydrogenase; BADH, CoA-dependent butyraldehyde dehydrogenase; BDH, butanol dehydrogenase (Jones and Woods 1986).

Alcohol production by hyperthermophilic microorganisms

Hyperthermophiles are a group of bacteria and archaea that have the ability to grow optimally at 80°C and above, or are capable of growing at 90°C and above (Blumer-Schuette et al. 2008). They possess various enzymes that can hydrolyze biomass into simple sugars, which can be metabolized by using either conventional or modified EM, ED, and/or PP pathways (Sieber and Schoenheit 2005). Many hyperthermophiles possess the ability to produce alcohol (Table 14.1), and possess/ utilize different types of ADHs (Ma and Tse 2015). It appears that the concentrations of alcohols produced are in sub-mM range, which may be due to the nature of key enzymes involved in the metabolic pathways. There have been no homolog sequences to either the commonly-known PDC or AlDH in hyperthermophilic genome sequences (Eram and Ma 2013), however, it is known that a two-step pathway is present in both hyperthermophilic bacteria and archaea (Ma et al. 1997; Eram et al. 2014; 2015). In this pathway, PDC is a bifunctional enzyme that also has POR activity catalyzing the oxidative decarboxylation of pyruvate (Ma et al. 1997; Eram et al. 2014; 2015). PDCs of the archaeal hyperthermophile *Pyrococcus furiosus* and *Thermococcus guaymasensis* are found to be 4.3 U/mg and 3.8 U/mg, respectively (Ma et al. 1997; Eram et al. 2014), which are much higher than those from bacterial hyperthermophiles *Thermotoga maritima* and *Thermotoga hypogea*, which are 1.4 U/mg and 1.9 U/mg, respectively (Eram et al. 2015). However, their much higher POR activities (~ 20–120 U/mg) may prevent them from having sufficient PDC activity to support higher alcohol production.

There are three hyperthermophilic archaea known to have the ability to produce butanol (Table 14.1), which are *Thermococcus paralvinellae* (formerly Strain ES1, Ma et al. 1995; Hensley et al. 2014), *Hyperthermus butylicus*, and *Pyrodictium abyssi* (Zillig et al. 1990; Pley et al. 1991). The genome sequences of *H. butylicum* and *T. paralvinellae* are available, however, no homolog of key enzymes involved in the butanol production pathway in *Clostridium* species can be found (Brugger et al. 2007). The genome for *P. abyssi* is yet to be sequenced. It is possible that a different pathway may be used to produce butanol at high temperatures.

Table 14.1 Major alcohol and hydrogen producing hyperthermophiles.

Organisms	Growth T$_{opt}$, °C (minimum-maximum)	Growth condition/ Substrate	Alcohol production (mM)	Hydrogen production (H$_2$ per glucose)	Reference
Thermotoga maritima	80 (55–90)	Batch/glucose	0.5[a]	4	Schroeder et al. 1994
Thermotoga hypogea	70–75 (56–90)	Batch/glucose	1	1.34	Fardeau et al. 1997
Thermotoga neapolitana	80 (55–90)	Batch/glucose	0.4[a]	4	van Ooteghem et al. 2004
Pyrococcus furiosus	100 (70–103)	Batch/maltose	1.5	2.6–3.5	Kengen and Stams 1994; Basen et al. 2014
Thermococcus kodakaraensis	85 (60–100)	Chemostat/starch	NA	3.33	Kanai et al. 2005
Thermococcus guaymasensis	88 (56–90)	Batch/glucose	0.8	28.7[b]	Ying and Ma 2011
Thermococcus paralvinellae	91 (81–91)	Batch/tryptone	0.05	4.5[b]	Ma et al. 1995; Hensley et al. 2014
Hyperthermus butylicus	95–106 (76–108)	Batch/tryptone	0.67[c]	0	Zillig et al. 1990
Pyrodictium abyssi	97 (80–110)	Batch/starch+yeast extract	0.0002[d]	0	Pley et al. 1991

[a] Ying and Ma unpublished data
[b] H$_2$ production in mM
[c] butanol production
[d] trace amount of butanol (0.4 nmol/10^8 cells, Pley et al. 1991)
NA, not available

A tungsten-containing aldehyde oxidoreductase (AOR) has been found to be present in a hyperthermophilic archaeon (Mukund and Adams 1995). This AOR in *P. furiosus* has been shown to catalyze the reduction of organic acids into corresponding aldehydes that can then be reduced into alcohols (Basen et al. 2014). When *P. furiosus* grows on maltose, added acetate or butyrate can be converted into corresponding ethanol and butanol, respectively. Reductants from the metabolism (or even added CO when a gene encoding a CO dehydrogenase is inserted) can be used as an electron donor for the reduction of organic acids to corresponding aldehydes that is then reduced to alcohol by ADH (Basen et al. 2014). This pathway may provide additional energy conservation benefits to the metabolism (Basen et al. 2014).

Hydrogen production

Microbial hydrogen production has been found to be mainly from dark fermentation using complex biomass (Classen et al 1990; van Niel et al. 2002; Kapdan and Kargi 2006; Kumar et al. 2019). Carbohydrates, including various simple sugars, are used as their substrates for producing hydrogen (Fig. 14.3). One glucose can be oxidized into 2 acetate, and 8 electrons released can be used for the reduction of 8 protons to generate 4 H_2. In the hyperthermophilic bacteria *Thermotoga*, a glyceraldehyde-3-phosphate dehydrogenase (GAPDH) catalyzes the oxidation of glyceraldehyde-3-phosphate (GAP), while a glyceraldehyde-3-phosphate ferredoxin oxidoreductase (GAPOR) is used to catalyze the same oxidation reaction in hyperthermophilic archaea *Pyrococcus*. The former will generate NAD(P)H and the latter will form reduced ferredoxin (Fd_{red}). Either an NAD(P)H- or a Fd_{red}-dependent or both NAD(P)H and Fd_{red}-dependent hydrogenase will catalyze the reduction of protons to produce hydrogen (Fig. 14.3, Schut and Adams 2009). Depending on the metal contents, there are [FeNi], [FeFe] and [Fe] hydrogenases (Lubitz et al. 2014; Peters et al. 2015), but [Fe] hydrogenase is present only in methanogens (Korbas et al. 2006).

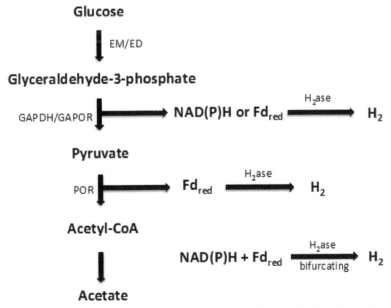

Figure 14.3 Hydrogen production by hyperthermophiles. EM/ED, Embden-Meyerhof pathway/Entner-Doudoroff pathway; GAPDH/GAPOR, glyceraldehyde-3-phosphate dehydrogenase/glyceraldehyde-3-phosphate ferredoxin oxidoreductase; H_2as, hydrogenase; POR, pyruvate ferredoxin oxidoreductase.

Hyperthermophilic bacteria *Thermotoga* species

Thermotoga species are gram negative, non-spore forming anaerobic bacteria. They can utilize complex biomass and produce hydrogen as an end-product. The most well-studied are *T. maritima* and *T. neapolitana*, which are able to produce up to 4 H_2 per glucose (Table 14.1) (Schroeder et al. 1994; van Ooteghem et al. 2002). *T. maritima* contains an [FeFe] hydrogenase, which is a bifurcating hydrogenase using both NAD(P)H and ferredoxin as an electron donor (Schut and Adams 2009; Peters et al. 2015). *Thermotoga* species use both EM (~ 85%) and ED (~ 15%) for catabolizing glucose (Selig et al. 1997), producing NAD(P)H. There is also a POR catalyzing the oxidation of pyruvate and generating Fd_{red}. This implies that both NAD(P)H and Fd_{red} can then be used for the production of hydrogen.

Hyperthermophilic archaea *Pyrococcus* species

Both *Pyrococcus* and *Thermococcus* species are hydrogen-producing archaea (Table 14.1). They have a modified EM pathway as there is no GAPDH, but GAPOR (Mukund and Adams 1995; Siebers and Schonheit 2005). *P. furiosus* is the most well-studied hydrogen-producing archaeon. POR catalyzes the oxidation of pyruvate to acetyl-CoA and the formation of Fd_{red}. Fd_{red} can be used for the reduction of NAD(P) to generate NAD(P)H, which is catalyzed by a ferredoxin:NADP oxidoreductase (Ma and Adams 1994). There are two cytoplasmic [NiFe]-hydrogenases using NADPH as an electron donor to produce hydrogen (Ma et al. 1993; Ma et al. 2000). However, there is also a membrane-bound [NiFe]-hydrogenase complex that can accept electrons from Fd_{red} directly, showing a very high ratio of H_2 evolution to H_2 uptake activity (Silva et al. 2000). This hydrogenase also functions as a proton pump, contributing to the formation of the cell's proton-motive force being coupled to the synthesis of ATP (Sapra et al. 2003).

Regulation of alcohol and hydrogen production

In general, hyperthermophiles produce more hydrogen compared to alcohol(s) (Table 14.1), which is expected as it is more thermodynamically favorable to produce H_2 at higher temperatures (Kengen et al. 2009). Naturally, less alcohol is produced in order to maximize energy conservation, however, this process may change depending on growth conditions. For example, it is observed that more ethanol is produced by *P. furiosus* when H_2 gas is added to the gas-phase in a batch culture (Yang and Ma unpublished results). Mesophilic *Enterobacter cloacae* can produce approximately 2 H_2 per glucose (Bisaillon et al. 2006), however, once its alcohol producing pathway is blocked by mutagenesis, its H_2 production yield increases to 3.4 H_2 per glucose (Kumar et al. 2001; Nath et al. 2006). Similarly, the mutant of *Thermococcus kodakarensis* shows an at least 5-fold increase in H_2 production (Santangebo et al. 2011). Such a principle may be applicable to enhancing the alcohol production by blocking the pathway of H_2 production in hyperthermophiles, meaning mutagenesis may be used for modulating the hydrogenase activity in achieving higher alcohol production. In reality, this may be more complex, for example, alcohol production is enhanced in *P. furiosus* by a single gene insertion (Basen et al. 2014), therefore, metabolic regulation and engineering warrant further investigation.

Concluding remarks

Many hyperthermophiles are excellent hydrogen producers by utilizing carbohydrates as substrates. Although they possess the metabolic pathways required for alcohol production, the yields are still quite low in comparison to other non-hyperthermophilic alcohol producing microorganisms. Further investigation is required to have a deeper understanding of their alcohol fermentation, especially butanol production, providing insight into developing a more efficient alcohol fermentation system at high temperatures. It is plausible to achieve higher yield of alcohol by blocking or modulating hydrogen production.

Acknowledgements

This work was supported by research grants from the Natural Sciences and Engineering Research Council (Canada) and the Canada Foundation for Innovation to KM.

References

Basen, M., G.J. Schut, D.M. Nguyen, G.L. Lipscomb, R.A. Benn, C.J. Prybol et al. 2014. Single gene insertion drives bioalcohol production by a thermophilic archaeon. PNAS 111: 17618–17623.
Bisaillon, A., J. Turcot and P.C. Hallenbeck. 2006. The effect of nutrient limitation on hydrogen production by batch cultures of *Escherichia coli*. Int. J. Hydrogen Energy 31: 1504–1508.
Blumer-Schuette, S.E., I. Kataeva, J. Westpheling, M.W. Adams and R.M. Kelly. 2008. Extremely thermophilic microorganisms for biomass conversion: status and prospects. Curr. Opin. Biotechnol. 19(3): 210–217.
Branduardi, P., V. Longo, N.M. Berterame, G. Rossi and D. Porro. 2013. A novel pathway to produce butanol and isobutanol in *Saccharomyces cerevisiae*. Biotechnol. Biofuels 6: 68–80.
Brugger, K., L. Chen, M. Stark, A. Zibat, P. Redder, A. Ruepp et al. 2007. The genome of *Hyperthermus butylicus*: a sulfur-reducing, peptide fermenting, neutrophilic Crenarchaeote growing up to 108°C. Archaea 2(2): 127–135.
Claassen, P.A.M., J.B. van, A.M. Lopez Contreras, E.W.J. van Niel, L. Sijtsma, A.J.M. Stams et al. 1999. Utilisation of biomass for the supply of energy carriers. Appl. Microbiol. Biotechnol. 52: 741–755.
Eram, M.S. and K. Ma. 2013. Decarboxylation of pyruvate to acetaldehyde for ethanol production by hyperthermophiles. Biomolecules 3: 578–596.
Eram, M.S., E. Oduaran and K. Ma. 2014. The Bifunctional pyruvate decarboxylase/pyruvate ferredoxin oxidoreductase from *Thermococcus guaymasensis*. Archaea 2014: Article ID 349379, 13 pages. doi:10.1155/2014/349379.
Eram, M.S., A. Wong, E. Oduaran and K. Ma. 2015. Molecular and biochemical characterization of bifunctional pyruvate decarboxylases and pyruvate ferredoxin oxidoreductases from *Thermotoga maritima* and *Thermotoga hypogea*. J. Biochem. 158(6): 459–66. doi: 10.1093/jb/mvv058.
Fardeau, M.-L., B. Ollivier, B.K.C. Patel, M. Magot, P. Thomas, A. Rimbault et al. 1997. *Thermotoga hypogea* sp. nov., a xylanolytic, thermophilic bacterium from an oil-producing well. Int. J. Syst. Bacteriol. 47: 1013–1019.
Garcia, V., J. Pakkila, H. Ojamo, E. Muurinen and R. Keiski. 2011. Challenges in biobutanol production: How to improve the efficiency? Renew. Sust. Energ. Rev. 15: 964–980.
Hensley, S.A., J.H. Jung, C.S. Park and J.F. Holden. 2014. *Thermococcus paralvinellae* sp. nov. and *Thermococcus cleftensis* sp. nov. of hyperthermophilic heterotrophs from deep-sea hydrothermal vents. Int. J. Syst. Evol. Microbiol. 64: 3655–3659.
Hoelzle, R.D., B. Virdis and D.J. Batstone. 2014. Regulation mechanisms in mixed and pure culture microbial fermentation. Biotechnol. Bioeng. 111(11): 2139–2154.
Jones, D.T. and D.R. Woods. 1986. Acetone-butanol fermentation revisited. Microbiol. Rev. 50: 484–524.
Kanai, T., H. Imanaka, A. Nakajima, K. Uwamori, Y. Omori, T. Fukui et al. 2005. Continuous hydrogen production by the hyperthermophilic archaeon, *Thermococcus kodakaraensis* KOD1. J. Biotechnol. 116: 271–282.
Kapdan, K. and F. Kargi. 2006. Bio-hydrogen production from waste materials. Enz. Microbiol. Technol. 38: 569–582.
Kengen, S.W.M. and A.J.M. Stams. 1994. Growth and energy conservation in batch cultures of *Pyrococcus furiosus*. FEMS Microbiol. Lett. 117: 305–310.
Kengen, S.W.M., H.P. Goorissen, M. Verhaart, A.J.M. Stams, E.W.J. van Niel and P.A.M. Claassen. 2009. Biological hydrogen production by anaerobic microorganisms. pp. 197–221. *In*: Soetaert, W. and E.J. Vandamme (eds.). Biofuels. John Wiley & Sons Ltd.
Korbas, M., S. Vogt, W. Meyer-Klaucke, E. Bill, E.J. Lyon, R.K. Thauer et al. 2006. The iron-sulfur cluster-free hydrogenase (Hmd) is a metalloenzyme with a novel iron binding motif. J. Biol. Chem. 281(41): 30804–30813.
Kour, D., K.L. Rana, N. Yadav, A.N. Yadav, A.A. Rastegari, C. Singh et al. 2019. Technologies for biofuel production: current development, challenges, and future prospects. pp. 1–50. *In*: Rastegari, A.A., A.N. Yadav and A. Gupta (eds.). Prospects of Renewable Bioprocessing in Future Energy Systems. Springer International Publishing, Cham. doi:10.1007/978-3-030-14463-0_1.
Kumar, N., A. Ghosh and D. Das. 2001. Redirection of biochemical pathways for the enhancement of H_2 production by *Enterobacter cloacae*. Biotechnol. Lett. 23: 537–541.
Kumar, S., S. Sharma, S. Thakur, T. Mishra, P. Negi, S. Mishra et al. 2019. Bioprospecting of microbes for biohydrogen production: Current status and future challenges. pp. 443–471. *In*: Molina, G., V.K. Gupta, B.N. Singh and N. Gathergood (eds.). Bioprocessing for Biomolecules Production. Wiley, USA.

Lee, J.M., Y.S. Jang, S.J. Choi, J.A. Im, H.H. Song, J.H. Cho et al. 2012. Metabolic engineering of *Clostridium acetobutylicum* ATCC 824 for isopropanol-butanol-ethanol fermentation. Appl. Environ. Microbiol. 78(5): 1416–1423.

Lee, S.Y., J.H. Park, S.H. Jang, L.K. Nielsen, J. Kim and K.S. Jung. 2008. Fermentative butanol production by *Clostridia*. Biotechnol. Bioeng. 101(2): 209–228.

Lubitz, W., H. Ogata, O. Ruediger and E. Reijerse. 2014. Hydrogenases. Chem. Rev. 114: 4081–4148.

Ma, K., R.N. Schicho, R.M. Kelly and M.W.W. Adams. 1993. Hydrogenase of the hyperthermophile *Pyrococcus furiosus* is an elemental sulfur reductase or sulfhydrogenase: Evidence for a sulfurreducing hydrogenase ancestor. Proc. Natl. Acad. Sci. USA 90: 5341–5344.

Ma, K. and M.W.W. Adams. 1994. Sulfide dehydrogenase from the hyperthermophilic archaeon *Pyrococcus furiosus:* a new multifunctional enzyme involved in the reduction of elemental sulphur. J. Bacteriol. 176: 6509–6517.

Ma, K., H. Loessner, J. Heider, M.K. Johnson and M.W.W. Adams. 1995. Effects of elemental sulfur on the metabolism of the deep sea hyperthermophilic archaeon, *Thermococcus* strain ES-1: Purification and characterization of a novel, sulfur-regulated, non-heme iron alcohol dehydrogenase. J. Bacteriol. 177: 4748–4756.

Ma, K., A. Hutchins, S.-J.S. Sung and M.W.W. Adams. 1997. Pyruvate ferredoxin oxidoreductase from the hyperthermophilic archaeon, *Pyrococcus furiosus*, functions as a coenzyme A-dependent pyruvate decarboxylase. Proc. Natl. Acad. Sci. USA 94: 9608–9613.

Ma, K., R. Weiss and M.W.W. Adams. 2000. Characterization of sulfhydrogenase II from the hyperthermophilic Archaeon *Pyrococcus furiosus* and assessment of its role in sulfur reduction. J. Bacteriol. 182: 1864–1871.

Ma, K. and C. Tse. 2015. Alcohol dehydrogenases and their physiological functions in hyperthermophiles. pp. 141–177. *In*: Li, F.-L. (ed.). Thermophilic Microorganisms. Caister Academic Press, Norfolk, UK.

Mukund, S. and M.W.W. Adams. 1995. Glyceraldehyde-3-phosphate ferredoxin oxidoreductase, a novel tungsten-containing enzyme with a potential glycolytic role in the hyperthermophilic archaeon, *Pyrococcus furiosus*. J. Biol. Chem. 270: 8389–8392.

Muller, V. 2008. Bacterial fermentation. pp. 1–8. *In*: Encyclopedia of Life Sciences (ELS). John Wiley & Sons, Ltd. Chichester. DOI: 10.1002/9780470015902.a0001415.pub2.

Nath, K., A. Kumar and D. Das. 2006. Effect of some environmental parameters on fermentative hydrogen production by *Enterobacter cloacae* DM11. Can. J. Microbiol. 52: 525–532.

Peters, J.W., G.J. Schut, E.S. Boyd, D.W. Mulder, E.M. Shepard, J.B. Broderick et al. 2015. [FeFe]- and [NiFe]-hydrogenase diversity, mechanism, and maturation. Biochimica et Biophysica Acta. 1853: 1350–1369.

Pley, U., J. Schipka, A. Gambacorta, H.W. Jannasch, H.R.R. Fricke and K.O. Stetter. 1991. *Pyrodictium abyssi* sp. nov. represents a novel heterotrophic marine archaeal hyperthermophile growing at 110°C. Syst. Appl. Microbiol. 14(3): 245–253.

Ranjan, A. and V.S. Moholkar. 2012. Biobutanol: science, engineering, and economics. Int. J. Energy Res. 36: 277–323.

Rastegari, A.A., A.N. Yadav and A. Gupta. 2019. Prospects of Renewable Bioprocessing in Future Energy Systems. Springer International Publishing, Cham.

Rodionova, M.V., R.S. Poudyal, I. Tiwari, R.A. Voloshin, S.K. Zharmukhamedov, H.G. Nam et al. 2017. Biofuel production: Challenges and opportunities. Int. J. Hydrog. Energy. 41: 8450–8461.

Santangebo, T.J., L. Cubonova and J.N. Reeve. 2011. Deletion of alternative pathways for reductant recycling in *Thermococcus kodakarensis* increases hydrogen production. Mol. Microbiol. 81(4): 897–911.

Sapra, R., K. Bagramyan and M.W.W. Adams. 2003. A simple energy conserving system: proton reduction coupled to proton translocation. Proc. Natl. Acad. Sci. USA 100: 7545–7550.

Savaliya, M.L., B. Dhorajiya, B.Z. Dholakiya. 2015. Recent advancement in production of liquid biofuels from renewable resources: A review. Res. Chem. Interm. 41: 475–509.

Schroeder, C., M. Selig and P. Schonheit. 1994. Glucose fermentation to acetate, CO_2 and H_2 in the anaerobic hyperthermophilic eubacterium *Thermotoga maritima* involvement of the Embden-Meyerhof pathway. Arch. Microbiol. 161: 460–470.

Schut, G.J. and M.W. Adams. 2009. The iron-hydrogenase of *Thermotoga maritima* utilizes ferredoxin and NADH synergistically: a new perspective on anaerobic hydrogen production. J. Bacteriol. 191(13): 4451–4457.

Selig, M., K.B. Xavier, H. Santos and P. Schonheit. 1997. Comparative analysis of Embden-Meyerhof and Entner-Doudoroff glycolytic pathways in hyperthermophilic archaea and the bacterium *Thermotoga*. Arch. Microbiol. 167: 217–32.

Siebers, B. and P. Schönheit. 2005. Unusual pathways and enzymes of central carbohydrate metabolism in Archaea. Curr. Opin. Microbiol. 8: 695–705.

Silva, P.J., E.C. Van Den Ban, H. Wassink, H. Haaker, B. de Castro, F.T. Robb et al. 2000. Enzymes of hydrogen metabolism in *Pyrococcus furiosus*. Eur. J. Biochem. 267: 6541–6551.

Tse, C. and K. Ma. 2016. Growth and metabolism of extremophilic microorganisms. pp. 1–46. *In*: Rampelotto, P.H. (ed.). Biotechnology of Extremophiles: Advances and Challenges. Springer.

van Niel, E.W.J., M.A.W. Budde, G.G. de Haas, F.J. Van Der Wal, P.A.M. Claassen and A.J.M. Stams. 2002. Distinctive properties of high hydrogen producing extreme thermophiles, *Caldicellulosiruptor saccharolyticus* and *Thermotoga elfii*. Int. J. Hydrogen. 27: 1391–1398.

van Ooteghem, S.A., S.K. Beer and P.C. Yue. 2002. Hydrogen production by the thermophilic bacterium *Thermotoga neapolitana*. Appl. Biochem. Biotechnol. 98–100(1-9): 177–189.

van Ooteghem, S.A., A. Jones, D. van der Lelie, B. Dong and D. Mahajan. 2004. H_2 production and carbon utilization by *Thermotoga neapolitana* under anaerobic and microaerobic growth conditions. Biotechnol. Lett. 26: 1223–1232.

Ying, X. and K. Ma. 2011. Characterization of a zinc-containing alcohol dehydrogenase with stereoselectivity from hyperthermophilic archaeon *Thermococcus guaymasensis*. J. Bacteriol. 193: 3009–3019.

Zillig, W., I. Holz, D. Janekovic, H. Klenk, E. Imsel, J. Trent et al. 1990. *Hyperthermus butylicus*, a hyperthermophilic sulfur-reducing archaebacterium that ferments peptides. J. Bacteriol. 172(7): 3959–3965.

Chapter 15

Microorganisms from Permafrost and their Possible Applications

A. Brouchkov,[1,2,3,]* V. Melnikov,[2,3] G.I. Griva,[2] E. Kashuba,[4,6] V. Kashuba,[4,5]
M. Kabilov,[11] O. Fursova,[10] V. Bezrukov,[7] Kh. Muradian,[8] V. Potapov,[8]
G. Pogorelko,[9,10] N. Fursova,[8] S. Ignatov,[7] V. Repin,[11] L. Kalenova,[3] A. Subbotin,[3]
Y.B. Trofimova,[3] E.V. Brenner,[11] S. Filippova,[12] V. Rogov,[1] V. Galchenko[12]
and A. Mulyukin[12]

Introduction

The isolation of microorganisms from ancient deposits, including permafrost, and the question that are they as old as the deposits are still areas of controversy. It is intriguing that many representatives of ancient bacteria are close to modern bacteria at the molecular level. In some cases geological data proves the old age of these organisms, while molecular data presents evidence of their modernity. It is unclear how bacteria stay viable for a long time. Despite the fact that it is unknown, if they are spores, individual cells surviving or a growing colony, *Bacillus anthracis* keeps viable for at least 105 years (Nicholson 2000). Bacteria from amber are reported to survive for 40 million and more years (Greenblatt et al. 2004). The discovery of bacteria living in salt deposits and apparently being of a huge age raises a number of questions. However, the individual findings did not insure the enormous life expectancy of bacteria, and the phenomenon has become obvious since permafrost was studied (Friedmann 1994).

Permafrost is defined as a lithosphere material (soil and sediment) that is permanently exposed to temperatures $\leq 0°C$ and remains frozen for at least two consecutive years, usually hundreds and thousands of years. It covers about 26% of terrestrial soil ecosystems in the Northern Hemisphere and can extend down to more than 1500 m into the subsurface (Steven et al. 2006). Permafrost is

---
[1] Lomonosov Moscow State University, Moscow, Russia.
[2] Tyumen Scientific Center, Siberian Branch of Russian Academy of Science, Tyumen, Russia.
[3] Tyumen State University, Tyumen, Russia.
[4] Department of Microbiology, Tumor and Cell Biology, Karolinska Institute, Stockholm, Sweden.
[5] Department of Molecular Oncogenetics, Institute of Molecular Biology and Genetics, Kyiv, Ukraine.
[6] R.E. Kavetsky Institute of Experimental Pathology, Oncology and Radiobiology, Kyiv, Ukraine.
[7] State Institute of Gerontology, Kyiv, Ukraine.
[8] State Research Center for Applied Microbiology and Biotechnology, Obolensk, Russia.
[9] NI Vavilov Institute of General Genetics, Russian Academy of Sciences, Moscow, Russia.
[10] Iowa State University, Ames IA, USA.
[11] Institute of Chemical Biology and Fundamental Medicine, Novosibirsk, Russia.
[12] Winogradsky Institute of Microbiology, Research Center of Biotechnology, Moscow, Russia.
* Corresponding author: brouchkov@hotmail.com

regarded as the natural depository of extant microorganisms that have survived for up to millions of years (Friedmann et al. 1994; Vorobjova 1997; Gilichinsky et al. 2007; Steven et al. 2008). Members of the major phyla (*Proteobacteria* and/or *Actinobacteria*) were found using culture-dependent and independent approaches in Alaskan, Canadian and Siberian permafrost of different ages and genesis (Shi et al. 1997; Vishnivetskya et al. 2000; 2006; Katayama et al. 2007; 2009; Steven et al. 2008; Yergau et al. 2010; Rivkina et al. 2015). Microorganisms have been recently found in ice-sediment communities in the surface layers of perennial and permanent lake ice (Psenner and Sattler 1998), in a high Arctic glacier (Skidmore et al. 2000), Greenland glacier (Sheridan et al. 2003) and sub glacial lake and rivers (Yadav et al. 2015a; Yadav et al. 2015b). Parkes (2000) reviewed convincing cases of bacteria in diverse environments which have remained viable over inordinate lengths of time.

Unfrozen water, held tightly by electrochemical forces onto the surfaces of mineral particles, occurs in even hard-frozen permafrost (Williams and Smith 1991). Bacterial cells are not frozen at temperatures of –2 and –4°C (Clein and Schimel 1995). The thin liquid layers provide a route for water flow, carrying solutes and small particles, possibly nutrients or metabolites, but its movement is extremely slow (Burt and Williams 1976). A bacterium of greater size (0.3 to 1.4 microns) than the thickness of the water layer (0.01–0.1 micron at temperatures –2 and –4°C) is unlikely to move, at least in ice (Ershov 1998). Therefore, microorganisms in permafrost have been isolated, trapped among mineral particles and ice, for orders of magnitude longer than a normal lifespan and maybe thousands and millions of years (Friedmann 1994). Their age in some cases is clearly proved by geological conditions, a history of freezing, radioisotope dating, as well as biodiversity. Here data obtained from old Siberian permafrost microbiological studies, and suggesting a possible explanation of controversial molecular data received from our and other studies of the ancient bacteria are presented.

'Modern' bacteria in ancient deposits: DNA stability and evolution rates

DNA in ancient bacteria is expected to decay due to number of reasons including thermal fluctuations, nucleotide deamination, radiation in the soil, etc., and the bacterium is supposed to lose its viability within about several hundred years (Lindahl 1993). Therefore, the nature of extreme longevity of microorganisms has no clear explanation. Proteins are far from being stable (Jaenicke 1996). Bacteria are affected by aging (Johnson and Mangel 2006). Genome is subjected to mutations, and reparations seem to be not effective enough to prevent damages (Cairns et al. 1994). Half-life of cytosine in water solution does not exceed a few hundred years (Levy and Miller 1998). Ancient DNA of mummies, mammoths, insects in amber and other organisms appears fragmented and destroyed (Willerslev and Cooper 2005). The major cause of DNA mutations was argued to be due to cytosine deamination, generating C→T and G→A miscoding lesions, but other studies show that damage may include adenine to hypoxanthine modifications resulting in A→G and T→C miscoding lesions (Binladen et al. 2006). Despite the similar nature of mutations, it is considered that a degree of variability of mutation rates is still an open question. It seems very unlikely that the rates are similar for different bacteria, for different environments, for different genes and for prokaryotes and eukaryotes.

At the same time it is a fact that younger genes tend to show accelerated evolutionary rates with respect to older genes. The 16S rRNA gene, for example, shows slow evolution, while other genes evolve more rapidly. The rate of substitution for 16S ribosomal DNA has been found to be remarkably uniform between about 1×10^8 and 5×10^8 substitutions per site per year (Clark et al. 1999). The phylogenetic substitution rates in mitochondria are approximately 0.5% per million years for avian protein-coding sequences and 1.5% per million years for primate protein-coding and d-loop sequences. However, it has been shown that for the α-Proteobacteria, the γ-Proteobacteria and the *Bacillus* group, the clock-like null hypothesis could not be rejected for only about 70% of the for several hundred analyzed orthologous sets, whereas the rest showed substantial anomalies (Novichkov et al. 2004). Two hundred and fifty million years old bacteria living in salt deposits was tested by Nickle et al. (2002) who performed relative rate tests using 16S rDNA with the same result; the branch leading to isolate 2-9-3 is not extraordinarily short, as would be expected of an organism that has not been evolving for millions of years. To explain the 16S rRNA gene studies of Permian bacteria results of Vreeland et al. (2000), it can be assumed that the rate of substitution in the 16S rRNA gene is about 5×10^{12} substitutions per site per year, i.e., a reduction of four orders of magnitude in comparison with the typical results. Bacteria from amber also show 'modern' genomic features (Greenblatt 2004; Gutiérrez and Marín 1998; Parkes 2000), and some permafrost microorganisms as well (Shi et al. 1997). Since DNA decays fast, and the rate estimations are so different, one needs to question how old are the ancient bacteria?

It is interesting that some bacteria show stable genomes and seem to have a recent origin, for example no sequence polymorphisms were detected in six gene fragments from 36 isolates from the three classical biovars, indicating that *Yersinia pestis* evolved from *Y. pseudotuberculosis* within the last 1,500–20,000 years (Achtman et al. 1999). Also, for the *Bacillus cereus* group of endospore-forming bacteria, representative of which will be discussed later in this chapter, including *Bacillus anthracis*, causative agent of anthrax, *B. cereus*, a saprophyte known for causing food poisoning, *Bacillus mycoides*, with rhizoidal colony formations, *Bacillus thuringiensis* containing parasporal crystal proteins, DNA studies (Nakamura et al. 1994) failed to clearly separate them, to the extent that they can be considered as single species. The rearrangement rate in the *Bu. aphidicola* was found close to zero during the last 100–150 years of evolution (Belda et al. 2005). It should be probably taken into account that the rates of spore forming bacteria evolution might be significantly slower. Estimates of the dormant state time vary from 10^2 to 10^4 years between times of growth (Nicholson et al. 2000). Some studies (Parkes et al. 2000) show that generation times of bacteria isolated from the sub seafloor sediments are of about thousands of years. Bacteria isolated from amber often show a high homology to the younger sequences, and it is sometimes explained by the fact that the amber samples were from different geoclimatic regions of the Earth (Veiga-Crespo et al. 2004). Contamination is always an issue; however, a number of studies show the same results: many isolated ancient bacteria have 'modern' genomes. The more similar the results the less possibility of mistakes, but it seems very unlikely that it is a contamination or error in all cases.

Microorganisms in permafrost

Permafrost microorganisms in comparison with ancient salt or amber isolates are widely distributed (Vishnivetskaya et al. 2006; Steven et al. 2008; Yergeau et al. 2010; Margesin and Miteva 2011). For more than a century there have been reports of living organisms in permafrost, some of which might certainly be millions of years old, if their age is similar to the age of permafrost itself. Living (or at least viable) bacteria apparently occur deep in solid-frozen ground (permafrost) in the cold regions (see the review by Gilichinsky and Wagener 1995). Rather than re-examining the individual cases reported, present-day knowledge of frozen ground and of the physiology of psychrophilic bacteria and other organisms (Morita 1997) has been considered. For example, are isolated bacteria as old as the permafrost itself or contamination with more recent bacteria has occurred? Do bacteria grow in the permafrost? And to what extent are 'normal' metabolic processes taking place? Or are they inactive and cryopreserved? An important characteristic of permafrost is that some water held tightly by electrochemical forces onto the surfaces of mineral particles or under the influence of capillary forces, occurs in even hard-frozen permafrost (Williams and Smith 1991; Brochkov and Williams 2002).

The thin liquid layers provide a route for water flow, which is normally from the warmer to the colder parts (Derjaguin and Churaev 1986). The water may carry solutes and small particles and perhaps may be bacteria, but its movement is extremely slow (Burt and Williams 1976): at a few degrees below °C it may take thousands of years to move a meter. A bacterium of greater size than the thickness of the water layer is likely to move much more slowly than the water. One can conclude that microorganisms in permafrost have been isolated, certainly from the ground surface, trapped among the mineral particles and ice. The longest, continuously frozen permafrost in the Northern Hemisphere is usually estimated as between one and three million years old (Ershov 1998). Abyzov's investigations at the Vostok station (Abyzov 1993) revealed bacteria, fungi, diatoms and other microorganisms were probably carried to the Antarctica by winds.

The ages of these individuals could be more than half a million years. Abyzov's (1993) showed the presence of viable bacteria in ice which was hundreds of thousands of years old and at a depth of thousand meters which could not have been contaminated from the surface or from below in recent times. Although most microorganisms do not grow at temperatures below 0°C, certain bacteria and fungi can be physiologically active and Friedmann (1994) noted metabolic activity in permafrost bacteria at –20°C. Others reporting evidence concerning bacterial activity in soils below 0°C, include Kalinina et al. (1994); and Clein and Schimel (1995). Water is a solvent for the molecules of life, and availability of water is a critical factor affecting the growth of all cells. But some particular water which is unfrozen in permafrost, although at less than 0°C and in the presence of ice, differs from 'ordinary' water. It is attached to the soil mineral particles surfaces. As the temperature falls to –2 or –3°C, the remaining water is in layers so thin that a bacterium could not be fitted in it. Metabolic activity and especially the ability of microorganisms to grow for a long time are greatly limited in the conditions of the environment within the permafrost.

The single bacterial cell is trapped and not even free to move or expand within the unfrozen water layer. Some microorganisms could probably grow only if a substantial degree of microbial activity is at temperatures below 0°C. But for the most part it appears unlikely. Microscopic pictures of frozen soils show single cells mostly (much fewer groups of a some cells), not colonies (Fig. 15.1), and this is another argument for dormancy microorganisms in permafrost (Melnikov et al. 2011).

One of the most unique, oldest and poorly studied permafrost environments are the permanently frozen alluvial Neogene sediments exposed in the Aldan River valley near the Mammoth Mountain in the Central Yakutia (Eastern Siberia). Intense cooling in the area began in the Late Pliocene (Tripati et al. 2008), and a small amount of temperature fluctuations did not affect the frozen state of these sediments (Lisiecki and Raymo 2005; Demezhko and Golovanova 2007). According to geological data, the age of the Mammoth Mountain-associated alluvial sediments is estimated to be no less than 3–3.5 Ma (Markov 1973; Velichko and Isavea 1992). Little is known about microorganisms in permafrost alluvium: relatively poor diversity of cultured heterotrophic bacteria and a low proportion between the numbers of colony-forming units and the direct viable counts (Zhang et al. 2013), while the structure of bacterial community remains unknown.

It is indeed an intriguing topic on the survival of bacteria belonging to different phylogenetic groups which could exist before freezing (Yadav et al. 2016; Yadav et al. 2018). On the other hand, due to the particular genesis of these sediments (Markov 1973) and contributions of different sources

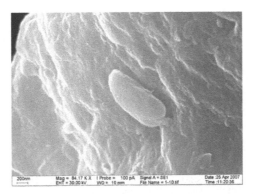

Figure 15.1 Electronic scanning microscope picture of bacteria in ancient Neogene permafrost, exposure of the Mammoth Mountain.

(such as relict plants as reported by Baranova et al. (1976), the microbial communities may differ from the other permafrost sediments. Next generation sequencing of 16S rRNA gene as a culture-independent approach is attractive due to the possibility to eliminate the clone library production step and to generate numerous sequences per sample (Rothberg and Leamon 2008; Caporaso et al. 2011) and especially to capture low-abundant microbial taxa in permafrost (Yang et al. 2012) and soils and deep sea water (Sogin et al. 2006; Roesch et al. 2007). One of possible biases in microbial community profiling from read abundance data could be associated with nucleic acid extraction (Amend et al. 2010), and different protocols were recommended to isolate total DNA from a studied sample (Feinstein et al. 2009; Brooks et al. 2015).

Case study

The sampling site was the ice-covered exposure (Figs. 15.2 and 15.3) of Mammoth Mountain (N62°56' E134°0.1') in the left bank of the Aldan River valley in Central Yakutia (Eastern Siberia). No specific permission was required for sampling and no endangered or protected species were involved in this study. The mean annual temperature near the exposure surface is presently about – 4°C. The exposure consisted of three sediment layers attributed to (1) Late Pleistocene (about 15 Ka–40 Ka-old Ice Complex); (2) Middle and Early Pleistocene (0.1–1 Ma old, frozen at the time of formation); (3) Neogene (frozen probably at the end of Neogene about 3–3.5 Ma ago) (Markov 1973).

Samples of alluvial Neogene sands were collected at 83-m altitude above the sea level (northern exposition) in a pit (about 100 × 100 cm) dug to a 1.5-m depth, 0.9–1.0 m deeper than the layer of seasonal thawing out. The pit surface was sterilized under a flame of a gas burner. Samples were transported in a frozen state in thermally controlled containers with cooling agents. The age of the permafrost exceeded 3 million years that was dated by paleoclimatic reconstructions (Bakulina and Spector 2000; Baranova et al. 1976). For surface sterilization, a sample of about 50 g was placed in a glass with 96% ethanol solution, then put on a burner flame, and packed into a sterile test tube.

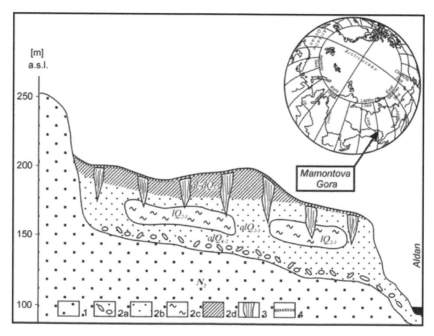

Figure 15.2 Profile of the exposure of the Mammoth Mountain: 1, Neogene sands; 2, Pleistocene sediment: a, pebbles in the ferrous sands; b, sands; c, lacustrine silt; d, silt; 3, ice wedge; 4, active layer.

Figure 15.3 Sampling site at Mammoth Mountain in the Aldan river valley (Central Yakutia, Eastern Siberia). Formations of Late Pleistocene (Ice Complex) (red), Pleistocene (green) and Neogene (yellow) are visible.

Control contamination of the processed pit by spreading 2-ml suspensions of *Yarrowia lipolytica* Y-3603 tagged with red fluorescent protein (Yuzbasheva et al. 2011) (10^8 cells/ml) onto 200-cm^2 surface was not detected by plating and direct microscopy examinations.

Sterile DNA-free 50-mL Cellstar® tubes were opened near the flame and immediately pushed into the exposed frozen sediments (– 4°C); the sampling procedure excluded any contact with exterior surfaces and used sampling devices. These samples were allotted for DNA isolation. In addition, frozen samples were taken from the same pit with extreme precautions using sterilized metal bores, forceps and scalpels were used for physicochemical and microscopy analyses. All samples were transported in a frozen state in a 36-L Coleman® cooler with saturated NaCl solution as a cooling agent and were stored at – 5°C that mimics the natural conditions.

Outer layers (2–3 cm) of stored frozen 4–5 kg samples were removed in the laboratory with sterile knives in a class II laminar hood and only internal parts were used. Composite samples were prepared by mixing of portions from three sediment samples. The grain-size distribution was determined using the common sieving technique; the water content was estimated after heating of wet samples at 105°C for 24 hours. Loss on Ignition (LOI) was measured as a difference before and after burning of dry samples at 430°C for 3 hours in a muffle furnace and served to estimate the organic carbon content (Shinner et al. 1996). The pH and Eh values of sediment suspensions (in 10 mM CaCl$_2$) were measured using a Thermo Scientific Orion Model 261S pH/mV/Temperature meter. Physicochemical analyses were performed in three independent sets, each in triplicates with calculations of the mean and SD using Excel software. Values of pH and Eh were measured in triplicates for 17 sub samples of the same sample.

Morphology and distribution of particles in permafrost alluvium were studied by light and electron microscopy of freeze fractures prepared as described (Rogov and Kurchatova 2013). Unthawed samples were broken at < 0° and the fracture surface was coated with formvar. After thawing, samples were examined under a light Reichert Microstar IV microscope and then under a Hitachi TM3000 electron microscope (voltage, 30 кV). Elemental compositions of particles were determined using a Swift ED3000 Energy Dispersive X-ray spectrometer (EDX). Experiments were also performed using a Nanoscope III multimode atomic force microscope (Veeco Instruments Inc., USA) in contact mode in the air. Commercial silicon CSC11 (spring constant 0.35 N/m) cantilevers

(MikroMasch) were used. The FemtoScan software (Filonov and Yaminsky 2007) was used for image processing. The scan rate was typically 2 Hz with 512 lines. The representative images are shown in Fig. 15.4.

All reagents used for DNA extraction and further analyses were molecular biology grade. Solutions and buffers were DNase-free and sterilized. All procedures were conducted in I class hoods. In these experiments, two counterpart sub samples from the same sediment sample that was collected by pushing in 50-ml sterile vials were used. Total community DNA was extracted from 0.25-g sub samples with DNA Spin Kit for Soil (MO-Bio) and QIAamp DNA Stool Mini Kit (Qiagen) according to the manufacturers' protocols. The V3–V4 region of the 16S rRNA genes was amplified with the primer pair 343F (5'-CTCCTACGGRRSGCAGCAG-3') and 806R (5'-GGACTACNVGGGTWTCTAAT-3') combined with Illumina adapter sequences, a pad and a linker of two bases, as well as barcodes on the primers. PCR amplification was performed in 50 μl reaction mixtures containing 0.7 U Phusion Hot Start II High-Fidelity and 1 × Phusion GC buffer (Thermo Fisher Scientific), 0.2 μM of each forward and reverse primers, 10 ng template DNA, 2.3 mM MgCl$_2$ (Sigma-Aldrich) and 0.2 mM of each dNTP (Life Technologies). Thermal cycling conditions were as follows: initial denaturation at 98°C for 1 minute, followed by 30 cycles of 98°C for 15 seconds, 62°C for 15 seconds, and 72°C for 15 seconds, with final extension at 72°C for 10 minutes.

A total of 200 ng PCR product from each sample was pooled together and purified using MinElute Gel Extraction Kit (Qiagen). Sample libraries for sequencing were prepared according to the MiSeq Reagent Kit Preparation Guide (Illumina) and the protocol described earlier (Caporaso et al. 2011). Sample denaturation was performed by mixing 4.5 μl of combined PCR products (4 nM) and 4.5 μl 0.2 M NaOH. Denatured DNA was diluted to 14 pM and 510 μl mixed with 90 μl of 14 pM Phix library. A total of 600 μl sample mixture, together with customized sequencing primers for forward, reverse and index reads, were loaded into the corresponding wells on the reagent cartridge of a Miseq 500 cycles kit and run on Miseq for 2 × 250 bp paired-ends sequencing

Figure 15.4 Images of the studied alluvial sediments as viewed by (A) light microscopy and (B, C) scanning electron microscopy. (A) Biomorphic particles (within a white frame) were examined using (B) EDX-SEM. A spectrum from a selected region is shown in the insert. (C) A biomorphic particle is adhered to minerals near the contact with ice. Designations: MP, mineral particles; BP, biomorphic particles.

(Illumina) at the SB RAS Genomics Core Facility (ICBFM SB RAS, Novosibirsk, Russia). The 16S read data were deposited in the NCBI database under accession SRP075638.

Raw sequences were analyzed with UPARSE pipeline (Edgar 2013) using USEARCH v8.1.1861. The UPARSE pipeline included merging of paired reads; read quality filtering; length trimming; merging of identical reads (dereplication); discarding singleton reads; removing chimeras and OTU clustering using the UPARSE-OTU algorithm. The OTU sequences were assigned a taxonomy using the RDP classifier 2.11 (Wang et al. 2007) and searched against the NCBI 16S database using BLASTN. Community structure analyses were based on the phylum, class and genus taxonomy levels. OTUs reads were aligned with the MUSCLE v.3.8.31 and phylogenetic analysis was performed using maximum likelihood phylogeny tool in CLC GW 8.5 (1000 bootstrap pseudoreplicates and GTR+G+T evolutionary model). Rarefaction curves of observed OTUs were obtained using USEARCH. Samples of different dilutions in sterile conditions were added to petri dishes containing YPD (BD, USA), MRS (BD, USA), and NA (BD, USA) media. Samples were also added into a liquid meat–peptone broth under anaerobic and aerobic conditions. As a result several bacterial strains were isolated from the frozen samples, one of them was *Bacillus* sp. strain *F* identified by Brouchkov et al. (2009).

Salmonella enterica var. Enteritidis and *Bacillus cereus* var. *toyoi* were received from the SRCAMB Culture Collection (#B-10130). All strains were cultivated routinely on Luria medium at 37°C, and stored by lyophilizate and at 70°C on Luria broth containing 20% glycerin. Bacterial suspensions for inoculation were prepared by growing microorganisms for 4 hours in nutrient broth at 37°C. The concentration of microorganisms was determined by spectrophotometer and appropriate dilutions were prepared in sterile 50 mM phosphate buffer solution (PBS) pH 7.0. The numbers of bacteria in the inocula were confirmed by plating onto nutrient agar.

Plasmid DNA was prepared using the highly efficient potassium hydroxide-based method providing a high yield of efficiently purified pDNA. 3 ml of *Bacillus* sp. strain *F* night culture grown at 37°C spin down for 30 seconds at 13400 rpm twice to collect all bacterial cells in one 1.5 ml vial. The pellet was resuspended in 0.2 ml of buffer 1 (50 mM Glucose, 25 mM Tris-HCl (pH = 8.0), 10 mM EDTA) and tubes were placed in ice followed by addition of 0.4 ml of buffer 2 (0.2 M NaOH, 1% SDS) and gently shaking several times. After incubation in ice for 5 minutes 0.3 ml of buffer 3 were added (7.5 M NH_4Ac) and incubated in ice for 5 minutes. The samples were then centrifuged for 5 minutes at 13400 rpm, and the supernatant moved to the new 1.5 ml tubes containing 0.6 ml of isopropanol and incubated for 5 minutes at room temperature. After centrifuging at 13400 rpm the supernatant was discarded completely and 0.1 ml of buffer 4 (2 M NH_4Ac) was added to the pellet followed by incubation in ice for 5 minutes. Samples were centrifuged for 2 minutes at 13400 rpm, and the supernatant was moved to the new 1.5 ml vial containing 0.1 ml of isopropanol and incubated for 5 minutes at room temperature. After centrifuging at 13400 rpm the pellet was washed once with 0.15 ml of 70% ethanol and dried at 65°C for 3 minutes and dissolved in 0.035 ml of distilled water; 3 µl of this suspension were used for electrophoresis.

DNA amplification was done using six pairs of arbitrary primers (Forw1: TAAACCGACAATGCGACTCCTCCGA and Rev1: AGTTGCATTTATTCCCATGACGTAC; Forw2: AATGCTGCATAATTCTCGGGGC and Rev2: AAGGCACGCAACGCCTACGACT; Forw3: GTCAGTTTCAGACATATTCAATGGC and Rev3: GTCATGGTGATACATTTTGTCAGTC; Forw4: GTCAGCGAAAGAGTGGTGCTCA and Rev4: TATCTGGATCACAAAACAGGG; Forw5: TTAGGATCCATGTTACGTCCTG and Rev5: TTTAGATCTTTGTTTGCCTCCC; Forw6: CACGGGGGACTCTAGAGGATCCGGGA and Rev6: CCATGATTACGGGCCCCCGATCTA) accordingly the DNA amplification program was followed: preliminary DNA denaturation at 95°C for 3 minutes; 35 cycles including DNA denaturation at 94°C for 15 seconds, annealing at 50°C for 40 seconds, and extension at 72°C for 2 minutes; and final extension at 72°C for 4 minutes. DNA sequencing of the PCR-products cloned into vector pGEM-Teasy (Promega) was done using M13-Forward and M13-Reverse primers.

Experiments were carried out on *Drosophila melanogaster*, whose population was kept in the State Institute of Gerontology, Kyiv, Ukraine for some years. A culture isolated from permafrost *Bacillus* F. sp. (1, 2, 5, 10, 25, 50, 100, 250, 500 million/ml) was added to the nutrient medium of *Drosophila melanogaster*. Heat shock was modeled by 30-minutes exposure of imago in a thermostat at 38°C. UV-radiation was applied for 60 minutes at 50 W in the survival experiments.

Six-week-old female Webster out bred mice was used in all experiments. The mice were kept in a cage of up to 10 per cage, and food and water were given *ad libitum*. All experiments were approved by the Animal Care and Ethics Committee of the SRCAMB.

Webster out bred mice was inoculated *per os* by *S. enterica* var. Enteritidis cells ($5 \cdot 10^6$ CFU/ mouse) for the modeling of infection. The dynamics of infection in mice was detected on the 3rd, 4th, 5th, 6th, and 7th days using CO_2-euthanase and post-mortem examination. The following characteristics were used for the estimation of mice organism (i) changing rate of internal organs that were increased in weight in 1.5–2.0 times; (ii) pathological changing in organs on 10–14th day after infection, that was expressed in cirrhosis and/or in forming of necrosis regions in the spleen and on the border of the liver; (iii) number of *Salmonella* cells in the liver and spleen that run up to 10^8–10^9 CFU/mouse on the 7th day after infection. Life duration of the infected mice was 7–10 days after infection on an average. Death specificity was confirmed by *Salmonella* isolation from homogenate of parenchymatous organs by growing on the selective medium SS-agar (SRCAMB, Russia).

Livers and spleens of mice were removed aseptically on specific days after infection as indicated in the results. Organs were weighed, and Mueller-Hinton broth was added to yield a 10% suspension (wt/vol) after homogenization in a Stomacher-80 (Thermo, UK). Serial 10-fold dilutions were plated in duplicate on the petri dishes filled by Mueller-Hinton agar. After incubation for 18 hours at 37°C, colonies were counted and the number of CFU per organ was calculated. Isolates from animal organs were identified as *S. enterica* var. Enteritidis by slide agglutination with the use of commercially available antisera to somatic (O) and flagellum (H) antigens (Immunoteks, Russia).

Bacillus cereus var. *toyoi* and *Bacillus* sp. strain F bacterial pathogenicity was studied using Webster out bred mice (weigh 20–24 g, age 10–12 weeks) by the intravenously, intraperitoneally, and intragastrically inoculation of 24 hours-culture at doses 10^7, 10^8, and 10^9 CFU per mouse. The number of mice in each group was 12. The observation time after inoculation was 21 days. The control group of mice was inoculated by PBS in the same scheme Pathogenicity was estimated using the following parameters: (i) survival of mice numbers at 21 days after infection; (ii) path morphology of euthanized mice organs on the 21th day after infection; (iii) the presence of *Bacillus* cells in blood, liver and spleen of the infected mice. All animal experiments and mice euthanasia was done accord to "Guide for Care and Use of Laboratory Animals".

Bacillus cereus var. *toyoi* and *Bacillus* sp. strain *F* bacilli were grown in 10 ml of Luria broth (Difco, USA) at 37°C overnight, the whole grown cells were transferred to 50 ml of the same broth and were further incubated at 37°C, when the cell growth reached the log phase, the cells were harvested by centrifugation and resuspended in PBS. The *Bacillus* sp. strain *F* bacteria at the dose of 1×10^9 CFU per mouse were administered orally for 7 consecutive days, while the mice in the control group were treated by PBS only. One day after the last *Bacillus* sp. strain *F* feeding the group of 10 mice was administrated with 5×10^6 CFU of *S. enterica* var. Enteritidis was equivalent to the LD50 dose. The number of the dead mice was recorded every day for 21 days. Alluvial sediments with the bulk density of 1.7 g cm^{-3} and porosity values of 45% were composed mainly from coarse sand fraction (0.2–2 mm, 84%) with pale quartz and feldspar as major minerals (Table 15.1).

The pH values of sediment suspensions (in 0.1 M $CaCl_2$) after thawing varied from 7.08 to 8.90; and the Eh was in a range +355 mV – +406 mV (17 measurements). Light microscopy of the sediment fractures demonstrated the cryogenic structure with crystals and thin streaks of ice with embedded biomorphic particles (Fig. 15.4A). SEM examinations with EDS analysis showed cell-like morphology of these propagules (6–8 μm in diameter) and the presence of C and O, as well as Cl and Ca as components of salts or encrusting material on the outer parts and surroundings (Fig. 15.4B, C). Tightly packed triangular-faceted particles of 3–30 μm in long axis were similar

Table 15.1 Physical and chemical properties of alluvial Neogene sediments (composite sample).

Major minerals	Quartz, feldspar
Grain size distribution	
particle size, 1–2 mm	3%
0.5–1 mm	20%
0.25–0.5 mm	61%
0.1–0.25 mm	10%
< 0.1 mm	6%
Density, r	1.72 gcm^{-3}
r_d	1.39 gcm^{-3}
r_s	2.51 gcm^{-3}
Porosity	45%
Water content	24%
Loss on ignition*	3.6%
Organic carbon*	1.6–2%
Total salinity*	< 0.05%
Content* NO_3	< 20 mg kg^{-1} dw
NO_2	< 10 mg kg^{-1} dw
total P	< 20 mg kg^{-1} dw
pH of suspension (1 : 1 in 0.01 M $CaCl_2$)	7.08–8.90
Eh	+355–+406 mV

Table shows the mean values; SD ± 5% are not indicated.
* Data were taken from a different study (Zhang et al. 2013).

in micromorphology to fragmented quartz (Mahaney 2002) to be attributed to salts or encrusting components. Some micron- and submicron-sized propagules resembled bacterial cells that were free or attached to minerals (Fig. 15.4). The absence of structures that are similar to dividing cells among biomorphic structures suggests that the conditions in the alluvial Neogene sediments were not supportive for growth and multiplication of many bacteria.

DNA was extracted from the same sample using DNA Spin Kit for Soil (MO-Bio) or QIAamp DNA Stool Mini Kit (Qiagen) with the yields 150 and 438 ng DNA per 0.25 mg sediment, respectively. A total of 31,239 (Qiagen kit) and 15,404 (Mo-Bio kit) 16S rRNA gene sequences (400–435 bp) were assigned to 61 OTUs. Only three phyla comprised more than 5% OTUs abundance: *Bacteroidetes* (71 and 52%, Qiagen and Mo-Bio kits), *Proteobacteria* (23 and 42%), and *Firmicutes* (5%) and jointly covered 99% of reads. The use of two DNA isolation kits caused no variations in the order of predominance at the phylum level: *Bacteroidetes* > *Proteobacteria* > *Firmucutes* (Fig. 15.5A). Rarefaction curves reached asymptotic values, thus indicating the sufficient read coverage for identification of taxa constituting the studied bacterial. Differences in OTU contributions for two DNA isolation procedures (Qiagen and Mo-Bio kits) were prominent for *Bacteroidetes* and *Proteobacteria*, not for *Firmicutes*.

Totally, members of 6 classes held 98% of the total reads and the top 11 families (and genera) shared a more than 90% contribution to the bacterial community from ancient permafrost alluvium (Fig. 15.5B, C). Representatives of the class *Sphingobacteria* dominated within *Bacteroidetes* and sustained 70.7% (Qiagen kit) and 51% (Mo-Bio kit) of the total reads (Fig. 15.5). Major contributors were members of the family *Chitinophagaceae* and the genus *Sediminibacterium* (65.9 and 48.2%

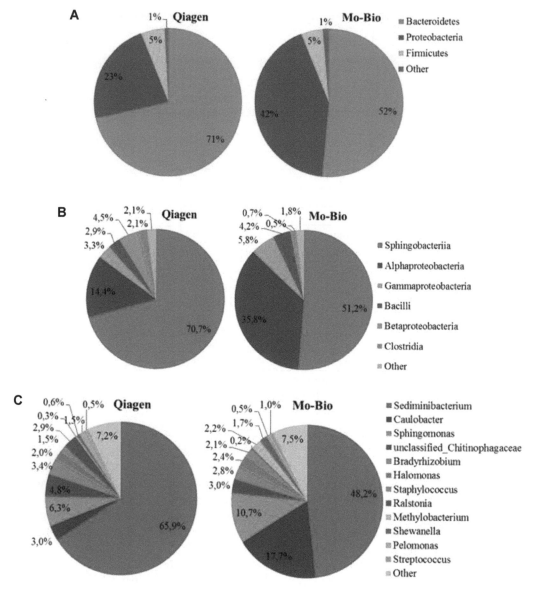

Figure 15.5 Composition of the bacterial community from 3–3.5 Ma permafrost alluvial sediments under Mammoth Mountain (Eastern Siberia). Relative read abundance of different bacterial phyla (A), classes (B), and genera (C).

for the two kits, respectively) and an unclassified group that possessed 92.5% similarity to the genus *Niabella*.

The second abundant lineage *Proteobacteria* was dominated by *Alphaproteobacteria* (35.8 and 14.4%) followed by *Gammaproteobacteria* (3.3 and 5.8%) and *Betaproteobacteria* (4.5 and 0.7%). The major groups (within *Alphaproteobacteria*) with > 1% abundance represented the following families and genera: *Caulobacteraceae* (the genus *Caulobacter*, the mean contribution 11.4%), *Sphingomonadaceae* (*Sphingomonas*, 8.5%), *Bradyrhizobiaceae* (*Bradyrhizobium*, 3.1%), and *Methylobacteriaceae* (*Methylobacterium*, 2.2%). Among *Gammaproteobacteria*, the most abundant were members of *Halomonadaceae* and *Shewanellaceae* and especially the genera *Halomonas* (2.2%) and *Shewanella* (1.2%). The *Betaproteobacteria* class was mainly represented by the families *Burkholderiaceae* and the genus *Ralstonia* (1.5%) and *Comamonadaceae*, the genus *Pelomonas* 1.1%.

The dominant sub-divisions within *Firmicutes* were the class *Bacilli* (especially *Staphylococcaceae*, *Streptococcaceae*, and *Lactobacillaceae*); members of the genera *Staphylococcus* and *Streptococcus* contributed with a mean 1.8 and 0.75% of the total reads, respectively (Fig. 15.5C). In general, variations of DNA extraction procedures produced no substantial interferences in the qualitative profile, except for the reliable detection of *Burkholderiaceae* with the Qiagen kit. However, differences in numerical contributions from most top divisions (especially, *Sediminibacterium*, *Caulobacter*, and *Sphingomonas*) were significant for two groups of data obtained using the two kits.

The other phyla contributed with ~ 1% to the total reads. *Actinobacteria* shared approximately 0.7%, among which members of the *Micrococcaceae* and *Brevibacteriaceae* families (closely relative to *Arthrobacter* and *Brevibacterium* genera) were detected by sequencing after DNA isolation with both the kits. Representatives of the other lineages (*Deinococcus-Thermus, Cyanobacteria/ Chloroplast, Fusobacteria,* and *Acidobacteria*) were rare and detected using only one of DNA isolation procedure, and some of them were assigned to unclassified groups. On the whole, the bacterial community from the studied permafrost sediments is composed from members of both major and minor groups with different relatedness to the known taxa (Fig. 15.6).

Analysis of phospolipid fatty acids (PLFA) demonstrated the dominance of bacteria over fungi; the analysis of fatty acids specific for Gram-positive and Gram-negative bacteria revealed an approximately twofold higher amount of Gram-negative bacteria compared to Gram-positive bacteria. Direct microbial counts after natural permafrost enrichment showed the presence of $(4.7 \pm 1.5) \times 10^8$ cells g^{-1} sediment dry mass. Viable heterotrophic bacteria were found at 0°C, 10°C and 25°C, but not at 37°C. Spore-forming bacteria were not detected. The numbers of viable fungi were low and were only detected at 0°C and 10°C. Selected culturable bacterial isolates were identified as representatives of *Arthrobacter phenanthrenivorans, Subtercola frigoramans* and *Glaciimonas immobilis*. Representatives of each of these species were characterized with regard to their growth temperature range, their ability to grow on different media, to produce enzymes, to grow in the presence of NaCl, antibiotics, and heavy metals and to degrade hydrocarbons. All strains could grow at −5°C; the upper temperature limit for growth in liquid culture was 25°C or 30°C. Sensitivity to rich media, antibiotics, heavy metals and salt increased with decrease in temperature (20°C > 10°C > 1°C). In spite of the ligninolytic activity of some strains, no biodegradation activity

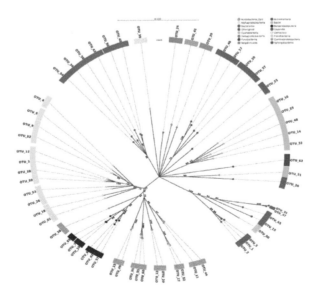

Figure 15.6 Maximum likelihood phylogenetic tree of the found ancient permafrost alluvium OTUs based on partial 16S rRNA gene sequences. Bootstrap values that were above 50% are shown at the nodes.

was detected in 16S metagenomics analysis which allowed us to find bacterial lineages in ancient permafrost alluvium, which were not captured by the culture-dependent approach, except for some members of *Actinobacteria* (Zhang et al. 2013). Moreover, variations of DNA isolation procedures (two kits for the same sample) were useful mainly to detect some minor groups, although they caused no changes in the compositional profile for major taxa at least. Taken together, obtained data showed some specific features of the bacterial component in permafrost alluvium. Strikingly, *Bacteroidetes* constitute the most abundant fraction in the ancient permafrost alluvial sediments under the study. Members of this lineage occur, but do not dominate over *Proteobacteria* in various unfrozen alluvial aquifers and sediments (Medihala and Lawrence 2012; Yergau et al. 2012; Gibbons et al. 2014; Missimer et al. 2014; Handley et al. 2015; Li et al. 2015), active and permafrost soils (Roesch et al. 2007; Yang et al. 2012), and are quite rare in some permafrost sediments (Yergau et al. 2010; Rivkina et al. 2015). Second, low abundance of *Actinobacteria* in the studied 3–3.5 Ma permafrost alluvium is in contrast with permafrost soils and lake sediments (Vishnivetskya et al. 2006; Yang et al. 2012; Rivkina et al. 2015), deep ground ice (Steven et al. 2008; Yergau et al. 2010) and is common in unfrozen alluvial water and sediments (Medihala and Lawrence 2012; Amaltifano et al. 2014; Missimer et al. 2014). The significant proportion of *Proteobacteria* as in our study and even their predominance is common as demonstrated in the cited works mentioned above.

Members of most bacterial families and genera in the studied community from nutritionally poor permafrost alluvial sediments are widespread in terrestrial and aquatic habitats. Thus, the closest relatives of almost all *Bacteroidetes* in the studied sample are related to the genus *Sediminibacterium* firstly described by Qu and Yuang (2008) who isolated the strain *S. salmoneum* from sediment of a eutrophic reservoir. Other species (*S. ginsengisoli* and *S. goheungense*) were found in soils and freshwater (Kim et al. 2013; Kang et al. 2014). Representatives of the genus *Caulobacter* inhabit oligotrophic environments (Abraham et al. 1999) and of the genus *Sphingomonas* are ubiquitous in a variety of aqueous and terrestrial habitats (Balkwill et al. 2006). The occurrence of *Bradyrhizobium* nitrogen-fixing plant root symbionts and *Staphylococcus* is not surprising since plant and animal remnants are present in studied sediments; staphylococci were also found among the isolates from some permafrost environments (Steven et al. 2008). It is noteworthy that even halophiles of the genus *Halomonas* (Dobson and Franzmann 1996) persisted in permafrost alluvium. These bacteria could enter there before freezing from neighboring saline environments that are currently and probably were widespread in this region, taking into account the prevalence of evaporation over precipitation. As a result of freezing, the salinity of unfrozen water microenvironments could increase and be suitable for halophilic bacteria.

Overall, the conditions in permafrost alluvial Neogene sediments are unlikely to support active growth of many bacteria based on absent dividing cells in the studied sample as shown using SEM. In this regard, an issue on the survival mechanisms becomes intriguing, especially for non-spore-forming bacteria that are predominant in the studied sediments and a variety of natural habitats. Specifically, little is known about the adaptation mechanisms of bacteria belonging to *Sediminibacterium* genus; a recent study has shown a flexible adaptation of closely related isolates to sustain stresses (Ayarza et al. 2015). It is quite plausible that most bacteria in ancient permafrost alluvium exist or even survive as uncultivable (or yet-to-be-cultivated) and viable non-culturable cells (Vartoukian et al. 2010; Puspita et al. 2012), common in natural habitats (Roszak and Colwell 1987; Amann et al. 1995). In particular, entering of readily cultivable non-spore-forming bacteria to a non-culturable state in response to nutrient starvation, temperature downshift and other stresses is well known for numerous species (Roszak and Colwell 1987; Kaprelyants et al. 1993; Oliver 2010; Puspita et al. 2012; Li et al. 2014). Moreover, non-spore-forming bacteria are able to produce dormant cyst-like cells intended for long-term survival, which were found in some ancient permafrost and unfrozen environments using direct electron microscopy (Mulyukin et al. 2003; Soina et al. 2004; Suzina et al. 2004). Specific procedures are often required for growth recovery in cultured and permafrost populations containing cyst-like cells (Mulyukin et al. 2009; Kriazhevskikh et al. 2012) and certain approaches are recommended to improve PCR-based detection of dormant cystous cells and spores (Mulyukin et al. 2013). Further studies are needed to clarify in what state microbial cells persist in permafrost alluvium and similar environments.

Hence, ancient permafrost Neogene alluvial sediments represent an environment with a particular compositional profile of bacterial community where diverse microorganisms survive for a long time. This study contributes to better knowledge of a conserved biological component to be subjected to thawing and spreading with water flows. Modern genomics allows rephrasing old questions of evolutionary biology. Now, it seems to be possible to consider the issue of mutation rates at the level of complete genomes. Our isolate is related to the *Bacillus* group, and *B. anthracis*, for example, represents one of the most molecularly monomorphic bacteria known. It is interesting that the chromosome *B. anthracis* is collinear with the closest near neighbor strains.

The molecular basis of thermal stability of biological materials is a significant, as yet unsolved, problem (Baker and Agard 1994; Jaenicke 1996). In general, the stability of a polymer is defined as the free energy change, G, for the reaction folded-unfolded under physiological conditions. Most proteins are characterized by values of G = 5–15 kcal/mol. In terms of thermodynamics, the fraction of residues in random coil regions divided by the fraction of residues in ordered regions (k) would appear as the following:

$$k = e^{-\frac{G}{RT}},$$

where G is the free energy change for the thermal transition of one residue from an ordered region to a random coil one, kcal/mol; T–temperature, °K; R – gas constant, ~ 0,001989, kcal/mol · °K.

Another similar expression could probably be written for an approximate estimation of time of existence of proteins or DNA:

$$t_e = t^* e^{-\frac{G}{RT}}$$

where t* – period of temperature fluctuations of molecules, normally about 10^{-12}–10^{-13} seconds.

Calculations show that even for G = 30 kcal/mol the time of existence of molecular bonds in the polymer chain is less than 300 years which is close to experimental data (Levy and Miller 1998). Maximal value of energy of activation described is about 45 kcal/mol; normally it is much less (Brouchkov et al. 2009). These calculations are very approximate and extreme, but they do show how unstable proteins and DNA are.

We can presume that live microorganisms in permafrost, if they are really old, apparently have special mechanisms of repair of cell structures (Brouchkov and Williams 2002; Johnson et al. 2007) or preserve them, as otherwise they are prone to collapse because of the duration of their existence. If so, their mutation rates can be extremely low. This explains ancient permafrost bacteria genome homology to modern strains. Moreover, the ability to keep the genome stable might be not only the permafrost microorganism's privilege, but also a feature of ancient isolates from salt and amber—a number of publications can be considered as being in agreement with this point of view (Greenblatt et al. 2004; Gutiérrez and Marín 1998; Nickle et al. 2002; Parkes 2000; Shi et al. 1997; Vishnivetskaya et al. 2006; Johnson et al. 2007; Hinsa-Leisure et al. 2010) as well as data presented here.

Melt fraction extracted from the frozen soil samples was sown on petri dishes containing LB agar. Bacterial cells resembling bacilli were obtained as a result after incubation of the petri dishes at 20°C, pure culture was prepared after several passages and named as *Bacillus* sp. strain *F*.

Believed to be as ancient as the permafrost sample from which this strain was isolated, *Bacillus* F exhibits a surprisingly high level of homology with modern *Bacillus cereus* strains, particularly with *Bacillus cereus* strain ATCC10987. The difference in chromosomal nucleotide sequences between these two strains does not exceed 1.5%, which is comparable to or even less than the difference between available chromosomal sequences of other *Bacillus cereus* strains. These observations may reflect the adaptability of *Bacillus cereus* for long-term survival and the evolution strategies of this organism in the permafrost environment.

The bacterial strain under study was sensitive to ampicillin with minimal inhibitory concentration (MIC) 100 mg/L, tetracycline – 25 mg/L, rifampicin – 50 mg/L, gentamicin – 25 mg/L, kanamycin – 50 mg/L, and resistant to spectinomycin with MIC 100 mg/L, and hygromycin – 50 mg/L.

Pure culture of *Bacillus F* was studied using Atomic Force Microscopy (Fig. 15.7). The morphology of the *Bacillus* sp. strain *F* vegetative cells look like typically rod-shaped cells with smooth surfaces. The dimension of these cells was around 1 μm height, and 3 μm length. Often the cells also showed many pili or fimbria structures.

Next, using the alkaline lysis from 10 ml of the overnight culture was isolated a 3 Kb plasmid (Fig. 15.8).

This plasmid DNA was used to transform into competent cells of *Escherichia coli* (XL1-Blue) which did not produce any resistance to hygromycin or spectinomycin colonies that we assumed is associated with the localization of the resistance gene not in the plasmid but in genomic DNA of isolated bacteria or with the absence of any recognition of replication initiation point by the *E. coli* replicative complex.

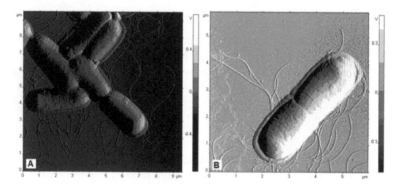

Figure 15.7 AFM images of *Bacillus* sp. strain *F* vegetative cells. (A) group of vegetative cells. (B) one bacterial cell. The imaging sizes are 6 × 6 μm.

Figure 15.8 Plasmid DNA extracted from ancient bacterial strain. (M) 1 kb + DNA ladder (Invitrogen), (1) isolated plasmid DNA.

PCR using nonspecific primers was used for further plasmid analysis. PCR using primers for relatively conservative bacterial plasmid sequences was carried out with deliberately low temperature annealing of primers to reduce the conservatism of the annealing sites and expand the range of targets. As a result, the 150 bp, 270 bp and 1200 bp amplicons were obtained (Fig. 15.9).

These PCR fragments were cloned to the pGEM-Teasy vector (Promega) and sequenced using standard M13 primers. Using the original short snippet sequenced DNA fragment as a template for the following specific primers, a complete sequencing of the plasmid was implemented by us. After whole sequencing comparative sequence analysis revealed that the plasmid shows 99 up to 100% identity to the plasmid pAH820_3 of *Bacillus cereus* AH820 (available at GenBank under accession number ACC NC_011657.1). The full compliance with the plasmid DNA isolated from bacterial cultures of the ancient with the modern *Bacillus cereus* was unique. The plasmid contains five Open Reading Frames (ORF) coding five unknown proteins amino acid sequences which were predicted by computational programs *in silico*. The analysis of protein databases by the BLAST program of NCBI server (http://blast.ncbi.nlm.nih.gov/Blast.cgi) were performed by us. The first, fourth and fifth ORFs coding peptides of 48, 76 and 44 amino acids length accordingly and do not contain any conserved domains and are not known to be present in any of the known proteins among the sequences in database. Second ORF encodes the large protein of 461 amino acid length which is 76% identical to the mobilization protein of *Bacillus thuriengiensis*. The program of the primary protein sequence analysis PFAM (http://pfam.sanger.ac.uk/) permitted to find in this protein conserved domain characterized for plasmid recombination enzymes. The third ORF encodes peptides of 75 amino acid length where the first 39 are identical to 46% of the TonB-dependent receptor domain protein of Robiginitalea biformata. The theoretical calculations propose 2% genomic variation per million years. It can be also described as $1 * 10^{-8}$ nucleotide substitution per nucleotide site per year for mitochondrial DNA. Practical data shows that specific mutations rate is exponential, after which a plateau is reached (the genome is saturated in the variable substitution sites). For some reason, the highest homology would lie among the older sequences. This phenomenon may be ascribed to the fact that the amber samples were from different geoclimatic regions of the planet (Veiga-Crespo et al. 2004).

Experimental estimation of the *Bacillus* sp. strain *F* strain pathogenicity was done in comparision with the same characteristics of the *Bacillus cereus* var. *toyoi* commercial strain. *Bacillus cereus* var. *toyoi* is pathogenic for mice with $LD_{50} = (8 \pm 2.5) \cdot 10^6$ CFU at an intravenous injection, and with $LD_{50} = (1 \pm 0.4) \cdot 10^7$ CFU with an intraperitoneal injection. Moreover, single instances of mice death were detected after intragastrically injection of *Bacillus cereus* var. *toyoi*. At the same conditions, *Bacillus* sp. strain *F* intravenously, intraperitoneally or intragastrically injection was not cause of the mice death (Fig. 15.10).

Both *Bacillus cereus* var. *toyoi* and *Bacillus* sp. strain *F* injection by intravenously and intraperitoneally routes were not caused by cutaneous edema and hyperemia after one day of

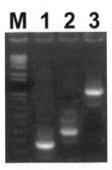

Figure 15.9 PCR products amplified from the ancient bacterial plasmid with the following sets of primer. (M) 1 kb (+) DNA Ladder (Invitrogen), (1) Forw1+Rev1+Forw2+Rev2 (2) Forw3+Rev3+Forw4+Rev4 (3) Forw5+Rev5+Forw6+Rev6.

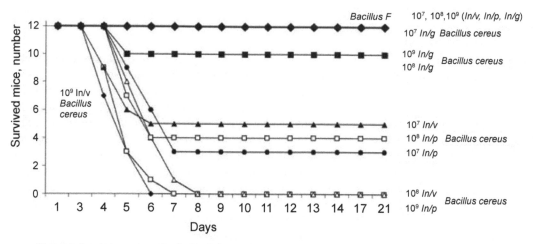

Note: *In/v - Intravenously; In/p - Intraperitoneal; In/g - Intragastrically*

Figure 15.10 Mice surveillance after *Bacillus cereus* var. *toyoi* and *Bacillus* sp. strain *F* intravenously, intraperitoneally and intragastrically injection (10^7, 10^8, and 10^8 CFU per mouse).

treatment. It was shown that the absence of the *Bacillus cereus* var. *toyoi* and *Bacillus* sp. strain *F* cells in the blood, liver and spleen of the survived mice on the 21th day of observation; and pathomorphological changes of euthanized mice' organs on the 21th day were not detected. So, on the base of the presented tests *Bacillus cereus* var. *toyoi* strain was determined as low pathogenic bacterial strain, and *Bacillus* sp. strain *F* – as a non-pathogenic bacterial strain. Low pathogenicity of the commercial probiotic strain *Bacillus cereus* var. *toyoi* (Bactisubtil) may be explained by the following facts: its cells was found to carry three genes *hblA*, *hblC*, and *hblD* that encode hemolysin BL (Hbl), the primary virulence factor in *B. cereus* diarrhea and the *bceT* gene, which encodes the single-component toxin enterotoxin T. Moreover, the cells of this probiotic strain have production of lecithinase (Granum and Lund 1997). It was proposed that non-pathogenic properties of the *Bacillus* sp. strain *F* are based on the absence of such pathogenic factors (data not shown).

The effect of relic microorganism *Bacillus F*, living in the severe environment of the Siberian permafrost during thousands and millions of years, on development and stress resistance of *Drosophila melanogaster* has been studied. In manipulating such objects with practically "eternal life span", molecular carriers of the unprecedented longevity potential and possibilities of their transmission to other biological objects should primarily be addressed. Here for the first time the influence of *Bacillus* F. application on development, survival, stress resistance and the gross physiological predictors of aging rate in *D. melanogaster* has been discussed. To establish optimal and toxic doses, a wide range of *Bacillus* sp. concentrations were tested (1–500 million cells of *Bacillus* sp. per 1 ml of the ß ies feeding medium). Surprisingly, no toxic effects of *Bacillus* sp. could be registered even on such a "sensitive" model as the developing larvae. In fact, the rate of development, survival and body mass gradually increased with elevation of *Bacillus* sp. concentration. The gain of higher body mass within shorter periods of development could indicate enhanced anabolic and/or declined catabolic effects of *Bacillus* sp. higher motor activity and gaseous exchange rates were observed in images developed on the mediums with *Bacillus* sp. application. Survival of the flies at the heat shock (30 minutes at 38°C) and ultraviolet irradiation (60 minutes, 50W UV lamp) was increased, indicating elevated stress resistance, apparently due to stimulation of DNA-repair and chaperone-mediated protection of macromolecules (Figs. 15.11 and 15.12). Further research is clearly warranted to identify more effective anti-stress and anti aging preparations and schemes of *Bacillus* sp. application on models of laboratory mammals and human cell cultures.

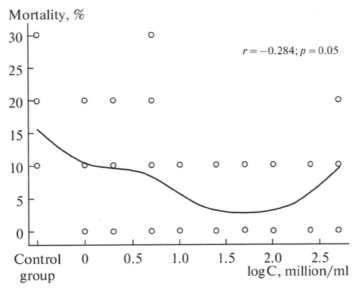

Figure 15.11 Mortality of *D. melanogaster* larvae during their development on the nutrient media with different *Bacillus* sp. concentrations (logarithmic concentration axis).

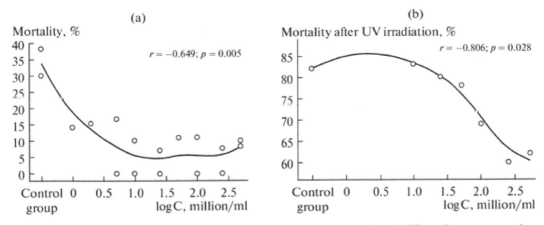

Figure 15.12 Mortality of *D. melanogaster* imagoes grown on the nutrient media with different B. sp. concentrations 24 hours after heat shock (38°C, 30 minutes), (a) and after UV irradiation (60 minutes under a 50 Watt UV lamp), (b) (logarithmic concentration axis).

The same bacterial culture was used for testing with mice. Mice F1 CBA/Black-6 of 18–20 g weight was taken, and each group included about 15 mice. All studies were done according the European convention for protection of vertebrate animals dated 18.03.1986. In the first series of tests 126 mice were used for study of influence of doses of culture on immune system of young animals (age 3–4 months). Bacterial culture was injected once in peritoneal $2.5 \cdot 10^3$; $5 \cdot 10^3$; $10 \cdot 10^3$; $20 \cdot 10^3$; $50 \cdot 10^3$; $500 \cdot 10^3$; $5 \cdot 10^6$ and $50 \cdot 10^6$ cells in physiological solution per animal. In the second series of tests an effect of bacterial culture on physiological and behavioral reactions of elderly mice was estimated (CBA, ♀ age of 17 months), and the culture was entered once in peritoneal $5 \cdot 10^3$ cells in 100 mkl of physiological solution. The control group was presented by animals of the same age. Standard techniques were used for timus and spleens activity (indexes – relation of weight of organ to weight of a body, %), activity of factors of nonspecific immune resistance on a level phagocytes (%) and metabolic (the HCT-TEST, %) activity spleens macrophages. Activity of cellular immunity was estimated on reaction of hypersensitivity of slowed down type (HST) *in vivo* on RCR. Behavioral reactions in the Open field test were studied. Life expectancy was estimated as well as muscular force of animals. Microbial culture of *Bacillus* sp., strain F in a dozen of 5000 cells promotes increase in indexes of timus and spleens. The culture in a small dose (5 · 10^3 cells) stimulates, and in average dozes (500 · 10^3 and 5 · 10^6 cells) suppresses phagocytes activity of spleens macrophages. Average doses (from 50 · 10^3 up to 5 · 10^6 м.т.) promote a prevalence of functional activity humoral, and a dose in 50 · 10^6 cells – cellular immunity.

The minimal lifespan of mice was 589 days in the control group, and maximal – 833 days. The minimal lifespan of mice was 836 days in the test group, and maximal – 897 (Fig. 15.13). The weight of a mice body of the test group was higher than the control group after 2 months of injection. Muscular force of test animals had increased to about 60% in comparison with the control group. Visiting of internal sectors of the open field was testified in the test group by increasing the ability of orientation and research activity, as well as an increase in number of vertical racks was observed. Therefore, the bacterial culture stimulates the immune system and improves emotional stability of laboratory animals. Increase in life expectancy of mice proves the presence of geroprotectors in the culture of *Bacillus F.* It needs to be to emphasized that studies of relic microorganisms are rather preliminary since the nature of the longevity of ancient microorganisms and their influence on higher organisms is unknown.

Bacterial strain *Bacillus* sp. strain *F* was assessed as nonpathogenic because during all experiments there were no detected mice deaths, any pathogen in animal organs or in the site of injection. The probiotic efficacy of *Bacillus* sp. strain *F* for an experimental model of *S. enterica* infection in mice was studied. Experimental murine salmonellosis is a widely used experimental model for acute systemic salmonellosis in humans (Ohl and Miller 2001). For mice model the strain *S. enterica enterica* serotype *typhi* (Cordeiro et al. 2000; Silva et al. 1999; Sukupolvi et al. 1997; Szabó et al. 2009) was used. In this chapter *S. enterica* model was described as a variant to cause a chronic carrier state in mice after oral inoculation as a model for a human carrier state. It is known that the maximum pathogenic bacteria in organs (spleen and livers) could be detected after 7–10 days of infection (Sukupolvi et al. 1997). In our experiment bacterial cultures from liver and spleen samples were performed at days 3, 4, 5, 6, and 7 post-infection. After 7 days of infection there is no

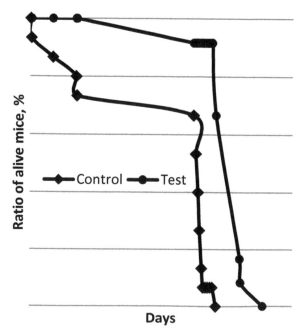

Figure 15.13 Test on longevity of 17th month mice, injection of 5000 bacterial cells.

difference in number of *S. enterica* in organs obtained from control mice and mice treatment with *Bacillus cereus* var. *toyoi*. But treatment with *Bacillus* sp. strain *F* decreased bacterial number in organs. This decrease was very important for mice survival (Fig. 15.14).

Therapy was started on hour 4 after *per os S. enterica* infection with an inoculum of the LD50 (5 * 10⁶ CFU/mouse). The bacterial suspension of *Bacillus* sp. strain *F* cells was administered to groups of 10 mice *per os* for 7 days. Figure 15.2 provides data of a representative experiment which indicate that *Bacillus* sp. strain *F* treatment of mice reduced their mortality compared with that of untreated animals and treated by *B. cereus* var. *toyoi*. *Salmonella* cells were present in the livers and spleens of dead mice. Enteritidis infection as described here was shown to be useful in the determining the effect of *Bacillus* sp. strain *F* on the host to the infection. The use of inbred mice with a stable genetic background ensures that the resistance to *S. enterica* var. Enteritidis can be clearly defined in all groups of animals (Plant et al. 1983). Ten mice each were used for treatment with *Bacillus* sp. strain *F*, treated with *Bacillus cereus* var. *toyoi* and for the control group. Mice were treated orally one time per day with *Bacillus* sp. strain *F* and *Bacillus cereus* var. *toyoi* for seven days after infection. The follow-up observation was performed for 21 days after infection. Ninety per cent of mice treated with *Bacillus* sp. strain *F* survived compared to 20% of the untreated mice or mice treated with *Bacillus cereus* var. *toyoi* (Fig. 15.15).

The following three basic mechanisms have been proposed for how orally ingested non indigenous bacteria can have a probiotic effect on a host: (i) immunomodulation, that is stimulation of the gut-associated lymphoid tissue, e.g., induction of cytokines; (ii) competitive exclusion of gastrointestinal pathogens, e.g., competition for adhesion sites; and (iii) secretion of antimicrobial compounds which suppress the growth of harmful bacteria (Duc et al. 2004). It was proposed that notable probiotic properties of the *Bacillus* sp. strain F as compared with the *Bacillus cereus* var. *toyoi* strain can be explained by the production of unknown bacteriocin-like inhibitory substances (unpublished data).

Here *S. enterica* var. Enteritidis model as a variant to cause a chronic carrier state in mice after oral inoculation as a model for a human carrier state has been described. Pathogenic mechanism of *S. enterica* var. Enteritidis action connected with gastrointestinal tract (Deng et al. 2007). This is the reason that oral infection and treatment of mice was used by us. To protect mice from *Salmonella* infection antibiotics or bacterial extract can be used (Cordeiro et al. 2000). It is known that live microorganisms such as yeast and bacterial species have also been used to protect from *Salmonella*

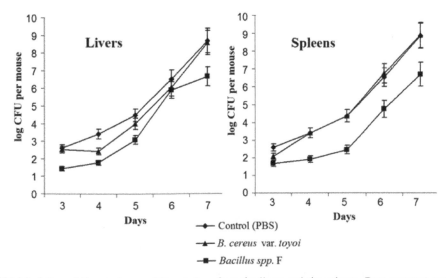

Figure 15.14 Isolation of *S. enterica* on different days from the livers and the spleens. Bars represent mean values (+/– standard deviation, n = 5).

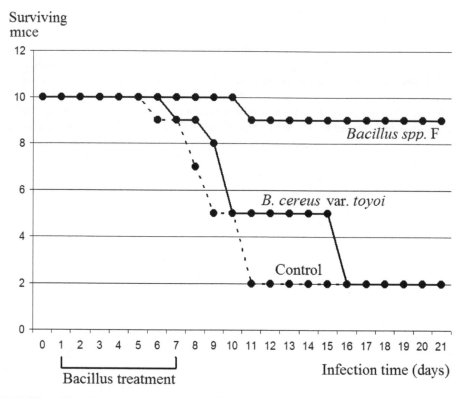

Figure 15.15 Effect of *Bacillus cereus* var. *toyoi* and *Bacillus* sp. strain *F* treatment on mortality in outbred mice after *per os* inoculation one time 5 · 10⁶ CFU/mouse (day number 0) with *S. enterica* var. Enteritidis (1 mice in each group).

infection (Silva et al. 1999; Szabó et al. 2009). In our case *Bacillus* sp. strain F protect mice from *Salmonella* infection. Probiotic bacteria reduced colonization by pathogens and decrease host defense mechanisms. Preliminary results for *Bacillus* sp. strain F showed increasing of humoral and cell immunity of mice.

Conclusion

Bacillus sp. Strain *F* from ancient permafrost has been isolated, and non-pathogenicity of these bacteria was found. Special animal model for mice was described as a variant to cause a chronic carrier state in mice. The possibilities for orally treating of the *S. enteric* var. Enteritidis mice infection by probiotic *Bacillus* sp. strain *F* obtained from the relic permafrost as well as other useful properties were clearly demonstrated here.

References

Abraham, W.R., C. Strömpl and H. Meyer. 1999. Phylogeny and polyphasic taxonomy of *Caulobacter* species. Proposal of *Maricaulis* gen. nov. with *Maricaulis maris* (Poindexter) comb. nov. as the type species, and emended description of the genera *Brevundimonas* and *Caulobacter*. Int. J. Syst. Bacteriol. 49: 1053–73.

Abyzov, S.S. 1993. Microorganisms in the Antarctic ice. pp. 265–295. *In*: Friedmann, E.I. (ed.). Antarctic Microbiology. Wiley-Liss, Inc. New York.

Achtman, M., K. Zurth, G. Morelli, G. Torrea, A. Guiyoule and E. Carniel. 1999. *Yersinia pestis*, the cause of plague, is a recently emerged clone of *Yersinia pseudotuberculosis*. Proc. Natl. Acad. Sci. USA 96: 14043–14048.

Amalfitano, S., A. Del Bon and A. Zoppini. 2014. Groundwater geochemistry and microbial community structure in the aquifer transition from volcanic to alluvial areas. Water Res. 65: 384–94.

Amann, R., W. Ludwig and K.H. Schleifer. 1995. Phylogenetic identification and in situ detection of individual microbial cells without cultivation. Microbiol. Rev. 59: 143–69.

Amend, A.S., K.A. Seifert and T.D. Bruns. 2010. Quantifying microbial communities with 454 pyrosequencing: does read abundance count? Mol. Ecol. 19: 5555–65.

Ashcroft, F. 2000. Life at the Extremes. Harper Collins 326p.

Ayala-del-Río, H.L., P.S. Chain, J.J. Grzymski, M.A. Ponder, N. Ivanova and P.W. Bergholz. 2010. The genome sequence of *Psychrobacter arcticu* 273-4, a psychroactive siberian permafrost bacterium, reveals mechanisms for adaptation to low-temperature growth. Appl. Environ. Microbiol. 76: 2304–2312.

Ayarza, J.M., M.A. Mazzella and L. Erijman. 2015. Expression of stress-related proteins in *Sedimini bacterium* sp. growing under planktonic conditions. J. Basic Microbiol. 55: 1134–40.

Baker, D. and D. Agard. 1994. Kinetics versus thermodynamics in protein folding. Biochemistry 33: 750509.

Bakulina, N.T. and V.B. Spector. 2000. Reconstruction of climatic parameters of Neogene of Yakutia based on palinology data. pp. 21–32. *In*: Maksimov, G.N. and A.N. Fedorov (eds.). Climate and Permafrost, Yakutsk: Permafrost Institute (in Russian).

Balkwill, D.L., J.K. Fredrickson and M.F. Romine. 2006. *Sphingomonas* and related genera. pp. 605–29. *In*: Dworkin, M. (eds.). The Prokaryotes: An Evolving Electronic Resource for the Microbiological Community. New York: Springer-Verlag.

Baranova, U.P., I.A. Il'inskay, V.P. Nikitin, G.P. Pneva, A.F. Fradkina and N.Y. Shvareva. 1976. Miocene of the Mammoth Mountain. In Works of Geological Institute of Russian Academy of Sciences Nauka Moscow: 284.

Belda, E., A. Moya and F.J. Silva. 2005. Genome rearrangement distances and gene order phylogeny in g-proteobacteria. Mol. Biol. Evol. 22: 1456–1467.

Binladen, J., C. Wiuf, M.T.P. Gilbert, M. Bunce, R. Barnett, G. Larson et al. 2006. Assessing the fidelity of ancient DNA sequences amplified from nuclear genes. Genetics 172: 733–741.

Brooks, J.P., D.J. Edwards and M.D. Harwich. 2015. The truth about metagenomics: Quantifying and counteracting bias in 16S rRNA studies. BMC Microbiol. 15: 66.

Brouchkov, A. and P. Williams. 2002. Could microorganisms in permafrost hold the secret of immortality? What does it mean? Contaminat. Freez. Ground 49–56.

Brouchkov, A., A.M. Peterson, E.V. Glinska, G.I. Griva and V.E. Repin. 2012. Biological Properties of bacteria isolated from permafrost in Central Yakutia. Proceedings of the Tenth International Conference on Permafrost. Salekhard, Yamal-Nenets Autonomous District, Russia June 25–29, The Northern Publisher Salekhard Salekhard, 2: 25–29.

Brouchkov, A.V., V.P. Mel'nikov, Iu.G. Sukhoveĭ, G.I. Griva, V.E. Repin, L.F. Kalenova et al. 2009. Relict microorganisms of cryolithozone as possible objects of gerontology. Adv. Gerontol. 22: 253–258.

Bunt, I.S. and C.C. Lee. 1970. Seasonal primary production in Antarctic sea ice at McMurdo Sound in 1967. J. mar. Res. 28: 304–320.

Burt, T.P. and P.J. Williams. 1976. Hydraulic conductivity in frozen soils. Earth Surface Processes 1: 349–360.

Cairns, J., J. Overbaugh and S. Miller. 1994. The origin of mutations. Nature 335: 142–145.

Cano, R.J. 1996. Analysing ancient DNA. Endeavour 20: 162–167.

Caporaso, J.G., C.L. Lauber, W.A. Walters, D. Berg-Lyons, C.A. Lozupone, P.J. Turnbaugh et al. 2011. Global patterns of 16S rRNA diversity at a depth of millions of sequences per sample. Proc. Natl. Acad. Sci. USA 108: 4516–22.

Clark, M.A., N.A. Moran and P. Bauman. 1999. Sequence evolution in bacterial endosymbionts having extreme base compositions. Mol. Biol. Evol. 16: 1586–1598.

Clein, J.S. and J.P. Schimel. 1995. Microbial activity of tundra and taiga soils at sub-zero temperatures. Soil Biol. Biochem. 27: 1231–1234.

Cordeiro, C., D.J. Wiseman, P. Lutwyche, M. Uh, J.C. Evans and B.B. Finlay. 2000. Antibacterial efficacy of gentamicin encapsulated in pH-sensitive liposomes against an *in vivo* Salmonella enterica serovar typhimurium intracellular infection model. Antimicrob. Agents Chemother. 44: 533–539.

Demezhko, D.Y. and I.V. Golovanova. 2007. Climatic changes in the Urals over the past millennium—an analysis of geothermal and meteorological data. Clim. Past. 3: 237–42.

Deng, S.X., A.C. Cheng, M.S. Wang and P. Cao. 2007. Gastrointestinal tract distribution of Salmonella enteritidis in orally infected mice with a species-specific fluorescent quantitative polymerase chain reaction. World J. Gastroenterol. 13: 6568–6574.

Derjaguin, B.V. and N.V. Churaev. 1986. Flow of nonfreezing water interlayers and frost heaving. Cold Reg. Sci. Technol. 12: 57–66.

Dobson, S.J. and P.D. Franzmann. 1996. Unification of the genera Deleya (Baumann et al. 1983), *Halomonas* (Vreeland et al. 1980), and *Halovibrio* (Fendrich 1988) and the species *Paracoccus halodenitrificans* (Robinson and Gibbons 1952) into a single genus, *Halomonas,* and placement of the *genus Zymobacter* in the family Halomonadaceae. Int. J. Syst. Evol. Microbiol. 46: 550–558.

Duc, L.H., H.A. Hong, T.M. Barbosa, A.O. Henriques and S.M. Cutting. 2004. Characterization of Bacillus Probiotics available for human use. Appl. Environ. Microbiol. 70: 2161–2171.

Eaton, K.A., A. Honkala, T.A. Auchtung and R.A. Britton. 2011. Probiotic *Lactobacillus reuteri* ameliorates disease due to enterohemorrhagic *Escherichia coli* in germfree mice. Infect. Immun. 79: 185–191.

Edgar, R.C. 2013. UPARSE: highly accurate OTU sequences from microbial amplicon reads. Nat. Methods 10: 996–98.

Ershov, E. 1998. Foundations of Geocryology. Moscow State University: Moscow, Russia 1–575.

Etienne-Mesmin, L., V. Livrelli, M. Privat, S. Denis, J.M. Cardot, M. Alric et al. 2011. Blanquet-Diot S effect of a new probiotic *Saccharomyces cerevisiae* strain on survival of *Escherichia coli* O157:H7 in a dynamic gastrointestinal model. Appl. Environ. Microbiol. 77: 1127–1131.

Feinstein, L.M., W.J. Sul and C.B. Blackwood. 2009. Assessment of bias associated with incomplete extraction of microbial DNA from soil. Appl. Environ. Microbiol. 75: 5428–33.

Filonov, A.S. and I.V. Yaminsky. 2007. Scanning Probe Microscopy Image Processing Software User's Manual FemtoScan. Advanced Technologies Center, Moscow. Available from URL.

Friedmann, E. 1994. Permafrost as microbial habitat. pp. 21–26. *In*: Gilichinskii, D.A. (ed.). Viable Microorganisms in Permafrost Russian Academy of Sciences Pushchino, Russia.

Fursova, O., O. Potapov, A. Pogorelko, A. Brouchkov, G. Griva, N. Fursova et al. 2013. Probiotic activity of a bacterial strain isolated from ancient permafrost against salmonella infection in mice. Probiotics and Antimicrobial Proteins. Springer Pub. Co. 4: 145–153.

Gibbons, S.M., E. Jones, A. Bearquiver, F. Blackwolf, W. Roundstone, N. Scott et al. 2014. Human and environmental impacts on river sediment microbial communities. PloS One 9.

Gilichinsky, D. and S. Wagener. 1995. Microbial life in permafrost: a historical review. Permafrost. Periglac. 6: 243–250.

Gilichinsky, D. 2002. Permafrost as a microbial habitat. pp. 932–56. *In*: Bitton, G. (ed.). Encyclopedia of Environmental Microbiology. New York: Wiley.

Gilichinsky, D.A., G.S. Wilson, E.I. Friedmann, C.P. Mckay, R.S. Sletten, E.M. Rivkina et al. 2007. Microbial populations in Antarctic permafrost: biodiversity, state, age, and implication for astrobiology. Astrobiology 7: 275–311.

Granum, P.E. and T. Lund. 1997. *Bacillus cereus* and its food poisoning toxins. FEMS Microbiol. Lett. 157: 223–228.

Greenblatt, C.L., J. Baum, B.Y. Klein, S. Nachshon, V. Koltunov and R.J. Cano. 2004. *Micrococcus luteus*—survival in Amber. Microb. Ecol. 48: 120–7.

Gutiérrez, G. and A. Marín. 1998. The most ancient DNA recovered from an amber-preserved specimen may not be as ancient as it seems. Mol. Biol. Evol. 15: 926–929.

Handley, K.M., K.C. Wrighton, C.S. Miller, M.J. Wilkins, R.S. Kantor, B.C. Thomas et al. 2015. Disturbed subsurface microbial communities follow equivalent trajectories despite different structural starting points. Environ. Microbiol. 17: 622–36.

Henwood, A. 1993. Recent plant resins and the taphonomy of organisms in amber: a review. Mod. Geol. 19: 35–59.

Hinsa-Leisure, S.M., L. Bhavaraju, J.L.M. Rodrigues, C. Bakermans, D.A. Gilichinsky, M. Hofreiter et al. 2001. DNA sequences from multiple amplifications reveal artifacts induced by cytosine deamination in ancient DNA. Nucleic Acids Res. 29: 4793–4799.

Ho, S.Y., M.J. Phillips, A. Cooper and A.J. Drummond. 2000. Time dependency of molecular rate estimates and systematic overestimation of recent divergence times. Mol. Biol. Evol. 22: 1561–1568.

Jaenicke, R. 1996. Stability and folding of ultrastable proteins: Eye lens crystallins and enzymes from thermophiles. FASEB J. 10: 84–92.

Johnson, L.R. and M. Mangel. 2006. Life histories and the evolution of aging in bacteria and other single-celled organisms. Mech. Ageing Dev. 127: 786–793.

Johnson, S.S., M.B. Hebsgaard, T.R. Christensen, M. Mastepanov, R. Nielsen, K. Munch et al. 2007. Ancient bacteria show evidence of DNA repair. Proc. Natl. Acad. Sci. USA 104: 14401–14405.

Joint, F. 2002. WHO Working Group Report on Drafting Guidelines for the Evaluation of Probiotics in Food. London, Ontario, Canada 30.

Kalinina, L.V., J.G. Holt and J.J. McGrath. 1994. Identity of bacterial from Siberian permafrost soils. In IUMS Congresses '94; 7th International Congress of Bacteriology and Applied Microbiology Division; 7th International Congress of Mycology Division, Prague, Czech Republic 3–8.

Kang, H., H. Kim and B.I. Lee. 2014. Sediminibacterium *goheungense* sp. nov., isolated from a freshwater reservoir. Int. J. Syst. Evol. Microbiol. 64: 1328–33.

Kaprelyants, A.S., J.C. Gottschal and D.B. Kell. 1993. Dormancy in non-sporulating bacteria. FEMS Microbiol. Rev. 104: 271–86.

Katayama, T., T. Kato and M. Tanaka. 2009. *Glaciibacter superstes* gen. nov., sp. nov., a novel member of the family Microbacteriaceae isolated from a permafrost ice wedge. Int. J. Syst. Evol. Microbiol. 59: 482–6.

Kim, Y.J., N.L. Nguyen and H.Y. Weon. 2013. Sediminibacterium *ginsengisoli* sp. nov., isolated from soil of a ginseng field, and emended descriptions of the genus Sediminibacterium and of Sediminibacterium *salmoneum*. Int. J. Syst. Evol. Microbiol. 63: 905–12.

Kriazhevskikh, N.A., E.V. Demkina and N.A. Manucharova. 2012. Reactivation of dormant and nonculturable bacterial forms from paleosols and subsoil permafrost. Microbiology 81: 435–45.

Lambert, J.B., G.O. Poinar and J.R. Amber. 2002. The organic gemstone. Acc. Chem. Res. 35: 628–636.

Levy, M. and S.L. Miller. 1998. The stability of the RNA bases: Implications for the origin of life. Biochemistry 95: 7933–7938.

Li, L., N. Mendis and H. Trigui. 2014. The importance of the viable but non-culturable state in human bacterial pathogens. Front. Microbiol. 5: 218.

Li, P., Y. Wang and X. Dai. 2015. Microbial community in high arsenic shallow groundwater aquifers in Hetao basin of Inner Mongolia, China. PLoS ONE 10.

Lindahl, T. 1993. Instability and decay of the primary structure of DNA. Nature 362: 709–715.

Lisiecki, L.E. and M.E. Raymo. 2005. A Pliocene-Pleistocene stack of 57 globally distributed benthic δ18O records. Paleoceanography 20.

Mahaney, W.C. 2002. Atlas of Sand Grain Surface Textures and Applications. Oxford–Toronto: Oxford University Press.

Margesin, R. and V. Miteva. 2011. Diversity and ecology of psychrophilic microorganisms. Res. Microbiol. 162: 346–361.

Markov, K.K. 1973. Cross-section of the Newest Sediments. Moscow, Russia: Moscow University Press.

Medihala, P.G. and J.R. Lawrence. 2012. Swerhone GDW Spatial variation in microbial community structure, richness, and diversity in an alluvial aquifer. Can. J. Microbiol. 58: 1135–51.

Melnikov, V.P., V.V. Rogov, A.N. Kurchatova, A.V. Brouchkov and G.I. Griva. 2011. Distribution of microorganisms in frozen soils. Earth Cryos. 4: 86–90.

Missimer, T.M., C. Hoppe-Jones and K.Z. Jadoon. 2014. Hydrogeology, water quality, and microbial assessment of a coastal alluvial aquifer in western Saudi Arabia: potential use of coastal wadi aquifers for desalination water supplies. Hydrogeol. J. 22: 1921–34.

Morita, R.Y. 1997. Bacteria in Oligotrphic Environments. Chapman and Hall, New York.

Mulyukin, A.L., V.S. Soina and E.V. Demkina. 2003. Formation of resting cells by non-spore-forming microorganisms as strategy of long-term survival in the environment. Proc. SPIE 4939: 208–18.

Mulyukin, A.L., E. Demkina and N.A. Kryazhevskikh. 2009. Dormant forms of *Micrococcus luteus* and *Arthrobacter globiformis* not platable on standard media. Microbiology 78: 407–18.

Mulyukin, A.L., N.E. Suzina and G.I. El-Registan. 2013. Effective PCR detection of vegetative and dormant bacterial cells due to a unified method for preparation of template DNA encased within cell envelopes. Microbiology 82: 295–305.

Nakamura, L.K. 1994. DNA relatedness among *Bacillus thuringiensis* serovars. Int. J. Syst. Bacteriol. 44: 125–129.

Nicholson, W.L., N. Munakata, G. Horneck, H.J. Melosh and P. Setlow. 2000. Resistance of *Bacillus* endospores to extreme terrestrial and extraterrestrial environments. Microbiol. Mol. Biol. Rev. 64: 548–572.

Nickle, D.C., G.H. Learn, M.W. Rain, J.I. Mullins and J.E. Mittler. 2002. Curiously modern DNA for a "250 Million-Year-Old" bacterium. J. Mol. Evol. 54: 134–137.

Novichkov, P.S., M.V. Omelchenko, M.S. Gelfand, A.A. Mironov, Y.I. Wolf, E.V. Koonin et al. 2004. Genome-wide molecular clock and horizontal gene transfer in bacterial evolution. J. Bacteriol. 186: 6575–6585.

Ochman, H. and A.C. Wilson. 1987. Evolution in bacteria: evidence for a universal substitution rate in cellular genomes. J. Mol. Evol. 26: 74–86.

Ohl, M.E. and S.I. Miller. 2001. Salmonella: a model for bacterial pathogenesis. Annu. Rev. Med. 52: 259–74.

Oliver, J.D. 2010. Recent findings on the viable but nonculturable state in pathogenic bacteria. FEMS Microbiol. Rev. 34: 415–25.

Pääbo, S., R.G. Higuchi and A.C. Wilson. 1989. Ancient DNA and the polymerase chain reaction. The emerging field of molecular archaeology. J. Biol. Chem. 264: 9709–9712.

Panieri, G., S. Lugli, V. Manzi, M. Roveri, B.C. Schreiber and K.A. Palinska. 1982. Ribosomal RNA gene fragments fromfossilized *cyanobacteria* identified in primary gypsumfromthe late Miocene. Ital. Geobiol. 8: 101–111.

Parkes, R.J. 2000. A case of bacterial immortality? Nature 407: 844–855.

Parkes, R.J., B.A. Cragg and P. Wellsbury. 2000. Recent studies on bacterial populations and processes in subseafloor sediments: a review. Hydrogeology 8: 11–28.

Plant, J.E., G.A. Higgs and C.S. Easmon. 1983. Effects of antiinflammatory agents on chronic *Salmonella typhimurium* infection in a mouse model. Infect Immun. 42: 71–75.

Poinar, G.O. 1994. The range of life in amber: significance and implications in DNA studies. Experientia 50: 536–542.

Psenner, R. and B. Sattler. 1998. Life at the freezing point. Science 280: 2073–2074.

Puspita, I.D., Y. Kamagata and M. Tanaka. 2012. Are uncultivated bacteria really uncultivable? Microbes Environ. 27: 356–66.

Qu, J.H. and H.L. Yuan. 2008. Sediminibacterium *salmoneum* gen. nov., sp. nov., a member of the phylum Bacteroidetes isolated from sediment of a eutrophic reservoir. Int. J. Syst. Evol. Microbiol. 58: 2191–94.

Rishi, P., S. Kaur, M.P.S. Bhalla, S. Preet and R.P. Tiwari. 2008. Selection of probiotic *Lactobacillus acidophilus* and its prophylactic activity against murine *Salmonellosis*. Int. J. Pro. Pre. 3: 89–98.

Rishi, P., S.K. Mavi, S. Bharrhan, G. Shukla and R. Tewari. 2009. Protective efficacy of probiotics alone or in conjunction with a prebiotic in Salmonella-induced liver damage. FEMS Microbiol. Ecol. 69: 222–230.

Rivkina, E., L. Petrovskaya and T. Vishnivetskaya. 2015. Metagenomic analyses of the late Pleistocene permafrost—additional tools for reconstruction of environmental conditions. Biogeosci. Discuss 12: 12091–119.

Roesch, L.F.W., R.R. Fulthorpe and A. Riva. 2007. Pyrosequencing enumerates and contrasts soil microbial diversity. ISME J. 1: 283–90.

Rogov, V.V. and A.N. Kurchatova. 2013. Method of manufacturing replica for analyses of microstructure of frozen rocks in scanning electron microscope. Patent RU 2 528 256.

Roszak, D.B. and R.R. Colwell. 1987. Survival strategies of bacteria in the natural environment. Microbiol. Rev. 51: 365–79.

Rothberg, J.M. and J.H. Leamon. 2008. The development and impact of 454 sequencing. Nat. Biotechnol. 26: 1117–24.

Sarkar, S. 2010. Approaches for enhancing the viability of probiotics: a review. British Food J. 112: 329–349.

Schinner, F., R. Öhlinger and E. Kandeler. 1996. Potential nitrification. Meth. Soil Biol. 146–149.

Selby, C.P. and A. Sancar. 2006. A cryptochrome/photolyase class of enzymes with single-stranded DNA-specific photolyase activity. Proc. Natl. Acad. Sci. U.S.A. 103: 17696–700.

Service, R.F. 1996. Just how old is that DNA, anyway? Science 272: 810.

Sheridan, P.P., V.I. Miteva and J.E. Brenchley. 2003. Phylogenetic analysis of anaerobic psychrophilic enrichment cultures obtained from a Greenland glacier ice core. Appl. Environ. Microbiol. 69: 2153–2160.

Shi, T., R.H. Reeves and D.A. Gilichinsky. 1997. Characterization of viable bacteria from Siberian permafrost by 16S rDNA sequencing. Microb. Ecol. 33: 169–79.

Silva, A.M., E.A. Bambirra, A.L. Oliveira, P.P. Souza, D.A. Gomes, E.C. Vieira et al. 1999. Protective effect of bifidus milk on the experimental infection with Salmonella enteritidis subsp. typhimurium in conventional and gnotobiotic mice. J. Appl. Microbiol. 86: 331–336.

Skidmore, M.L., J.M. Foght and J. Martin. 2000. Microbial life beneath a high arctic glacier sharp. Appl. Environ. Microbiol. 66: 3214–3220.

Sogin, M.L., H.G. Morrison and J.A. Huber. 2006. Microbial diversity in the deep sea and the underexplored rare biosphere. Proc. Natl. Acad. Sci. USA 103: 12115–20.

Soina, V.S., A.L. Mulyukin and E.V. Demkina. 2004. The structure of resting bacterial populations in soil and subsoil permafrost. Astrobiology 4: 345–58.

Steven, B., R. Léveillé and W.H. Pollard. 2006. Microbial ecology and biodiversity in permafrost. Extremophiles 10: 259–267

Steven, B., W.H. Pollard, C.W. Greer and L.G. Whyte. 2008. Microbial diversity and activity through a permafrost/ground ice core profile from the Canadian high Arctic. Environ. Microbiol. 10: 3388–3403.

Sukupolvi, S., A. Edelstein, M. Rhen, S.J. Normark and J.D. Pfeifer. 1997. Development of a murine model of chronic Salmonella infection. Infect Immun. 65: 838–842.

Suzina, N.E., A.L. Mulyukin and A.N. Kozlova. 2004. Ultrastructure of resting cells of some non-spore-forming bacteria. Microbiology 73: 435–47.

Szabó, I., L.H. Wieler, K. Tedin, L. Scharek-Tedin, D. Taras, A. Hensel et al. 2009. Influence of probiotic strain of *Enterococcus faecium* on *Salmonella enterica* serovar *Typhimurium* DT104 infection in a porcine animal infection mode. Appl. Environ. Microbiol. 75: 2621–2628.

Tiedje, J.A. 2010. Characterization of a bacterial community from a northeastern Siberian seacost permafrost sample. FEMS Microbiol. Ecol. 74: 103–113.

Tripati, A.K., R.A. Eagle and A. Morton. 2008. Evidence for glaciation in the Northern Hemisphere back to 44 Ma from ice-rafted debris in the Greenland Sea. Earth Planet Sci. Lett. 265: 112–22.

Vartoukian, S.R., R.M. Palmer and W.G. Wade. 2010. Strategies for culture of 'unculturable' bacteria. FEMS Microbiol. Lett. 309: 1–7.

Veiga-Crespo, P., M. Poza, M. Prieto-Alcedo and T.G. Villa. 2004. Ancient genes of *Saccharomyces cerevisiae*. Microbiology 150: 2221–2227.

Velichko, A.A. and L.L. Isavea. 1992. Landscape types during the Last Glacial Maximum. *In*: Frenzel, B., B. Pecsi and A.A. Velichko (eds.). Atlas of Palaeoclimates and Palaeoenvironments of the Northern Hemisphere. Budapest: INQUA/ Hungarian Academy of Sciences.

Venkateswaran, K., M. Kempf, F. Chen, M. Satomi, W. Nicholson and R. Kern. 2003. *Bacillus nealsonii* sp. nov., isolated from a spacecraft-assembly facility, whose spores are gamma-radiation resistant. Int. J. Syst. Evol. Microbiol. 53: 165–172.

Vishnivetskaya, T., S. Kathariou, J. McGrath, D. Gilichinsky and J.M. Tiedje. 2000. Low-temperature recovery strategies for the isolation of bacteria from ancient permafrost sediments. Extremophiles 4: 165–173.

Vishnivetskaya, T.A., M.A. Petrova, J. Urbance, M. Ponder, C.L. Moyer, D.A. Gilichinsky et al. 2006. Bacterial community in ancient Siberian permafrost as characterized by culture and culture-independent methods. Astrobiology 6: 400–414.

Vishnivetskaya, T.A., M.A. Petrova and J. Urbance. 2006. Bacterial community in ancient Siberian permafrost as characterized by culture and culture-independent methods. Astrobiology 6: 400–14.

Vorobyova, E., V. Soina and M. Gorlenko. 1997. The deep cold biosphere: facts and hypothesis. FEMS Microbiol. Rev. 20: 277–90.

Wang, Q., G.M. Garrity and J.M. Tiedje. 2007. Naïve Bayesian classifier for rapid assignment of rRNA sequences into the new bacterial taxonomy. Appl. Environ. Microbiol. 73: 5261–67.

Willerslev, E. and A. Cooper. 2005. Ancient DNA. Proc. Roy. Soc. 272: 3–16.

Williams, P.J. and M.W. Smith. 1991. The Frozen Earth. Cambridge University Press. 311p.

Yadav, A.N., S.G. Sachan, P. Verma and A.K. Saxena. 2015a. Prospecting cold deserts of north western Himalayas for microbial diversity and plant growth promoting attributes. J. Biosci. Bioeng. 119: 683–693.

Yadav, A.N., S.G. Sachan, P. Verma, S.P. Tyagi, R. Kaushik and A.K. Saxena. 2015b. Culturable diversity and functional annotation of psychrotrophic bacteria from cold desert of Leh Ladakh (India). World J. Microbiol. Biotechnol. 31: 95–108.

Yadav, A.N., S.G. Sachan, P. Verma and A.K. Saxena. 2016. Bioprospecting of plant growth promoting psychrotrophic Bacilli from cold desert of north western Indian Himalayas. Indian J. Exp. Biol. 54: 142–150.

Yadav, A.N., P. Verma, S.G. Sachan, R. Kaushik and A.K. Saxena. 2018. Psychrotrophic microbiomes: molecular diversity and beneficial role in plant growth promotion and soil health. pp. 197–240. *In*: Panpatte, D.G., Y.K. Jhala, H.N. Shelat and R.V. Vyas (eds.). Microorganisms for Green Revolution-Volume 2: Microbes for Sustainable Agro-ecosystem. Springer, Singapore. doi:10.1007/978-981-10-7146-1_11.

Yang, S., X. Wen and H. Jin. 2012. Pyrosequencing investigation into the bacterial community in permafrost soils along the China-Russia crude oil pipeline (CRCOP). PLoS ONE 7.

Yergeau, E., H. Hogues, L.H. Whyte and C.W. Greer. 2010. The functional potential of high Arctic permafrost revealed by metagenomic sequencing, qPCR and microarray analyses. ISME J. 4: 1206–14.

Yergeau, E., J.R. Lawrence and S. Sanschagrin. 2012. Next-generation sequencing of microbial communities in the Athabasca River and its tributaries in relation to oil sands mining activities. Appl. Environ. Microbiol. 78: 7626–37.

Yuzbasheva, E.Y., T.V. Yuzbashev and I.A. Laptev. 2011. Efficient cell surface display of Lip2 lipase using C-domains of glycosylphosphatidylinositol-anchored cell wall proteins of *Yarrowia lipolytica*. Appl. Microbiol. Biotechnol. 91: 645–54.

Zhang, D.C., A. Brouchkov, G. Griva, F. Schinner and R. Margesin. 2013. Isolation and characterization of bacteria from ancient siberian permafrost sediment. Biology 2: 85–106.

Chapter 16

Biodiversity and Biotechnological Applications of Extremophilic Microbiomes

Current Research and Future Challenges

Ajar Nath Yadav,[1],* Tanvir Kaur,[1] Rubee Devi,[1] Divjot Kour[1] and Neelam Yadav[2]

|||||||||||||||||||||||||||||||||||||||||||||||||||||||||||||||||||||||||||||||||||||||||||||||||||||||||||||||||||||||||||||||||||||||||||||||||||||||||||||||||||||||||

Introduction

This chapter contains current knowledge about extreme microbiomes and their biotechnological applications. The diverse groups of extremophilic microbes are emerging as an important and promising tool for sustainable agriculture and environment. The extremophilic microbes from different extreme habitats have the ability to promote plant growth directly or indirectly, by releasing plant growth regulators; solubilization of phosphorus, potassium and zinc; biological nitrogen fixation or by producing siderophores, ammonia, HCN and other secondary metabolites which are antagonistic against pathogenic microbes. These PGP microbes could be used as biofertilizers/bioinoculants in place of chemical fertilizers for sustainable agriculture. These stressed adaptive microbes could be used for mitigation of different abiotic stresses such as saline, heat, cold, alkalinity, acidic and drought stress. The chapter encompasses current knowledge of extreme microbiomes and their potential biotechnological applications. It will be very useful to the faculty, researchers and students associated with microbiology, biotechnology, agriculture, molecular biology, environmental biology and related subjects.

Minute forms of life 'microbes' exist in environments with a broad range of conditions and an extreme state is one where they could exist. The microbes adapted to physiological conditions such moderate temperature, salinity and pressure is very well described but life in extreme conditions is rarely discussed. So, keeping in this in view this chapter contains the current knowledge about the microbiome of extremophilic conditions. Extremophilic microbes are those that survive in harsh environments where other moderate life cannot live. Microbes, the earth's oldest life forms are known to have developed robust metabolic functions which are a survival secret that helps them to shelter in harsh conditions and have been reported worldwide such as acidophilic (below 4.0 pH) (Verma et al. 2013), alkalophilic (over 9.0 pH) (Yadav et al. 2015e), barophilics (> 0.5 MPa pressure)

[1] Department of Biotechnology, Dr. KSG Akal College of Agriculture, Eternal University, Baru Sahib, Sirmour-173101, Himachal Pradesh, India.
[2] Gopi Nath P.G. College, Veer Bahadur Singh Purvanchal University, Ghazipur-275201, Uttar Pradesh, India.
* Corresponding author: ajarbiotech@gmail.com

(Verma et al. 2017), halophilic (> 1 M NaCl concentration) (Yadav et al. 2019a; Yadav et al. 2015c), hyperthermophiles (60–120°C temperature), metalophilic (Verma et al. 2016), psychrophilic (below 0°C temperature) (Yadav et al. 2015a; Yadav et al. 2016; Yadav et al. 2015b), thermophilic (45–60°C temperature) (Kumar et al. 2014; Sahay et al. 2017), radiophilic and xerophiles (< 0.85 water activity) (Kour et al. 2020a; Verma et al. 2019). There are different culturable and unculturable techniques for exploration, characterization, identification and potential applications in diverse sectors (Fig. 16.1). The three known taxonomic forms of life called archaea, bacteria and eukarya are known to be supported by extreme territories, whereas, archaea is the largest and main life form which is found in extreme conditions that is followed by bacteria and eukarya. Archaea are mostly found in the acidophilic, alkalophilic, halophilic and hyperthermophilic conditions for example *Methanopyrus kandleri* are known to grow at the temperature above 122°C (Gaba et al. 2017; Saxena et al. 2016). Cynobacteria is a very well known bacteria that can survive in hypersaline, high metal concentration and xerophilic conditions. Among, eukarya, fungi are the most ecological

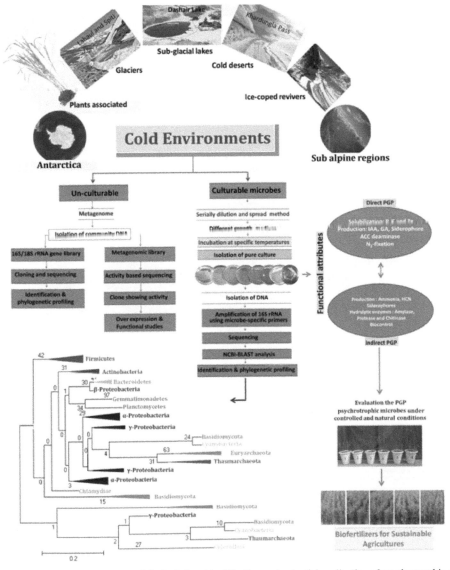

Figure 16.1 A schematic representation of the isolation, identification and potential application of psychrotrophic microbes from different habitats of extreme cold environments. Adapted with permission from Yadav et al. (2018a).

successful organisms that live in hot and cold deserts, alkaline and acidic, metal enriched water, hypersalinity and the deep oceans conditions (Rampelotto 2013).

Extremophilic microbiome has a wide range of applications from bioremediation of toxic elements to production of biomolecules for industrial and medical purposes to agricultural application. Microbes survive in heavy metal concentration that can be used in bioremediation of several pollutants that are toxic to environment (Johnson 2014), whereas radiophiles has an application in managing the pollutants based on nuclear waste (Appukuttan et al. 2006). Extremophiles survive in low water concentration, thermophilic and psychrophilic conditions that can be used in agricultural fields to help crop plants grow optimally in the rainfed, high or low temperature conditions (Yadav 2017). The potential applications of extremophiles have formed a great impact in the field of biotechnology which has been described. The stressed adaptive microbiomes could be utilized for plant growth promotion and mitigation of abiotic stresses conditions (Fig. 16.2).

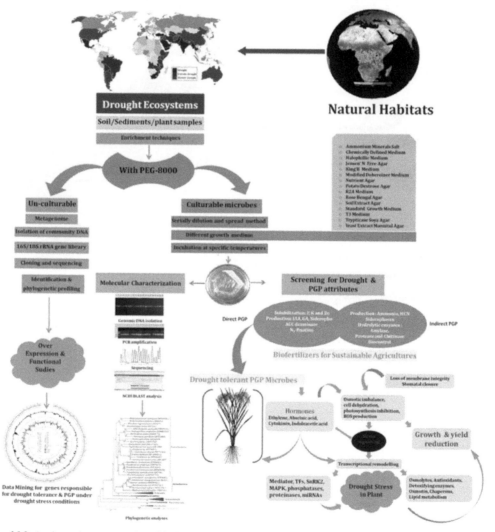

Figure 16.2 A schematic representation of the isolation, characterization, identification, and potential application of drought-tolerant microbes. Adapted with permission from Kour et al. (2019b).

Current perspective of extremophile microbiome

Recently extremophilic microbes have been drawing great interest, due to their biological activity. There are fundamental studies dedicated to their taxonomy and biodiversity, specific features of bacterial geographic distribution in different types of soil and their relationship between composition of microbial community's structure, genetic diversity and the saline soil have been shown (Dobrovol'skaya et al. 2015; Dobrovol'skii and Chernov 2011). Extremophilic microbes have evolved several mechanisms for survival under extreme conditions in the environment. In this study these adaptive mechanisms and the distinctive perspective on the fundamental characteristics of life have been described.

Extremophiles microbiome has been providing an extensive and versatile metabolic diversity with astonishing physiological capabilities to colonize extreme environments. Yet there are many types of strategies including molecular, biochemical and physiological extremophiles that are not understood and recognized completely. These environments host exceptional microbiome (extremophiles) and has a complex genetic and physiological adaptation that enable their survival, growth and development under harsh conditions (Bhowal and Chakraborty 2015; Singh et al. 2019). Recently according to frontier work extremophiles are channeled to consider their peculiar electrology, survival mechanism as well as physiology (Rothschild and Mancinelli 2001). In addition, their novel biomolecules and enzymes have been synthesized to extreme conditions and pursued for biotechnological application (Selvarajan et al. 2017). Natural environments with extreme acidity ($pH \leq 5$) are widespread and can be produced by various natural processes including volcanic activity, nitrification, fermentation and anaerobic respiration (Lukhele et al. 2019).

Since the last two decades, the research data on extremophiles has increased exponentially as the enzymes extracted from the extremophilic microbiome have been shown in various industries such as biotechnological industries, diary textiles, detergent, food and beverages, leather pulp and paper and pharma medicines (Fig. 16.3). Currently analysis encapsulates the awareness of numerous extremophiles and their potential medical and biotechnological applications. As the name suggests this microbiome lives in extreme conditions. These extreme parameters are mainly high radiation, heavy metals, high salt concentrations, high or low temperatures, low nutrient availability, low water content and pH (Parihar and Bagaria 2019; Yadav et al. 2017).

The rapidly developed molecular technologies have contributed to the accumulation of unparalleled precision of information about existence of extremophiles by exposing an unexpectedly

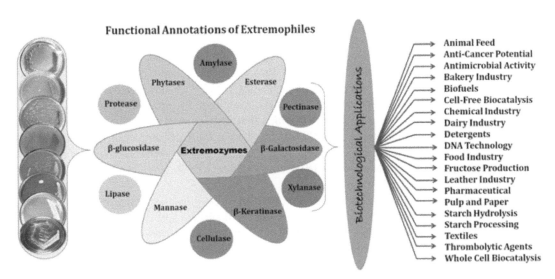

Figure 16.3 Extremozymes producing microbes and its biotechnological application in food and allied sectors. Adapted with permission from Kour et al. (2019a).

high level of diversity and complexity, the breakthrough of DNA sequencing technology has fundamentally revolutionized the microbiology of extreme environments. Consequently, the combinatorial approach will greatly improve the understanding of how microbiome has been surviving in the extreme environments (Rampelotto 2013; Singh et al. 2019).

Recently an industry has been producing hydrogen and chemical catalyst drivers. However after the discovery of thermophilic bacteria *Thermotoga elfill* and *Caldicellulosiruptor saccharolyticus* both microbiome have been widely used for large-scale of hydrogen production (Das and Veziroğlu 2001; De Vrije et al. 2002). The application of the strain including *Aeropyrum, Caldicellulosiruptor, Pyrococcus* and *Thermoanaero bacterium* that has been shown immense potential in the development of hydrogen by anaerobic fermentation and hydrogenase (Nishimura and Sako 2009; Ren et al. 2008). Thermophilic clostridium is used by green biologics for the production of biobutanol from corn stock (Coker 2016).

Biodiversity of extremophilic microbiomes

The extremophilic microbiome has been evolved to grow optimally under high pressure and temperature in extreme environments that are beneficial for sustainable environment (Yadav et al. 2019c). This microbiome has been reported from one of the most diverse extreme environments such as acidophilic, alkaliphile, barophiles, halophilic, metallotolerant, psychrophilic, polyextremophilic, thermophilic and xerophilic. It can be concluded that, all three domains of life like archaea, bacteria and eukarya have found a different extreme environments across the globe. Cherif et al. (2018) isolated four halophilic bacterial species *Alphaproteobacteria, Betaproteobacteria, Gammaproteobacteria, Firmicutes* and *Actinobacteria* from desert salt adapted and also halotolerant bacteria *Cladosporium, Alternaria, Aspergillus, Penicillium, Ulocladium,* and *Engyodontium*. Kadnikov et al. (2016) reported Acidophilic *Gallionaceae* sp. They also isolated *Sulfurisphaera javensis* sp. from an Indonesian hot spring (Tsuboi et al. 2018). It was also reported by Falagán et al. (2019) that *Acidithiobacillus sulfuriphilus* sp. was separated from the neutral pH environment and (Mukherjee et al. 2019) also isolated *Aspergillus niger* from the dumping area of Midnapore town, West Bengal. O'Dell et al. (2018) isolated bacteria *Salinisphaera* sp. from Lake Brown, Western Australia, and other *Sulfobacillus thermosulfidooxidans* isolated from hot spring for the biosorption of heavy metal ions. Panyushkina et al. (2018) reported two bacterial strains *Leptospirillum ferriphilum* and *Acidithiobacillus thiooxidans* which were isolated from sulfur- and iron-oxidizing organisms and also *Chromulina freiburgensis Dof.,* isolated from Berkeley Pit Lake (Mitman 2019). In another investigation Parihar and Bagaria (2019) isolated *Methanococcoides burtonii* from perennially cold, anoxic hypolimnion of Ace Lake, Antarctica and other bacteria have been isolated *Bacillus, Arthrobacter, Pseudomonas, Pseudoalteromonas, Vibrio, Penicillium Halorubrum, Methanogenium, Cladosporium, Crystococcus* and *Candida* from ice zones like the Arctic and Antarctica ice beds, polar zones, glaciers. Singh et al. (2019) reported one psychrophiles bacterial strain *Pseudomonas* sp. isolated from Arctic tundra soils, and two Barophiles bacterial strain *Moritella yayanosii* and *Shewanella benthica* that was isolated from Mariana Trench have also been reported (Marteinsson et al. 1999). Takai et al. (2000) isolated from the deep-sea and other identified bacterial strain *Palaeococcus ferrophilus, Moritella japonica* (Nogi et al. 1998), *Colwellia echini* sp. (Christiansen et al. 2018), *Colwellia ponticola* sp. (Park et al. 2019), *Rubritalea profundi* (Song et al. 2019).

There are some metallotolerant microbiome such as, *Aspergillus niger, Aspergillus fumigates,* and *pencillium rubens* that were isolated from the top soil/surface soil (Khan et al. 2019). Satoh et al. (2018) reported many types of radioresistant *Deinococcus aerius* that were isolated from orange-pigmented, nonmotile, desiccation-tolerant, UV- and gammaresistant and coccoid bacterium from the upper troposphere in Japan and some others are *Firmicutes, Proteobacteria, Deinococcus-Thermus, Actinobacteria* sp., Sajjad et al. (2018) and *Rubrobacter xylanophilus* Tomariguchi and Miyazaki (2019) were isolated from desert soil and Arima Onsen in Japan. In an investigation

Varshney et al. (2018) reported some thermophilic bacteria such as *Asterarcys quadricellulare* and *Chlorella sorokiniana* that they isolated from green algae and others isolated from agricultural waste composting in Vietnam which identified the bacteria are *Bacillus* sp. As well as some other strains that include *Bacillus cytotoxicus* and *Bacillus licheniformis* (Cavello et al. 2018), *Anoxybacillus kamchatkensis, Anoxybacillus salavatliensis, Anoxybacillus flavithermus, Aeribacillus pallidus, Geobacillus toebii, Geobacillus galactosidasius, Brevibacillus thermoruber* and *Bacillus thuringiensis* (Yadav et al. 2018b), *Bacillus haynesii* (Rehman et al. 2019), *Bacillus subtilis* (Siu-Rodas et al. 2018), *Coelastrella* sp. (Narayanan et al. 2018), *Geobacillus yumthangensis* sp. (Najar et al. 2018). There have been some reports of polyextremophilic microbes *Aspergillus protuberus* that were isolated from marine water by Corral et al. (2018) and other microbes *Aspergillus* sp. have been isolated from marine fungus from Barents Sea (Sartorio et al. 2020) and *Acinetobacter* sp. were isolated from high-altitude Andean lakes (Sartorio et al. 2020) as well as *Natronolimnobius aegyptiacus* sp., was isolated from Athalassohaline Wadi An Natrun, Egypt (Zhao et al. 2018) with *Halorubrum* sp. isolated from Salar de Uyuni, Bolivia (DasSarma et al. 2019).

Potential biotechnological applications of extremophiles

Extremophiles have adapted to survive and flourish in the extreme ecological niches that cannot be inhabited by other organisms such as deep-sea hydrothermal vents, hot and cold deserts, hot springs, inland saline systems, soda lakes, sulfataric fields, solar salterns and environment highly contaminated with nuclear waste or heavy metals (Raddadi et al. 2015). As a result of the extreme environmental conditions at these locations, the microorganisms surviving here have developed biological adaptations, mechanisms and strategies to remain metabolically active (Schultz and Rosado 2019). Thus, due to their unique properties, which enable them to withstand such environmental conditions, extremophiles have great biotechnological potential and are the main targets for constant research and discovery of new products (Horikoshi and Bull 2011). Owing to fact of their unique physiological properties, their applications range from the bioremediation of toxic pollutants, production of different biomolecules of medical and industrial importance to management of nuclear-waste-polluted environments (Johnson 2014; Navarro et al. 2013).

Many extremozymes have been reported from extremophiles including amylases from *Engyodontium album* (Ali et al. 2014a; Ali et al. 2019), *Aspergillus gracilis* (Ali et al. 2014b), proteases from *Aspergillus flavus* (Sa et al. 2012) and cellulases from *Aspergillus terreus* (Gunny et al. 2014), pectinases from *Geomyces* sp. (Poveda et al. 2018; Rafiq et al. 2019), chitinase from *Glaciozyma antarctica* (Yusof et al. 2017) and lipases from *Penicillium* sp. (Pandey et al. 2016). The extremophilic microbes and their secondary metabolites could be utilized in industry, agriculture and allied sectors (Rana et al. 2019; Rastegari et al. 2019a; Rastegari et al. 2019b; Yadav et al. 2019b) (Fig. 16.4).

Extremophiles are also a very good source of amphipathic organic molecules known as surfactants (Pacwa-Płociniczak et al. 2011). Biosurfactants find applications in the oil, pharmaceutical and detergent industries. Biosurfactants have the capability to reduce surface and interfacial tension and strengthen the interaction between pollutants and microorganisms, increase the solubility and availability of hydrophobic organic compounds, and, consequently, the removal of pollutants (Liu et al. 2017). Biosurfactants are also known to have antifungal, antitumor and antiviral activities (Balan et al. 2017). Biosurfactants are also used in the detergent industry, which is in fact one of the largest consumers of biosurfactants (Banat et al. 2010). The detergent industry also requires biosurfactants with stability at alkaline pH and thermophilic temperatures because of the washing process and performance (Grbavčić et al. 2011). Biosurfactants producing microbes have been reported from extreme environments including *Bacillus, Burkholderia, Pseudomonas, Rhodococcus* and *Sphingomonas* from low temperature (Gesheva et al. 2010; Janek et al. 2013); *Achromobacter, Bacillus, Citrobacter, Lysinibacillus, Ochrobactrum* and *Pseudomonas* from desert soils and hot springs (Joy et al. 2017); *Acidithiobacillus thiooxidans* and *Virgibacillus salaries* from extreme

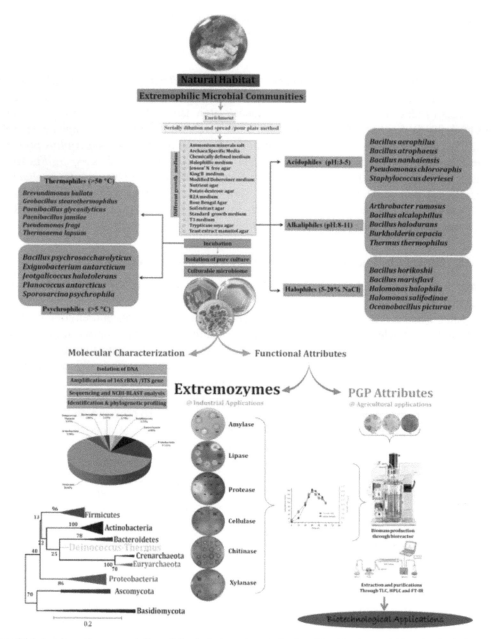

Figure 16.4 A schematic representation of the isolation, characterization, identification and biotechnological potential application of extremophilic microbes. Adapted with permission from Kour et al. (2019a).

pH environments (Elazzazy et al. 2015; Karwowska et al. 2015); *Bacillus asahii, B. detrensis, Fictibacillus barbaricus*, and *Paenisporosarcina indica* from saline and hypersaline environments (de Almeida Couto et al. 2015).

Extremophiles are also known to play a role in bioremediation (Kumar et al. 2019). *Mortierella* sp. from Antarctica has been reported as a good candidate for bioremediation of hydrocarbon spill (Hughes et al. 2007). Halophilic fungi *Debaryomyces hansenii* and *Debaryomyces subglobosus* and Antartic *Aspergillus fumigatus* have been reported for their ability to degrade phenol (Gerginova et al. 2013; Jiang et al. 2016). Mansour (2017) showed remediation of halite on sandstones by halophilic fungi *Wallemia sebi*. Bano et al. (2018) showed the removal of heavy metals by halophiles including *Aspergillus flavus, Aspergillus gracilis, Aspergillus penicillioides, Aspergillus restrictus* and *Sterigmatomyces halophiles*.

Many clinically important products have been produced from extremophiles. The oxygenated polyketides, penilactones A and B, with clinical importance has been produced from *Penicillium crustosum* isolated from deep sea of Antarctic (Wu et al. 2012). Eremofortine, eremophilane sesquiterpenes and chloro-trinoreremophilane sesquiterpene have been reported to show cytotoxic activity against cancer cell lines which were recovered from an Antarctic *Penicillium* sp. (Wu et al. 2013). Cladosin C showing activity against influenza A H1N1 virus has been obtained from deep-sea *Cladosporium sphaerospermum* (Wu et al. 2014). *Penicillium chrysogenum*, a psychrophilic and halophilic obtained from Vestfold Hills' saline lake produced bis-anthraquinone with insecticidal and medicinal properties (Sumarah et al. 2005). Furthermore, carotenoid and many fatty acids like heptadecanoic, linoleic, linolenic, myristic, palmitic and stearic have been reported from *Thelebolus microspores*, a cold-tolerant fungus (Singh et al. 2014).

Extremophiles also have the potential to act as biofertilizers (Kour et al. 2020a; Kour et al. 2020b; Kour et al. 2020c; Kour et al. 2020d). *Aspergillus niger* isolated from the tundra in Arctic Archipelago of Svalbard, showed phosphate solubilization ability. P-solubilizing archeae including *Haloarcula, Halobacterium, Halococcus, Haloferax, Halolamina, Halosarcina, Halostagnicola, Haloterrigena, Natrialba, Natrinema* and *Natronoarchaeum* has been isolated from sediments, water as well as the rhizospheric soil from Rann of Kutch, Gujarat (Yadav et al. 2015d).

There are various success stories for application of the extremophiles as well as their products. There are still other possibilities which could be explored for the application of the extremophiles. One such emerging field is the production of the polyhydroxyalkanoates by extremophilic microbes which will be useful for the production of bioplastics. As there is growing demand of the industries, researchers may focus on unexplored extreme environments and make best efforts to discover novel microbial communities with capabilities to yield products of biotechnological importance.

Conclusion and future prospect

Extremophilic microbes that survive in unique and extreme conditions have been considered to have various biotechnological applications in the environment as well as in agriculture. Microbial strains from extreme environment that have multifarious plant growth promoting potential can be used as bioinoculants in agriculture for the improvement of plant growth in the various abiotic conditions. These microbes in the environment can also be used for bioremediation of toxic elements or chemicals. In conclusion, extremophilic microbes are considered as sustainable resources that can be exploited in various biotechnological sectors that can be used for the development of economy.

References

Ali, I., A. Akbar, M. Anwar, B. Yanwisetpakdee, S. Prasongsuk, P. Lotrakul et al. 2014a. Purification and characterization of extracellular, polyextremophilic α-amylase obtained from halophilic *Engyodontium album*. Iran J. Biotechnol. 12: 35–40.

Ali, I., A. Akbar, B. Yanwisetpakdee, S. Prasongsuk, P. Lotrakul and H. Punnapayak. 2014b. Purification, characterization, and potential of saline waste water remediation of a polyextremophilic α-amylase from an obligate halophilic *Aspergillus gracilis*. BioMed Res. Int. 2014.

Ali, I., S. Khaliq, S. Sajid and A. Akbar. 2019. Biotechnological applications of halophilic fungi: past, present, and future. pp. 291–306. *In*: Tiquia-Arashiro, S.M. and M. Grube (eds.). Fungi in Extreme Environments: Ecological Role and Biotechnological Significance. Springer.

Appukuttan, D., A.S. Rao and S.K. Apte. 2006. Engineering of *Deinococcus radiodurans* R1 for bioprecipitation of uranium from dilute nuclear waste. Appl. Environ. Microbiol. 72: 7873–7878.

Balan, S.S., C.G. Kumar and S. Jayalakshmi. 2017. Aneurinifactin, a new lipopeptide biosurfactant produced by a marine *Aneurinibacillus aneurinilyticus* SBP-11 isolated from Gulf of Mannar: purification, characterization and its biological evaluation. Microbiol. Res. 194: 1–9.

Banat, I.M., A. Franzetti, I. Gandolfi, G. Bestetti, M.G. Martinotti, L. Fracchia et al. 2010. Microbial biosurfactants production, applications and future potential. Appl. Microbiol. Biotechnol. 87: 427–444.

Bano, A., J. Hussain, A. Akbar, K. Mehmood, M. Anwar, M.S. Hasni et al. 2018. Biosorption of heavy metals by obligate halophilic fungi. Chemosphere 199: 218–222.

Bhowal, S. and R. Chakraborty. 2015. Microbial diversity of acidophilic heterotrophic bacteria: An overview. pp. 157–173. *In*: Jha, P. (ed.). Biodiversity, Conservation and Sustainable Development: Issues and Approaches. New Academic Publishere.

Cavello, I., M.S. Urbieta, A.B. Segretin, A. Giaveno, S. Cavalitto and E.R. Donati. 2018. Assessment of keratinase and other hydrolytic enzymes in thermophilic bacteria isolated from geothermal areas in *Patagonia Argentina*. Geomicribiol. J. 35: 156–165.

Cherif, H., M. Neifar, H. Chouchane, A. Soussi, C. Hamdi, A. Guesmi et al. 2018. Extremophile diversity and biotechnological potential from desert environments and saline systems of southern Tunisia. pp. 33–64. *In*: Extremophiles. CRC Press.

Christiansen, L., P.K. Bech, M. Schultz-Johansen, H.J. Martens and P. Stougaard. 2018. *Colwellia echini* sp. nov., an agar- and carrageenan-solubilizing bacterium isolated from sea urchin. Int. J. Syst. Evol. Microbiol. 68: 687–691.

Coker, J.A. 2016. Extremophiles and biotechnology: current uses and prospects. F1000Research 5.

Corral, P., F.P. Esposito, P. Tedesco, A. Falco, E. Tortorella, L. Tartaglione et al. 2018. Identification of a sorbicillinoid-producing *Aspergillus* strain with antimicrobial activity against *Staphylococcus aureus*: a new polyextremophilic marine fungus from Barents Sea. Mar. Biotechnol. 20: 502–511.

Das, D. and T.N. Veziroğlu. 2001. Hydrogen production by biological processes: a survey of literature. Int. J. Hydrogen Energ. 26: 13–28.

DasSarma, P., B.P. Anton, S. DasSarma, V.J. Laye, D. Guzman, R.J. Roberts et al. 2019. Genome sequence and methylation patterns of *Halorubrum* sp. strain BOL3-1, the first haloarchaeon isolated and cultured from Salar de Uyuni, Bolivia. Microbiol Resour Announc 8: e00386–00319.

de Almeida Couto, C.R., V.M. Alvarez, J.M. Marques, D. de Azevedo Jurelevicius and L. Seldin. 2015. Exploiting the aerobic endospore-forming bacterial diversity in saline and hypersaline environments for biosurfactant production. BMC Microbiol. 15: 240.

De Vrije, T., G. De Haas, G. Tan, E. Keijsers and P. Claassen. 2002. Pretreatment of Miscanthus for hydrogen production by *Thermotoga elfii*. Int. J. Hydrogen Energ. 27: 1381–1390.

Dobrovol'skaya, T., D. Zvyagintsev, I.Y. Chernov, A. Golovchenko, G. Zenova, L. Lysak et al. 2015. The role of microorganisms in the ecological functions of soils. Eurasian Soil Sci. 48: 959–967.

Dobrovol'skii, G. and I.Y. Chernov. 2011. Rol'pochvy v formirovanii i sokhranenii biologicheskogo raznoobraziya [The Role of Soil in the Formation and Conservation of Biological Diversity]. Moscow: Publishing House of the Association of Scientific Publications KMK.

Elazzazy, A.M., T. Abdelmoneim and O. Almaghrabi. 2015. Isolation and characterization of biosurfactant production under extreme environmental conditions by alkali-halo-thermophilic bacteria from Saudi Arabia. Saudi J. Biol. Sci. 22: 466–475.

Falagán, C., A. Moya-Beltrán, M. Castro, R. Quatrini and D.B. Johnson. 2019. *Acidithiobacillus sulfuriphilus* sp. nov.: an extremely acidophilic sulfur-oxidizing chemolithotroph isolated from a neutral pH environment. Int. J. Syst. Evol. Microbiol. 69: 2907–2913.

Gaba, S., R.N. Singh, S. Abrol, A.N. Yadav, A.K. Saxena and R. Kaushik. 2017. Draft genome sequence of *Halolamina pelagica* CDK2 isolated from natural salterns from Rann of Kutch, Gujarat, India. Genome Announc. 5: 1–2.

Gerginova, M., J. Manasiev, H. Yemendzhiev, A. Terziyska, N. Peneva and Z. Alexieva. 2013. Biodegradation of phenol by Antarctic strains of *Aspergillus fumigatus*. Z. Naturforsch. C 68: 384–393.

Gesheva, V., E. Stackebrandt and E. Vasileva-Tonkova. 2010. Biosurfactant production by halotolerant rhodococcusfascians from casey station, Wilkes Land, Antarctica. Curr. Microbiol. 61: 112–117.

Grbavčić, S., D. Bezbradica, L. Izrael-Živković, N. Avramović, N. Milosavić, I. Karadžić et al. 2011. Production of lipase and protease from an indigenous *Pseudomonas aeruginosa* strain and their evaluation as detergent additives: compatibility study with detergent ingredients and washing performance. Bioresour. Technol. 102: 11226–11233.

Gunny, A.A.N., D. Arbain, R.E. Gumba, B.C. Jong and P. Jamal. 2014. Potential halophilic cellulases for in situ enzymatic saccharification of ionic liquids pretreated lignocelluloses. Bioresour. Technol. 155: 177–181.

Horikoshi, K. and A.T. Bull. 2011. Prologue: definition, categories, distribution, origin and evolution, pioneering studies, and emerging fields of extremophiles. In: Extremophiles Handbook.

Hughes, K.A., P. Bridge and M.S. Clark. 2007. Tolerance of Antarctic soil fungi to hydrocarbons. Sci. Total Environ. 372: 539–548.

Janek, T., M. Łukaszewicz and A. Krasowska. 2013. Identification and characterization of biosurfactants produced by the Arctic bacterium *Pseudomonas putida* BD2. Colloids Surf B Biointerfaces 110: 379–386.

Jiang, Y., Y. Shang, K. Yang and H. Wang. 2016. Phenol degradation by halophilic fungal isolate JS4 and evaluation of its tolerance of heavy metals. Appl. Microbiol. Biotechnol. 100: 1883–1890.

Johnson, D.B. 2014. Biomining—biotechnologies for extracting and recovering metals from ores and waste materials. Curr. Opin. Biotechnol. 30: 24–31.

Joy, S., P.K. Rahman and S. Sharma. 2017. Biosurfactant production and concomitant hydrocarbon degradation potentials of bacteria isolated from extreme and hydrocarbon contaminated environments. Chem. Eng. J. 317: 232–241.

Kadnikov, V., D. Ivasenko, A. Beletskii, A. Mardanov, E. Danilova, N. Pimenov et al. 2016. A novel uncultured bacterium of the family Gallionellaceae: description and genome reconstruction based on metagenomic analysis of microbial community in acid mine drainage. Microbiology 85: 449–461.

Karwowska, E., M. Wojtkowska and D. Andrzejewska. 2015. The influence of metal speciation in combustion waste on the efficiency of Cu, Pb, Zn, Cd, Ni and Cr bioleaching in a mixed culture of sulfur-oxidizing and biosurfactant-producing bacteria. J. Hazard Mater. 299: 35–41.

Khan, I., M. Aftab, S. Shakir, M. Ali, S. Qayyum, M.U. Rehman et al. 2019. Mycoremediation of heavy metal (Cd and Cr)–polluted soil through indigenous metallotolerant fungal isolates. Environ. Monit. Assess. 191: 585.

Kour, D., K.L. Rana, T. Kaur, B. Singh, V.S. Chauhan, A. Kumar et al. 2019a. Extremophiles for hydrolytic enzymes productions: biodiversity and potential biotechnological applications. pp. 321–372. *In*: Molina, G., V.K. Gupta, B. Singh and N. Gathergood (eds.). Bioprocessing for Biomolecules Production. doi:10.1002/9781119434436.ch16.

Kour, D., K.L. Rana, A.N. Yadav, N. Yadav, V. Kumar, A. Kumar et al. 2019b. Drought-tolerant phosphorus-solubilizing microbes: biodiversity and biotechnological applications for alleviation of drought stress in plants. pp. 255–308. *In*: Sayyed, R.Z., N.K. Arora and M.S. Reddy (eds.). Plant Growth Promoting Rhizobacteria for Sustainable Stress Management : Volume 1: Rhizobacteria in Abiotic Stress Management. Springer Singapore, Singapore. doi:10.1007/978-981-13-6536-2_13.

Kour, D., K.L. Rana, T. Kaur, I. Sheikh, A.N. Yadav, V. Kumar et al. 2020a. Microbe-mediated alleviation of drought stress and acquisition of phosphorus in great millet (*Sorghum bicolour* L.) by drought-adaptive and phosphorus-solubilizing microbes. Biocatal. Agric. Biotechnol. 23: 101501. doi:https://doi.org/10.1016/j.bcab.2020.101501.

Kour, D., K.L. Rana, I. Sheikh, V. Kumar, A.N. Yadav, H.S. Dhaliwal et al. 2020b. Alleviation of drought stress and plant growth promotion by *Pseudomonas libanensis* EU-LWNA-33, a drought-adaptive phosphorus-solubilizing bacterium. Proc. Natl. Acad. Sci. India Sec. B Biol. Sci. doi:10.1007/s40011-019-01151-4.

Kour, D., K.L. Rana, A.N. Yadav, I. Sheikh, V. Kumar, H.S. Dhaliwal et al. 2020c. Amelioration of drought stress in Foxtail millet (*Setaria italica* L.) by P-solubilizing drought-tolerant microbes with multifarious plant growth promoting attributes. Environ. Sustain. 3: 23–34. doi:10.1007/s42398-020-00094-1.

Kour, D., K.L. Rana, A.N. Yadav, N. Yadav, M. Kumar, V. Kumar et al. 2020d. Microbial biofertilizers: Bioresources and eco-friendly technologies for agricultural and environmental sustainability. Biocatal. Agric. Biotechnol. 23: 101487. doi:https://doi.org/10.1016/j.bcab.2019.101487.

Kumar, M., A.N. Yadav, R. Tiwari, R. Prasanna and A.K. Saxena. 2014. Deciphering the diversity of culturable thermotolerant bacteria from Manikaran hot springs. Ann. Microbiol. 64: 741–751.

Kumar, M., R. Saxena, P.K. Rai, R.S. Tomar, N. Yadav, K.L. Rana et al. 2019. Genetic diversity of methylotrophic yeast and their impact on environments. pp. 53–71. *In*: Yadav, A.N., S. Singh, S. Mishra and A. Gupta (eds.). Recent Advancement in White Biotechnology Through Fungi: Volume 3: Perspective for Sustainable Environments. Springer International Publishing, Cham. doi:10.1007/978-3-030-25506-0_3.

Liu, G., H. Zhong, Y. Jiang, M.L. Brusseau, J. Huang, L. Shi et al. 2017. Effect of low-concentration rhamnolipid biosurfactant on Pseudomonas aeruginosa transport in natural porous media. Water Resour. Res. 53: 361–375.

Lukhele, T., R. Selvarajan, H. Nyoni, B.B. Mamba and T.A. Msagati. 2019. Acid mine drainage as habitats for distinct microbiomes: current knowledge in the era of molecular and omic technologies. Curr. Microbiol. 1–18.

Mansour, M.M. 2017. Effects of the halophilic fungi *Cladosporium sphaerospermum*, *Wallemia sebi*, *Aureobasidium pullulans* and *Aspergillus nidulans* on halite formed on sandstone surface. Int. Biodeterior. Biodegrad. 117: 289–298.

Marteinsson, V.T., J.-L. Birrien, A.-L. Reysenbach, M. Vernet, D. Marie, A. Gambacorta et al. 1999. Thermococcus barophilus sp. nov., a new barophilic and hyperthermophilic archaeon isolated under high hydrostatic pressure from a deep-sea hydrothermal vent. Int. J. Syst. Evol. Microbiol. 49: 351–359.

Mitman, J.E.M. 2019. Growth, Lipid Production and Biodiesel Potential of *Chromulina freiburgensis* Dofl., An Acidophilic Chrysophyte Isolated from Berkeley Pit Lake. Montana Tech of The University of Montana.

Mukherjee, R., T. Paul, J.P. Soren, S.K. Halder, K.C. Mondal, B.R. Pati et al. 2019. Acidophilic α-amylase production from *Aspergillus niger* RBP7 using potato peel as substrate: a waste to value added approach. Waste Biomass Valorization 10: 851–863.

Najar, I.N., M.T. Sherpa, S. Das, K. Verma, V.K. Dubey and N. Thakur. 2018. *Geobacillus yumthangensis* sp. nov., a thermophilic bacterium isolated from a north-east Indian hot spring. Int. J. Syst. Evol. Microbiol. 68: 3430–3434.

Narayanan, G.S., S. Seepana, R. Elankovan, S. Arumugan and M. Premalatha. 2018. Isolation, identification and outdoor cultivation of thermophilic freshwater microalgae *Coelastrella* sp. FI69 in bubble column reactor for the application of biofuel production. Biocatal. Agric. Biotechnol. 14: 357–365.

Navarro, C.A., D. von Bernath and C.A. Jerez. 2013. Heavy metal resistance strategies of acidophilic bacteria and their acquisition: importance for biomining and bioremediation. Biolo. Res. 46: 363–371.

Nishimura, H. and Y. Sako. 2009. Purification and characterization of the oxygen-thermostable hydrogenase from the aerobic hyperthermophilic archaeon *Aeropyrum camini*. J. Biosci. Bioeng. 108: 299–303.

Nogi, Y., C. Kato and K. Horikoshi. 1998. *Moritella japonica* sp. nov., a novel barophilic bacterium isolated from a Japan Trench sediment. J. Gen. Appl. Microbiol. 44: 289–295.

O'Dell, K.B., E.A. Hatmaker, A.M. Guss and M.R. Mormile. 2018. Complete genome sequence of *Salinisphaera* sp. strain LB1, a moderately halo-acidophilic bacterium isolated from Lake Brown, Western Australia. Microbiol. Resour. Announc. 7: e01047–01018.

Pacwa-Płociniczak, M., G.A. Płaza, Z. Piotrowska-Seget and S.S. Cameotra. 2011. Environmental applications of biosurfactants: recent advances. Int. J. Mol. Sci. 12: 633–654.

Pandey, N., K. Dhakar, R. Jain and A. Pandey. 2016. Temperature dependent lipase production from cold and pH tolerant species of *Penicillium*. Mycosphere 7: 1533–1545.

Panyushkina, A., I. Tsaplina, T. Kondrat'Eva, A. Belyi and A. Bulaev. 2018. Physiological and morphological characteristics of acidophilic bacteria *Leptospirillum ferriphilum* and *Acidithiobacillus thiooxidans*, members of a chemolithotrophic microbial consortium. Microbiology 87: 326–338.

Parihar, J. and A. Bagaria. 2019. The extremes of life and extremozymes: diversity and perspectives. Acta Sci. Microbiol. 3.

Park, S., J.-M. Park and J.-H. Yoon. 2019. *Colwellia ponticola* sp. nov., isolated from seawater. Int. J. Syst. Evol. Microbiol. 69: 3062–3067.

Poveda, G., C. Gil-Durán, I. Vaca, G. Levicán and R. Chávez. 2018. Cold-active pectinolytic activity produced by filamentous fungi associated with Antarctic marine sponges. Biol. Res. 51: 28.

Raddadi, N., A. Cherif, D. Daffonchio, M. Neifar and F. Fava. 2015. Biotechnological applications of extremophiles, extremozymes and extremolytes. Appl. Microbiol. Biotechnol. 99: 7907–7913.

Rafiq, M., N. Hassan, M. Rehman and F. Hasan. 2019. Adaptation mechanisms and applications of psychrophilic fungi. pp. 157–174. *In*: Fungi in Extreme Environments: Ecological Role and Biotechnological Significance. Springer.

Rampelotto, P.H. 2013. Extremophiles and Extreme Environments. Multidisciplinary Digital Publishing Institute.

Rana, K.L., D. Kour, I. Sheikh, A. Dhiman, N. Yadav, A.N. Yadav et al. 2019. Endophytic fungi: biodiversity, ecological significance and potential industrial applications. pp. 1–62. *In*: Yadav, A.N., S. Mishra, S. Singh and A. Gupta (eds.). Recent Advancement in White Biotechnology through Fungi: Volume 1: Diversity and Enzymes Perspectives. Springer, Switzerland.

Rastegari, A.A., A.N. Yadav and N. Yadav. 2019a. Genetic manipulation of secondary metabolites producers. pp. 13–29. *In*: Gupta, V.K. and A. Pandey (eds.). New and Future Developments in Microbial Biotechnology and Bioengineering. Elsevier, Amsterdam. doi:https://doi.org/10.1016/B978-0-444-63504-4.00002-5.

Rastegari, A.A., A.N. Yadav, N. Yadav and N. Tataei Sarshari. 2019b. Bioengineering of secondary metabolites. pp. 55–68. *In*: Gupta, V.K. and A. Pandey (eds.). New and Future Developments in Microbial Biotechnology and Bioengineering. Elsevier, Amsterdam. doi:https://doi.org/10.1016/B978-0-444-63504-4.00004-9.

Rehman, S., B.R. Jermy, S. Akhtar, J.F. Borgio, S. Abdul Azeez, V. Ravinayagam et al. 2019. Isolation and characterization of a novel thermophile; *Bacillus haynesii*, applied for the green synthesis of ZnO nanoparticles. Artificial Cells, Nanomedicine, and Biotechnology 47: 2072–2082.

Ren, N., G. Cao, A. Wang, D.-J. Lee, W. Guo and Y. Zhu. 2008. Dark fermentation of xylose and glucose mix using isolated *Thermoanaerobacterium thermosaccharolyticum* W16. Int. j. Hydrogen Energ. 33: 6124–6132.

Rothschild, L.J. and R.L. Mancinelli. 2001. Life in extreme environments. Nature 409: 1092–1101.

Sa, A., A. Singh, S. Garg, A. Kumar and H. Kumar. 2012. Screening, isolation and characterisation of protease producing moderately halophilic microorganisms. Asian Jr. of Microbiol. Biotech. Env. Sc. 14: 603–612.

Sahay, H., A.N. Yadav, A.K. Singh, S. Singh, R. Kaushik and A.K. Saxena. 2017. Hot springs of Indian Himalayas: Potential sources of microbial diversity and thermostable hydrolytic enzymes. 3 Biotech 7: 1–11.

Sajjad, W., S. Khan, M. Ahmad, M. Rafiq, M. Badshah, S. Zada et al. 2018. Effects of ultra-violet radiation on cellular proteins and lipids of radioresistant bacteria isolated from desert soil. Folia Biol. 66: 41–52.

Sartorio, M.G., G.D. Repizo and N. Cortez. 2020. Catalases of the polyextremophilic Andean isolate *Acinetobacter* sp. Ver 3 confer adaptive response to H_2O_2 and UV radiation. FEBS J. 1–15.

Satoh, K., H. Arai, T. Sanzen, Y. Kawaguchi, H. Hayashi, S.-i. Yokobori et al. 2018. Draft genome sequence of the radioresistant bacterium *Deinococcus aerius* TR0125, isolated from the high atmosphere above Japan. Genome Announc. 6: e00080–00018.

Saxena, A.K., A.N. Yadav, M. Rajawat, R. Kaushik, R. Kumar, M. Kumar et al. 2016. Microbial diversity of extreme regions: An unseen heritage and wealth. Indian J. Plant Genet. Resour. 29: 246–248.

Schultz, J. and A.S. Rosado. 2019. Extreme environments: a source of biosurfactants for biotechnological applications. Extremophiles 1–18.

Selvarajan, R., T. Sibanda, M. Tekere, H. Nyoni and S. Meddows-Taylor. 2017. Diversity analysis and bioresource characterization of halophilic bacteria isolated from a South African saltpan. Molecules 22: 657.

Singh, P., K. Jain, C. Desai, O. Tiwari and D. Madamwar. 2019. Microbial community dynamics of extremophiles/extreme environment. pp. 323–332. *In*: Microbial Diversity in the Genomic Era. Elsevier.

Singh, S.M., P.N. Singh, S.K. Singh and P.K. Sharma. 2014. Pigment, fatty acid and extracellular enzyme analysis of a fungal strain Thelebolus microsporus from Larsemann Hills, Antarctica. Polar Rec. 50: 31–36.

Siu-Rodas, Y., M. de los Angeles Calixto-Romo, K. Guillén-Navarro, J.E. Sánchez, J.A. Zamora-Briseno and L. Amaya-Delgado. 2018. *Bacillus subtilis* with endocellulase and exocellulase activities isolated in the thermophilic phase from composting with coffee residues. Rev. Argent. Microbiol. 50: 234–243.

Song, J., I. Kang, Y. Joung, S. Yoshizawa, R. Kaneko, K. Oshima et al. 2019. Genome analysis of *Rubritalea profundi* SAORIC-165 T, the first deep-sea verrucomicrobial isolate, from the northwestern Pacific Ocean. J. Microbiol. 57: 413–422.

Sumarah, M.W., J.D. Miller and G.W. Adams. 2005. Measurement of a rugulosin-producing endophyte in white spruce seedlings. Mycologia 97: 770–776.

Takai, K., A. Sugai, T. Itoh and K. Horikoshi. 2000. *Palaeococcus ferrophilus* gen. nov., sp. nov., a barophilic, hyperthermophilic archaeon from a deep-sea hydrothermal vent chimney. Int. J. Syst. Evol. Microbiol. 50: 489–500.

Tomariguchi, N. and K. Miyazaki. 2019. Complete genome sequence of *Rubrobacter xylanophilus* strain AA3-22, isolated from Arima Onsen in Japan. Microbiol. Res. Announc. 8: e00818–00819.

Tsuboi, K., H.D. Sakai, N. Nur, K.M. Stedman, N. Kurosawa and A. Suwanto. 2018. *Sulfurisphaera javensis* sp. nov., a hyperthermophilic and acidophilic archaeon isolated from Indonesian hot spring, and reclassification of Sulfolobus tokodaii Suzuki et al. 2002 as Sulfurisphaera tokodaii comb. nov. Int. J. Syst. Evol. Microbiol. 68: 1907–1913.

Varshney, P., J. Beardall, S. Bhattacharya and P.P. Wangikar. 2018. Isolation and biochemical characterisation of two thermophilic green algal species-Asterarcys quadricellulare and *Chlorella sorokiniana*, which are tolerant to high levels of carbon dioxide and nitric oxide. Algal Res. 30: 28–37.

Verma, P., A.N. Yadav, S.K. Kazy, A.K. Saxena and A. Suman. 2013. Elucidating the diversity and plant growth promoting attributes of wheat (*Triticum aestivum*) associated acidotolerant bacteria from southern hills zone of India. Natl. J. Life Sci. 10: 219–227.

Verma, P., A.N. Yadav, K.S. Khannam, S. Kumar, A.K. Saxena and A. Suman. 2016. Molecular diversity and multifarious plant growth promoting attributes of Bacilli associated with wheat (*Triticum aestivum* L.) rhizosphere from six diverse agro-ecological zones of India. J. Basic Microbiol. 56: 44–58.

Verma, P., A.N. Yadav, K.S. Khannam, S. Mishra, S. Kumar, A.K. Saxena et al. 2019. Appraisal of diversity and functional attributes of thermotolerant wheat associated bacteria from the peninsular zone of India. Saudi J. Biol. Sci. 26: 1882–1895. doi:https://doi.org/10.1016/j.sjbs.2016.01.042.

Verma, P., A.N. Yadav, V. Kumar, D.P. Singh and A.K. Saxena. 2017. Beneficial plant-microbes interactions: biodiversity of microbes from diverse extreme environments and its impact for crop improvement. pp. 543–580. *In*: Singh, D.P., H.B. Singh and R. Prabha (eds.). Plant-Microbe Interactions in Agro-Ecological Perspectives: Volume 2: Microbial Interactions and Agro-Ecological Impacts. Springer Singapore, Singapore. doi:10.1007/978-981-10-6593-4_22.

Wu, G., A. Lin, Q. Gu, T. Zhu and D. Li. 2013. Four new chloro-eremophilane sesquiterpenes from an Antarctic deep-sea derived fungus, *Penicillium* sp. PR19N-1. Mar. Drugs 11: 1399–1408.

Wu, G., H. Ma, T. Zhu, J. Li, Q. Gu and D. Li. 2012. Penilactones A and B, two novel polyketides from Antarctic deep-sea derived fungus *Penicillium crustosum* PRB-2. Tetrahedron 68: 9745–9749.

Wu, G., X. Sun, G. Yu, W. Wang, T. Zhu, Q. Gu et al. 2014. Cladosins A–E, hybrid polyketides from a deep-sea-derived fungus, *Cladosporium sphaerospermum*. J. Nat. Prod. 77: 270–275.

Yadav, A.N. 2017. Agriculturally important microbiomes: biodiversity and multifarious PGP attributes for amelioration of diverse abiotic stresses in crops for sustainable agriculture. Biomed. J. Sci. Tech. Res. 1: 1–4.

Yadav, A.N., S. Gulati, D. Sharma, R.N. Singh, M.V.S. Rajawat, R. Kumar et al. 2019a. Seasonal variations in culturable archaea and their plant growth promoting attributes to predict their role in establishment of vegetation in Rann of Kutch. Biologia 74: 1031–1043. doi:10.2478/s11756-019-00259-2.

Yadav, A.N., D. Kour, K.L. Rana, N. Yadav, B. Singh, V.S. Chauhan et al. 2019b. Metabolic engineering to synthetic biology of secondary metabolites production. pp. 279–320. *In*: Gupta, V.K. and A. Pandey (eds.). New and Future Developments in Microbial Biotechnology and Bioengineering. Elsevier, Amsterdam. doi:https://doi.org/10.1016/B978-0-444-63504-4.00020-7.

Yadav, A.N., R. Kumar, S. Kumar, V. Kumar, T. Sugitha, B. Singh et al. 2017. Beneficial microbiomes: biodiversity and potential biotechnological applications for sustainable agriculture and human health. J. Appl. Biol. Biotechnol. 5: 45–57.

Yadav, A.N., S.G. Sachan, P. Verma and A.K. Saxena. 2015a. Prospecting cold deserts of north western Himalayas for microbial diversity and plant growth promoting attributes. J. Biosci. Bioeng. 119: 683–693.

Yadav, A.N., S.G. Sachan, P. Verma and A.K. Saxena. 2016. Bioprospecting of plant growth promoting psychrotrophic Bacilli from cold desert of north western Indian Himalayas. Indian J. Exp. Biol. 54: 142–150.

Yadav, A.N., S.G. Sachan, P. Verma, S.P. Tyagi, R. Kaushik and A.K. Saxena. 2015b. Culturable diversity and functional annotation of psychrotrophic bacteria from cold desert of Leh Ladakh (India). World J. Microbiol. Biotechnol. 31: 95–108.

Yadav, A.N., D. Sharma, S. Gulati, S. Singh, R. Dey, K.K. Pal et al. 2015c. Haloarchaea endowed with phosphorus solubilization attribute implicated in phosphorus cycle. Sci. Rep. 5: 12293.

Yadav, A.N., D. Sharma, S. Gulati, S. Singh, R. Dey, K.K. Pal et al. 2015d. Haloarchaea endowed with phosphorus solubilization attribute implicated in phosphorus cycle. Scientific Reports 5: 12293.

Yadav, A.N., S. Singh, S. Mishra and A. Gupta. 2019c. Recent Advancement in White Biotechnology Through Fungi. Volume 3: Perspective for Sustainable Environments. Springer International Publishing, Cham.

Yadav, A.N., P. Verma, M. Kumar, K.K. Pal, R. Dey, A. Gupta et al. 2015e. Diversity and phylogenetic profiling of niche-specific Bacilli from extreme environments of India. Ann. Microbiol. 65: 611–629.

Yadav, A.N., P. Verma, S.G. Sachan, R. Kaushik and A.K. Saxena. 2018a. Psychrotrophic microbiomes: molecular diversity and beneficial role in plant growth promotion and soil health. pp. 197–240. *In*: Panpatte, D.G., Y.K. Jhala, H.N. Shelat and R.V. Vyas (eds.). Microorganisms for Green Revolution-Volume 2: Microbes for Sustainable Agro-ecosystem. Springer, Singapore. doi:10.1007/978-981-10-7146-1_11.

Yadav, P., S. Korpole, G.S. Prasad, G. Sahni, J. Maharjan, L. Sreerama et al. 2018b. Morphological, enzymatic screening, and phylogenetic analysis of thermophilic bacilli isolated from five hot springs of Myagdi, Nepal. J. App. Biol. Biotech. 6: 1–8.

Yusof, N., M.F.M. Raih, N. Mahadi, R. Illias, F.D.A. Bakar and A.M.A. Murad. 2017. *In silico* analysis and 3D structure prediction of a chitinase from psychrophilic yeast *Glaciozyma antarctica* PI12. Malaysian Applied Biology 46: 117–123.

Zhao, B., Q. Hu, X. Guo, Z. Liao, F. Sarmiento, N.M. Mesbah et al. 2018. *Natronolimnobius aegyptiacus* sp. nov., an extremely halophilic alkalithermophilic archaeon isolated from the athalassohaline Wadi An Natrun, Egypt. International Journal of Systematic and Evolutionary Microbiology 68: 498–506.

Index

||||||||||||||||||||||||||||||||||||||||||||||||||||||||||||||||||||||||||||||||||||||||||||||||||||||||||||||||||||||||||||||||||||||||||||||||||||||||

Editors

Ajar Nath Yadav is an Assistant Professor (Senior Scale) and Assistant Controller of Examination at Eternal University, Baru Sahib, Himachal Pradesh, India. He has 5 years of teaching and 11 years of research experience in the field of Microbial Biotechnology, Extreme Microbiology and Plant-Microbe-Interactions. Dr. Yadav obtained his doctorate degree in Microbial Biotechnology, jointly from IARI, New Delhi and BIT, Mesra, Ranchi, India; M.Sc. (Biotechnology) from Bundelkhand University and B.Sc. (CBZ) from University of Allahabad, India. Dr. Yadav has 150 publications with h-index of 45, i10-index of 110, and 5125 citations (Google Scholar). Dr. Yadav is editor of 14 Springer-Nature, 02 CRC Press Taylor & Francis and 02 Elsevier books. In his credit one granted patent "Insecticidal formulation of novel strain of *Bacillus thuringiensis* AK 47. Dr. Yadav has published 115 research communications in different international and national conferences. Dr. Yadav has got 12 Best Paper Presentation Awards, and 01 Young Scientist Award (NASI-Swarna Jyanti Purskar). Dr. Yadav received "Outstanding Teacher Award" in 6th Annual Convocation 2018 by Eternal University, Baru Sahib, Himachal Pradesh. Dr. Yadav has a long standing interest in teaching at the UG, PG and Ph.D. level and is involved in taking courses in microbiology, and microbial biotechnology. Dr. Yadav is currently handling two projects. Dr. Yadav guided 2 scholor for Ph.D. degree and one for M.Sc. dissertation. Presently he is guiding 3 scholars for PhD degree in Microbial Biotechnology. He has been serving as an editor/editorial board member and reviewer for different national and international peer-reviewed journals. He has lifetime membership of Association of Microbiologist in India, and Indian Science Congress Council, India. https://sites.google.com/site/ajarbiotech/ for more details.

Ali Asghar Rastegari is currently working as an assistant professor in the Faculty of Biological Science, Department of Molecular and Cellular Biochemistry, Falavarjan Branch, Islamic Azad University, Isfahan, I.R. Iran. He has 13 years of experience in the field of Enzyme Biotechnology, Nanobiotechnology, Biophysical Chemistry, Computational Biology and Biomedicine. Dr. Rastegari gained a Ph.D. in Molecular Biophysics in 2009, the University of Science and Research, Tehran Branch, Iran; M.Sc. (Biophysics), in 1994 from Institute of Biochemistry and Biophysics, the University of Tehran, and B.Sc. (Microbiology) in 1990 from the University of Isfahan, Iran. To his credit are 39 publications in various international, national journals and publishers. He is editor of 2 books published by CRC Press, Taylor & Francis, 2 books by Elsevier, and 6 books by Springer. He has issued 12 abstracts in different conferences/symposiums/workshops. He has presented 2 papers at national and international conferences/symposiums. Dr. Rastegari is a reviewer of different national and international journals. He has a lifetime membership of Iranian Society for Trace Elements Research (ISTER), The Biochemical Society of I.R. IRAN, Member of Society for Bioinformatics in Northern Europe (SocBiN), Membership of Boston Area Molecular Biology Computer Types (BAMBCT), Bioinformatics/Computational Biology Student Society

(BIMATICS Membership), Ensemble genome database and Neuroimaging Informatics Tools and Resources Clearinghouse (NITRC).

Neelam Yadav is currently working on microbial diversity from diverse sources and their biotechnological applications agriculture and allied sectors. She obtained her post graduation degree from Veer Bahadur Singh Purvanchal University, Uttar Pradesh, India. She has research interests in the area of beneficial microbiomes and their biotechnological application in agriculture, medicine environment and allied sectors. Dr. Yadav has 55 publications with h-index of 20, i10-index of 34, and 1179 citations (Google Scholar). Dr. Yadav is editor of 2 books published by CRC Press, Taylor & Francis, 2 books by Elsevier and 8 books by Springer-Nature. She is Editor/ associate editor/reviewer of different international and national journals including Plos One, Extremophiles, Annals of Microbiology, Journal of Basic Microbiology, Advance in Microbiology and Biotechnology. She has the lifetime membership of Association of Microbiologists in India, Indian Science Congress Council, India and National Academy of Sciences, India. https://sites. google.com/site/neelamanyadav/ for more details.